Phased Array
Antennas

To Eli,

Best compliments from

Arun Bhattacharyya

5/9/2007.

Phased Array Antennas

Floquet Analysis, Synthesis, BFNs, and Active Array Systems

ARUN K. BHATTACHARYYA

WILEY-
INTERSCIENCE

A JOHN WILEY & SONS, INC., PUBLICATION

Library of Congress Cataloging-in-Publication Data:

Bhattacharyya, Arun.
  Phased array antennas: floquet analysis, synthesis, BFNs, and active array systems /
  by Arun K. Bhattacharyya.
      p. cm.
  "A Wiley-Interscience publication."
  Includes bibliographical references and index.
  ISBN-13: 978-0-471-72757-6
  ISBN-10: 0-471-72757-1
  1. Phased array antennas. 2. Electronics—Mathematics. I. Title.

  TK6590.A6B45 2005
  621.382′4—dc22   2005049360

Printed in the United States of America

10 9 8 7 6 5 4 3 2 1

For my Teachers

Contents

Preface

The purpose of this book is to present in a comprehensive manner the analysis and design of phased array antennas and systems. The book includes recent analytical developments in the phased array arena published in journals and conference proceedings. Efforts have been made to develop the concept in a logical manner starting from fundamental principles. Detailed derivations of theorems and concepts are provided to make the book as self-contained as possible. Several design examples and design guidelines are included in the book. The book should be useful for antenna engineers and researchers, especially those involved in the detailed design of phased arrays. The reader is assumed to have a basic knowledge of engineering mathematics and antenna engineering at a graduate level. The book can be used either as a text in an advanced graduate-level course or as a reference book for array professionals.

The book contains 14 chapters that may be broadly divided into three sections. The first section, which includes Chapters 1–6, is mostly devoted to the development of the Floquet modal-based approach of phased array antennas starting with an introductory chapter. The second section, which includes Chapters 7–10, presents applications of the approach to important phased array structures. The third section, which includes Chapters 11–14, is not directly related to the Floquet modal analysis as such; however, it covers several important aspects of a phased array design. This section includes beam array synthesis, array beam forming networks, active phased array systems, and statistical analysis of phased arrays. Several practice problems are included at the end of each chapter to provide a reader an interactive experience. Information on the solution manual and selective software may be available at **http://hometown.aol.com/painta9/**.

Chapter 1 presents a brief discussion on phased array fundamentals. There are two goals of this chapter. First, it gives a basic overview of phased array characteristics. Second, the limitations of conventional first-order analysis of phased array antennas are spelled out. In this way, the reader can comprehend the limitations of the traditional approach and appreciate the need for a higher order analysis (which is developed in the next few chapters). The chapter begins with the definition of the element pattern followed by the array pattern and array factor. The maximum-gain theorem of a general array antenna is presented next. Scan characteristics of a pencil beam array is discussed in light of gain, grating lobe, and beam-width considerations. The phase-quantization effects are discussed. Prevalent array synthesis procedures for pencil beam arrays are presented. The scope and limitation of this first-order approach is discussed at the end.

Chapter 2 initiates the development of Floquet modal analysis for array antennas, which is one of the main objectives of the book. Using simple analytic means, we show that Floquet modal expansion evolves from Fourier expansion. The relation-ship between a Floquet mode and observable antenna parameters, such as radiation direction, is then established. Derivation of Floquet modal functions for an arbitrary array grid structure is presented next. Finally, the coupling of a Floquet mode with a guided mode is considered and interesting consequences are discussed.

In Chapter 3 expressions for normalized Floquet modal functions are deduced. The Floquet modal expansion method is illustrated through an example of an infinite array of electric current sources. Two parameters of an array antenna are defined, namely the Floquet impedance and active element pattern, and their significance is discussed in the context of array performance. A detailed discussion of array blindness is also presented. A scattering matrix formulation for an infinite array of rectangular horn apertures is presented.

Chapter 4 demonstrates that the "results of an infinite array analysis" can be utilized to analyze a finite array with arbitrary excitation. In the first part, important theorems and concepts relevant for the development of finite array analysis are presented. It is well known that the most important factor to consider in a finite array analysis is the mutual coupling. It is demonstrated that the mutual coupling between the elements can be determined using infinite array data obtained through a Floquet modal analysis. Next, the active impedances of the elements and the radiation pattern of a finite array with respect to arbitrary amplitude taper are obtained. The chapter concludes with an alternate approach for finite array analysis. This alternate approach relates infinite array characteristics to finite array characteristics through convolution.

It is often beneficial to divide the entire array into a number of identical groups. Such a group, called a subarray, consists of few elements that are excited by a single feed. In Chapter 5, a systematic procedure is presented to analyze an array of subarrays. Using matrix theory it is shown that a subarray impedance matrix can be constructed from the Floquet impedance of a single element. Important characteristic features of an array of subarrays are presented.

Chapter 6 presents the generalized scattering matrix (GSM) approach to analyze multilayer array structures. The chapter begins with the definition of a GSM fol-lowed by the cascading rule of two GSMs. Advantages of the GSM approach over a transmission matrix approach are discussed from a numerical stability standpoint. Using the method of moments and modal matching, the GSMs of several important "building blocks" are deduced. The stationary character of the present approach is then established. Convergence of the solution is discussed in detail. The chapter concludes with a discussion of advantages and disadvantages of the GSM approach.

Chapter 7 applies the GSM approach to analyze probe-fed and slot-fed multi-layer patch arrays. The radiation and impedance characteristics of patch arrays are presented. The analysis of an electromagnetically coupled (EMC) patch array and stripline-fed patch array with mode suppressing vias are discussed. Results of finite arrays are presented at the end.

Chapter 8 primarily focuses on horn radiators as array elements. It begins with the linearly flared rectangular horns. It is shown that under certain conditions a horn array structure may support surface and leaky waves. The dips and nulls present in active element patterns are explained via surface and leaky wave coupling. The wide-angle impedance matching aspect of a horn array is discussed. Characteristics of step horns, which may be used for enhanced radiation efficiency, are presented. Design guidelines of "high efficiency horns" are presented at the end.

In Chapter 9, the analyses of three important passive printed array structures are presented. They include frequency-selective surfaces (FSSs), screen polarizers, and printed reflect arrays. In the first part of the chapter, features of a single- and two-layer FSS are presented in terms of return loss, copolar, and cross-polar characteristics. The analysis of a horn antenna loaded with an FSS is considered. In the second part, the analysis of a meander line polarizer screen is presented. The chapter concludes with an analysis of a printed reflect-array antenna for linear and circular polarizations. The gain enhancement method of reflect-array antennas is also discussed.

Chapter 10 presents a method for analyzing complex multilayer array structures that have many useful applications. The analysis is very general to handle layers with different lattice structures, periodicities, and axes orientations. In order to perform the analysis, a mapping relation between the modes associated with a local lattice and the global lattice is established. The mapping relation in conjunction with the GSM cascading rule is utilized to obtain the overall GSM of the structure. The methodology is demonstrated by considering two examples of multilayer structures that have practical applications.

Chapter 11 presents various synthesis methods for shaped beam array antennas. The chapter begins with an analysis of a linear array to study the effects of array size and element size on the array pattern. This is followed by Woodward's beam superposition method for obtaining a shaped beam. Next, different optimization algorithms that are commonly employed for beam shaping are presented. Examples of shaped beams using various optimization schemes are provided.

One of the most important tasks in a phased array design is designing the beam forming network (BFN). Chapter 12 presents comprehensive treatments of the most common types of BFNs. The chapter begins with the simplest type of BFN using passive power divider circuits. The Butler matrix BFN that operates on the principle of the FFT (fast Fourier transform) algorithm is considered. Implementations of Butler BFNs using power dividers and hybrids are shown. The operational principle of a Blass matrix BFN is presented next, followed by the Rotman lens design and analysis. The chapter ends with a discussion of digital beam formers and optical beam formers, including their principles of operations, advantages, and limitations.

In Chapter 13, the basic structures and subsystems of active array antennas are presented. The chapter begins with a generic block diagram of an active array. Typical circuit configurations of each block are considered. Important system parameters are defined and discussed in the context of antenna performance. The intermodulation (IM) products caused by the amplifier nonlinearity are studied. Locations and power levels of IM beams are obtained. To aid the array system analysis, the noise temperature and noise figure of active array components are deduced. A typical example of an array system analysis is presented. Various active array calibration methods are presented at the end.

Chapter 14 presents a statistical analysis of an array antenna with respect to the amplitude and phase uncertainties of the amplifiers and phase shifters. The analysis is developed from the fundamental principles of probability theory. The first part deduces the statistics of the array factor in terms of the amplitude and phase errors. To that end, the statistical parameters of the real and imaginary parts of the array factor are obtained. This is followed by a deduction of the probability density function for the far-field intensity. Also obtained are the simplified closed-form expressions of the probability density functions for the far-field intensity at the beam peak, null, and peak side-lobe locations. Approximate expressions for the 95% confidence boundaries are deduced. Finally, the effect of element failure on the array statistics is introduced. The numerical results with and without element failure are presented. Effects of amplitude and phase uncertainties and element failure on side-lobe levels and null depths are shown.

I express my sincere appreciation to the antenna professionals and friends Owen Fordham, Guy Goyette, Eng Ha, Philip Law, Stephen Kawalko, Gordon Rollins, James Sor, Murat Veysoglu, and Paul Werntz for devoting their personal time to review the chapters. Their expert comments and constructive suggestions were invaluable for improving the text. I express my sincere thanks to Kai Chang for his encouragement and recommendation for publication of the book. Thanks are also due to anonymous reviewers for their suggestions. I am thankful to Rachel Witmer and George Telecki of the editorial department and the production editor Lisa VanHorn for their cooperations during the course of the project. I am grateful to my wife, Arundhuti, and daughters, Atreyi and Agamoni, for their encouragement and moral support in writing the book and for their help in proofreading. I express my profound love and gratitude to my mother, who has been a constant inspiration for me. I am indebted to my teachers for teaching me the fundamentals at various stages of my career. This book is dedicated to them as a token of my appreciation of their love and devotions for teaching.

ARUN K. BHATTACHARYYA

El-Sugendo, California
January 2006

CHAPTER ONE

Phased Array Fundamentals: Pattern Analysis and Synthesis

1.1 INTRODUCTION

In this chapter we present an overview of the basic electromagnetic properties of phased array antennas. We begin with the basics and develop useful concepts for the analysis of phased arrays. Important array terminology is introduced to familiarize the reader with the topic. The element pattern, array pattern, and array factor are introduced, leading to a discussion of how element gain and array gain are tied together. Copolarization and cross-polarization as proposed by Ludwig are presented, and the maximum-gain theorem of an array antenna is deduced. Scan characteristics of a pencil beam array are presented with consideration to gain, grating lobes, and beam deviation due to different types of phase shifters. Prevalent aperture synthesis procedures for sum and difference patterns are presented with theoretical details. Emphasis is placed on the foundation and conceptual development of these methods, and a few words on the scope and limitation of this chapter are included at the end.

1.2 ARRAY FUNDAMENTALS

A phased array antenna consists of an array of identical radiating elements in regular order, as shown in Figure 1.1. In a typical array antenna, all the elements radiate coherently along a desired direction. This is particularly true for a pencil beam array where a linear phase progression between the elements is set to accomplish

Phased Array Antennas. By Arun K. Bhattacharyya
© 2006 John Wiley & Sons, Inc.

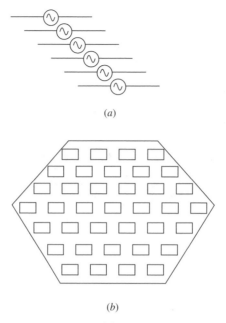

(a)

(b)

FIGURE 1.1 (*a*) Linear array of dipoles. (*b*) Two-dimensional array of rectangular horns in triangular lattice.

this coherent radiation. For a shaped beam array, however, all the elements do not radiate coherently at a given direction. The shape of the beam decides the amplitude and phase distribution of the array, which is usually nonlinear. In the following sections we will define a few terms that are necessary to understand the radiation of an array antenna.

1.2.1 Element Pattern, Directivity, and Gain

In order to estimate the radiated power of an array antenna in the far-field region, one needs to understand the radiated field intensity of an element in the far-field region. The element pattern is defined as the field intensity distribution of a radiating element as a function of two far-field coordinates, while the radial distance remains constant. In a spherical coordinate system the radiated electric field in the far-field location can be expressed as

$$\vec{E}(r, \theta, \phi) = A \frac{\exp(-jk_0 r)}{r} \vec{e}(\theta, \phi) \tag{1.1}$$

In the above A is a constant which is related to the input excitation of the antenna, $\vec{e}(\theta, \phi)$ is the element pattern, (r, θ, ϕ) is the spherical coordinates of the far-field point, also known as the observation point, and k_0 is the wave number in free space. It should be pointed out that unless otherwise stated, the radiation pattern is defined at the far-field region where $r \gg \lambda_0$, λ_0 being the wavelength in free space. In (1.1) $\vec{e}(\theta, \phi)$ is a complex vector function having components along $\hat{\theta}$- and $\hat{\phi}$-directions only. The radial component does not exist in the far-field region.

The complex directive pattern of an element is the far-field intensity pattern normalized with respect to the square root of the average radiated power per unit solid angle. The total radiated power is determined by integrating the Poynting vector on a spherical surface covering the antenna element as

$$P_r = \iint_\Omega \frac{|\vec{E}|^2}{\eta} r^2 d\Omega = \frac{|A|^2}{\eta} \int_0^{2\pi} \int_0^\pi |\vec{e}(\theta, \phi)|^2 \sin\theta \, d\theta \, d\phi \qquad (1.2)$$

In (1.2) η represents free-space impedance, which is equal to 120π (or 377) ohms. The average power per unit solid angle is

$$P_r^{\mathrm{av}} = \frac{P_r}{4\pi} \qquad (1.3)$$

Thus the complex directive pattern of the element becomes

$$\vec{D}(\theta, \phi) = \sqrt{\frac{1}{\eta}} \sqrt{\frac{4\pi}{P_r}} A \vec{e}(\theta, \phi) \qquad (1.4)$$

Observe that in order to make $\vec{D}(\theta, \phi)$ dimensionless the factor $\sqrt{1/\eta}$ is introduced in (1.4). Substituting the expression of P_r from (1.2) in (1.4), we obtain

$$\vec{D}(\theta, \phi) = \frac{\sqrt{4\pi} \, \vec{e}(\theta, \phi)}{\sqrt{\int_0^{2\pi} \int_0^\pi |\vec{e}(\theta, \phi)|^2 \sin\theta \, d\theta \, d\phi}} \qquad (1.5)$$

In deducing (1.5) we assume $|A| = A$, which we justify by accommodating the complex exponent part of A into the element pattern $\vec{e}(\theta, \phi)$.

The directivity of an element signifies the relative power flux per solid angle with respect to that of an isotropic radiator that radiates an equal amount of power. The directivity at (θ, ϕ) is the square of the magnitude of $\vec{D}(\theta, \phi)$. Usually the directivity is expressed in dBi, where i stands for isotropic. The complex gain pattern of an element is the far-field intensity pattern normalized with respect to the incident power at the antenna input instead of the total radiated power. Thus the complex gain pattern with respect to the field intensity can be expressed as

$$\vec{G}(\theta, \phi) = A\sqrt{\frac{4\pi}{\eta P_{inc}}}\; \vec{e}(\theta, \phi) \qquad (1.6)$$

In (1.6) P_{inc} is the incident power and A is related to P_{inc}. The gain at a far-field point is given by $|\vec{G}(\theta, \phi)|^2$ [however, at places we use the word "gain" to indicate either $\vec{G}(\theta, \phi)$ or $|\vec{G}(\theta, \phi)|^2$, which can be understood from the context]. Since the total radiated power is reduced by the antenna mismatch loss and other radio-frequency (RF) losses, the gain does not exceed the directivity.

1.2.2   Copolarization and Cross-Polarization

As mentioned before, the radiated field emanating from a radiating source has two mutually orthogonal components along $\hat{\theta}$ and $\hat{\phi}$, respectively. For linearly polarized radiation, the copolarization vector essentially is the preferred electric field vector. The cross-polarization vector is orthogonal to both the copolarization vector and the direction of radiation. One can define the preferred polarization direction according to one's preference. However, we will follow *Ludwig's third definition* [1] as it is well accepted by the antenna community and has practical significance. According to this definition, the copolarization and cross-polarization components at a far-field point (θ, ϕ) of an electric current source (or electric field source for an aperture antenna) polarized along the x-direction are defined as

$$e_{co}(\theta, \phi) = \vec{e}(\theta, \phi) \cdot [\hat{\theta}\cos\phi - \hat{\phi}\sin\phi] \qquad (1.7a)$$

$$e_{cr}(\theta, \phi) = \vec{e}(\theta, \phi) \cdot [\hat{\theta}\sin\phi + \hat{\phi}\cos\phi] \qquad (1.7b)$$

where $\vec{e}(\theta, \phi)$ represents the far electric field pattern of the source. The above definition is consistent with the standard pattern measurement method where aperture of the antenna under test (AUT) is placed on the $z = 0$ plane and the probe antenna is mounted on a rigid rod such that the probe always remains perpendicular to the rod. The probe end of the rod is free to move along the great circles on ϕ-constant planes (Figure 1.2) without any twist, while the other end is fixed at a reference point on the AUT. For the copolarization measurement, the probe must be aligned with the principal polarization direction when it is situated at the bore sight ($\theta = 0$) of the AUT. Under these conditions, the orientation of the probe for an arbitrary probe location defines the copolarization vector. The cross-polarization vector is perpendicular to the copolarization vector and the radius vector.

The copolarization gain is of practical importance because it is used for estimating the amount of power received by a receiving antenna, which is polarized at the same sense of the transmitting antenna. The copolarization gain can be calculated by taking the copolarization component of $\vec{e}(\theta, \phi)$ in (1.6). The cross-polarization gain

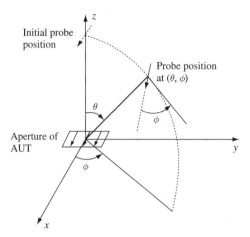

FIGURE 1.2 Antenna copolar pattern measurement scheme following Ludwig's third definition. Notice that the probe maintains an angle ϕ with the tangent of the circle (θ-direction) shown by the dotted line, which is consistent with (1.7a). The principal polarization of the AUT is along x.

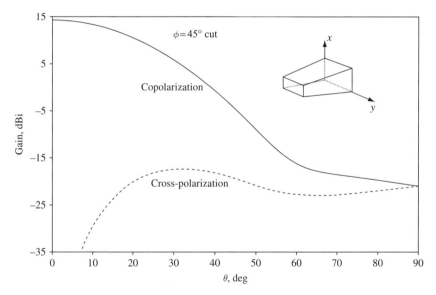

FIGURE 1.3 Copolarization and cross-polarization patterns of a square horn of length $3\lambda_0$ and of aperture size $1.6\lambda_0 \times 1.6\lambda_0$. The input waveguide dimension is $0.6\lambda_0 \times 0.3\lambda_0$ and is excited by the TE_{01} mode.

can be obtained similarly. Figure 1.3 shows the copolarization and cross-polarization gain patterns of a linearly flared horn. The cross-polarization field intensity vanishes along the two principal plane cuts (E- and H-plane cuts); therefore we choose the diagonal plane cut for the plot.

It is worth mentioning that there is no preferred direction for a circularly polarized radiation. The right and left circular polarization unit vectors (rcp and lcp) for $\exp(j\omega t)$ time dependence are defined as

$$\hat{e}_{\mathrm{rcp}}(\theta, \phi) = \frac{1}{\sqrt{2}}[\hat{\theta} - j\hat{\phi}] \tag{1.8a}$$

$$\hat{e}_{\mathrm{lcp}}(\theta, \phi) = \frac{1}{\sqrt{2}}[\hat{\theta} + j\hat{\phi}] \tag{1.8b}$$

Notice, the polarization vectors are mutually orthogonal because $\hat{e}_{\mathrm{rcp}} \cdot \hat{e}_{\mathrm{lcp}}^* = 0$.

1.2.3 Array Pattern

The radiated field of an array essentially is the summation of the individual element fields. It can be shown that the far-field pattern of an array of identical elements can be represented by a product of two quantities, namely the *element pattern* and the *array factor*. The element pattern signifies the radiation behavior of an individual element and the array factor signifies the arraying effect, including array architecture and relative excitations of the elements. To establish this relation we consider a linear array of N identical elements along the x-axis with element spacing a as shown in Figure 1.4. The excitation coefficient $A_n (n = 1, 2, ..., N)$ is assumed to be a complex number incorporating amplitude and phase in a single entity. Invoking (1.1) and applying superposition of the individual fields, the array field can be expressed as

$$\vec{E}_{\mathrm{array}} = A_1 \frac{\exp(-jk_0 r_1)}{r_1} \vec{e}(\theta, \phi) + A_2 \frac{\exp(-jk_0 r_2)}{r_2} \vec{e}(\theta, \phi) + \cdots$$
$$+ A_N \frac{\exp(-jk_0 r_N)}{r_N} \vec{e}(\theta, \phi) \tag{1.9}$$

In (1.9) $r_n (n = 1, 2, ..., N)$ is the distance from the nth element to the observation point located at the far field. For $r_n \gg Na$ we can have the following approximation:

$$r_n \approx r_1 - (n-1)a \sin\theta \cos\phi \qquad n = 1, 2, \ldots, N \tag{1.10}$$

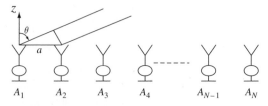

FIGURE 1.4   Linear array of N elements. The elements are situated along the x-axis.

The above approximation should be used for the r_n that lies inside the argument of the complex exponential function, because complex exponential functions are highly oscillatory. For the r_n in the denominator, however, a more crude approximation, namely $r_n \approx r_1$, is admissible because $1/r$ is a slow-varying function for large r. Thus (1.9) becomes

$$\vec{E}_{array} = \vec{e}(\theta, \phi)\frac{\exp(-jk_0 r_1)}{r_1}[A_1 + A_2 \exp(jk_0 a \sin\theta \cos\phi) + \cdots$$

$$+ A_N \exp\{jk_0(N-1)a \sin\theta \cos\phi\}] \qquad (1.11)$$

The quantity inside the square bracket is known as the array factor (AF) for a linear array. Ignoring the radial dependence part we observe that the array pattern reduces to a product of the element pattern and the array factor. For a two-dimensional array a similar development results in the following array factor:

$$AF(\theta, \phi) = \sum_{n=1}^{N} A_n \exp[jk_x x_n + jk_y y_n] \qquad (1.12)$$

In (1.12), (x_n, y_n) represents the coordinate of the nth element and k_x, k_y are

$$k_x = k_0 \sin\theta \cos\phi \qquad k_y = k_0 \sin\theta \sin\phi \qquad (1.13)$$

Figure 1.5 depicts a typical pattern of a linear array. For this plot 20 elements that are uniformly exited and spaced at $1\lambda_0$ apart are used. The plot shows the relative

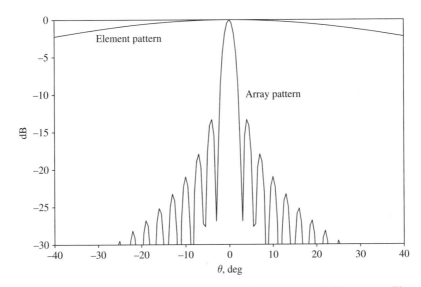

FIGURE 1.5 Radiation pattern of a uniformly excited linear array with 20 elements. Element spacing λ_0.

field intensity (with respect to the peak field intensity) versus θ on $\phi = 0°$ plane. To examine the arraying effect, the element pattern is also plotted. Evidently, the array beam width is much smaller than the element beam width, implying a high level of beam focusing capability of the array.

The array pattern deduced in (1.11) assumes identical element patterns for all the elements. This is not rigorously valid because even if the elements have identical shapes they have different surroundings.[1] For instance, the edge element of an array will have a different radiation pattern than that of the center element. However, in most applications the *identical radiation pattern* assumption may be reasonable, particularly in the bore-sight region where small variations in the element patterns do not introduce an error beyond the acceptable limit.

1.2.4 Array Gain

The gain of an array can be determined after normalizing the field intensity in (1.11) with respect to the total incident power of the array. It is convenient to use the element gain pattern as the element pattern in (1.11) because the field intensity is already normalized with respect to its incident power. The array pattern thus can be written as

$$\vec{E}_{\text{array}} = \vec{G}(\theta, \phi)[A_1 + A_2 \exp(jk_0 a \sin \theta \cos \phi) + \cdots$$
$$+ A_N \exp\{jk_0(N - 1)a \sin \theta \cos \phi\}] \qquad (1.14)$$

In (1.14) the element gain pattern, $\vec{G}(\theta, \phi)$, is normalized so that

$$\int_0^{2\pi} \int_0^{\pi} |\vec{G}(\theta, \phi)|^2 \sin \theta \, d\theta \, d\phi = 4\pi(1 - L) \qquad (1.15)$$

where L is the antenna loss factor. We must emphasize here that for the array gain determination we consider A_n as the normalized incident voltage or current [not to be confused with the A in (1.6)] for the nth element such that $|A_n|^2$ becomes its incident power. Thus the total incident power of the array is given by

$$P_{\text{inc}} = \sum_{n=1}^{N} |A_n|^2 \qquad (1.16)$$

Thus the complex gain pattern of the array can be written as

$$\vec{G}_{\text{array}}(\theta, \phi) = \vec{G}(\theta, \phi) \frac{\sum_{n=1}^{N} A_n \exp(jk_0 na \sin \theta \cos \phi)}{\sqrt{\sum_{n=1}^{N} |A_n|^2}} \qquad (1.17)$$

---

[1] The radiation pattern of an element, including the effects of surrounding elements, is known as the *active element pattern*. Analysis of the active element pattern will be considered in a later chapter.

The co- and cross-polarization gain patterns simply are the corresponding components of $\vec{G}_{\text{array}}(\theta, \phi)$.

The above array gain pattern expression is rigorously valid if the following two conditions are satisfied: (a) The element gain pattern $\vec{G}(\theta, \phi)$ is measured or computed in array environment with all other elements match terminated, that is $\vec{G}(\theta, \phi)$ represents active element gain pattern, and (b) elements have identical active element patterns. In reality condition (b) is not generally satisfied because the active element pattern differs from element to element. For such situations (1.17) can be modified as

$$\vec{G}_{\text{array}}(\theta, \phi) = \frac{\sum_{n=1}^{N} A_n \vec{G}_n(\theta, \phi) \exp(jk_0 na \sin\theta \cos\phi)}{\sqrt{\sum_{n=1}^{N} |A_n|^2}} \tag{1.18}$$

In (1.18) $\vec{G}_n(\theta, \phi)$ represents the active element gain pattern for the nth element of the array. Figure 1.6 is a pictorial definition of the active element pattern.

1.2.5 Maximum-Array-Gain Theorem

The maximum-gain theorem yields the optimum excitation condition of the array to achieve maximum gain along a desired direction. The statement and the proof of the theorem follow.

Statement The maximum gain (magnitude) of an array at a desired direction occurs if the incident voltage (or current) is proportional to the complex conjugate of the active element gain along the desired direction.

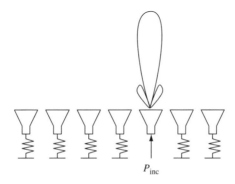

FIGURE 1.6 Pictorial definition of the active element pattern of an element in a seven-element array. The element under consideration is excited while the other elements are match terminated.

Proof   In order to prove the above theorem we consider (1.18). Without loss of generality we can assume that the total incident power is unity, implying

$$\sum_{n=1}^{N} |A_n|^2 = 1 \tag{1.19}$$

Then the array gain pattern becomes

$$\vec{G}_{\text{array}}(\theta, \phi) = \sum_{n=1}^{N} A_n \vec{G}_n(\theta, \phi) \exp(jk_0 na \sin\theta \cos\phi) \tag{1.20}$$

We will consider the copolar gain at (θ, ϕ); therefore we extract the copolar component in both sides of (1.20), leaving

$$G_{\text{array}}^{\text{co}}(\theta, \phi) = \sum_{n=1}^{N} A_n G_n^{\text{co}}(\theta, \phi) \exp(jk_0 na \sin\theta \cos\phi) \tag{1.21}$$

From triangular inequality [2] we know that the magnitude of the sum of complex numbers is less than or equal to the sum of their magnitudes. Applying this in (1.21) we write

$$|G_{\text{array}}^{\text{co}}(\theta, \phi)| = |\sum_{n=1}^{N} A_n G_n^{\text{co}}(\theta, \phi) \exp(jk_0 na \sin\theta \cos\phi)|$$

$$\leq \sum_{n=1}^{N} |A_n G_n^{\text{co}}(\theta, \phi) \exp(jk_0 na \sin\theta \cos\phi)| = \sum_{n=1}^{N} |A_n||G_n^{\text{co}}(\theta, \phi)| \tag{1.22}$$

The equality holds if

$$A_n G_n^{\text{co}}(\theta, \phi) \exp(jk_0 na \sin\theta \cos\phi) = |A_n| \cdot |G_n^{\text{co}}(\theta, \phi)| \tag{1.23}$$

for all n. To satisfy (1.23), the phase of A_n must be negative of the phase of $G_n^{\text{co}}(\theta, \phi) \exp(jk_0 na \sin\theta \cos\phi)$. Symbolically,

$$\angle A_n = -\angle G_n^{\text{co}} - k_0 na \sin\theta \cos\phi \tag{1.24}$$

If (1.24) is satisfied, then the magnitude of the copolar gain of the array along the (θ, ϕ) direction becomes

$$|G_{\text{array}}^{\text{co}}(\theta, \phi)| = \sum_{n=1}^{N} |A_n| \cdot |G_n^{\text{co}}(\theta, \phi)| \tag{1.25}$$

Equation (1.24) yields the phase of the excitations; however, the magnitude of A_n for maximum gain is yet to be found. Toward that effort we maximize (1.25) subject to the constraint in (1.19). We invoke Lagrange's multiplier method[2] to maximize the array gain with respect to $|A_n|$. Accordingly we write

$$\frac{\partial |G^{co}_{array}|}{\partial |A_n|} + \mu \frac{\partial}{\partial |A_n|} \sum_{i=1}^{N} |A_i|^2 = 0 \tag{1.26}$$

and deduce

$$|G^{co}_n(\theta, \phi)| + 2\mu |A_n| = 0 \tag{1.27}$$

In (1.26) and (1.27) μ is Lagrange's multiplier. From (1.27) we notice that $|A_n|$ is proportional to $|G^{co}_n(\theta, \phi)|$. The phase of A_n is already deduced in (1.24). Now, $G^{co}_n(\theta, \phi) \exp(jk_0 na \sin\theta \cos\phi)$ can be considered as the complex copolar gain of nth element, $\tilde{G}^{co}_n$, which includes the additional phase for the path difference of the radiated field. Thus, we obtain

$$\tilde{G}^{co}_n(\theta, \phi) = G^{co}_n(\theta, \phi) \exp(jk_0 na \sin\theta \cos\phi) \tag{1.28}$$

Combining (1.24), (1.27), and (1.28) we write

$$A_n = C \tilde{G}^{co}_n{}^*(\theta, \phi) \tag{1.29}$$

where C is a constant and the asterisk indicates the complex conjugate. This proves that the highest gain of an array can be accomplished if the element's incident voltage (or current) is proportional to the complex conjugate of its active element gain.

Equation (1.29) implies that for identical radiation patterns of the elements, the amplitude distribution should be uniform in order to have maximum array gain. However, the amplitude distribution should not be uniform for dissimilar elements; thus the maximum-gain theorem has very important significance in designing a pencil beam array with dissimilar elements. An array of elements feeding a parabolic reflector can be treated as an array of dissimilar elements because the secondary patterns produced by the feeds differ from each other. Another important application of the maximum-gain theorem is designing a conformal array, for instance, an array on a spherical surface. The bore sights of the elements differ from each other; hence the element patterns are considered to be different, though the elements look alike physically (see problem 1.4).

---

[2] See, e.g., A. Mizrahi and M. Sullivan, *Calculus and Analytic Geometry*, 3rd ed., Wadsworth Publishing Company, 1990.

1.2.6 Array Taper Efficiency

The array taper efficiency (ATE) signifies the influence of the "taper distribution" on the total aperture efficiency of an array. For an array with elements that have 100% aperture efficiency, the ATE becomes equal to the array aperture efficiency. For a planar or linear array of N elements, the ATE is defined as

$$\text{ATE} = \frac{\{\sum_n |A_n|\}^2}{N \sum_n |A_n|^2} \times 100\% \tag{1.30}$$

For a uniform taper, the ATE is 100%. For a nonuniform taper it is less than 100%. The array aperture efficiency is the product of the ATE and the element aperture efficiency in the array environment. Thus, the ATE has a very important role in deciding the overall aperture efficiency of the full array. The ATE will be considered for a few cases in Section 1.4.

1.3 PENCIL BEAM ARRAY

A pencil beam array is generally referred to as an array that produces narrow beams. Thus, a pencil beam array exhibits maximum possible gain in a desired direction. Typically, the amplitude taper of such an array is decided by the side-lobe requirement; thus only the element phase is adjusted to maximize the array gain. The element phase corresponding to a maximum gain is deduced in (1.24). In this section we employ (1.24) to study pencil beams and address important aspects concerning beam scanning. For a linear array of N elements with element spacing a, the radiation pattern at the $\phi = 0$ plane can be expressed as [per (1.14)]

$$\vec{E}_{\text{array}} = \vec{G}(\theta, 0)[A_1 + A_2 \exp(jk_0 a \sin \theta) + \cdots$$
$$+ A_N \exp\{jk_0(N-1)a \sin \theta\}] \tag{1.31}$$

The complex excitation coefficients for the nth element can be expressed as

$$A_n = B_n \exp(j\psi_n) \tag{1.32a}$$

where B_n and ψ_n, respectively, are the magnitude and phase of the excitation coefficient. According to the maximum-gain theorem, the maximum gain at a scan angle θ_0 occurs if

$$\psi_n = -k_0(n-1)a \sin \theta_0 \tag{1.32b}$$

The corresponding array pattern becomes

$$\vec{E}_{\text{array}} = \vec{G}(\theta, 0)[B_1 + B_2 \exp\{jk_0 a(\sin \theta - \sin \theta_0)\} + \cdots$$
$$+ B_N \exp\{jk_0(N-1)a(\sin \theta - \sin \theta_0)\}] \tag{1.33}$$

In (1.33), $\vec{E}_{\text{array}}$ represents the scan array pattern associated with the scan angle θ_0.

1.3.1 Scan Loss and Beam Broadening

For determining the scan array gain, the total power should be normalized. We assume $\sum_{n=1}^{N} B_n^2 = 1$, so that $\vec{E}_{\text{array}}$ in (1.33) becomes the gain pattern of the array. The copolar gain is given by the copolar component of $\vec{E}_{\text{array}}$ in (1.33). At the scan location $\theta = \theta_0$ the copolar gain becomes

$$G_{\text{array}}^{\text{co}} = G^{\text{co}}(\theta_0, 0)[B_1 + B_2 + \cdots + B_N] \tag{1.34}$$

Observe that the copolar array gain at a scan location is proportional to the copolar element gain at that location. The element spacing has no direct influence[3] on the scan gain. Typically, the element gain peaks at the bore sight, implying that the scan gain decreases as the scan angle moves away from the bore sight. The scan loss is defined as the relative gain loss in decibels with respect to the bore-sight scan. Thus from (1.34) we obtain the scan loss as

$$\text{Scan loss at } \theta_0 = 20 \log \left| \frac{G^{\text{co}}(0,0)}{G^{\text{co}}(\theta_0,0)} \right| \qquad \text{dB} \tag{1.35}$$

Obviously, the larger the element size, the higher is the scan loss because a larger element has a sharp roll-off pattern. Figure 1.7 shows the scan loss for three elements of different sizes.

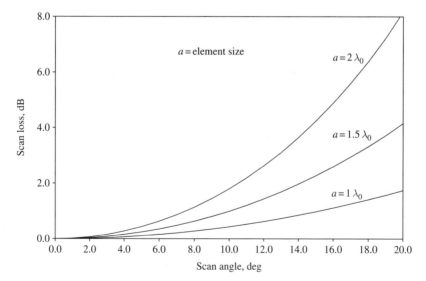

FIGURE 1.7 Scan loss versus scan angle for three elements (λ_0 = wavelength).

---

[3] Unless the *active element pattern* is used in the analysis, which is dependent upon the element spacing. However, for a first-order analysis we ignore mutual coupling; thus we can assume that the element pattern is independent of element spacing.

The beam width increases as the scan angle increases. This is consistent with the scan loss behavior because a low gain corresponds to a wide beam. This can also be understood from (1.33). We notice that the array factor is an explicit function of $(\sin\theta - \sin\theta_0)$. Therefore, the shape of the array factor remains unchanged with the scan angle in the $\sin\theta$ space as depicted in Figure 1.8. If the element gain does not have a rapid variation within the scan region, then the beam width of the scan beam remains unaffected if it is measured in the $\sin\theta$ scale. Mathematically, we can write

$$\Delta\sin\theta = \delta u \qquad (1.36)$$

where δu, the beam width in the $\sin\theta$ scale, is independent of θ. For a narrow beam (1.36) leads to

$$\Delta\theta \approx \frac{\delta u}{\cos\theta} \approx \frac{\delta u}{\cos\theta_0} \qquad (1.37)$$

The above relation holds in the vicinity of the beam peak. Equation (1.37) thus implies that the beam width increases by the factor $1/\cos\theta_0$, where θ_0 is the scan angle.

1.3.2 Scan Array Design Consideration

The scan loss has a very important role in designing a scan beam array. For instance, suppose we need to design an array to scan within 10° off bore sight for a minimum

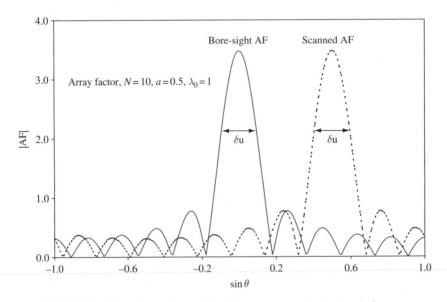

FIGURE 1.8 Plot of array factors for different scan angles in the $\sin\theta$ space.

gain of 40 dBi. The design essentially includes determining the number of elements and the array aperture size that would yield the desired gain while satisfying the other constraints.[4] To aid the design trade, we must understand the relation between the array aperture size and the total number of elements that would satisfy the gain requirement. Toward that pursuit we first consider a uniform excitation. The effects of the amplitude taper will be considered later. From (1.34) we find that for uniform excitation, the array gain at the scan angle θ_0 can be expressed as

$$G_{\text{array}}^{\text{co}} = G^{\text{co}}(\theta_0, 0)\sqrt{N} \tag{1.38}$$

because $B_n = 1/\sqrt{N}$ for all n. For 100% aperture efficient elements, the element gain pattern can be expressed as

$$G^{\text{co}}(\theta_0, 0) = \sqrt{\frac{4\pi a^2}{\lambda_0^2}} \frac{\sin[(k_0 a \sin \theta_0)/2]}{(k_0 a \sin \theta_0)/2} \tag{1.39}$$

where a is the element size and λ_0 is the wavelength in free space. We consider square elements for simplicity. Using (1.39) in (1.38) we obtain

$$G_{\text{array}}^{\text{co}} = \sqrt{\frac{4\pi A}{\lambda_0^2}} \frac{\sin[(k_0 \sqrt{A/N} \sin \theta_0)/2]}{(k_0 \sqrt{A/N} \sin \theta_0)/2} \tag{1.40}$$

In (1.40) $A = Na^2$ is the total aperture area of the array. We assumed that the elements are touching each other, leaving no interelement spacing.

Equation (1.40) can be rearranged to obtain A in terms of N. The result is

$$A = \frac{4N}{k_0^2 \sin^2 \theta_0} \left[\sin^{-1} \left(\sqrt{\frac{\pi}{4N}} G_{\text{array}}^{\text{co}} \sin \theta_0 \right) \right]^2 \tag{1.41}$$

Using the scan gain at $\theta_0 = 10°$ as a parameter, we have plotted A versus N in Figure 1.9. Observe that the array aperture size decreases as the number of elements increases. To minimize the cost, the number of elements should be as low as possible. On the other hand, a smaller number of elements increases the aperture size, which may have packaging problem, particularly for space applications. The acceptable cost and the packaging constraint decide the operating point on the constant-gain curve.

It is interesting to observe that the minimum number of elements can be found from (1.41), because the argument of the sine-inverse function cannot exceed unity.

---

[4] That includes packaging, cost, etc.

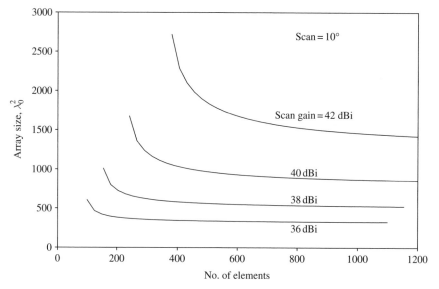

FIGURE 1.9 Array aperture size versus number of elements with the scan gain as a parameter.

Thus for a given scan gain at θ_0, the theoretical limit for the minimum number of elements becomes

$$N_{min} = \frac{1}{4}\pi[G^{co}_{array}\sin\theta_0]^2 \qquad (1.42)$$

The corresponding element size can be obtained from (1.41) as

$$a = \sqrt{\frac{A}{N}} = \frac{\lambda_0}{2\sin\theta_0} \qquad (1.43)$$

The above element size results in 3.92 dB lower scan gain at θ_0 than that at the bore-sight scan. These results generally differ for nonuniform array taper and other array elements.

1.3.3 Grating Lobes

A phased array may have several radiating lobes of peak intensities comparable with that of the desired beam. Such undesired radiating lobes are known as grating lobes. The grating lobes should not be confused with the side lobes. Usually, a side lobe has much lower peak intensity than that of a grating lobe. For quantitative understanding of grating lobes, let us consider the following array factor of a liner array [see (1.33)]:

$$AF(\theta) = B_1 + B_2 \exp\{jk_0 a(\sin\theta - \sin\theta_0)\} + \cdots$$
$$+ B_N \exp\{jk_0(N-1)a(\sin\theta - \sin\theta_0)\} \quad (1.44)$$

The above array factor is associated with the desired main beam radiation at a scan angle $\theta = \theta_0$. Notice that $AF(\theta)$ is a periodic function of $k_0 a \sin\theta$ of period 2π. To illustrate, we have plotted $AF(\theta)$ versus $\sin\theta$ for $a = 1.5\lambda_0, N = 10$ and $\theta_0 = 17.5°$ ($\sin\theta_0 = 0.3$) in Figure 1.10. The periodicity in the $\sin\theta$ space is $2\pi/(k_0 a) = \lambda_0/a$. Thus the first grating lobe peak adjacent to the main lobe occurs at $\sin\theta = \sin\theta_0 + 2\pi/(k_0 a)$. In general, the peaks occur at $\sin\theta = \sin\theta_0 \pm 2m\pi/(k_0 a)$, where m is a positive integer. In θ-space, the grating lobe peaks are located at

$$\theta_{gm} = \sin^{-1}\left(\sin\theta_0 \pm \frac{2m\pi}{k_0 a}\right) \quad m = 1, 2, 3, \cdots \quad (1.45)$$

Notice that all peaks have identical magnitudes. For the actual radiation pattern of the array the array factor must be multiplied by the element pattern as given in (1.33). Thus the grating lobe peaks will be modified by the element gain at the grating lobe location. The radiation pattern of the linear array is illustrated in Figure 1.11.

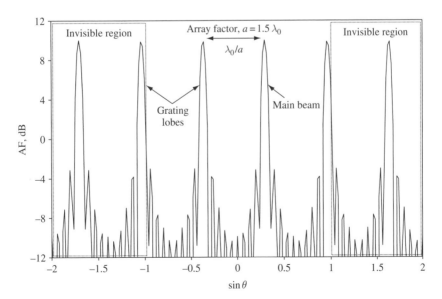

FIGURE 1.10 Array factor versus $\sin\theta$ plot to show grating lobes.

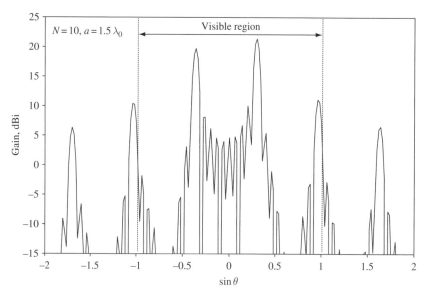

FIGURE 1.11   Radiation pattern of an array showing the effect of the element pattern.

It is interesting to observe that in the $\sin\theta$-space several grating lobes may exist outside the range of $\sin\theta$, where $|\sin\theta| > 1$. These grating lobes are not responsible for energy radiation in the visible space because the values of θ are not real. However, invoking Fourier spectrum analysis it can be shown that the lobes outside the visible space are related to the reactive energy (as opposed to the radiating energy) of the array. The radiating energy and the reactive energy together decide the active impedance of the array sources. Thus the grating lobes in the invisible region have significance in regard to impedance match [3].

From (1.45) we see that no real solution for θ_{gm} can be found if $2\pi/(k_0 a) > 2$ for all real values of θ_0. It follows that a scan array with element spacing less than $\frac{1}{2}\lambda_0$ does not have any grating lobe in the visible region.

Planar Array   For a two-dimensional planar array, the array factor can be expressed as

$$\mathrm{AF}(\theta, \phi) = \sum_{n=1}^{N} A_n \exp[jk_0 \sin\theta\{x_n \cos\phi + y_n \sin\phi\}] \qquad (1.46)$$

where (x_n, y_n) represents the coordinate of the nth element and A_n is its complex excitation coefficient. For the general lattice structure shown in Figure 1.12a, we can express the coordinate of the nth element as

$$x_n = pa + \frac{qb}{\tan\gamma} \qquad y_n = qb \qquad (1.47)$$

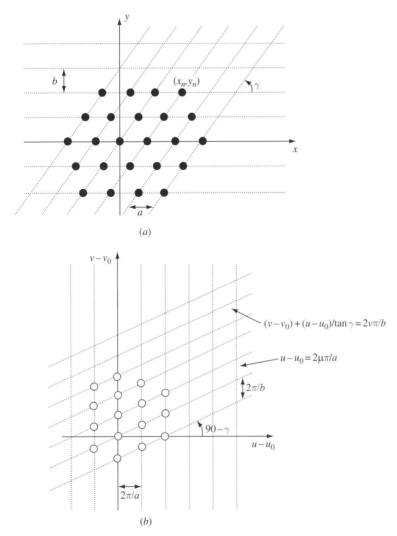

(a)

(b)

FIGURE 1.12 (a) A general lattice structure of an array represented by the symbol $[a, b, \gamma]$. (b) The Grating lobe locations of array in uv-plane.

where p and q are two integers representing that the nth element is situated at the pth oblique column and qth horizontal row. The unit cell area is $a \times b$ and γ denotes the cell angle. For brevity of representation, we will use the symbols $[a, b, \gamma]$ to represent such a cell. Notice, $\gamma = 90°$ for a rectangular lattice.

For simplicity of presentation we denote $k_0 \sin\theta \cos\phi$ as u and $k_0 \sin\theta \sin\phi$ as v. Thus the two-dimensional array factor can be expressed as

$$\text{AF}(u, v) = \sum_{n=1}^{N} A_n \exp[j\{x_n u + y_n v\}] \qquad (1.48)$$

In order to have a beam peak at (θ_0, ϕ_0), the complex excitation coefficient should be

$$A_n = B_n \exp[-j\{x_n u_0 + y_n v_0\}] \tag{1.49}$$

where B_n is a positive real number and (u_0, v_0) are given by

$$u_0 = k_0 \sin\theta_0 \cos\phi_0 \qquad v_0 = k_0 \sin\theta_0 \sin\phi_0 \tag{1.50}$$

The array factor thus can be written as

$$\mathrm{AF}(u,v) = \sum_{n=1}^{N} B_n \exp[j\{x_n(u-u_0) + y_n(v-v_0)\}] \tag{1.51}$$

Substituting the expressions of x_n and y_n from (1.47) into (1.51), we have

$$\mathrm{AF}(u,v) = \sum_{n=1}^{N} B_n \exp\left[j\left\{\left(pa + \frac{qb}{\tan\gamma}\right)(u-u_0) + qb(v-v_0)\right\}\right] \tag{1.52}$$

Rearranging (1.52) we obtain

$$\mathrm{AF}(u,v) = \sum_p \sum_q B_n \exp[jpa(u-u_0)]\exp\left[jqb\left\{(v-v_0) + \frac{u-u_0}{\tan\gamma}\right\}\right] \tag{1.53}$$

Notice, $\mathrm{AF}(u,v)$ will have peaks if the following two conditions satisfy simultaneously.

$$a(u-u_0) = 2\mu\pi \qquad b\left\{(v-v_0) + \frac{u-u_0}{\tan\gamma}\right\} = 2\nu\pi \tag{1.54}$$

where μ and ν are integers.

The above equations represent two sets of parallel lines on the u–v plane, as shown in Figure 1.12b. The intersection points represent the grating lobe locations. Notice that the grating lobes also lie on a triangular lattice. The minimum separation between the main lobe and the nearest grating lobes is the smaller number between $2\pi/b$ and $2\pi/(a\sin\gamma)$. The location of the nearest grating lobe determines the maximum scan angle of an array.

To illustrate this point further, let us consider an array with $\gamma = 60°$ and $b = a\sin\gamma$. This corresponds to an array with equilateral triangular grid. Suppose the array is required to scan within $\pm\theta_s$ in all planes with no grating lobe within the scan region. This implies that no grating lobe should exist inside the circle on the u–v plane representing the scan region (Figure 1.13). For an extreme situation (scanning to the edge of the circle on the $\phi = 90°$ plane, for instance), the nearest grating

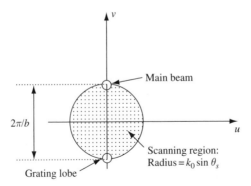

FIGURE 1.13 Main beam and grating lobe locations of an optimum array scanning at the edge-of-coverage region.

lobe on the scan plane must be situated at least at the diametrically opposite end of the circle. It follows that

$$2k_0 \sin \theta_s \leq \frac{2\pi}{b} = \frac{2\pi}{a \sin \gamma} \tag{1.55}$$

Thus, for an equilateral triangular grid, the distance between two elements (equal to a side of the equilateral triangle) must satisfy the following condition:

$$a \leq \frac{\lambda_0/\sin \gamma}{2 \sin \theta_s} = \frac{1.155\lambda_0}{2 \sin \theta_s} \tag{1.56}$$

For a square grid, $\gamma = 90°$; therefore for nonexistence of a grating lobe inside the scanning region, the interelement spacing should satisfy

$$a \leq \frac{\lambda_0/\sin \gamma}{2 \sin \theta_s} = \frac{\lambda_0}{2 \sin \theta_s} \tag{1.57}$$

We thus see that for satisfying the grating lobe constraint (no grating lobe inside the scan region) a triangular grid array allows larger element spacing (hence larger element size) than that of a square grid array. The former grid structure is preferable over the latter, because the former employs about 15% fewer elements for a given array aperture size.

1.3.4 Fixed-Value Phase Shifter Versus True Time Delay Phase Shifter

We have established that to produce a beam at a scan angle θ_s, the phase of the nth element of a linear array should be

$$\psi_n = -k_0(n-1)a \sin \theta_s = -\frac{2\pi f_0}{c}(n-1)a \sin \theta_s \tag{1.58}$$

where f_0 is the frequency of operation and c is the speed of light. In practice, a phase shifter is attached with each element for providing the required phase. Notice that the required phase is proportional to the operating frequency. Thus, for an ideal operation of a scanning beam array in a frequency band, the phase of the phase shifter should vary linearly with frequency. However, for a fixed-value phase shifter the phase does not change with the frequency. To minimize the error, the element phase is set with respect to the center frequency of the band. Thus at the center frequency the peak of the array beam occurs at the desired scan angle, but it deviates from the desired scan angle at other frequencies of the band. This effect is shown in Figure 1.14. This beam deviation can be estimated using (1.58). Suppose θ is the scan angle corresponding to the frequency f, which is different from the center frequency f_0. Then, for a fixed-phase shifter we can write

$$\psi_n = \frac{2\pi f_0}{c}(n-1)a\sin\theta_s = \frac{2\pi f}{c}(n-1)a\sin\theta \qquad (1.59)$$

yielding

$$\theta = \sin^{-1}\left(\frac{f_0}{f}\sin\theta_s\right) \qquad (1.60)$$

Writing $f = f_0 + \Delta f$ and $\theta = \theta_s + \Delta\theta$ in (1.60) and for a small frequency deviation, we deduce the beam deviation $\Delta\theta$ as

$$\Delta\theta \approx -\frac{\Delta f}{f_0}\tan\theta_s \qquad (1.61)$$

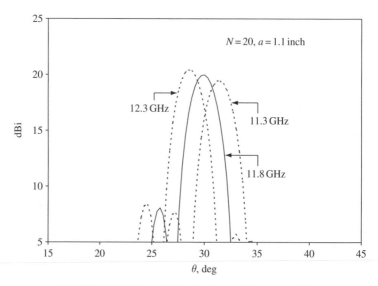

FIGURE 1.14   Beam deviation due to frequency shift.

Equation (1.61) implies that the beam deviation increases with frequency shift and with the scan angle as well. For a wide beam and a small scan angle this beam deviation may not pose much of a problem, but for a narrow beam and/or a wide scan angle, the gain reduces significantly with the frequency at the desired scan location. This gain variation with frequency causes signal distortion for a signal of finite bandwidth. For a uniform distribution the gain variation with frequency at a desired scan location can be estimated. For an array of N elements, the array factor is given by

$$\text{AF}(k, \theta) = \frac{\sin[Na(k\sin\theta - k_0\sin\theta_s)/2]}{\sin[a(k\sin\theta - k_0\sin\theta_s)/2]} \tag{1.62}$$

where k is the wave number corresponding to the frequency f. Setting $f = f_0 + \Delta f$, the magnitude of the array factor at the desired scan location is

$$\text{AF}(k, \theta_s) = \frac{\sin[Na\sin\theta_s\,\pi\,\Delta f/c]}{\sin[a\sin\theta_s\,\pi\,\Delta f/c]} \tag{1.63}$$

The gain loss at the desired scan location is therefore

$$\text{Gain loss in dB} = -20\log\left|\frac{\text{AF}(k, \theta_s)}{\text{AF}(k_0, \theta_s)}\right| = -20\log\left|\frac{\sin[Na\sin\theta_s\,\pi\,\Delta f/c]}{N\sin[a\sin\theta_s\,\pi\,\Delta f/c]}\right| \tag{1.64}$$

In the above gain loss expression, the frequency-independent element pattern is assumed. In Figure 1.15 we have plotted the gain loss versus scan angle for a 20-element linear array with element spacing of $1\lambda_0$. We used frequency deviations of 3 and 5%, respectively, above the center frequency. Expectedly, the more the frequency deviation, the more is the beam deviation and subsequent gain loss. Also, the gain loss increases with increasing array size. The gain loss limits the bandwidth of a scan beam array. For $N > 7$, the 3-dB-gain-loss bandwidth can be shown to be

$$\frac{2\,\Delta f}{f_0} \approx \frac{0.88\lambda_0}{L\sin\theta_s} \tag{1.65}$$

where $L = Na$ is the length of the array.

The above beam deviation and subsequent gain loss can be prevented if true time delay phase shifters are used. In a true time delay phase shifter, the phase shift is proportional to the operating frequency. Such phase shifters may be realized using nondispersive delay lines, for example transverse electromagnetic (TEM) lines.

1.3.5 Phase Quantization

A typical phase shifter has a finite number of quantized phase states. This is particularly true for a digitally controlled phase shifter employing electronic switches. For an N-bit phase shifter, the number of quantized phase states is 2^N; thus the discrete phase interval is $\Delta\psi = 2\pi/2^N$. Such phase shifters introduce phase quantization error because they cannot implement intermediate-phase values. In a scan beam

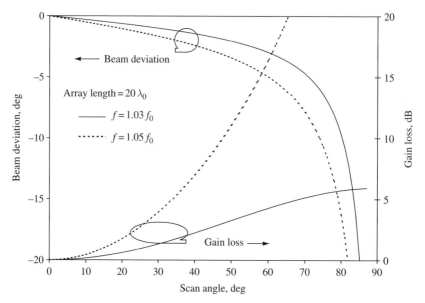

FIGURE 1.15 Gain loss due to frequency deviation from the designed frequency versus scan angle; $N = 20$, $a = \lambda_0$.

array, phase quantization introduces undesired quantization beams and affects gain and side-lobe performances.

To illustrate the phase quantization effect, consider a uniformly distributed continuous source with linearly progressed phase. The source function can be expressed as

$$f(x) = A \exp(-jKx) \qquad |x| \leq \frac{1}{2}L \tag{1.66}$$

The radiation pattern of the source is given by

$$F(\theta) = \int_{-L/2}^{L/2} f(x) \exp(jk_0 x \sin \theta)\,dx \tag{1.67}$$

Substituting (1.66) in (1.67) and performing the integration, we obtain

$$F(\theta) = \frac{AL \sin[(K - k_0 \sin \theta)L/2]}{(K - k_0 \sin \theta)L/2} \tag{1.68}$$

The beam peak occurs at $\theta_0 = \sin^{-1}(K/k_0)$ and no grating lobe exists. If the source phase is quantized (as shown in Figure 1.16) with a quantization interval $2\pi/2^N$, then the continuous source appears to be an array of discrete sources. The length of a discrete source that has a uniform phase distribution (as per Figure 1.16) is

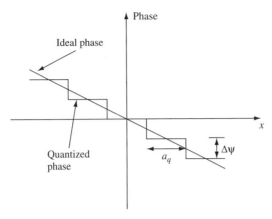

FIGURE 1.16 Phase quantization of a linearly progressed phase.

given by $a_q = 2\pi/(K2^N)$. We now make an assumption that the source length L is divisible by the length a_q. With this assumption we can say that the phase quantization modifies the continuous source to an array of L/a_q elements. The phase difference between two adjacent elements is $2\pi/2^N$. The radiation pattern for such an array can be represented in terms of the array factor and the element pattern. The array may produce grating lobes in the visible region. These grating lobes, primarily resulting from the phase quantization, are often called quantization lobes. Typically, the grating lobe peak nearest to the main beam peak occurs at $\theta_g = \sin^{-1}[\sin\theta_0 - 2\pi/k_0 a_q]$. The main lobe peak occurs at θ_0. The intensity at the grating lobe peak, which is proportional to the element pattern intensity at that location, can be represented as

$$G(\theta_g) = \frac{B\sin[k_0 a_q \sin\theta_g/2]}{k_0 a_q \sin\theta_g/2} \tag{1.69}$$

The intensity at the main beam peak at θ_0 is

$$G(\theta_0) = \frac{B\sin[k_0 a_q \sin\theta_0/2]}{k_0 a_q \sin\theta_0/2} \tag{1.70}$$

The relative intensity of the grating (quantization) lobe thus becomes

$$\frac{G(\theta_g)}{G(\theta_0)} = -\frac{\sin\theta_0}{\sin\theta_g} = -\frac{\sin\theta_0}{\sin\theta_0 - 2\pi/k_0 a_q} = \frac{1}{2^N - 1} \tag{1.71}$$

The gain loss due to phase quantization is proportional to the relative element gain at the beam peak. The gain loss is given by

$$\text{Gain loss} = \frac{\sin(k_0 a_q \sin\theta_0/2)}{k_0 a_q \sin\theta_0/2} = \frac{2^N}{\pi}\sin\left(\frac{\pi}{2^N}\right) \tag{1.72}$$

For a 3-bit phase shifter, the gain loss is about 0.2 dB. At this point it should be mentioned that the theory presented above is applicable for a quantized phase of a continuous source with uniform amplitude taper. For a nonuniform amplitude taper, the phase quantization results in an array of dissimilar element aperture fields; thus the above theory does not apply rigorously. Nevertheless, the quantization beams and gain loss still occur and the results can be used with some acceptable errors, particularly for moderate-amplitude tapers. Furthermore, the results can also be used for an array (as opposed to a continuous source) where phase quantization results in a larger a_q as compared with the actual array element spacing.

The phase quantization also causes scan angle quantization, leading to beam pointing error, as shown in Figure 1.17. The pointing error is prominent in the vicinity of a scan angle for which the phase quantization error is zero. For instance, for the bore-sight scan, all the elements may be set at zero phase. The phase of a phase shifter does not change its phase state unless the desired phase value changes by at least one-half of the quantization interval. Thus, the beam remains at the bore sight unless the phase of the edge elements changes by at least one-half of the quantization interval. It follows that the scan locations and the corresponding maximum beam pointing error can be estimated from the following equations:

$$k_0 a \sin \theta_n = \frac{n2\pi}{2^N} \qquad k_0 L(\Delta \sin \theta_n) = \frac{2\pi}{2^N} \qquad (1.73)$$

In Figure 1.17, we used $a = 0.6\lambda_0$, $N = 3$, and $L = 6\lambda_0$. Observe that the maximum beam pointing error occurs near 0 and 12° scan angles within the plotting range.

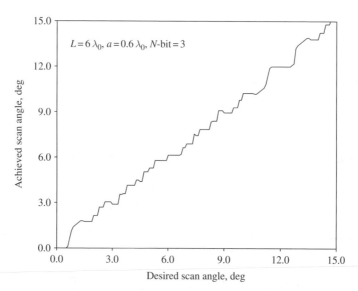

FIGURE 1.17 Desired versus achieved scan angle in an array with phase quantization.

The estimated beam pointing error is about $1.2°$ (only one-half of the error is shown near $0°$ scan), as can be observed from Figure 1.17.

1.4   LINEAR ARRAY SYNTHESIS

An array synthesis essentially involves determination of the amplitude and phase of the elements that will produce a desired beam pattern. In this section we present commonly used array synthesis methods. The first part deals with linear array synthesis, which is followed by a continuous line source synthesis. For a linear array it is convenient to express the array factor in terms of a simple polynomial function introduced by Schelkunoff [4]. The polynomial representation of the array factor enables us to understand several important aspects of a phased array antenna, as we shall see shortly.

1.4.1   Array Factor: Schelkunoff's Polynomial Representation

We have established that for a linear array of N elements the array factor is given by

$$\mathrm{AF}(z) = A_0 + A_1 z + A_2 z^2 + \cdots + A_{N-1} z^{N-1} \qquad (1.74)$$

where $A_0, A_1, A_2, \ldots, A_{N-1}$ are complex excitation coefficients and z is the complex exponential factor given by

$$z = \exp(jk_0 a \sin \theta) \qquad (1.75)$$

Observe that the array factor is now represented by a polynomial of z, known as Schelkunoff's polynomial. For a real value of $\sin \theta$, $|z| = 1$, which represents a unit circle in the complex z-plane, known as Schelkunoff's unit circle (Figure 1.18). On the circumference of the unit circle, a segment exists where $|\sin \theta| < 1$. This segment is known as the visible region and any point lying outside the segment,

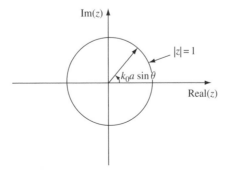

FIGURE 1.18   Schelkunoff's unit circle.

whether on the circumference or not, belongs to the invisible region. At the invisible region, θ does not assume a real value.

The polynomial in (1.74) can be expressed in terms of a product of $N-1$ factors as

$$\text{AF}(z) = A_{N-1}(z - z_1)(z - z_2) \cdots (z - z_{N-1}) \tag{1.76}$$

where $z_1, z_2, \ldots, z_{N-1}$ are the zeros of the polynomial. Expanding the right-hand side of (1.76) and comparing with (1.74), one obtains the relative amplitudes of the array elements in terms of the zeros of the polynomial. The zeros play an important role in the array synthesis procedure.

A zero is associated with a null location of the array. A side lobe exists between two successive zeros in the visible segment of the unit circle. The locations of the zeros on the unit circle determine the side-lobe levels. For a low side-lobe level, the two successive zeros should be close to each other but far from the main beam location. This can be understood from the array factor expression in (1.76). The magnitude of the array factor is given by

$$|\text{AF}(z)| = |A_{N-1}| \cdot |z - z_1| \cdot |z - z_2| \cdots |z - z_{N-1}| \tag{1.77}$$

Suppose the main beam is at $\theta = 0°$, that is, at $z = 1$. The side-lobe peak between the two successive zeros z_1 and z_2, for example, is located approximately at $z \approx (z_1 + z_2)/2$. Thus the side-lobe ratio (SLR) becomes

$$\text{SLR} = \frac{|\text{AF}((z_1 + z_2)/2)|}{|\text{AF}(0)|} \approx \frac{|z_1 - z_2|^2 \cdot |(z_1 + z_2)/2 - z_3| \cdots |(z_1 + z_2)/2 - z_{N-1}|}{|(1 - z_1) \cdot (1 - z_2) \cdot (1 - z_3) \cdots (1 - z_{N-1})|} \tag{1.78}$$

From (1.78) we observe that the SLR in decibels becomes large[5] if $|z_1 - z_2|$ is small, that is, the zeros are closely spaced on the unit circle. Furthermore, the zeros must be far from the bore-sight point on the unit circle in order to have larger magnitude of the denominator. Thus, a *large SLR is achieved at the expense of beam broadening*. This information would be useful in the synthesis process we consider next.

1.4.2 Binomial Array

For a binomial array of N elements, Schelkunoff's polynomial has only one zero. The array factor can be expressed as

$$\text{AF}(z) = (z - z_1)^{N-1} \tag{1.79}$$

---

[5] The SLR in decibels is defined as $\text{SLR} = 20 \log |G(\theta_0)/G(\theta_{\text{SL}})|$, where $G(\theta_0) =$ gain at the beam peak and $G(\theta_{\text{SL}}) =$ gain at the side-lobe peak.

A binomial series expansion for the right-hand side of (1.79) yields

$$\mathrm{AF}(z) = z^{N-1} - {}^{N-1}C_1\, z^{N-2} z_1 + {}^{N-1}C_2\, z^{N-3} z_1^2 - \cdots + (-1)^{N-1}\, {}^{N-1}C_{N-1} z_1^{N-1} \quad (1.80)$$

The excitation coefficients are readily obtained from the coefficients of the polynomial. Notice that the array factor has only one null at $z = z_1$; thus the array has no visible side lobe. Consider an array of element spacing $a = \lambda_0/2$. This particular array does not have any grating lobe and the entire circumference of the unit circle pertains to the visible region. We assume the beam peak at $\theta = 0°$ and set the null location at $\theta = 90°$. The zero of the polynomial is given by

$$z_1 = \exp(jk_0 a \sin 90°) = -1 \quad (1.81)$$

The relative excitation coefficient of the nth element ($1 \leq n \leq N$) becomes

$$A_n = {}^{N-1}C_{N-n} = \frac{(N-1)!}{(N-n)!((n-1)!} \quad (1.82)$$

Figure 1.19 shows the excitation coefficients of a 10-element binomial array. Figure 1.20 shows the corresponding array pattern. As can be seen, the amplitude distribution is highly tapered. Evidently, the array pattern does not have any side lobe, as expected. The ATE as defined in (1.30) is

$$\mathrm{ATE} = \frac{|\sum_{n=1}^{N} A_n|^2}{N \sum_{n=1}^{N} |A_n|^2} = \frac{2^{2N-2}}{N^{2N-2} C_{N-1}} \quad (1.83)$$

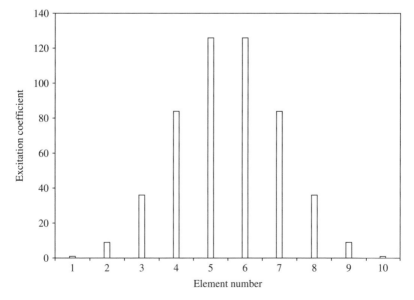

FIGURE 1.19 Excitation coefficients of a binomial array with 10 elements.

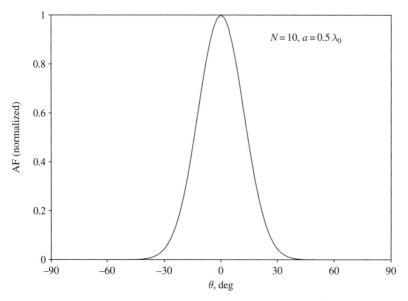

FIGURE 1.20   Array factor (AF) plot of the binomial array of Figure 1.19.

For a large array, Starling's approximation[6] for the factorial of a large number can be employed to replace the factorial terms in (1.83). The approximate ATE becomes

$$\text{ATE} \approx \sqrt{\frac{\pi}{N}\left(1-\frac{1}{N}\right)} \tag{1.84}$$

In Figure 1.21 we have plotted the ATE versus the number of elements, N, deduced in (1.83). Also plotted is the ATE using the approximate expression in (1.84). As noted, the approximation matches well within 1% for $N > 15$. The ATE is over 80% when $N < 5$ and decreases monotonically as N increases. For $N > 11$, the ATE becomes lower than 50%. Thus, from the array aperture efficiency point of view, a binomial array is not preferable, unless the array size is sufficiently small.

It must be pointed out that the ATE deduced above assumes $\lambda_0/2$ element spacing. However, the ATE expression in (1.83) remains unchanged for a different element spacing if $z_1 = -1$ and the beam peak is set at $\theta = 0°$. For element spacing greater than $\lambda_0/2$, nulls appear in the visible region and grating lobes emerge. For element spacing less than $\lambda_0/2$, the nulls occur in the invisible region and no grating lobes exist in the visible region.

---

[6] For a large M, Starling's approximation is given by $M! \approx \sqrt{2\pi M}\exp(-M)M^M$.

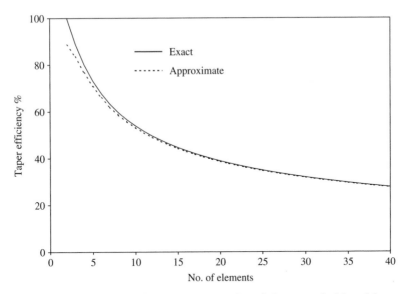

FIGURE 1.21 Array taper efficiency versus number of elements of a binomial array.

1.4.3 Dolph–Chebyshev Array

The Dolph–Chebyshev (D–C) array synthesis was developed by C. L. Dolph, who employed a Chebyshev polynomial function as the array factor [5]. This synthesis procedure allows designing an array with a desired SLR and with an improved array taper efficiency. A distinguished feature of a D–C array lies on its uniform side-lobe levels. To conduct such an array synthesis, we must understand important properties of the Chebyshev polynomials. The Chebyshev polynomial of order m, where m is an integer, is defined as [6]

$$T_m(x) = \begin{cases} \cos(m\cos^{-1}x) & |x| \leq 1 \\ \cosh(m\cosh^{-1}x) & x > 1 \\ (-1)^m\cosh(m\cosh^{-1}|x|) & x < -1 \end{cases} \qquad (1.85)$$

It can be verified using simple trigonometry that $T_m(x)$ is a polynomial of x of order m and the following recurrence relation holds:

$$T_{m+1}(x) = 2xT_m(x) - T_{m-1}(x) \qquad (1.86)$$

The first few polynomials are given by

$$T_0(x) = 1 \qquad T_1(x) = x \qquad T_2(x) = 2x^2 - 1 \qquad T_3(x) = 4x^3 - 3x \qquad (1.87)$$

From (1.85) we see that the magnitude of a Chebyshev polynomial lies between zero and unity if $|x| < 1$ and exceeds unity if $|x| > 1$, as depicted in Figure 1.22.

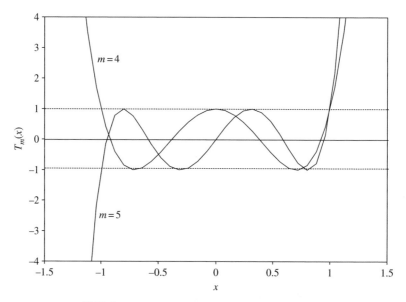

FIGURE 1.22   Chebyshev polynomial functions.

This property of the polynomial is utilized to construct the array factor. Toward that end, the variable x is defined in such a way that $|x| < 1$ corresponds to the side-lobe region of the array, while the bore sight corresponds to a point where $|x| > 1$. A D–C array has symmetrical amplitude distribution.

Consider the following array factor of a linear array with N elements of Element spacing a:

$$\text{AF}(\theta) = \sum_{-(N-1)/2}^{(N-1)/2} A_n \exp(jk_0 na \sin \theta) \qquad (1.88)$$

Without loss of generality we can assume a bore-sight beam. For a scan beam, the phase factors can be accommodated inside the arguments of the exponential functions. For symmetric excitations, $A_n = A_{-n}$, so that (1.88) reduces to

$$\text{AF}(\theta) = A_0 + 2 \sum_{n=1}^{(N-1)/2} A_n \cos(n\psi) \qquad (1.89)$$

with $\psi = k_0 a \sin \theta$. We assume odd number of elements; however, the procedure is similar for even number of elements. The above array factor is now equated with the Chebyshev polynomial of order $N-1$. Thus, a correspondence between the variable x and the array factor variable ψ needs to be found. Inspecting (1.85) and (1.89), $x = \cos(\psi/2)$ could be a possible choice because this relation transforms the array factor in (1.89) into a polynomial of x of order $N-1$. To verify, we substitute $\psi = 2\cos^{-1} x$ in (1.89) and obtain

$$\mathrm{AF}(x) = A_0 + 2 \sum_{n=1}^{(N-1)/2} A_n \cos(2n\cos^{-1} x) \qquad (1.90)$$

The right-hand side of (1.90) is a sum of N Chebyshev polynomials with the highest order $N-1$; thus the sum is a polynomial of x of order $N-1$.

At this point it is important to realize that the array factor should be represented by a single Chebyshev polynomial; otherwise the simplicity of the synthesis procedure would be lost. The order of the polynomial should be $N-1$. Notice that the entire visible region is defined by $|x| < 1$ because ψ is real in the visible region. Therefore, if we attempt to express the array factor by a single Chebyshev polynomial, then a scaling of the variable would be necessary, because at the bore sight the argument of the Chebyschev polynomial must be greater than unity. A simple choice is introducing a new variable $x' = cx$, where c is a constant greater than 1. Thus we write:

$$\mathrm{AF}(x) = A_0 + 2 \sum_{n=1}^{(N-1)/2} A_n \cos(2n\cos^{-1} x) = BT_{N-1}(cx) \qquad (1.91)$$

In (1.91), B is a constant coefficient and c is determined from the SLR. The main-beam location corresponds to $x = 1$ and the side-lobe region corresponds to $|cx| < 1$. In the region $|cx| < 1$, $T_{N-1}(cx)$ becomes a cosine function with real argument, implying that the side lobes have uniform peaks of magnitude B. Thus the SLR of the array, R, is given by

$$R = \left| \frac{BT_{N-1}(c)}{B} \right| = T_{N-1}(c) = \cosh[(N-1)\cosh^{-1} c] \qquad (1.92)$$

The above equation can be solved for c and the solution is[7]

$$c = \frac{1}{2}[(R + \sqrt{R^2-1})^{1/(N-1)} + (R - \sqrt{R^2-1})^{1/(N-1)}] \qquad (1.93)$$

The next task is to determine the excitation coefficients A_n's. This can be done by expressing Chebyshev functions in (1.91) as polynomials of x and then comparing term by term. This could be a tedious task, particularly for a large N. An alternate approach is using the basic form of the array factor and then applying a Fourier expansion method. Toward that end, we rewrite (1.91) as

$$\sum_{n=-(N-1)/2}^{(N-1)/2} A_n \exp(jn\,\psi) = BT_{N-1}\{c\cos(\psi/2)\} \qquad (1.94)$$

---

[7] This can be shown using $\cosh^{-1} R = \ln[R + \sqrt{R^2-1}]$.

Recall that $\psi = k_0 a \sin\theta$ and $A_{-n} = A_n$. The left-hand side of (1.94) is a Fourier series expansion with A_n's as Fourier coefficients. The Fourier coefficients are given by the following integral:

$$A_m = \frac{B}{2\pi} \int_0^{2\pi} T_{N-1}\left(c\cos\frac{\psi}{2}\right)\exp(-jm\psi)\,d\psi \qquad (1.95)$$

Observe that the integral in (1.95) represents the Fourier inverse of the array pattern. We also know that this integral vanishes if $|m| \geq (N+1)/2$ because the array consists of only N elements. Thus, the Fourier inverse of the array pattern is space limited. According to Shannon's sampling theorem [7] discrete samples of the array pattern will construct the array excitations without error (see Section A.1 in the Appendix). In regard to the Fourier integral in (1.95), the sampling interval should be determined from $\Delta m\,\Delta\psi \leq 2\pi$. Because $\Delta m = N$, the largest possible sampling interval should be $\Delta\psi = 2\pi/N$. Using this sampling interval we discretize the integrand in (1.95) and obtain

$$A_m = \frac{B}{N}\sum_{i=1}^{N} T_{N-1}\left(c\cos\frac{\psi_i}{2}\right)\exp(-jm\psi_i) \qquad (1.96)$$

where $\psi_i = 2\pi i/N$. It can be shown that the imaginary part of the right-hand side of (1.96) vanishes, leaving

$$A_m = \frac{B}{N}\sum_{i=1}^{N} T_{N-1}\left(c\cos\frac{\psi_i}{2}\right)\cos(m\psi_i) \qquad (1.97)$$

Ignoring the unimportant constant factor outside the summation sign of (1.97), we write

$$A_m = \sum_{i=1}^{N} T_{N-1}\left(c\cos\frac{\psi_i}{2}\right)\cos(m\psi_i) \qquad \text{for } N \text{ odd} \qquad (1.98)$$

For even number of elements, A_m turns out to be

$$A_m = \sum_{i=1}^{N} T_{N-1}\left(c\cos\frac{\psi_i}{2}\right)\cos\left[\left(m-\frac{1}{2}\right)\psi_i\right] \qquad \text{for } N \text{ even} \qquad (1.99)$$

Another approach of determining A_n's is finding the zeros of $T_{N-1}(cx)$ and using them to determine the zeros for Schelkunoff's polynomial. The zeros then can be used to construct the array factor in the form of Schelkunoff's polynomial and hence the excitation coefficients can be obtained directly.

The amplitude distribution of an 11-element D–C array is shown in Figure 1.23. The SLR for this array was considered as 20 dB. The radiation pattern of the array with element spacing $0.5\lambda_0$ is shown in Figure 1.24. Expectedly, uniform side lobes are observed. The nulls are not evenly spaced in θ.

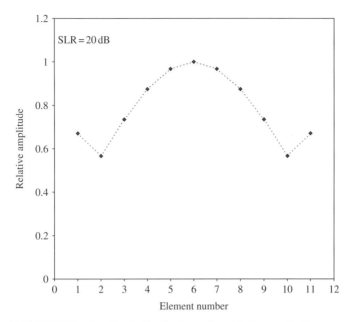

FIGURE 1.23 Amplitude distributions of an 11-element D–C array.

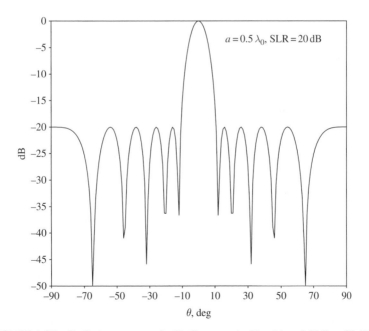

FIGURE 1.24 Radiation pattern of a D–C array win $N = 11$ and SLR $= 20$ dB.

It is worth pointing out that all the nulls in a D–C array may not lie inside the visible region if $a < \lambda_0/2$. For $a < \lambda_0/2$, the number of visible nulls is approximately given by the number $2(N-1)a/\lambda_0 - 1$. The SLR has an effect also on the number of visible nulls. This number decreases with an increase in the SLR. Thus a smaller element spacing does not necessarily improve the quality of the array factor in a regular D–C array. A modified version of the D–C array synthesis procedure for smaller element spacing can be found in [8].

Beam Width and Array Taper Efficiency   The 3-dB beam width of a D–C array can be determined by solving the following equation:

$$T_{N-1}\left(c\cos\frac{\psi}{2}\right) = \frac{R}{\sqrt{2}} \tag{1.100}$$

In (1.100), $\psi = k_0 a \sin\theta'$, where $2\theta'$ is the 3-dB beam width. For $R > \sqrt{2}$, $\sin\theta'$ is given by

$$\sin\theta' = \frac{1}{k_0 a} 2\cos^{-1}\left\{\frac{1}{c}\cosh\left[\frac{\cosh^{-1}(R/\sqrt{2})}{N-1}\right]\right\} \tag{1.101}$$

Using the expression of c from (1.93), (1.101) can be simplified to

$$\sin\theta' = \frac{2}{k_0 a}\cos^{-1}\left[2^{-1/\{2(N-1)\}}\frac{\{R+\sqrt{R^2-2}\}^{1/(N-1)}+\{R-\sqrt{R^2-2}\}^{1/(N-1)}}{\{R+\sqrt{R^2-1}\}^{1/(N-1)}+\{R-\sqrt{R^2-1}\}^{1/(N-1)}}\right] \tag{1.102}$$

From (1.102) we observe that the term inside the square bracket decreases as R increases. Thus, the inverse-cosine function increases with R, that is, the beam width increases with the SLR. This effect is shown in Figure 1.25 for two arrays.

The ATE versus SLR for two arrays with $N = 11$ and $N = 41$, respectively, are plotted in Figure 1.26. It is observed that ATE initially increases, reaches to a peak, and then monotonically decreases with the SLR. The ATE peaks at two different SLR values for the two arrays. For $N = 11$, ATE peaks at a smaller SLR than that with $N = 41$. This effect can be explained in the following way. The ATE depends on two factors, namely the power in the main lobe and the beam width. The power in the main lobe increases as the SLR increases. On the other hand, the beam width also increases with the SLR. These two factors have opposing effects on the ATE. At the lower range of SLR the main-lobe power increases rapidly with SLR. Therefore at this range the main-beam power has the dominating effect on ATE. At the upper range of SLR the main-lobe power almost reaches to saturation; thus ATE is mostly decided by the beam width. Since the beam width increases

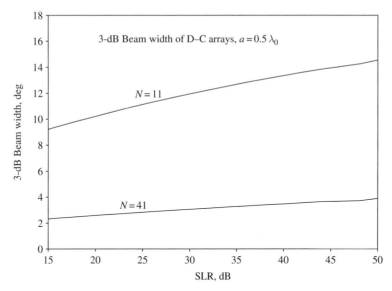

FIGURE 1.25 Dolph–Chebyshev arrays showing that larger SLR corresponds to a wider beam.

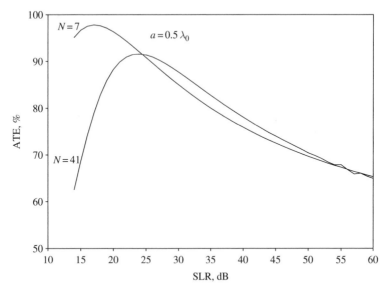

FIGURE 1.26 Array taper efficiency versus SLR for two D–C arrays.

with SLR, the ATE decreases. Furthermore, the power in the side lobes is smaller for a smaller N than that with a larger N. Therefore in the case of a smaller array, the main-beam power controls ATE only within a limited range of SLR, implying that the ATE peaks at a smaller SLR.

1.4.4 Taylor Line Source Synthesis

Taylor's line source synthesis concerns synthesis of continuous one-dimensional source (line source) distribution. We know that for a continuous line source of length l and source distribution function $f(x)$, the radiation pattern is

$$F(\theta) = C \int_{-l/2}^{l/2} f(x) \exp(jk_0 x \sin\theta)\, dx \qquad (1.103)$$

In (1.103) C is a constant. The integral of (1.103) can be expressed in a simple form by introducing two variables, p and u, that are related to x and $\sin\theta$, respectively, as

$$p = \frac{2\pi x}{l} \qquad u = \frac{k_0 l \sin\theta}{2\pi} \qquad (1.104)$$

The radiation pattern, expressed as a function of u, simplifies to

$$F(u) = B \int_{-\pi}^{\pi} g(p) \exp(jpu)\, dp \qquad (1.105)$$

with $B = Cl/2\pi$ and $g(p) = f(pl/2\pi)$. The above representation is very convenient to use for synthesizing a continuous line source.

Dolph–Chebyshev Pattern for Continuous Source Taylor's synthesis is evolved from D–C synthesis, which has been developed in the previous section. We have seen that the D–C array factor for an N-element array is represented by the Chebyshev polynomial of order $N - 1$. In the following development we will determine the radiation pattern of a continuous line source that evolves from the D–C array factor. Toward that effort we will apply the limiting conditions that N approaches infinity and the element spacing, a, approaches zero keeping the line source length $l(= Na)$ finite. Recall that the D–C array factor is given by

$$F(\psi) = T_{N-1}\left(c \cos\frac{\psi}{2}\right) \qquad (1.106)$$

with $\psi = k_0 a \sin\theta$. Using $Na = l$ we obtain $\psi = 2\pi u/N$, where u is defined in (1.104). The SLR is equal to $T_{N-1}(c)$. We define a parameter A such that $T_{N-1}(c) = \cosh(\pi A)$. This yields

$$c = \cosh\left(\frac{\pi A}{N-1}\right) \approx \cosh\left(\frac{\pi A}{N}\right) \qquad (1.107)$$

because N is infinitely large. Using (1.107) in (1.06) we express the array factor as a function of u as below:

$$F(u) = T_{N-1} \left\{ \cosh\left(\frac{\pi A}{N}\right) \cos\left(\frac{\pi u}{N}\right) \right\}$$

$$= \cos\left[(N-1)\cos^{-1}\left\{ \cosh\left(\frac{\pi A}{N}\right) \cos\left(\frac{\pi u}{N}\right) \right\} \right] \qquad (1.108)$$

We now expand the functions inside the second bracket in Taylor's series and retain the dominant terms. Then using the limiting condition that N approaches infinity, we deduce

$$F(u) = \cos(\pi\sqrt{u^2 - A^2}) \qquad |u| > A \qquad (1.109a)$$

For $|u| < A$, $F(u)$ becomes

$$F(u) = \cosh(\pi\sqrt{A^2 - u^2}) \qquad |u| < A \qquad (1.109b)$$

Source Distribution   The source distribution of the continuous source can be obtained by taking the inverse Fourier transform of the far-field pattern $F(u)$. From (1.105), the source distribution $g(p)$ is obtained as

$$g(p) = \frac{1}{2\pi B} \int_{-\infty}^{\infty} F(u)\exp(-jpu)\,du \qquad (1.110)$$

Recall that $g(p)$ exists only in the finite range of p from $-\pi$ to π. Thus $g(p)$ can be constructed from the samples of $F(u)$ with Shannon's sampling interval of $\Delta u = 2\pi/(2\pi) = 1$. Applying Shannon's sampling theorem and ignoring the unimportant $1/2\pi B$ factor, we obtain

$$g(p) = \sum_{n=-\infty}^{\infty} F(n)\exp(-jnp) \qquad (1.111)$$

The source distribution expression in (1.111) is valid for any finite source distribution that has a radiation pattern $F(u)$. For a symmetric radiation pattern, $F(n) = F(-n)$; thus the right-hand side of (1.111) simplifies to

$$g(p) = F(0) + 2\sum_{n=1}^{\infty} F(n)\cos(np) \qquad (1.112)$$

In the case of the D–C pattern in (1.109), the continuous source distribution becomes

$$g(p) = \cosh(\pi A) + 2\sum_{n=1}^{\infty} \cos(\pi\sqrt{n^2 - A^2})\cos(np) \qquad (1.113)$$

It is interesting to see that the infinite series in (1.113) is oscillatory and thus nonconvergent. This indicates that *a Dolph–Chebyshev pattern can never be realized by a continuous source distribution*. This can also be understood from Parseval's theorem [9]. According to the theorem, the following relation must hold (equivalent to power conservation):

$$B^2 \int\limits_{-\pi}^{\pi} g^2(p)\,dp = \frac{1}{2\pi} \int\limits_{-\infty}^{\infty} |F(u)|^2\,du \qquad (1.114)$$

For $F(u) = \cos(\pi\sqrt{u^2 - A^2})$, the right-hand side of (1.114) is a divergent integral. This indicates that in order to satisfy (1.114), $g(p)$ must be unbounded at least at some finite number of points within the interval $-\pi < p < \pi$, which contradicts with the definition of a bounded continuous source distribution. On the other hand, a finite array with discrete number of elements satisfies (1.114) because a discrete element is represented by a Dirac delta function, which is unbounded. Thus, in principle, a D–C pattern can be realized only by an array of discrete sources, as we have seen in the previous section.

In reality, however, a discrete source (a Dirac delta source) is replaced by an element that has a finite aperture length because a Dirac delta source can never be implemented due to its infinite input impedance (or admittance). The nonisotropic element pattern (due to nonzero aperture size) modifies the ideal D–C array pattern of an array. Thus, an ideal D–C pattern is not realizable using practical source elements. For that matter, this argument is equally valid for other types of arrays, such as binomial arrays.

Taylor Line Source   We have seen that the D–C pattern function $F(u)$ defined in (1.109) is not realizable by a continuous and bounded source because the integral on the right-hand side of (1.114) is nonconvergent. In order to satisfy the convergence criterion, T. T. Taylor modified the D–C array pattern. Taylor developed a new pattern function such that it closely resembles a scaled D–C pattern function at the bore sight and near-in side-lobe region and closely resembles $\sin(\pi u)/(\pi u)$ in the far-out region including the invisible region [10]. To implement this novel idea, the first few zeros are taken from the zeros of the D–C array pattern, and the remaining zeros are taken from zeros of the $\sin(\pi u)/(\pi u)$ function. In order to make a smooth transition between two different sets of zeros, Taylor defined two parameters σ and $\bar{n}$, where σ is the scale factor for the scaled D–C pattern and $\bar{n}$ is the number of "equal" side lobes. The scaled D–C pattern function used by Taylor is

$$F_1(u) = \cos\left[\pi\sqrt{\left(\frac{u}{\sigma}\right)^2 - A^2}\right] \qquad (1.115)$$

Accordingly, the zeros are

$$u_n = \sigma\sqrt{\frac{(2n-1)^2}{4} + A^2} \qquad n = 1, 2, \ldots, \bar{n} \qquad (1.116)$$

The remaining zeros are the zeros of $\sin(\pi u)/(\pi u)$; thus

$$u_n = n \qquad n > \bar{n} \tag{1.117}$$

Taylor considered that $u_{\bar{n}} = \bar{n}$ in order to have the zeros spaced at a regular order. This relation yields

$$\sigma = \frac{\bar{n}}{\sqrt{A^2 + (\bar{n} - 1/2)^2}} \tag{1.118}$$

With the zeros, the pattern function of a symmetrical array becomes

$$F(u) = \prod_{n=1}^{\bar{n}-1}\left(1 - \frac{u^2}{u_n^2}\right) \prod_{n=\bar{n}}^{\infty}\left(1 - \frac{u^2}{n^2}\right) \tag{1.119}$$

Now utilizing the relation $\sin(\pi u)/(\pi u) = \prod_{n=1}^{\infty}(1 - u^2/n^2)$, we obtain

$$F(u) = \frac{\sin(\pi u)}{\pi u} \prod_{n=1}^{\bar{n}-1} \frac{1 - u^2/u_n^2}{1 - u^2/n^2} \tag{1.120}$$

The source distribution $g(p)$ can be obtained invoking the relation in (1.112) because $g(p)$ exists only within the interval $-\pi \leq p \leq \pi$ (space limited). This can be ascertained by performing the complex integration of the inverse transformation using appropriate contour deformation (see problem 1.16).

Notice that $F(n)$ is zero for $n \geq \bar{n}$. Thus the series in (1.112) will have only a finite number of terms, given by

$$g(p) = F(0) + 2 \sum_{n=1}^{\bar{n}-1} F(n)\cos(np) \tag{1.121}$$

Figure 1.27 depicts Taylor's line source distribution for $\bar{n} = 6$ and SLR of 30 dB. The edge taper was 11.6 dB. In this particular case the distribution is monotonically decreasing from the center to the end but may not be such for other cases. Figure 1.28 is the Taylor radiation pattern corresponding to the source distribution of Figure 1.27. Observe that first five side lobes have *almost uniform* heights because in this region the pattern is dominated by the D–C pattern function. The influence of the $\sin(\pi u)/(\pi u)$ function restrains from being exactly uniform. The SLR is very close to 30 dB, as desired.

Beam Width and Aperture Efficiency   We have seen that the main-lobe region is controlled by the scaled D–C pattern in (1.115); thus the beam width of a Taylor pattern would be approximately σ times the beam width of the D–C array of identical length and SLR. It is found numerically that the value of σ varies from 1.13 to 1.03 for $\bar{n}$ varying from 2 to 10 and SLR varying from 20 to 50 dB. This

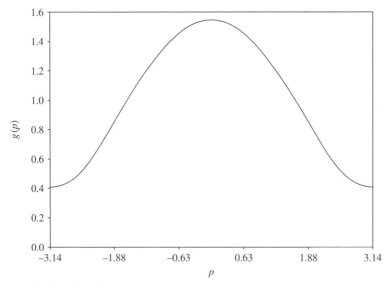

FIGURE 1.27   Taylor's line source distribution with $\bar{n} = 6$, SLR=30 dB.

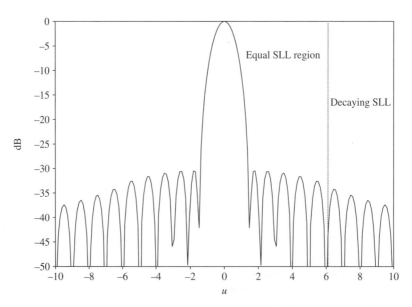

FIGURE 1.28   Taylor's pattern with $\bar{n} = 6$, SLR $= 30$ dB.

implies that the Taylor pattern is 13–3% wider than the D–C pattern of comparable size and SLR. For instance, a Taylor pattern with 25 dB SLR realized with $\bar{n} = 6$ is about 7% wider than the D–C pattern of comparable SLR. The larger the value of $\bar{n}$, the closer are the beam widths.

Following the definition of the ATE of a discrete array the aperture efficiency (AE) of a continuous source is defined as

$$AE = \frac{[\int_{-\pi}^{\pi} g(p)\,dp]^2}{2\pi \int_{-\pi}^{\pi} g^2(p)\,dp} = \frac{1}{1 + 2\sum_{n=1}^{\bar{n}-1} F^2(n)} \qquad (1.122)$$

In Figure 1.29 we have plotted AE versus SLR with $\bar{n} = 4, 8, 40$. The AE monotonically decreases with SLR for a small $\bar{n}$; however, for a large $\bar{n}$ the efficiency is not monotonic. This behavior has a similar explanation as that of D–C array taper efficiency described in Section 1.4.3.

1.4.5 Bayliss Difference Pattern Synthesis

We have seen that Taylor's line source yields a sum beam pattern, that is, a symmetric beam pattern with respect to the beam peak. Unlike a Taylor pattern, the Bayliss pattern [11] is a difference beam pattern that has a null at the boresight direction. The pattern function is fabricated using a similar concept as that of Taylor. For controlling the SLR, the zeros are set such that the difference pattern closely follow the function $q(u)\sin[\pi\sqrt{(u/\sigma)^2 - A^2}]$ in the vicinity of $|u| = 0$, where $q(u)$ is an odd function of u. Because the function $q(u)$ has an influence on

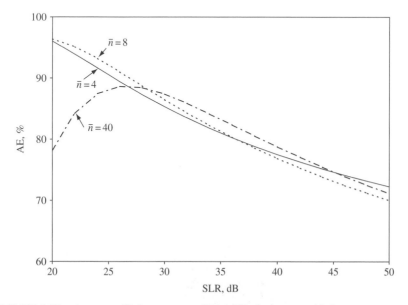

FIGURE 1.29 Aperture efficiency versus SLR of Taylor arrays with $\bar{n}$ as a parameter.

the location of the beam peak and on the SLR, an analytical relation between A and SLR cannot be found (this is contrary to Taylor's pattern function); hence the near-in zeros cannot be determined analytically. Bayliss determined the near-in zeros numerically. The Bayliss pattern function $F(u)$ is given by

$$F(u) = \frac{u \cos \pi u \prod_{n=1}^{\bar{n}-1} [1 - u^2/u_n^2]}{\prod_{n=0}^{\bar{n}-1} [1 - u^2/(n+1/2)^2]} \tag{1.123}$$

Observe that $F(0) = 0$, that is, the pattern has a null at $u = 0$. Furthermore, $F(u)$ becomes proportional to $\cos(\pi u)/u$ as u approaches infinity, implying that the power integral at the right-hand side of (1.114) is convergent, as required for a realizable source function. Also, the Fourier inverse of $F(u)$ can be proven to be space limited.

Bayliss obtained the first four zeros numerically and aligned the following $\bar{n} - 5$ zeros with the zeros of $\sin[\pi\sqrt{(u/\sigma)^2 - A^2}]$. Thus, the first $\bar{n} - 1$ zeros are given by

$$u_n = \begin{cases} \sigma \xi_n & \text{for } n = 1, 2, 3, 4 \\ \sigma \sqrt{A^2 + n^2} & \text{for } n = 5, 6, \ldots, \bar{n} - 1 \end{cases} \tag{1.124}$$

The remaining zeros are set as the zeros of $\cos(\pi u)$. To match the zero at $n = \bar{n}$ the scale factor σ was selected as

$$\sigma = \frac{\bar{n} + 1/2}{\sqrt{A^2 + \bar{n}^2}} \tag{1.125}$$

In (1.124) and (1.125) A is related to the side-lobe ratio. The numerical data obtained by Bayliss for the parameters A and ξ_n are given in Table 1.1. A Bayliss difference pattern with 25 dB SLR and $\bar{n} = 8$ is shown in Figure 1.30. The side-lobe characteristics are very similar with that of Taylor sum patterns.

TABLE 1.1 Parameters for Computing Zeros of Bayliss Differences Patterns at Selected SLRs

Parameters	15 dB	20 dB	25 dB	30 dB	35 dB	40 dB
A	1.0079	1.2247	1.4355	1.6413	1.8431	2.0415
ξ_1	1.5124	1.6962	1.8826	2.0708	2.2602	2.4504
ξ_2	2.2561	2.3698	2.4943	2.6275	2.7675	2.9123
ξ_3	3.1693	3.2473	3.3351	3.4314	3.5352	3.6452
ξ_4	4.1264	4.1854	4.2527	4.3276	4.4093	4.4973

Source: From [12].

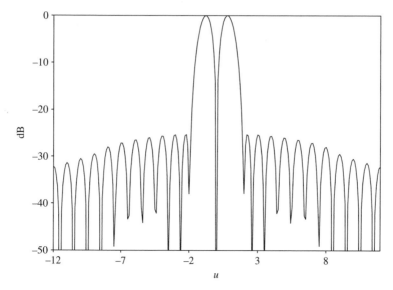

FIGURE 1.30 Bayliss difference pattern with SLR $= 25\,$dB and $\bar{n} = 8$.

The source distribution $g(p)$ is determined using (1.111). However, unlike Taylor's case, the infinite series at the right-hand side of (1.111) does not terminate because $F(n)$ does not vanish even for $|n| \geq \bar{n}$. However, if we use the samples of $F(u)$ at every $u = n + \dfrac{1}{2}$ instead of every $u = n$, then the infinite series terminates to a finite number of terms because $F\left(n + \dfrac{1}{2}\right) = 0$, for $|n| \geq \bar{n}$. Thus we can have $g(p)$ as

$$g(p) = \sum_{n=-\bar{n}}^{\bar{n}-1} F\left(n + \frac{1}{2}\right) \exp\left\{-jp\left(n + \frac{1}{2}\right)\right\} \tag{1.126}$$

Since $F(u)$ is an odd function of u, we further simplify (1.126) as

$$g(p) = \sum_{n=0}^{\bar{n}-1} F\left(n + \frac{1}{2}\right) \sin\left\{p\left(n + \frac{1}{2}\right)\right\} \tag{1.127}$$

In (1.127), the constant factor is suppressed. Figure 1.31 shows the source distribution corresponding to the Bayliss difference pattern of Figure 1.30. The source distribution is asymmetric with zero intensity at the center point. The edge taper is about 5.8 dB below the peak for this case.

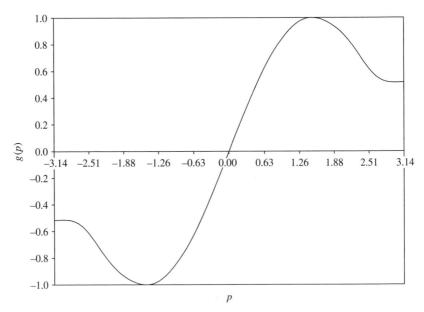

FIGURE 1.31   Bayliss source distribution for SLR $= 25$ dB and $\bar{n} = 8$.

1.5   PLANAR APERTURE SYNTHESIS

The previous section was devoted to the syntheses of linear arrays and continuous line sources. In this section we will consider syntheses of planar sources, in particular circular sources. For a circularly symmetric source distribution, the pattern function can be expressed in terms of two-dimensional Fourier transform as below:

$$F(k_x, k_y) = \int_0^a \int_0^{2\pi} f(\rho) \exp(jk_x\rho\cos\alpha + jk_y\rho\sin\alpha)\, \rho\, d\rho\, d\alpha \qquad (1.128)$$

The source lies on the xy-plane as shown in Figure 1.32. In (1.128) ρ is the radial distance of a source point from the origin, α is the angular distance measured from the x-axis, and k_x and k_y are spatial parameters given by

$$k_x = k_0 \sin\theta\cos\phi \qquad k_y = k_0 \sin\theta\sin\phi \qquad (1.129)$$

Substituting the expressions for k_x and k_y in (1.128), we derive

$$F(k_x, k_y) = \int_0^a \int_0^{2\pi} f(\rho) \exp[jk_0\rho\sin\theta\cos(\alpha - \phi)]\, \rho\, d\rho\, d\alpha \qquad (1.130)$$

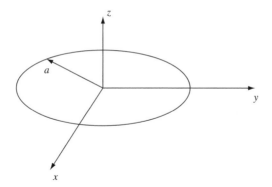

FIGURE 1.32   Circular source and the coordinate system.

Utilizing the identity [9]

$$2\pi J_0(x) = \int_0^{2\pi} \exp(jx\cos\beta)d\beta \tag{1.131}$$

we simplify (1.130) as

$$F(k_x, k_y) = 2\pi \int_0^a f(\rho)J_0(k_0\rho\sin\theta)\rho\,d\rho \tag{1.132}$$

Observe that the radiation pattern is a function of a single variable θ. We now make the following substitutions in (1.132):

$$u = \frac{a}{\pi}\sqrt{k_x^2 + k_y^2} = \frac{k_0 a}{\pi}\sin\theta \qquad p = \frac{\pi}{a}\rho \tag{1.133}$$

Then (1.132) transforms to

$$F(u) = \int_0^\pi pg(p)J_0(up)dp \tag{1.134}$$

with

$$g(p) = \frac{2a^2}{\pi}f\left(\frac{ap}{\pi}\right) \tag{1.135}$$

Equation (1.134) yields the relation between the source distribution and the far-field pattern of a circular aperture where the source is rotationally symmetric. The integral at the right-hand side of (1.134) is known as a Fourier–Bessel transform or

Hankel transform of $g(p)$. The inverse Hankel transformation [9] yields the source function $g(p)$:

$$g(p) = \int_0^{\infty} uF(u)J_0(up)\, du \qquad (1.136)$$

Equations (1.134) and (1.136) relate the source distribution to its far-field pattern and vice versa for a rotationally symmetric source.

1.5.1   Taylor's Circular Aperture Synthesis

Following a development similar to the Taylor line source synthesis, the radiation pattern of a circular source with controlled SLR was obtained as [13]

$$F(u) = \frac{2J_1(\pi u)}{\pi u} \prod_{n=1}^{\bar{n}-1} \frac{1 - u^2/u_n^2}{1 - u^2/\beta_n^2} \qquad (1.137)$$

with

$$J_1(\pi\beta_n) = 0 \qquad \beta_n \neq 0 \text{ for } n \neq 0$$

$$u_n = \sigma\sqrt{A^2 + \left(n - \frac{1}{2}\right)^2} \qquad \sigma = \frac{\beta_{\bar{n}}}{\sqrt{A^2 + \left(\bar{n} - \frac{1}{2}\right)^2}} \qquad (1.138)$$

The SLR is given by $\cosh(\pi A)$. In formulating the radiation pattern in (1.137), the following important points were considered:

1. The first $\bar{n} - 1$ zeros are the zeros of $\cos[\pi\sqrt{(u/\sigma)^2 - A^2}]$, so that $F(u)$ resembles $\cos[\pi\sqrt{(u/\sigma)^2 - A^2}]$ in the main-lobe and first few side-lobe regions, ensuring the SLR to be about $\cosh(\pi A)$.
2. The far-out side lobes must decay with u, so that the power integral becomes finite, making the source realizable. This justifies the factor $1/u$ in $F(u)$.
3. The factor $J_1(\pi u)$ is imported in order to make the source function of finite radius π and also to make the source function expressible by a finite series, as we shall see shortly. Also, $J_1(\pi u)$ helps converge the power integral.

The source distribution function $g(p)$ can be found using the inverse Hankel transformation given in (1.136). Employing the contour integration method, it can be shown that $g(p)$ is zero if $p > \pi$, that is, the pattern function $F(u)$ is realizable with a finite source function.

In order to determine $g(p)$ from $F(u)$, the sampling theorem, though applicable, may not be very beneficial because (a) two-dimensional samples (both k_x and k_y samples) need to be used and (b) infinite series representing $g(p)$ may not terminate to a finite number of terms. Instead, we first express $g(p)$ in terms of the following infinite series of orthogonal functions and then obtain the unknown coefficients by comparing the Hankel transforms:

$$g(p) = \begin{cases} \sum_{m=0}^{\infty} C_m J_0(\beta_m p) & 0 \le p \le \pi \\ 0 & p > \pi \end{cases} \tag{1.139}$$

Recall, (1.137) does not have a term with β_0. For (1.139) we set $\beta_0 = 0$. To obtain C_m's we obtain the Hankel transform of the right-hand side of (1.139) and compare that with $F(u)$ in (1.137). We invoke the identity [2]

$$\int_0^{\pi} J_0(\beta_m p) J_0(up) p \, dp = \frac{\pi u J_0(\beta_m \pi) J_1(\pi u)}{u^2 - \beta_m^2} \tag{1.140}$$

and obtain

$$F(u) = \frac{2J_1(\pi u)}{\pi u} \prod_{n=1}^{\bar{n}-1} \frac{1 - u^2/u_n^2}{1 - u^2/\beta_n^2} = \sum_{m=0}^{\infty} C_m \frac{\pi u J_0(\beta_m \pi) J_1(\pi u)}{u^2 - \beta_m^2} \tag{1.141}$$

The term at the extreme right essentially is a partial fraction representation of $F(u)$. From the expression of $F(u)$ represented by the middle term we observe that the number of partial fractions should be equal to $\bar{n}$ because only $\bar{n}$ factors exist in the denominator. This indicates that the C_m is nonzero only for $m < \bar{n}$. To determine C_m's we eliminate $J_1(\pi u)$, multiply both sides by $u^2 - \beta_i^2$, and then take the limiting value that u approaches β_i. This leaves

$$C_i \pi \beta_i J_0(\pi \beta_i) = \lim_{u \to \beta_i} \left[F(u) \frac{u^2 - \beta_i^2}{J_1(\pi u)} \right] \tag{1.142}$$

Using L'Hospital's rule, we deduce

$$C_i \pi \beta_i J_0(\pi \beta_i) = F(\beta_i) \frac{2\beta_i}{\pi J_0(\pi \beta_i)} \tag{1.143}$$

Extracting C_i from (1.143) and then substituting in (1.139), we finally obtain $g(p)$ as

$$g(p) = \begin{cases} \dfrac{2}{\pi^2} \sum_{m=0}^{\bar{n}-1} \dfrac{F(\beta_m)}{J_0^2(\pi \beta_m)} J_0(\beta_m p) & 0 \le p \le \pi \\ 0 & p > \pi \end{cases} \tag{1.144}$$

Figure 1.33 shows the far-field pattern $F(u)$ versus u with $\bar{n} = 4$ and 25 dB SLR. Three near-in side lobes are almost uniform, as evident from the plot. The far-out side lobes decay with u, as expected. Figure 1.34 presents the source distributions $g(p)$ corresponding to three different SLRs. The larger the SLR, the larger is the

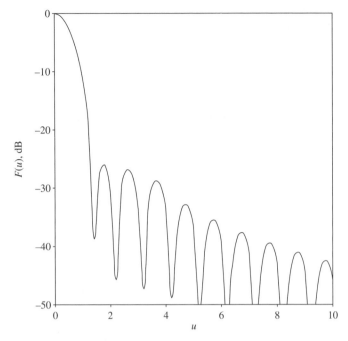

FIGURE 1.33   Taylor's pattern for circular source with SLR $= 25$ dB, $\bar{n} = 4$.

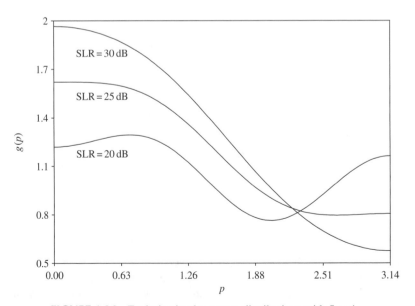

FIGURE 1.34   Taylor's circular source distributions with $\bar{n} = 4$.

edge taper. The aperture efficiency is computed using the standard definition, which, for a circular source, becomes

$$\text{AE} = \frac{2}{\pi^2} \frac{[\int_0^\pi g(p)pdp]^2}{\int_0^\pi g^2(p)pdp} \tag{1.145}$$

Note, $\text{AE} = 100\%$ for $g(p) = 1$. Substituting the expression for $g(p)$ from (1.144) in (1.145) and invoking orthogonality, we deduce

$$\text{AE} = \left[1 + \sum_{n=1}^{\bar{n}-1} \frac{F^2(\beta_n)}{J_0^2(\pi\beta_n)}\right]^{-1} \tag{1.146}$$

Figure 1.35 shows AE versus SLR with $\bar{n}$ as a parameter. For a given $\bar{n}$, the aperture efficiency peaks at a particular SLR. For larger $\bar{n}$, the peak occurs at larger SLR. This behavior is very similar to that of a D–C array, which has been explained in Section 1.4.3.

1.5.2 Bayliss Difference Pattern Synthesis

The Bayliss difference pattern function for a circular source was obtained following the very similar steps used for a line source [11]. The pattern function is

$$F(u, \phi) = \frac{(\cos\phi)uJ_1'(\pi u)\prod_{n=1}^{\bar{n}-1}[1 - u^2/u_n^2]}{\prod_{n=0}^{\bar{n}-1}[1 - u^2/\beta_n^2]} \tag{1.147}$$

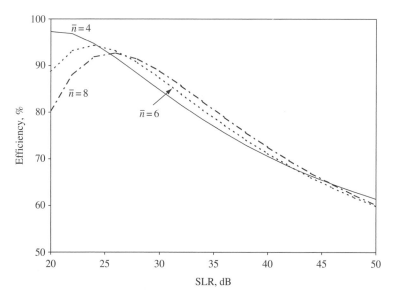

FIGURE 1.35 Aperture efficiency versus SLR of Taylor's circular source.

with

$$J_1'(\pi\beta_n) = 0 \qquad n = 0, 1, 2, \ldots$$

$$u_n = \begin{cases} \sigma\xi_n & \text{for } n = 1, 2, 3, 4 \\ \sigma\sqrt{A^2 + n^2} & \text{for } n = 5, 6, \ldots, \bar{n} - 1 \end{cases} \qquad (1.148)$$

In (1.147), $\pi\beta_0$ is the first zero of $J_1'(x)$, which is 1.841.... The values of ξ_n are tabulated in Table 1.1 for different SLRs. Observe that $F(u, \phi)$ satisfy the conditions necessary for a realizable source function. The source distribution can be determined using the inverse Fourier transform of $F(u, \phi)$. Ignoring the constant factor, the Fourier transform of $f(p)\cos\alpha$ can be expressed as $F(u)\cos\phi$ (see problem 1.18), where $F(u)$ is the Hankel transform of $f(p)$. Furthermore, noting the identity

$$\int_0^\pi J_1(\beta_m p)J_1(up)p\,dp = -\frac{\pi u J_1(\pi\beta_m)J_1'(\pi u)}{u^2 - \beta_m^2} \qquad (1.149)$$

one comprehends that $g(p)$ is expressible in terms of a Bessel function series because the denominator of $F(u, \phi)$ in (1.147) comprises the factor $u^2 - \beta_n^2$. Following the procedure in the previous section, we deduce

$$g(p) = \begin{cases} -\dfrac{2}{\pi^2}\displaystyle\sum_{m=0}^{\bar{n}-1}\frac{F(\beta_m, 0)}{J_1(\pi\beta_m)J_1''(\pi\beta_m)}J_1(\beta_m p) & 0 \le p \le \pi \\ 0 & p > \pi \end{cases} \qquad (1.150)$$

Figures 1.36 and 1.37 illustrate a circular Bayliss pattern and the corresponding source distribution. The source function in this case is $g(p)\cos\alpha$; therefore the source distribution undergoes $180°$ phase reversal from negative x to positive x.

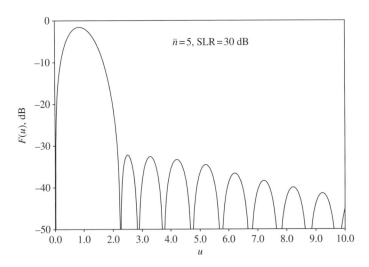

FIGURE 1.36 Bayliss difference pattern for a circular source.

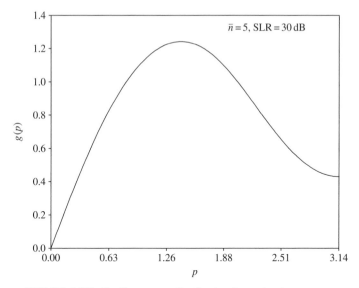

FIGURE 1.37 Bayliss source distribution for a circular source.

1.6 DISCRETIZATION OF CONTINUOUS SOURCES

Taylor's synthesis yields a continuous source distribution to satisfy desired patterns. To implement Taylor's source distribution in an array, one needs to discretize or sample the continuous source function to a finite number of samples. The number of samples is equal to the number of array elements. An element amplitude corresponds to the sample height at the element location. Figure 1.38 depicts the sampling of a continuous source. It is obvious that the radiation pattern of the sampled source will differ from the radiation pattern of the continuous source. This section is concerned with the effects of sampling on the radiation pattern.

For a continuous source function $f(p)$, the associated "sampled source function" in Figure 1.38 can be represented as

$$f_s(p) = \left[f(p) \sum_{n=-\infty}^{\infty} \delta(p - n\Delta) \right] \otimes e(p) \tag{1.151}$$

where Δ represents the sampling interval (or element spacing for the array) and $e(p)$ represents the aperture distribution of an element. The symbol $\otimes$ represents the convolution operation. The far-field pattern of the sampled source is the Fourier Transform (FT) of $f_s(p)$. According to the convolution theorem, we have

$$F_s(u) = \text{FT} \left\{ f(p) \sum_{n=-\infty}^{\infty} \delta(p - n\Delta) \right\} \times \text{FT}\{e(p)\} \tag{1.152}$$

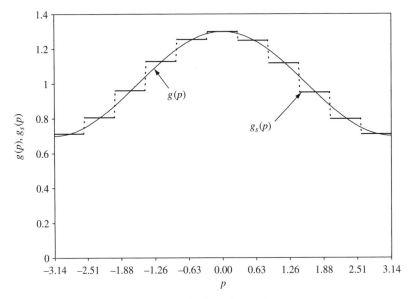

FIGURE 1.38   Discretization of a continuous source.

The FT of $e(p)$ is the element pattern. The first term at the right-hand side is

$$FT\left\{f(p)\sum_{n=-\infty}^{\infty}\delta(p-n\Delta)\right\}=\frac{1}{\Delta}\sum_{m=-\infty}^{\infty}F\left(u-\frac{2\pi m}{\Delta}\right) \qquad (1.153)$$

because the infinite Dirac delta series can be expressed as (see Section A.2)

$$\sum_{n=-\infty}^{\infty}\delta(p-n\Delta)=\frac{1}{\Delta}\sum_{m=-\infty}^{\infty}\exp\left(-\frac{j2\pi mp}{\Delta}\right) \qquad (1.154)$$

In (1.153) $F(u)=FT\{f(p)\}$, is the radiation pattern of the continuous source function $f(p)$. Thus we obtain

$$F_s(u)=\frac{1}{\Delta}\sum_{m=-\infty}^{\infty}F\left(u-\frac{2m\pi}{\Delta}\right)\times E(u) \qquad (1.155)$$

In (1.155), $E(u)=FT\{e(p)\}$. Observe that $F_s(u)$ is a superposition of the continuous source pattern $F(u)$ and an infinite number of shifted patterns $F(u-2m\pi/\Delta)$ multiplied by the element pattern $E(u)$ and a constant factor $1/\Delta$. A shifted pattern function $F(u-2m\pi/\Delta)$ is associated with a grating beam of a discrete array. Ignoring the element pattern factor temporarily, we observe that the sampling results in infinite number of shifted patterns. Thus the change in the radiation pattern due to sampling is equal to the total contributions of the shifted patterns. This is rigorously true if $E(u)$ is uniform, which corresponds to delta element sources. For finite element size, $E(u)$ will be non uniform; hence the element pattern has an effect on this pattern change. However, the effect is insignificant, particularly near the bore sight, because most element patterns are stationary near the bore sight.

Figure 1.39 shows three patterns (original and two shifted) of a continuous source. Typically, contributions of the shifted pattern near the bore sight are very small, particularly for a small sampling interval. Thus the near-in side lobes are not very much affected by the shifted patterns. In other words, the main beam and near-in side lobes of a sampled source do not differ significantly from that of the continuous source. Table 1.2 shows the first side-lobe level of a sampled source versus number of samples. In most cases, the side-lobe degradation is minimal. It is worth mentioning that the change in the radiation pattern due to sampling is a function of the number of samples only. For instance, a $20\lambda_0$ long source with 10 samples will have the same side-lobe degradation as that of a $40\lambda_0$ long source with 10 samples. The source distributions (and also the element aperture distributions), however, should be identical.

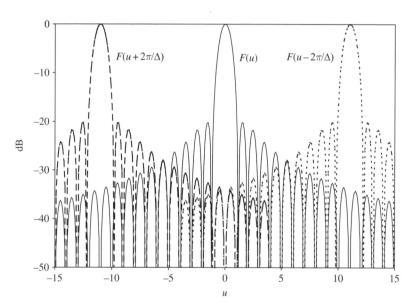

FIGURE 1.39 Radiation pattern of a sampled source.

TABLE 1.2 Variation of Peak SLR Versus Array Element (N)

N	SLR, dB	N	SLR, dB
101	20.12	15	20.05
51	20.11	11	20.0
26	19.58	7	20.0
21	20.08	5	19.71

Note: The source distribution $f(p) = 1 + 0.3\cos(p)$, $|p| < \pi$. Element size $2\pi/N$.

1.7 SUMMARY

The primary objective of this chapter was to introduce the fundamental radiation principle of phased array antennas, define various technical terms in phased arrays, and present aperture synthesis procedures. We covered most commonly used synthesis, yet many other array synthesis are available in the literature. An interested reader may be directed to the comprehensive text by Elliott [12].

The array analysis presented in this chapter does not consider mutual coupling effects categorically. However, mutual coupling effects can be incorporated rigorously if the *active element pattern* is considered instead of the *isolated element pattern*. Except for the edge elements, the active element patterns are practically invariant from element to element. Thus the theory presented in this chapter may be applicable if mutual coupling between the elements needs to be considered, particularly in a large array.

To an array designer, estimating the driving point impedances of the elements in the array environment is a major issue in order to ensure impedance match between the radiating elements and the components behind the elements, such as power amplifiers. In the following few chapters we will develop a systematic procedure to determine the active element pattern and active driving point impedance based on Floquet modal expansion theory.

REFERENCES

[1] A. C. Ludwig, "The Definition of Cross Polarization," *IEEE Trans. Antennas Propagat.*, Vol. AP-21, pp. 116–119, Jan. 1973.

[2] M. Abramowitz and I. E. Stegun, *Handbook of Mathematical Functions*, Applied Mathematics Series 55, National Bureau of Standards, Washington, DC, June 1964.

[3] D. R. Rhodes, "On the Stored Energy and Reactive Power of Planar Apertures," *IEEE Trans.*, Vol. AP-14, pp. 676–683, 1966.

[4] S. A. Schelkunoff, "A Mathematical Theory of Linear Arrays," *Bell Syst. Techn. J.*, Vol. 22, pp. 80–107, 1943.

[5] C. L. Dolph, "A Current Distribution for Broadside Arrays which Optimizes the Relationship Between Beamwidth and Sidelobe Level," *Proc. IRE*, Vol. 34, pp. 335–348, 1946.

[6] N. N. Lebedev, *Special Functions and Their Applications*, Dover, New York 1972.

[7] B. P. Lathi, *Modern Digital and Analog Communication Systems*, 2nd ed., Holt, Rinehart and Winston, New York, 1989.

[8] R. C. Hansen, *Phased Array Antennas*, Wiley, New York, 1998.

[9] J. Mathews and R. L. Walker, *Mathematical Methods of Physics*, 2nd ed., W. A. Benjamin, Reading, MA 1964.

[10] T. T. Taylor, "Design of Line Source Antennas for Narrow Beamwidth and Low Side Lobes," *IRE Trans. Antennas Propagat.*, Vol. AP-3, pp. 16–28, 1955.

[11] E. T. Bayliss, "Design of Monopulse Antenna Difference Patterns with Low Side Lobes," *Bell Syst. Tech. J.*, Vol. 47, pp. 623–650, 1968.

[12] R. S. Elliott, *Antenna Theory and Design*, Prentice-Hall, Englewood Cliffs, NJ, 1981.

[13] T. T. Taylor, "Design of Circular Apertures for Narrow Beamwidth and Low Side Lobes," *IRE Trans. Antennas Propagat.*, Vol. AP-8, pp. 17–22, 1960.

BIBLIOGRAPHY

Amitay, N., Galindo, V. and C. P. Wu, *Theory and Analysis of Phased Array Antennas*, Wiley-Interscience, New York, 1972.

Bach, H., and J. E. Hansen, "Uniformly Spaced Arrays," in R. E. Collin and F. J. Zucker (Eds.) *Antenna Theory*, Part 1, McGraw-Hill, New York, 1969, Chapter 5.

Cheng, D. K., and M. T. Ma, "A New Mathematical Approach for Linear Array Analysis," *IRE Trans., Antennas Propagat.*, Vol. AP-8, pp. 255–259, May 1960.

Elliott, R. S., "Design of Circular Apertures for Narrow Beamwidth and Asymmetric Sidelobes," *IEEE Trans. Antennas Propagat.*, Vol. AP-23, pp. 523–527, 1975.

Elliott, R. S., "Design of Line-Source for Sum Patterns with Sidelobes of Individually Arbitrary Heights," *IEEE Trans. Antennas Propagat.*, Vol. AP-24, pp. 76–83, 1976a.

Elliott, R. S., "Design of Line Source Antennas for Difference Patterns with Sidelobes of Individually Arbitrary Heights," *IEEE Trans. Antennas Propagat.*, Vol. AP-24, pp. 310–316, 1976b.

Elliott, R. S., "The Theory of Antenna Arrays," in R. C. Hansen (Ed.), *Microwave Scanning Antennas*, Vol. 2, Academic New York, 1966, Chapter 1.

Harrington, R. F., "Antenna Excitation for Maximum Gain," *IEEE Trans. Antennas Propagat.*, Vol. AP-13, pp. 896–903, Nov. 1965.

Ishimaru, A., "Theory of Unequally Spaced Arrays," *IRE Trans. Antennas Propagat.*, Vol. AP-10, pp. 691–702, 1962.

King, R. W. P., and S. S. Sandler, "The Theory of Broadside Arrays," *IEEE Trans. Antennas Propagat*, Vol. AP-12, pp. 269–275, 1964.

Ksienski, A. "Equivalence Between Continuous and Discrete Radiating Arrays," *Can. J. Phys.*, Vol. 39, pp. 335–349, Feb. 1961.

Mailloux, R. J., *Phased Array Antenna Handbook*, Artech House, Boston 1994.

Rhodes, D. R., "The Optimum Linear Array for Single Main Beam," *Proc. IRE*, Vol. 41, pp. 793–794, June 1953.

Schell, A. C., and A. Ishimaru, "Antenna Pattern Synthesis," in R. E. Collin and F. J. Zucker (Eds.), *Antenna Theory*, Part 1, McGraw-Hill, New York, 1969, Chapter 7.

Stegen, R. J., "Excitation Coefficients and Beamwidth of Tchebyscheff Arrays," *Proc. IRE*, Vol. 41, pp. 1671–1674, Nov. 1953.

Villeneuve, A. T., "Taylor Patterns for Discrete Arrays," *IEEE Trans. Antennas Propagat.*, Vol. AP-32, pp. 1089–1093, 1984.

Woodward, P. M. and J. D. Lawson, "The Theoretical Precision with Which an Arbitrary Radiation Pattern May be Obtained from a Source of Finite Size," *J. IEE*, Vol. 95, Pst. III, pp. 363–370, 1948.

PROBLEMS

1.1 Determine the copolarization and cross-polarization vectors for an antenna with principal polarization along the y-direction.

1.2 The radiation pattern of a circular horn excited by the TE_{11} mode with principal polarization along the x-direction can be expressed as

$$\vec{e}_x(\theta, \phi) = \hat{\theta} f(\theta) \cos \phi - \hat{\phi} g(\theta) \sin \phi$$

with $f(0) = g(0)$. (a) Express the linear copolarization and cross-polarization components. (b) If the horn is excited with two orthogonal TE_{11} modes (with equal amplitude) such that the y-polarization lags in phase by 90°, then determine the lcp and rcp components as a function of θ and ϕ.

1.3 Verify (1.10).

1.4 Figure P1.4 shows a four-element array along the circumference of a cylinder. The radiation pattern of the nth element is given by

$$\vec{e}_n(\theta, \phi) = A\hat{\theta} \cos[\alpha(\theta - \tfrac{1}{2}\pi)] \cos[\beta(\phi - \phi_n)]$$

where α and β are constants and ϕ_n is the angular position of the n th element, which is given by $\phi_n = (n - 0.5)\pi/6$ with $n = -1, 0, 1, 2$. Determine the relative amplitude and phase distributions that would produce a beam at $\theta = \pi/2$ and $\phi = 0$ with maximum possible array gain. The radius of the cylinder is $1.5\lambda_0$.

1.5 Assume that the gain pattern of a circular element is given by

$$G(\theta) = \left(\frac{4\pi r}{\lambda_0}\right) \frac{J_1(k_0 r \sin \theta)}{k_0 r \sin \theta}$$

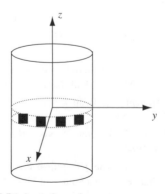

FIGURE P1.4 A four-element array on a cylinder.

where r is the radius of the aperture and $J_1(x)$ is the first-order Bessel function of x. Show that in order to have minimum number of array elements, the element diameter should be

$$2r = \frac{1.17\lambda_0}{2\sin\theta_0}$$

for a scan beam array of maximum scan angle θ_0. Assume a square grid.

1.6 Obtain the maximum possible element spacing for a linear array scanning to $60°$ with no grating lobe in the visible region.

1.7 Given that the solution for the equation $\sin x / x = 1/\sqrt{2}$ is $x = 1.39$, prove the bandwidth relation in (1.65).

1.8 For $a = \lambda_0/4$, show that the visible region lies on the right-half segment of the Schelkunoff's unit circle.

1.9 Locate the zeros of Schelkunoff's polynomial on the unit circle for a uniform distribution and with $N = 5$, $a = 0.6\lambda_0$. From the null locations, estimate the side-lobe peak locations and the corresponding SLRs.

1.10 Consider a five-element array with $\lambda_0/2$ element spacing. Consider that the zeros are $z_1 = \exp(j0.6\pi)$, $z_2 = \exp(j0.87\pi)$, $z_3 = \exp(j1.13\pi)$, and $z_4 = \exp(j1.4\pi)$, respectively (notice that the zeros are closely spaced than that of the previous problem). Estimate the SLR corresponding to the highest side lobe.

1.11 Show that $\sum_{n=1}^{N} A_n^2 = {}^{2N-2}C_{N-1}$; hence prove the relation in (1.83) for the ATE of a binomial array.

1.12 In a Dolph–Chebyshev array of N elements show that the maximum possible number of nulls inside the visible region $0 < \theta < \pi/2$ is given by the equation

$$\frac{1}{c}\cos\left[\frac{(2n+1)\pi}{2(N-1)}\right] = \cos(k_0 a/2)$$

where $T_{N-1}(c)$ is the SLR. For small a/λ_0 and c near unity, show that n can be approximated as

$$n \approx (N-1)\frac{a}{\lambda_0} - \frac{1}{2}$$

Thus, the total number of nulls in the region $-\pi/2 < \theta < \pi/2$ is $2n \approx 2(N-1)a/\lambda_0 - 1$.

1.13 Design a seven-element Dolph–Chebyshev array with SLR of 25 dB. How many zeros of Schelkunoff's polynomial lie inside the visible region of Schelkunoff's circle if (a) element spacing $= \lambda_0/2$, and (b) element spacing $= \lambda_0/5$.

1.14 Using the expression for $g(p)$ in (1.111), for $-\pi < p < \pi$ and zero elsewhere, show that the Fourier transform of $g(p)$ is equal to $2\pi F(u)$. *Hint:* Use contour integration method for summing the infinite series after taking the FT of $g(p)$.

1.15 Show that the term inside the summation sign of (1.113) approaches $(-1)^n \cos(np)$ as n becomes very large compared to A. Thus for a given value of p, the series does not converge.

1.16 Show that if $p > \pi$ and z has a negative imaginary part (could be very small), then $\int_{-\infty}^{\infty} \exp[j(\pi - p)u]/(u+z)\, du = 0$. Hence convince yourself that $g(p)$, the Fourier inverse of $F(u)$ in (1.120), vanishes for $|p| > \pi$.

1.17 Show that the radiation pattern normalized with respect to the bore-sight field of a uniform, circular planar source is given by

$$F_{\text{uni}}(u) = \frac{2J_1(\pi u)}{\pi u}$$

1.18 The two-dimensional FT of the function $f(p)\cos\alpha$ is given by

$$F(u, \phi) = \int_0^{2\pi} \int_0^{\pi} f(p)\cos\alpha \exp[jup\cos(\phi - \alpha)]p\,dp\,d\alpha$$

where u and p are defined in (1.133). Show that $F(u, \phi)$ can be represented as

$$F(u, \phi) = 2\pi j\cos\phi \int_0^{\pi} f(p)J_1(up)p\,dp$$

1.19 Deduce (1.150).

CHAPTER TWO

Introduction to Floquet Modes in Infinite Arrays

2.1 INTRODUCTION

The primary objective of this chapter is to introduce the modal functions associated with infinite array antennas. The rationale for undertaking such infinite array analysis is as follows:

(a) Elements in the central region of an electrically large array have similar active impedance characteristics as that of an element in an infinite array.

(b) It will be shown in Chapter 4 that the performance of a finite array can be determined accurately utilizing infinite array results.

(c) The infinite array results are applied to predict the mutual coupling between the elements in an array environment.

(d) The embedded element pattern that includes mutual coupling effects can be determined directly from the infinite array results.

There are several other reasons for analyzing an infinite array, as will be apparent in the chapters that follow.

Mathematically, an infinite array antenna is equivalent to an array of infinite source functions. The solution of Maxwell's equations under the "infinite source function" excitation essentially provides the radiation characteristics of the associated infinite array antenna. The Fourier transform method is a very powerful method to obtain solutions for Maxwell's equations pertaining to the radiation problems. For that reason, we begin with the definition of the Fourier transform and conduct Fourier analysis of source functions. In order to investigate the Fourier spectrum

Phased Array Antennas. By Arun K. Bhattacharyya
© 2006 John Wiley & Sons, Inc.

of an array of sources, we apply superposition of an infinite number of identical sources placed at a regular interval. It is observed that the Fourier integral evolves to a series of Dirac delta functions; consequently the continuous Fourier spectrum of a single source evolves to discrete spectral lines. We deduce the Floquet series expansion for an array of source functions, where the amplitude and phase have two different periodicities. Such source functions are relevant for a scanned beam array antenna analysis. The basis functions that constitute the Floquet series are called Floquet modal functions. This implies that an infinite array of sources can be represented alternatively as a superposition of Floquet modal functions. The advantage of using Floquet modal functions is that the solution of Maxwell's equations is obtained almost by inspection. The linearity of Maxwell's equation allows constructing the complete solution of an infinite array of sources.

To illustrate the Floquet modal analysis, we first consider a one-dimensional infinite array antenna and determine the electromagnetic fields in terms of Floquet modes. We establish the correspondence between a Floquet modal function and the direction of radiation. We then deduce the Floquet modal functions associated with a planar infinite array with an arbitrary lattice structure and study the results in light of antenna radiation. The last part of the chapter deals with the interaction of a Floquet mode and a guided mode and discusses the interesting consequences of such interaction.

2.2 FOURIER SPECTRUM AND FLOQUET SERIES

In this section we present the Fourier analysis and Floquet series representations of periodic functions, starting from the basic definition of the Fourier transform. We begin with the Fourier transform of a nonperiodic function and use superposition to derive appropriate series expansion formulas relevant to an infinite array.

2.2.1 Fourier Transform

The one-dimensional Fourier transform of a function $f(x)$[1] is defined as

$$\tilde{f}(k_x) = \frac{1}{2\pi} \int_{-\infty}^{\infty} f(x) \exp(jk_x x) \, dx \tag{2.1}$$

where $\tilde{f}(k_x)$ is the Fourier transform or Fourier spectrum or simply the spectrum of $f(x)$. The variable k_x is the spectral frequency, analogous to the variable ω for a time-domain function. The function $f(x)$ can be recovered from its Fourier transform using the following inverse transformation integral:

$$f(x) = \int_{-\infty}^{\infty} \tilde{f}(k_x) \exp(-jk_x x) \, dk_x \tag{2.2}$$

---

[1] $f(x)$ could be a real or a complex function of a real variable x.

The relation in (2.2) can be established by direct substitution of (2.1) in (2.2) and using the following identity:

$$\delta(x - x_0) = \frac{1}{2\pi} \int_{-\infty}^{\infty} \exp\{\pm jk_x(x - x_0)\} \, dk_x \tag{2.3}$$

In (2.3) $\delta(x)$ represents the Dirac delta function of x. The above identity can be proven by employing a contour integration technique on the complex k_x-plane (see problem 2.1).

2.2.2 Periodic Function: Fourier Series

Let us consider a periodic function $g(x)$ with periodicity a, where $g(x)$ is given by

$$g(x) = \sum_{n=-\infty}^{\infty} f(x - na) \tag{2.4}$$

Notice, $g(x) = g(x + ma)$, where m is an integer. Following the definition in (2.1) the Fourier transform of $g(x)$ can be expressed as

$$\tilde{g}(k_x) = \frac{1}{2\pi} \int_{-\infty}^{\infty} g(x) \exp(jk_x x) \, dx \tag{2.5}$$

Substituting $g(x)$ from (2.4) in (2.5) and then interchanging the summation and integral signs, one obtains

$$\tilde{g}(k_x) = \frac{1}{2\pi} \sum_{n=-\infty}^{\infty} \int_{-\infty}^{\infty} f(x - na) \exp(jk_x x) \, dx \tag{2.6}$$

A substitution of $x^{/}$ for $x - na$ in (2.6) results in the following infinite series (shift theorem):

$$\tilde{g}(k_x) = \tilde{f}(k_x) \sum_{n=-\infty}^{\infty} \exp(jnk_x a) \tag{2.7}$$

The infinite series in (2.7) can be replaced by the following infinite series of Dirac delta functions (see the Section A.2 in the Appendix):

$$\sum_{n=-\infty}^{\infty} \exp(jnk_x a) = \frac{2\pi}{a} \sum_{n=-\infty}^{\infty} \delta\left(k_x - \frac{2n\pi}{a}\right) \tag{2.8}$$

The above identity can also be proven by directly applying Poisson's summation formula [1] (see problem 2.3). Using (2.8) in (2.7) we obtain

$$\tilde{g}(k_x) = \frac{2\pi}{a} \tilde{f}(k_x) \sum_{-\infty}^{\infty} \delta\left(k_x - \frac{2n\pi}{a}\right) \tag{2.9}$$

Since the magnitude of the Dirac-delta function $\delta(\alpha)$ is nonzero at $\alpha = 0$ and zero elsewhere, $\tilde{g}(k_x)$ would be zero everywhere except at discrete points $k_x = 0, \pm(2\pi/a), \pm(4\pi/a), \ldots$. Thus, the periodic function $g(x)$ has a set of *discrete* spectral lines as opposed to a continuous spectrum of the nonperiodic function $f(x)$. Observe that the heights of the spectral lines for $g(x)$ follow the envelope of the spectrum for $f(x)$. The spectra of $f(x)$ and $g(x)$ are depicted in Figure 2.1. We will investigate the basic reason for such a discrete spectrum. By definition, the spectrum of $g(x)$ is equal to the phasor summation of the individual spectra of the shifted functions $f(x - na)$, where n varies from $-\infty$ to ∞, as in (2.6). By virtue of the shift theorem [2], the spectrum of a shifted function is equal to the spectrum of the nonshifted function times an exponential phase factor associated with the shift. These infinite phase factors associated with infinite number of shifted functions add coherently only at discrete points and cancel elsewhere, as given in (2.8).

Using (2.9) we express the periodic function $g(x)$ in an alternative format. From the general relation in (2.2), the function $g(x)$ can be expressed as

$$g(x) = \int_{-\infty}^{\infty} \tilde{g}(k_x) \exp(-jk_x x) \, dk_x \tag{2.10}$$

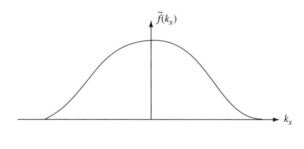

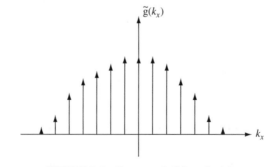

FIGURE 2.1   Spectra of $f(x)$ and $g(x)$.

Substituting the expression of $\tilde{g}(k_x)$ from (2.9) to (2.10) and after performing the integration, we obtain

$$g(x) = \frac{2\pi}{a} \sum_{n=-\infty}^{\infty} \tilde{f}\left(\frac{2n\pi}{a}\right) \exp\left(-\frac{j2n\pi x}{a}\right) \qquad (2.11)$$

The right-hand side of (2.11) is the Fourier-series expansion of the periodic function $g(x)$. This alternate representation of a periodic function helps explain many physical phenomena that occur in the arena of electromagnetism.

2.2.3 Floquet Series

In the previous section we deduced an alternate format of a periodic function $g(x)$. It was assumed that the magnitude and phase of $g(x)$ have the same periodicity, a. Now let us consider a more general type of a complex function $h(x)$ such that both the magnitude and the phase are periodic but with different periodicities. These functions have useful applications in scanned beam arrays. A direct application of the Fourier expansion is not very handy for such functions, because the overall periodicity may be much larger than the amplitude or phase periodicity[2]. However, a function of this type can also be expressed in terms of a Fourier-like series called a Floquet series. We will articulate the Floquet series expansion method in this section.

The function under consideration can be expressed as

$$h(x) = \sum_{n=-\infty}^{\infty} f(x - na) \exp(-jn\varphi) \qquad (2.12)$$

where $f(x)$ is a complex function of the real variable x and φ is a real constant. Notice that the magnitude of $h(x)$ is a periodic function of periodicity a, because

$$h(x + a) = \exp(-j\varphi)h(x) \qquad (2.13)$$

and the phase of $h(x)$ has, in general, a different periodicity. Furthermore, every $\Delta x = a$ interval, the phase of $h(x)$ decreases by a constant amount φ, as depicted in Figure 2.2.

Now we will try to find a Fourier-like series representation for $h(x)$. Toward that end we first find the Fourier transform of $h(x)$, which is given by

$$\tilde{h}(k_x) = \frac{1}{2\pi} \sum_{-\infty}^{\infty} \exp(-jn\varphi) \int_{-\infty}^{\infty} f(x - na) \exp(jk_x x) dx \qquad (2.14)$$

---

[2] Strictly speaking, for some functions, the Fourier method may not be applicable if one periodicity (magnitude or phase) is an irrational number. Example: $h(x) = \sum f(x - n) \exp(-jn)$.

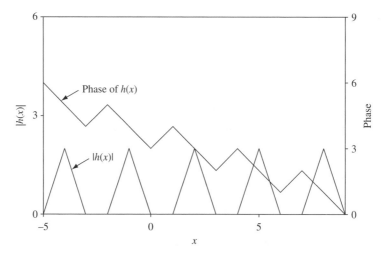

FIGURE 2.2   Magnitude and phase of $h(x)$.

Substituting x' for $x - na$ we obtain

$$\tilde{h}(k_x) = \tilde{f}(k_x) \sum_{-\infty}^{\infty} \exp\{jn(k_x a - \varphi)\} \tag{2.15}$$

Using the identity given in (2.8) we write

$$\tilde{h}(k_x) = \frac{2\pi}{a} \tilde{f}(k_x) \sum_{-\infty}^{\infty} \delta\left(k_x - \frac{2n\pi}{a} - \frac{\varphi}{a}\right) \tag{2.16}$$

The right-hand side of (2.16) is the Fourier transform of $h(x)$. Notice, the spectrum of $h(x)$ exists in discrete intervals, as in the case of a fully periodic function $g(x)$. However, locations of the spectral lines of $h(x)$ deviate from that of $g(x)$ by an amount φ/a. This is pictorially shown in Figure 2.3. Using (2.16) we can write $h(x)$ in terms of a series of complex exponential functions as

$$h(x) = \frac{2\pi}{a} \sum_{n=-\infty}^{\infty} \tilde{f}\left(\frac{2n\pi + \varphi}{a}\right) \exp\left(-\frac{j(2n\pi + \varphi)x}{a}\right) \tag{2.17}$$

The right-hand side of (2.17) is the Floquet series expansion of $h(x)$. Substitute $\varphi = 0$ in $h(x)$, the Floquet series turns into a Fourier series, as expected. The Floquet series expansion in (2.17) can also be established by directly applying Poisson's summation formula [1].

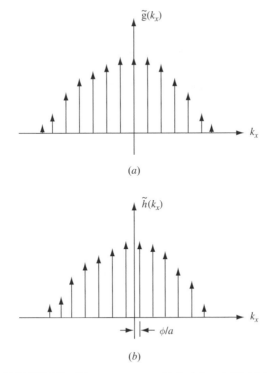

FIGURE 2.3 Discrete spectra of (a) $g(x)$ and (b) $h(x)$.

2.2.4 Two-Dimensional Floquet Series

Consider a complex function $h(x, y)$ of two independent variables x and y defined as

$$h(x, y) = \sum_m \sum_n f(x - x_{mn}, y - y_{mn}) \exp(-jk_{x0}x_{mn} - jk_{y0}y_{mn})$$

$$-\infty < m < \infty \qquad -\infty < n < \infty \tag{2.18}$$

where (x_{mn}, y_{mn}) are discrete grid points on the xy-plane; k_{x0} and k_{y0} are two constants that determine the discrete phase shift between the adjacent cells. For rectangular grids, $x_{mn} = ma$ and $y_{mn} = nb$, where m and n are integers running from $-\infty$ to ∞. With respect to a general grid shown in Figure 2.4, x_{mn} and y_{mn} can be expressed as

$$x_{mn} = ma + \frac{nb}{\tan \gamma} \qquad y_{mn} = nb \tag{2.19}$$

where γ is the grid angle. Notice, for rectangular grids $\gamma = 90°$.

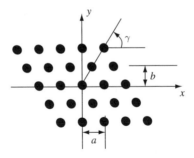

FIGURE 2.4 General grid structure of a planar periodic array antenna.

In order to obtain the Floquet series expansion of $h(x, y)$, we first find the Fourier transform of $h(x, y)$. The two-dimensional Fourier transform is defined as

$$\tilde{h}(k_x, k_y) = \frac{1}{4\pi^2} \int_{-\infty}^{\infty} \int_{-\infty}^{\infty} h(x, y) \exp(jk_x x + jk_y y) dx\, dy \qquad (2.20)$$

As for the one-variable function, the two-variable function $h(x, y)$ can be recovered from its Fourier transform using the following integral relation:

$$h(x, y) = \int_{-\infty}^{\infty} \int_{-\infty}^{\infty} \tilde{h}(k_x, k_y) \exp(-jk_x x - jk_y y) dk_x dk_y \qquad (2.21)$$

Substituting $h(x, y)$ from (2.18) into (2.20) we obtain

$$\tilde{h}(k_x, k_y) = \frac{1}{4\pi^2} \sum_m \sum_n \{ \exp(-jk_{x0} x_{mn} - jk_{y0} y_{mn})$$

$$\times \iint_{x\, y} f(x - x_{mn}, y - y_{mn}) \exp(jk_x x + jk_y y) dx\, dy \} \qquad (2.22)$$

Substituting $x' = x - x_{mn}$ and $y' = y - y_{mn}$ we obtain

$$\tilde{h}(k_x, k_y) = \tilde{f}(k_x, k_y) \sum_m \sum_n \exp\{ jx_{mn}(k_x - k_{x0}) + jy_{mn}(k_y - k_{y0}) \} \qquad (2.23)$$

Temporarily let us focus on the term with a double summation. Using the expressions for x_{mn} and y_{mn} from (2.19) the double-summation term becomes

$$S = \sum_m \sum_n \exp\{ jx_{mn}(k_x - k_{x0}) + jy_{mn}(k_y - k_{y0}) \}$$

$$= \sum_{m=-\infty}^{\infty} \exp\{ jma(k_x - k_{x0}) \} \sum_{n=-\infty}^{\infty} \exp\left[jnb \left\{ (k_y - k_{y0}) + \frac{k_x - k_{x0}}{\tan \gamma} \right\} \right] \qquad (2.24)$$

Applying (2.8) we deduce

$$S = \frac{2\pi}{a} \sum_{m=-\infty}^{\infty} \delta\left(k_x - k_{x0} - \frac{2m\pi}{a}\right) \frac{2\pi}{b} \sum_{n=-\infty}^{\infty} \delta\left(k_y - k_{y0} + \frac{k_x - k_{x0}}{\tan\gamma} - \frac{2n\pi}{b}\right)$$

(2.25)

Because the Delta functions are non zero only at discrete points, we can substitute $k_x = 2m\pi/a + k_{x0}$ into the second summation, yielding

$$S = \frac{2\pi}{a} \sum_{m=-\infty}^{\infty} \left[\delta\left(k_x - k_{x0} - \frac{2m\pi}{a}\right) \frac{2\pi}{b} \sum_{n=-\infty}^{\infty} \delta\left(k_y - k_{y0} + \frac{2m\pi}{a\tan\gamma} - \frac{2n\pi}{b}\right) \right]$$

(2.26)

Using the above expression for the summation part in (2.23), we finally obtain

$$\tilde{h}(k_x, k_y) = \frac{4\pi^2}{ab} \tilde{f}(k_x, k_y) \sum_m \sum_n \delta\left(k_x - k_{x0} - \frac{2m\pi}{a}\right) \delta\left(k_y - k_{y0} + \frac{2m\pi}{a\tan\gamma} - \frac{2n\pi}{b}\right)$$

(2.27)

Thus the Fourier spectrum for $h(x, y)$ exists only at discrete points in the $k_x k_y$-plane. These points are given by

$$k_x = k_{xmn} = k_{x0} + \frac{2m\pi}{a} \qquad k_y = k_{ymn} = k_{y0} - \frac{2m\pi}{a\tan\gamma} + \frac{2n\pi}{b}$$

(2.28)

The Floquet series of $h(x, y)$ is obtained by substituting $\tilde{h}(k_x, k_y)$ from (2.27) into (2.21). This yields

$$h(x, y) = \frac{4\pi^2}{ab} \sum_m \sum_n \tilde{f}(k_{xmn}, k_{ymn}) \exp(-jk_{xmn}x - jk_{ymn}y)$$

(2.29)

where k_{xmn} and k_{ymn} are defined in (2.28). We now consider the following two special cases.

Rectangular Grid For a rectangular grid we set $\gamma = 90°$ in (2.28) to obtain the spectral points. The spectral points are shown in Figure 2.5a. The spectral points are spaced at regular intervals of $2\pi/a$ and $2\pi/b$ along the k_x- and k_y-directions, respectively, where $a \times b$ is the unit cell size. In particular, for a square grid the spectral points lie on a square grid.

Equilateral Triangular Grids For equilateral triangular grids we set $\gamma = 60°$ and $b = a\sin(60°) = a\sqrt{3}/2$ in (2.28), where a represents the length of a side of the triangle. The spectral points are given by

$$k_{xmn} = k_{x0} + \frac{2m\pi}{a} \qquad k_{ymn} = k_{y0} - \frac{2m\pi}{a\sqrt{3}} + \frac{4n\pi}{a\sqrt{3}}$$

(2.30)

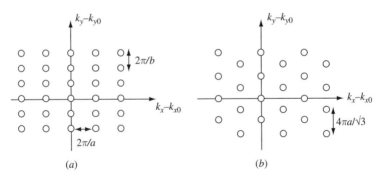

(a)   (b)

FIGURE 2.5 Spectral points of (a) rectangular grid of cell size $a \times b$ and (b) equilateral triangular grid with side a.

Substituting integer values for m and n we compute the spectral points. Figure 2.5b shows the discrete spectral points on the $k_x k_y$-plane. The spectral points are on an equilateral triangular lattice. The distance between two adjacent spectral points is $4\pi/a\sqrt{3}$. Interestingly, the spectral points are further apart as compared to a square grid of side a.

2.3 FLOQUET EXCITATIONS AND FLOQUET MODES

In the previous section we derived Floquet series expansions of a special type of periodic functions, which are expressible in the form given by (2.12) and (2.18) for one- and two-dimensional cases, respectively. We will now examine the electromagnetic fields produced by an array of current sources represented by the above equations. Such sources are called *Floquet sources* because the electromagnetic fields produced by them can be represented in terms of Floquet modal functions, as we shall see shortly. We first consider the one-dimensional case. Suppose the current vector is y-directed (see Figure 2.6) and the current distribution is uniform along y (this uniform distribution assumption is necessary in order to make the problem one dimensional). Also we assume the source current to be a surface current located on the plane $z = 0$. The surface current excitation function can be expressed as

$$\vec{I} = \hat{y} \sum_{n=-\infty}^{\infty} f(x - na) \exp(-jn\varphi) \qquad (2.31)$$

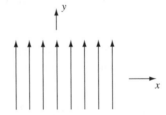

FIGURE 2.6 Linear infinite array of y-directed current elements.

The above surface current will produce the TM_y fields that can be obtained from a vector potential [3] $\vec{A} = \hat{y}A_y$, where A_y satisfies the following scalar Helmholtz equation:

$$\nabla^2 A_y + k_0^2 A_y = -J_y \qquad (2.32a)$$

In (2.32a) J_y is the y-component of the volume current density (amperes per square meter). The surface current density in (2.31) can be expressed in terms of the volume current density by introducing Dirac delta functions as below:

$$\nabla^2 A_y + k_0^2 A_y = -\delta(z) \sum_{n=-\infty}^{\infty} f(x - na) \exp(-jn\varphi) \qquad (2.32b)$$

In the above equation k_0 is the wave number in free space given by

$$k_0^2 = \omega^2 \mu_0 \varepsilon_0 \qquad (2.33)$$

where ω is the angular frequency and μ_0 and ε_0 are the permeability and permittivity of free space. A time dependent factor $\exp(j\omega t)$ is assumed throughout the analysis.

In order to find a solution for A_y in (2.32b) we expand the right-hand side of (2.32b) into a Floquet series as in (2.17). This yields

$$\nabla^2 A_y + k^2 A_y = -\delta(z) \frac{2\pi}{a} \sum_{n=-\infty}^{\infty} \tilde{f}\left(\frac{2n\pi + \varphi}{a}\right) \exp\left(-\frac{j(2n\pi + \varphi)x}{a}\right) \qquad (2.34)$$

Inspecting the right-hand side of (2.34) we can assume a solution of A_y in the following format:

$$A_y = \sum_{n=-\infty}^{\infty} F_n(z) \exp\left(-\frac{j(2n\pi + \varphi)x}{a}\right) \qquad (2.35)$$

In the above, no y-dependent term is included in A_y because the forcing function in (2.34) is independent of y. Substituting A_y from (2.35) into (2.34) and then comparing term by term, one obtains

$$\frac{\partial^2 F_n(z)}{\partial z^2} + k_{zn}^2 F_n(z) = -\delta(z) \frac{2\pi}{a} \tilde{f}(k_{xn}) \qquad (2.36)$$

with

$$k_{xn} = \frac{2n\pi + \varphi}{a} \qquad k_{zn}^2 = k_0^2 - k_{xn}^2 \qquad (2.37)$$

A solution for $F_n(z)$ that satisfies (2.36) is given by

$$F_n(z) = \begin{cases} \dfrac{\pi}{jak_{zn}} \tilde{f}(k_{xn}) \exp(-jk_{zn}z) & z > 0 \\[2ex] \dfrac{\pi}{jak_{zn}} \tilde{f}(k_{xn}) \exp(jk_{zn}z) & z < 0 \end{cases} \qquad (2.38)$$

The above solution is valid if the source current is symmetrically located in free space. If that is not the case (e.g., if a ground plane exists at $z = -d$), then the solution would have a different form. The final solution for A_y in the $z > 0$ region thus becomes

$$A_y = \frac{\pi}{ja} \sum_{n=-\infty}^{\infty} \frac{\tilde{f}(k_{xn})}{k_{zn}} \exp(-jk_{xn}x - jk_{zn}z) \qquad z > 0 \qquad (2.39)$$

The electromagnetic field components can be found using the following equations:

$$\vec{H} = \nabla \times (\hat{y}A_y) \qquad \vec{E} = \frac{1}{j\omega\varepsilon_0} \nabla \times \vec{H} \qquad (2.40)$$

From (2.40) we find $E_x = E_z = 0$, and $E_y = -j\omega\mu_0 A_y$. The nonzero electric field component produced by the current source is then

$$E_y = -\frac{\pi\omega\mu_0}{a} \sum_{n=-\infty}^{\infty} \frac{\tilde{f}(k_{xn})}{k_{zn}} \exp(-jk_{xn}x - jk_{zn}z) \qquad z > 0 \qquad (2.41)$$

Thus, the electromagnetic fields produced by the current source in (2.31) are expressed in terms of an infinite series. Each term of the above series satisfies the wave equation. If the time factor $\exp(j\omega t)$ is included, the solution for E_y becomes

$$E_y(x, y, z, t) = -\frac{\pi\omega\mu_0}{a} \sum_{n=-\infty}^{\infty} \frac{\tilde{f}(k_{xn})}{k_{zn}} \exp(j\omega t - jk_{xn}x - jk_{zn}z) \qquad z > 0 \quad (2.42)$$

We will now try to interpret the results in light of antenna radiation. The exponential term inside the summation sign is known as a Floquet modal function or a Floquet mode. Thus a Floquet mode is associated with a plane wave. The following vector represents the propagation direction of the nth Floquet mode:

$$\vec{p}_n = \hat{x}k_{xn} + \hat{z}k_{zn} \qquad (2.43)$$

Figure 2.7 shows the direction of propagation of the nth Floquet mode.

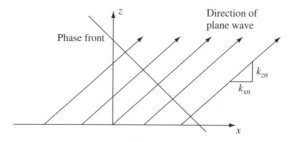

FIGURE 2.7 Direction of plane wave associated with a Floquet mode.

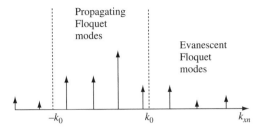

FIGURE 2.8 Floquet modal amplitudes.

2.3.1 Main Beam and Gratings

As stated before, each term under the summation sign in (2.42) represents a plane wave. The propagation vector associated with the nth Floquet mode makes an angle θ_n with the z-axis where

$$\tan \theta_n = \frac{k_{xn}}{k_{zn}} = \frac{k_{x0} + 2n\pi/a}{\sqrt{k_0^2 - (k_{x0} + 2n\pi/a)^2}} \tag{2.44}$$

with $k_{x0} = \varphi/a$. The Floquet mode corresponding to $n = 0$ is considered as the dominant mode because this mode propagates along the intended scan angle given by $\theta_0 = \sin^{-1}(\varphi/k_0 a)$. The other modes that have real k_{zn} are the grating beams. A Floquet mode becomes a grating beam if $k_0 \geq |(\varphi + 2n\pi)/a|$, $n \neq 0$. For an imaginary k_{zn}, the modal field components decay with z. Such a mode is known as an evanescent Floquet mode. For an evanescent Floquet mode, the propagation angle θ_n is imaginary. Figure 2.8 shows a conceptual sketch of the various Floquet modes. The relative amplitudes of E_y are given in (2.42). Typically, only a finite number of Floquet modes are propagating and the rest are evanescent.

2.4 TWO-DIMENSIONAL FLOQUET EXCITATION

For a two-dimensional Floquet excitation, the surface current should have the following form:

$$\vec{I}(x, y) = \hat{y} \sum_m \sum_n f(x - x_{mn}, y - y_{mn}) \exp(-jk_{x0}x_{mn} - jk_{y0}y_{mn}) \tag{2.45}$$

For a general grid structure, x_{mn} and y_{mn} are given by [see (2.19)]

$$x_{mn} = ma + \frac{nb}{\tan \gamma} \qquad y_{mn} = nb \tag{2.46}$$

where γ is the grid angle. Expanding the above surface current source into a Floquet series, we obtain [see (2.29)]

$$\vec{I}(x, y) = \hat{y} \frac{4\pi^2}{ab} \sum_m \sum_n \tilde{f}(k_{xmn}, k_{ymn}) \exp(-jk_{xmn}x - jk_{ymn}y) \tag{2.47}$$

where

$$k_{xmn} = k_{x0} + \frac{2m\pi}{a} \qquad k_{ymn} = k_{y0} - \frac{2m\pi}{a \tan \gamma} + \frac{2n\pi}{b} \qquad (2.48)$$

Following the same procedure used in Section 2.3, we obtain the radiated electric field components for the $z > 0$ region as

$$E_x = \frac{2\pi^2}{ab\omega\varepsilon_0} \sum_m \sum_n \frac{k_{xmn}k_{ymn}}{k_{zmn}} \tilde{f}(k_{xmn}, k_{ymn}) \exp\{j(-k_{xmn}x - k_{ymn}y - k_{zmn}z)\} \quad (2.49a)$$

$$E_y = -\frac{2\pi^2}{ab\omega\varepsilon_0} \sum_m \sum_n \frac{k_0^2 - k_{ymn}^2}{k_{zmn}} \tilde{f}(k_{xmn}, k_{ymn})$$

$$\times \exp\{j(-k_{xmn}x - k_{ymn}y - k_{zmn}z)\} \qquad (2.49b)$$

$$E_z = \frac{2\pi^2}{ab\omega\varepsilon_0} \sum_m \sum_n k_{ymn} \tilde{f}(k_{xmn}, k_{ymn}) \exp\{j(-k_{xmn}x - k_{ymn}y - k_{zmn}z)\} \qquad (2.49c)$$

The (m, n) terms in the above infinite series are associated with the TM_{ymn} (transverse magnetic) Floquet mode. If k_{zmn} is real, the associated Floquet mode becomes simply a plane wave that propagates along a direction parallel to the vector $\vec{p}_{mn} = \hat{x}k_{xmn} + \hat{y}k_{ymn} + \hat{z}k_{zmn}$. The corresponding radiation angle (θ_{mn}, ϕ_{mn}) in the spherical coordinate system is given by

$$k_{xmn} = k_0 \sin \theta_{mn} \cos \phi_{mn}$$
$$k_{ymn} = k_0 \sin \theta_{mn} \sin \phi_{mn} \qquad (2.50a)$$
$$k_{zmn} = k_0 \cos \theta_{mn}$$

In particular the $(0,0)$ Floquet mode, considered as the dominant mode, radiates along the intended angle (θ_0, ϕ_0). Thus, from (2.48) and (2.50a) and setting $m = n = 0$, we obtain

$$k_{x0} = k_0 \sin \theta_0 \cos \phi_0$$
$$k_{y0} = k_0 \sin \theta_0 \sin \phi_0 \qquad (2.50b)$$

Equation (2.50b) yields the relation between the phase progression factors (k_{x0}, k_{y0}) and the intended radiation direction (θ_0, ϕ_0).

A Floquet mode becomes a propagating plane wave only if the following condition is satisfied:

$$k_{xmn}^2 + k_{ymn}^2 \leq k_0^2 \qquad (2.51)$$

If the above condition is not satisfied, then the corresponding Floquet mode is an evanescent mode that decays along the z-direction.

2.4.1 Circle Diagram: Rectangular Grids

In a two-dimensional infinite array, the number of simultaneous propagating Floquet modes and their directions of propagation can be determined graphically. Using $\gamma = 90°$ in (2.48) for a rectangular grid we obtain the mode numbers for the (m, n) Floquet mode as

$$k_{xmn} = k_{x0} + \frac{2m\pi}{a} \qquad k_{ymn} = k_{y0} + \frac{2n\pi}{b} \tag{2.52}$$

Combining (2.52) and (2.50b) we obtain for the (m, n) Floquet mode

$$\left(k_{xmn} - \frac{2m\pi}{a} \right)^2 + \left(k_{ymn} - \frac{2n\pi}{b} \right)^2 = k_0^2 \sin^2 \theta_0 \leq k_0^2 \tag{2.53}$$

Equation (2.53) represents a family of circular regions of radii k_0, as depicted in Figure 2.9. Each circle represents a Floquet mode. The center of the (m, n)-mode circle is located at $(2m\pi/a, 2n\pi/b)$. The shaded circle centered at the origin corresponds to the dominant Floquet mode. The circular area inside the dominant-mode circle represents the visible scan region because all the points inside the circle satisfy (2.51). A higher order propagating Floquet mode manifests as a grating beam. A higher order mode turns into a grating beam if the (k_{xmn}, k_{ymn}) point lies inside the dominant-mode circle. A necessary condition for this to happen is that the higher order mode circles must intersect with the dominant-mode circle. From Figure 2.9 one observes that mode circles do not intersect if $k_0 < \pi/a$ and $k_0 < \pi/b$. Thus, an array will have only one beam in the entire visible region if the following conditions are satisfied simultaneously:

$$a \leq \frac{1}{2}\lambda_0 \qquad b \leq \frac{1}{2}\lambda_0 \tag{2.54}$$

In (2.54) λ_0 represents the wavelength in free space. Thus for nonexistence of a grating beam the maximum allowable element spacing is $\lambda_0/2$, regardless of the

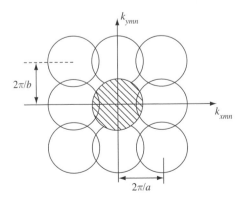

FIGURE 2.9 Circle diagram for rectangular grid.

scan angle of the desired beam. For some applications, existence of grating beams may be allowed subject to the constraint that no grating beam must fall inside the scan region. In this situation, element spacing may be increased to a larger value than $\lambda_0/2$. To quantify the optimum element spacing, consider that $\theta = \theta_{max}$ be the maximum scan angle from the bore sight. Therefore, the higher order mode circles must not intersect the k_{xmn}- or k_{ymn}-axis within a distance $|k_0 \sin \theta_{max}|$ from the origin (see Figure 2.9). This condition is satisfied if the following relation holds:

$$k_0 \sin \theta_{max} = \frac{2\pi}{a} - k_0 \tag{2.55}$$

The above equation yields the optimum value of the element spacing a as

$$a = \frac{\lambda_0}{1 + \sin \theta_{max}} \tag{2.56}$$

A similar relation holds for the element spacing b along the y-direction. Notice, for a special case with $\theta_{max} = 90°$, we obtain $a = b = \lambda_0/2$ as in (2.54).

2.4.2 Circle Diagram: Isosceles Triangular Grids

For an isosceles triangular grid we have $\tan \gamma = 2b/a$ (see Figure 2.4). Substituting this in (2.48) we have

$$k_{xmn} = k_{x0} + \frac{2m\pi}{a} \qquad k_{ymn} = k_{y0} + \frac{(2n-m)\pi}{b} \tag{2.57}$$

For propagating Floquet modes, we must have

$$\left(k_{xmn} - \frac{2m\pi}{a}\right)^2 + \left(k_{ymn} - \frac{(2n-m)\pi}{b}\right)^2 \leq k_0^2 \tag{2.58}$$

The Floquet mode circles associated with (2.58) are shown in Figure 2.10. Observe that the centers of the circles, in this case, lie on triangular grid points. The procedure of the previous section can be followed to determine the optimum spacing for a desired maximum scan angle, θ_{max}. For a symmetrical scan region, the centers of the immediate neighboring circles of the main beam circle must be symmetrically located. This happens if

$$\frac{2\pi}{b} = \sqrt{\left(\frac{2\pi}{a}\right)^2 + \left(\frac{\pi}{b}\right)^2} \tag{2.59}$$

Simplifying, we obtain

$$\frac{2b}{a} = \sqrt{3} \tag{2.60}$$

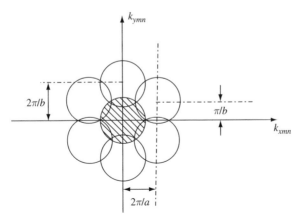

FIGURE 2.10 Circle diagram for isosceles triangular grid.

Equation (2.60) signifies that a symmetrical scanning region is possible if (2.60) holds, that is, the array elements are arranged in equilateral triangular grids. The maximum scan angle in such a situation is given by

$$k_0 \sin \theta_{\max} = \frac{2\pi}{b} - k_0 \qquad (2.61)$$

Substituting a from (2.60) we obtain the optimum element spacing a as

$$a = \frac{2\lambda_0}{\sqrt{3}\,(1 + \sin \theta_{\max})} = \frac{1.155\lambda_0}{1 + \sin \theta_{\max}} \qquad (2.62)$$

Comparing (2.62) with (2.56), we conclude that a similar scanning range can be achieved with about 15% larger element spacing if equilateral triangular grids are used. Therefore, from a grating lobe perspective, an equilateral triangular grid array utilizes about 15% fewer elements than that with a square grid array of identical aperture areas.

2.5 GRATING BEAMS FROM GEOMETRICAL OPTICS

Thus far we have applied Fourier transform techniques to study the behavior of an infinite array source. While the Fourier transform method is vital for a precise analysis of an infinite array, some of the important features can be derived using simple concepts of geometrical optics. This simple procedure does not add any new information; nevertheless it helps in understanding some of the fundamental properties of an array from a different perspective. For instance, the relation between the element spacing and the grating beam directions can be appreciated using simple geometry and fundamental principles of electromagnetic radiation.

To illustrate, consider the excitation function in (2.45). The total radiation field is the superposition of the individual fields emanating form the sources represented by the right-hand side of (2.45). The fields add coherently if the phases are equal or differ by an integral multiple of 2π. For a large r the distance traveled by the field emanating from the (m, n) element at $R(r, \theta, \phi)$ is given by (see Figure 2.11)

$$r_{mn} \approx r - \hat{r} \cdot \vec{\rho}_{mn} \tag{2.63}$$

where

$$\hat{r} = \hat{x} \sin\theta \cos\varphi + \hat{y} \sin\theta \sin\phi + \hat{z} \cos\theta$$
$$\vec{\rho}_{mn} = \hat{x} x_{mn} + \hat{y} y_{mn} \tag{2.64}$$

In (2.63) and (2.64), $\hat{x}$, $\hat{y}$, $\hat{z}$, and $\hat{r}$ are the unit vectors. Substituting (2.64) in (2.63), we obtain

$$r_{mn} \approx r - x_{mn} \sin\theta \cos\phi - y_{mn} \sin\theta \sin\phi \tag{2.65}$$

The phase of the radiated field at R produced by the (m, n) element is the phase delay due to the path length plus the phase lag of the source, which is given by

$$\begin{aligned}\psi_{mn} &= -k_0 r_{mn} - k_{x0} x_{mn} - k_{y0} y_{mn} \\ &= -k_0 r + (k_0 \sin\theta \cos\phi - k_{x0}) x_{mn} + (k_0 \sin\theta \sin\phi - k_{y0}) y_{mn}\end{aligned} \tag{2.66}$$

Substituting the expressions for x_{mn} and y_{mn} from (2.46) we obtain

$$\psi_{mn} = -k_0 r + (k_0 \sin\theta \cos\phi - k_{x0}) \left(ma + \frac{nb}{\tan\gamma} \right) + (k_0 \sin\theta \sin\phi - k_{y0}) nb \tag{2.67}$$

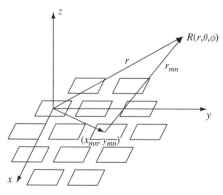

FIGURE 2.11 Array geometry and the coordinate system.

For field coherence, the phase difference between the fields emanating from two adjacent elements should be an integral multiple of 2π. Accordingly, we set

$$\psi_{10} - \psi_{00} = 2p\pi \qquad \psi_{01} - \psi_{00} = 2q\pi \tag{2.68}$$

where p and q are integers. Using (2.67) in (2.68) we obtain

$$(k_0 \sin\theta \cos\phi - k_{x0})a = 2p\pi$$
$$\frac{(k_0 \sin\theta \cos\phi - k_{x0})b}{\tan\gamma} + (k_0 \sin\theta \sin\phi - k_{y0})b = 2q\pi \tag{2.69}$$

The above equations yield

$$k_0 \sin\theta \cos\phi = k_{x0} + \frac{2p\pi}{a} \qquad k_0 \sin\theta \sin\phi = k_{y0} - \frac{2p\pi}{a \tan\gamma} + \frac{2q\pi}{b} \tag{2.70}$$

Now, $k_0 \sin\theta \cos\phi$ and $k_0 \sin\theta \sin\phi$ are the wave numbers along the x- and y-directions, respectively, for a plane wave propagating along the (θ, ϕ)-direction. Denoting the wave numbers as k_{xpq} and k_{ypq} we write

$$k_{xpq} = k_{x0} + \frac{2p\pi}{a} \qquad k_{ypq} = k_{y0} - \frac{2p\pi}{a \tan\gamma} + \frac{2q\pi}{b} \tag{2.71}$$

Equations (2.71) are identical with Equations (2.48) except the dummy variables p and q are replaced by m and n, respectively. By definition, the main beam corresponds to $p = q = 0$. Thus in reference to (2.68) we observe that the main beam is produced at a location where the element fields arrive at the same phase. Contrary to this, the grating beams are produced where most of the element fields, if not all, differ in phase by integral multiples of 2π.

2.6 FLOQUET MODE AND GUIDED MODE

If an infinite array structure is capable of supporting a guided wave, then under certain Floquet excitations, the guided mode may be strongly excited. Depending on the coupling mechanism between the radiating elements and the guided mode, the input voltage across each source element becomes either infinitely large or zero or the voltage and the current are in phase quadrature. The above three voltage–current relations lead to infinite, zero, or reactive input impedance; hence the array ceases to radiate. This effect is known as scan blindness of the array. A patch array is a typical example of such an array structure where the scan blindness occurs. The guided wave mode in this case is the surface wave mode that is supported by the dielectric substrate layer. Each patch excites at least one surface wave mode. If the propagation constant of a surface wave mode coincides with that of a Floquet mode, then these two modes couple strongly, leading to a resonance and subsequent scan blindness.

The possibility of strong surface wave coupling can be examined graphically from the Floquet circle diagram associated with an infinite array. It will be seen in the following development that a surface wave corresponds to a circle in the $k_x k_y$-plane similar to that of a Floquet mode. An intersection of a surface wave mode circle and a Floquet mode circle indicates a possibility of coupling between the surface wave mode and the respective Floquet mode.

To determine the surface wave mode loci (circles) in the $k_x k_y$-plane, we assume that the resultant surface wave produced by an infinite array is a plane wave with propagation constant β_{su}. The field intensity of the surface wave can be expressed as

$$\vec{E}_{su} = A \exp\{-j\beta_{su}^x x - j\beta_{su}^y y\}\vec{f}_{su}(z) \tag{2.72}$$

where $\vec{f}_{su}(z)$ represents the z-variation of the surface wave mode and β_{su}^x and β_{su}^y are the wave numbers of the surface wave mode along x- and y-directions, respectively. The relation between β_{su}^x, β_{su}^y and β_{su} is given by

$$\beta_{su}^{x\,2} + \beta_{su}^{y\,2} = \beta_{su}^2 \tag{2.73}$$

From (2.72) the equiphase surface is given by the following plane parallel to the z-axis:

$$x\beta_{su}^x + y\beta_{su}^y = C \tag{2.74}$$

To establish a surface wave resonant condition (or scan blindness condition), we assume that the propagation vector makes an angle α with the x-axis of the array coordinate, as shown in Figure 2.12. For simplicity we assume isosceles triangular

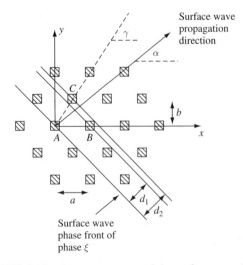

FIGURE 2.12   Array geometry and the surface wave front.

grid of grid angle γ. The surface wave phase front should be perpendicular to the direction of propagation. For simplicity, we assume that at time $t = 0$, the phase front passes through the origin. We also assume that the phase difference between the phase front through A and the antenna element A is ξ. For coherent coupling, the phase of the two elements on the phase front line differs by an integral multiple of 2π. Suppose the phase of the element at A is zero. Therefore, the phase of the phase front at A is ξ. The phase of the phase front through B would be $\xi - \beta_{su} d_1 - 2p\pi$, where d_1 is the perpendicular distance between the two phase fronts at the two locations as shown in Figure 2.12 and p is an integer. Suppose $-\psi_1$ is the phase of the element at B. At B the phase difference between the element and the phase front passing through it must be ξ. Therefore,

$$\xi = \xi - \beta_{su} d_1 - 2p\pi + \psi_1 \tag{2.75}$$

Similarly, for the element at C, the above phase relation becomes

$$\xi = \xi - \beta_{su} d_2 - 2q\pi + \psi_2 \tag{2.76}$$

where $-\psi_2$ is the phase of the element at C, q is an integer and d_2 is the distance between the phase fronts through A and C. From Figure 2.12, we write

$$\tan \gamma = \frac{2b}{a} \qquad d_1 = a\cos\alpha \qquad d_2 = b\frac{\cos(\gamma - \alpha)}{\sin\gamma} \tag{2.77}$$

For a desired scan direction (θ_0, ϕ_0), ψ_1 and ψ_2 are given by

$$\psi_1 = ak_{x0} \qquad \psi_2 = \frac{bk_{x0}}{\tan\gamma} + bk_{y0} \tag{2.78}$$

with

$$k_{x0} = k_0 \sin\theta_0 \cos\phi_0 \qquad k_{y0} = k_0 \sin\theta_0 \sin\phi_0 \tag{2.79}$$

Using (2.75)–(2.78) and eliminating α we finally derive

$$\left(k_{x0} - \frac{2p\pi}{a}\right)^2 + \left(k_{y0} - \frac{(2q-p)\pi}{b}\right)^2 = \beta_{su}^2 \tag{2.80}$$

We now invoke (2.57) to eliminate k_{x0} and k_{y0} and introduce the mode numbers for the (m, n) Floquet mode. This yields

$$\left(k_{xmn} - \frac{2(p+m)\pi}{a}\right)^2 + \left(k_{ymn} - \frac{\{2(q+n)-(p+m)\}\pi}{b}\right)^2 = \beta_{su}^2$$

Since p, q, m, and n are all integers, we can write

$$\left(k_{xmn} - \frac{2m'\pi}{a}\right)^2 + \left(k_{ymn} - \frac{\{2n'-m'\}\pi}{b}\right)^2 = \beta_{su}^2 \tag{2.81}$$

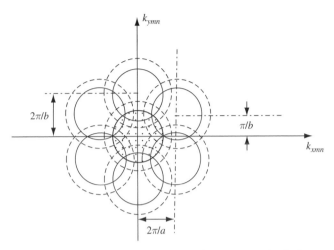

FIGURE 2.13 Floquet mode and surface wave mode circles.

where m' and n' are also integers. If at a given scan angle (2.81) is satisfied, then the surface wave is very strongly excited, leading to a scan blindness. Equation (2.81) represents a family of circles, called guided wave mode circles. Comparing with (2.58) we observe that the guided mode circles are concentric with Floquet mode circles. The broken-line circles in Figure 2.13 represent surface wave modes. The surface wave mode circles have larger radii than Floquet mode circles, because $\beta_{su} > k_0$. However, for some structures, the guided wave mode may have a smaller propagation constant than k_0. Scan blindness at the (θ_0, ϕ_0) scan direction occurs only if a surface wave mode circle passes through a point $(k_0 \sin \theta_0 \cos \phi_0, k_0 \sin \theta_0 \sin \phi_0)$. From Figure 2.13 we observe that a blind angle exists in every scan plane. Scan blindness will not occur if a surface wave mode circle does not intersect a Floquet mode circle. This happens if $\pi/a > \beta_{su}$ and $\pi/b > \beta_{su}$ for a rectangular grid of cell size $a \times b$.

The array structure considered above supports a guided wave mode that can propagate in all directions of the array plane. However, it is possible to have array structures that support a guided wave only in some specified directions. An example is a printed patch array on a grooved dielectric substrate. In such cases, the guided wave mode will have different shaped loci (e.g., periodically modulated lines) than that of concentric circles.

2.7 SUMMARY

In this chapter we introduced Floquet modes and Floquet modal analysis of infinite arrays starting from the basic Fourier transformation. Characteristics of Floquet modes in light of antenna radiation are explored. We introduced Floquet mode circle diagrams to determine the existence of grating beams and their respective radiation directions graphically. Coupling between a Floquet mode and a guided wave mode and related scan blindness phenomenon are discussed. The following chapter deals with Floquet modes in vector form, orthogonality, and its application for detailed array analysis, including active impedance and embedded element patterns.

REFERENCES

[1] R. E. Collin, *Field Theory of Guided Waves* 2nd ed., IEEE, New York, 1991, p. 820.
[2] J. Mathews and R. L. Walker, *Mathematical Methods of Physics*, 2nd ed., W. A. Benjamin, Reading, MA, 1970.
[3] R. F. Harrington, *Time-Harmonic Electromagnetic Fields*, McGraw-Hill, New York, 1961.

PROBLEMS

2.1 Using a contour integration method in the complex k_x-plane, prove that

$$\delta(x) = \frac{1}{2\pi} \int_{-\infty}^{\infty} \exp(jk_x x)\, dk_x$$

where $\delta(x)$ is the Dirac delta function. [*Hints*: For $x \neq 0$, use a closed contour formed by an infinite semicircle at the upper half (or the lower half, as appropriate) of the complex k_x-plane and the real axis and show that the integral vanishes. Now integrate the right-hand side with respect to x in the range $-\varepsilon < x < \varepsilon$ and then integrate with respect to k_x to show the final result equal to unity.]

2.2 Obtain the Fourier spectra of
   (a) $f(x) = u(x+a/2) - u(x-a/2)$
   (b) $f(x) = \cos(m\pi x/a)\,[u(x+a/2) - u(x-a/2)]$
where $u(x)$ is the unit step function.

2.3 One form of Poisson's sum formula can be expressed as

$$\sum_{n=-\infty}^{\infty} f(na) = \frac{2\pi}{a} \sum_{n=-\infty}^{\infty} \tilde{f}\left(\frac{2n\pi}{a}\right)$$

Using this, prove that

$$\sum_{n=-\infty}^{\infty} \exp(jnk_x a) = \frac{2\pi}{a} \sum_{n=-\infty}^{\infty} \delta\left(k_x - \frac{2n\pi}{a}\right)$$

2.4 Obtain the Fourier series expansions for the following periodic functions:

$$\text{(a)} \quad g(x) = \sum_{n=-\infty}^{\infty} \delta(x - na)$$

$$\text{(b)} \quad g(x) = \sum_{n=-\infty}^{\infty} f(x - na)$$

where $f(x) = \cos(2\pi x/a)$ for $|x| < a/4$, and $f(x) = 0$ for $|x| > a/4$.

2.5 A function $g(x)$ is defined as

$$g(x) = \sum_{n=-\infty}^{\infty} \delta(x - n) \exp\left(-\frac{jn\pi}{2}\right)$$

(a) Find the periodicity of $g(x)$ and then expand $g(x)$ into a Fourier series.

(b) Expand $g(x)$ into a Floquet series. Show that Floquet series and Fourier series expansions are identical.

2.6 A function $f(x)$ is defined as

$$f(x) = \frac{\sin \alpha x}{\alpha x} \qquad -\infty < x < \infty$$

(a) Obtain the Fourier Transform of $f(x)$ and show that the spectrum is band-limited of bandwidth 2α.

(b) Define $h(x) = \sum_{n=-\infty}^{\infty} f(na)\delta(x - na)$. Show that within the band of $f(x)$ the spectra of $f(x)$ and $h(x)$ differ by a constant multiplier if $a \leq \pi/\alpha$ (this is known as Shannon's sampling theorem). *Hints*: First prove that

$$\tilde{h}(k_x) = \frac{1}{a}\tilde{f}(k_x) \otimes \sum_{n=-\infty}^{\infty} \delta\left(k_x - \frac{2n\pi}{a}\right)$$

and then carry out the convolution integral to prove the results.

(c) Set $\alpha = k_0$ in (b), where k_0 is the wave number in free space. Then show that a continuous current source and its samples have identical radiation patterns if the sampling interval is less than one-half of the wavelength (this is known as Woodward's sampling theorem in antenna theory).

2.7 A function $f(x, y)$ is defined as

$$f(x, y) = \begin{cases} 1 & \text{if } |x| \leq a/2, |y| \leq b/2 \\ 0 & \text{otherwise} \end{cases}$$

If $h(x, y)$ is defined as

$$h(x, y) = \sum_{m=-\infty}^{\infty} \sum_{n=-\infty}^{\infty} f(x - x_{mn}, y - y_{mn}) \exp(-jk_{x0}x_{mn} - jk_{y0}y_{mn})$$

then obtain the Floquet series for $h(x, y)$ for (a) a rectangular grid and (b) a triangular grid. Use $a = b$ in both cases. Set $k_{x0} = k_{y0} = 0$ and then determine the locations of the first nine spectral points (increasing the distance from the origin) in the $k_x k_y$-plane for both cases.

2.8 Repeat problem 2.7 if

$$f(x, y) = \begin{cases} 1 & \sqrt{x^2 + y^2} \le a/2 \\ 0 & \text{otherwise} \end{cases}$$

2.9 Show that the function $F_n(z)$ in (2.38) is a solution of the differential equation (2.36).

2.10 Find the solution of the differential equation

$$\frac{\partial^2 F}{\partial z^2} + k_z^2 F = -\delta(z)$$

if $F(z) = 0$ at $z = -d$ and $F(z)$ is continuous at $z = 0$. Assume $F(z)$ represents a propagating wave in the $z > 0$ region.

2.11 An array of magnetic surface current elements is expressed as

$$\vec{M} = \hat{x} \sum_{m=-\infty}^{\infty} \sum_{n=-\infty}^{\infty} f(x - x_{mn}, y - y_{mn}) \exp(-jk_{x0}x_{mn} - jk_{y0}y_{mn})$$

where (x_{mn}, y_{mn}) is a coordinate point on an equilateral triangular lattice. The length of a side is $1.5\lambda_0$ and k_{x0} and k_{y0} are two constants. The function $f(x, y)$ is defined as

$$f(x, y) = \begin{cases} 1 & |x| \le \lambda_0/2, |y| \le \lambda_0/2 \\ 0 & \text{otherwise} \end{cases}$$

(a) Obtain the direction of the main beam in terms of k_{x0} and k_{y0}.
(b) Find the relative intensity and the direction of radiation of the nearest grating lobe.
(c) Find the power radiated by a single element in the array environment.
(d) If the array is intended to radiate along $\theta = 10°$, $\phi = 45°$, then determine the location of the grating lobe nearest to the main beam (*Hints:* Use a circle diagram.)

2.12 An infinite array of probe-fed patches is printed on a dielectric substrate of dielectric constant 2.56. The element spacing of the array is 2.0 cm with square lattice. The frequency of operation is 20 GHz. The dielectric substrate supports a surface wave mode with a propagation constant of $1.15k_0$, where k_0 is the free-space wave number. Obtain the blind angles of the array for the E- and H-plane scans using a circle diagram.

2.13   An infinite array of slot-coupled patches is printed on a substrate of dielectric constant of 2.56. The array elements are in an equilateral triangular lattice of side 2.0 cm. The frequency of operation is 20 GHz. Asymmetric striplines excite the slots, as shown in Figure P2.13. Because of two ground planes, the stripline region supports a parallel-plate mode that has a propagation constant of $1.02k_0$, where k_0 is the wave number in free space. The patch-side substrate supports a surface wave mode of propagation constant $1.15k_0$. Obtain the blind angles in the E, H, and diagonal planes.

2.14   An array of probe-fed patches is printed on air–dielectric substrate. An array of microstriplines is printed between patch rows as shown in Figure P2.14. The microstriplines support the TEM mode (because of the air–dielectric substrate). The element spacing of the array (square grid) is 3.0 cm, and the wavelength of operation is 2.0 cm. (a) Draw the circle diagram for the Floquet modes of the array. (b) Assume that the propagation constants of the microstrip transmission lines do not change with the patch excitations; plot k_y versus k_x for the microstrip transmission lines. (c) Find the blind angles in the E, H, and diagonal planes (assume that the transmission lines are parallel to the E-plane of the patches). (*Hint*: The propagation constant of a microstripline under Floquet excitation can be considered as $\beta = \pm k_0 \pm 2m\pi/a$, where m is a positive integer.)

2.15   A slotted rectangular waveguide that supports only the TE_{10} mode is excited periodically by probes as shown in Figure P2.15. The slots are cut on the broad wall with a small offset from the line of symmetry. The distance

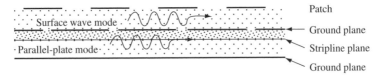

FIGURE P2.13   Side view of a slot-coupled patch array fed by asymmetric striplines.

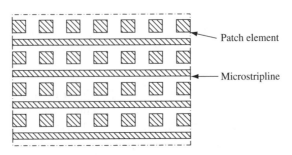

FIGURE P2.14   Top view of a portion of an infinite array of patches with microstriplines printed on the same plane.

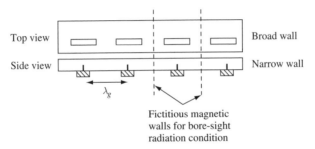

FIGURE P2.15 Top and side views of a waveguide slot array. A probe underneath feeds a slot.

between the two probes is equal to one guide wavelength of the TE_{10} mode. Assume that the slots do not significantly affect the propagation constant of the waveguide.

(a) Prove from guided mode coupling point of view that this one-dimensional infinite array has blindness in the bore-sight direction.

(b) From the image theory point of view observe that with respect to the bore-sight radiation a feed probe in an infinite array environment is equivalent to a situation of being inside a waveguide cavity with magnetic walls one-half wavelength apart on both sides. Thus the probe admittance becomes negligibly small; hence the slot radiation is negligibly small due to the mismatch problem.

CHAPTER THREE

Floquet Modal Functions

3.1 INTRODUCTION

For analytical simplicity it is often preferred to work with orthogonal modal functions that are normalized with respect to the modal power. This necessitates normalization of Floquet modal functions that are pertinent to an infinite array aperture analysis. This chapter derives the normalized Floquet modal functions. The normalization is done following a similar procedure as in the case of waveguide modes associated with the waveguide circuit analysis [1]. The normalized TE_z (transverse electric to z) and TM_z (transverse magnetic to z) vector modal functions are derived with respect to an arbitrary array grid structure. The application of Floquet modal expansion is demonstrated through two examples, namely a printed dipole array and a waveguide aperture array. In addition, several characteristic features of the arrays are presented. The mechanism of array blindness of a printed array is explained in light of surface wave coupling. A graphical method for determining the blind angles is presented. A general expression for the active element pattern, also known as the embedded element pattern, of an array is derived. It is found that the amplitudes of the dominant Floquet modes completely determine the active element pattern. The significance of the active element pattern in regard to mutual coupling is explained. Numerical results of the arrays under consideration revealing important electrical characteristics are shown.

3.2 TE_Z AND TM_Z FLOQUET VECTOR MODAL FUNCTIONS

For infinite array analysis it is useful to have a complete set of orthogonal functions that can be utilized to express the radiating fields. Several sets of orthogonal modal functions are possible to construct. However, for multilayer structure analysis,

Phased Array Antennas. By Arun K. Bhattacharyya
© 2006 John Wiley & Sons, Inc.

layered along the z-direction, TE_z and TM_z types of modal decomposition are very suitable because of the following reasons:

- The TE_z and TM_z modes retain their individual modal characteristics as the fields propagate from one medium to another medium, provided the media are homogeneous and isotropic. In other words a TE_z mode will not produce a TM_z mode while passing through the interface of a layered media and vice versa.

- The generalized scattering matrices (GSMs) of the interface of two media become diagonal, which has computational advantages, because these modes do not couple unless there is an obstacle.

In this section we deduce the mathematical expressions for the normalized modal fields.

3.2.1 TE_z Floquet Modal Fields

By definition, the TE_z modal fields have no electric field component along z. Therefore the TE_z modal field components can be generated from the electric vector potential $\vec{F}$ which has only the z-component [2, Ch. 4]. In terms of Floquet harmonics, $\vec{F}$ can be expressed as

$$\vec{F} = \hat{z} A \exp(-jk_{xmn}x - jk_{ymn}y - jk_{zmn}z) \qquad (3.1)$$

where A is a constant and k_{xmn}, k_{ymn} are the mode numbers for the TE_{zmn} Floquet mode[1]. For a general grid structure, these quantities are given by [see (2.48)]

$$k_{xmn} = k_{x0} + \frac{2m\pi}{a} \qquad k_{ymn} = k_{y0} - \frac{2m\pi}{a\tan\gamma} + \frac{2n\pi}{b} \qquad (3.2)$$

where a, b, and γ are the grid parameters shown in Figure 3.1. Such a grid structure is denoted symbolically by $[a,b,\gamma]$. In (3.2) k_{x0} and k_{y0} are phase constants related to the intended direction of radiation. If (θ, ϕ) is the intended radiation direction in the spherical coordinate system, then the phase constants are given by

$$k_{x0} = k_0 \sin\theta\cos\phi \qquad k_{y0} = k_0 \sin\theta\sin\phi \qquad (3.3a)$$

In (3.3a) k_0 is the wave number in free space. The wave number along z is obtained as

$$k_{zmn} = \sqrt{k^2 - k_{xmn}^2 - k_{ymn}^2} \qquad (3.3b)$$

---

[1] At places we use the symbol "(m, n)" to represent either the TE_{zmn} or TM_{zmn} Floquet mode.

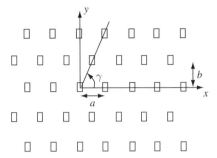

FIGURE 3.1 General array grid structure denoted by $[a, b, \gamma]$.

where k is the wave number in the medium. The wave number along z is real for a propagating mode and imaginary for an evanescent mode. The electric field components are determined using the relation $\vec{E} = -\nabla \times \vec{F}$, which yields

$$\vec{E} = jA \left(\hat{x} k_{ymn} - \hat{y} k_{xmn} \right) \exp(-jk_{xmn}x - jk_{ymn}y - jk_{zmn}z) \tag{3.4}$$

As expected the electric field vector is transverse to the z-direction. The magnetic field components are obtained from the Maxwell equation

$$\nabla \times \vec{E} = -j\omega\mu\vec{H} \tag{3.5}$$

yielding

$$\vec{H} = jAY_{mn}^{TE} \left(\hat{x} k_{xmn} + \hat{y} k_{ymn} - \hat{z} \frac{k^2 - k_{zmn}^2}{k_{zmn}} \right) \exp(-jk_{xmn}x - jk_{ymn}y - jk_{zmn}z) \tag{3.6}$$

where Y_{mn}^{TE} is the modal admittance for the TE$_{zmn}$ Floquet mode, which is given by

$$Y_{mn}^{TE} = \frac{k_{zmn}}{\omega\mu} \tag{3.7}$$

Now we normalize the modal fields by selecting the constant term A such that the complex power through a unit cell becomes equal to the complex conjugate of the modal admittance of the TE$_{zmn}$ mode. Mathematically

$$\iint\limits_{\text{unit cell}} \vec{E} \times \vec{H}^* \cdot \hat{z} \, dx \, dy = Y_{mn}^{TE^*} \tag{3.8}$$

Using (3.4) and (3.6) in (3.8) one obtains a solution for A as

$$A = \frac{1}{j\sqrt{ab(k^2 - k_{zmn}^2)}} \tag{3.9}$$

We substitute the above expression for A in (3.4) and (3.6) to obtain the final expressions for the normalized field vectors. For brevity of notation, usually the normalized field vectors are expressed in terms of the vector modal functions as follows:

$$\vec{E} = \vec{e}_{mn}^{\mathrm{TE}} \exp(-jk_{zmn}z) \tag{3.10}$$

where

$$\vec{e}_{mn}^{\mathrm{TE}} = \frac{\hat{x}k_{ymn} - \hat{y}k_{xmn}}{\sqrt{ab(k^2 - k_{zmn}^2)}} \exp(-jk_{xmn}x - jk_{ymn}y) \tag{3.11}$$

The normalized transverse magnetic field vector (with the z-component suppressed) is given by

$$\vec{H}_t = Y_{mn}^{\mathrm{TE}} \vec{h}_{mn}^{\mathrm{TE}} \exp(-jk_{zmn}z) \tag{3.12}$$

where $\vec{h}_{mn}^{\mathrm{TE}}$ is obtained as

$$\vec{h}_{mn}^{\mathrm{TE}} = \frac{\hat{x}k_{xmn} + \hat{y}k_{ymn}}{\sqrt{ab(k^2 - k_{zmn}^2)}} \exp(-jk_{xmn}x - jk_{ymn}y) \tag{3.13}$$

The vector functions $\vec{e}_{mn}^{\mathrm{TE}}$ and $\vec{h}_{mn}^{\mathrm{TE}}$ are independent of z, known as electric and magnetic vector modal functions, respectively. The z-component can be easily recovered from the transverse magnetic field vector using the Maxwell equation $\nabla \cdot \vec{H} = 0$. If V_{mn}^{TE} denotes the modal voltage (or modal amplitude) of the TE$_{zmn}$ mode, then the corresponding electromagnetic field components can be expressed in terms of the modal functions as

$$\vec{E}_{mn}^{\mathrm{TE}} = V_{mn}^{\mathrm{TE}} \vec{e}_{mn}^{\mathrm{TE}} \exp(-jk_{zmn}z)$$

$$\vec{H}_{mn}^{\mathrm{TE}} = V_{mn}^{\mathrm{TE}} Y_{mn}^{\mathrm{TE}} \left(\vec{h}_{mn}^{\mathrm{TE}} + \frac{\hat{z}\nabla \cdot \vec{h}_{mn}^{\mathrm{TE}}}{jk_{zmn}} \right) \exp(-jk_{zmn}z) \tag{3.14a}$$

If the fields propagate along the negative z-direction, the corresponding fields would be

$$\vec{E}_{mn}^{\mathrm{TE}} = V_{mn}^{\mathrm{TE}} \vec{e}_{mn}^{\mathrm{TE}} \exp(jk_{zmn}z)$$

$$\vec{H}_{mn}^{\mathrm{TE}} = -V_{mn}^{\mathrm{TE}} Y_{mn}^{\mathrm{TE}} \left(\vec{h}_{mn}^{\mathrm{TE}} - \frac{\hat{z}\nabla \cdot \vec{h}_{mn}^{\mathrm{TE}}}{jk_{zmn}} \right) \exp(jk_{zmn}z) \tag{3.14b}$$

The complex power passing through a unit cell is

$$P_{mn}^{\mathrm{TE}} = |V_{mn}^{\mathrm{TE}}|^2 Y_{mn}^{\mathrm{TE}*} \tag{3.15}$$

Notice that the following relations hold for the modal vectors:

$$\vec{e}_{mn}^{\,\text{TE}} = -\hat{z} \times \vec{h}_{mn}^{\,\text{TE}} \qquad \vec{h}_{mn}^{\,\text{TE}} = \hat{z} \times \vec{e}_{mn}^{\,\text{TE}}$$

$$\iint\limits_{\text{unit cell}} \vec{e}_{mn}^{\,\text{TE}} \times \vec{h}_{ij}^{\,\text{TE}*} \cdot \hat{z}\, dx\, dy = \delta_{mi}\delta_{nj} \tag{3.16a}$$

where δ_{pq} is the Kronecker delta. The integral relation in (3.16a) is equivalent to

$$\iint\limits_{\text{unit cell}} \vec{e}_{mn}^{\,\text{TE}} \cdot \vec{e}_{ij}^{\,\text{TE}*}\, dx\, dy = \delta_{mi}\delta_{nj} \tag{3.16b}$$

3.2.2 TM$_z$ Floquet Modal Fields

The TM$_z$ Floquet modal fields can be constructed from the magnetic vector potential, $\vec{A}$, that has only the z-component [2]. The magnetic vector potential can be expressed as

$$\vec{A} = \hat{z}B \exp(-jk_{xmn}x - jk_{ymn}y - jk_{zmn}z) \tag{3.17}$$

where B is a constant. The electric and magnetic field components can be found from the following relations:

$$\vec{H} = \nabla \times \vec{A} \qquad \nabla \times \vec{H} = j\omega\varepsilon\vec{E} \tag{3.18}$$

Because $\vec{A}$ has only the z-component, the magnetic field vector is transverse to z (TM$_z$). Following the procedure in the foregoing section, the electric and magnetic vector modal functions for the TM$_{zmn}$ mode are obtained as

$$\vec{e}_{mn}^{\,\text{TM}} = \frac{\hat{x}k_{xmn} + \hat{y}k_{ymn}}{\sqrt{ab(k^2 - k_{zmn}^2)}} \exp(-jk_{xmn}x - jk_{ymn}y) \tag{3.19a}$$

$$\vec{h}_{mn}^{\,\text{TM}} = -\frac{\hat{x}k_{ymn} - \hat{y}k_{xmn}}{\sqrt{ab(k^2 - k_{zmn}^2)}} \exp(-jk_{xmn}x - jk_{ymn}y) \tag{3.19b}$$

The modal admittance is

$$Y_{mn}^{\text{TM}} = \frac{\omega\varepsilon}{k_{zmn}} \tag{3.20}$$

Interestingly, we notice

$$\vec{e}_{mn}^{\,\text{TM}} = \vec{h}_{mn}^{\,\text{TE}} \qquad \vec{h}_{mn}^{\,\text{TM}} = -\vec{e}_{mn}^{\,\text{TE}} \tag{3.21}$$

The electric and magnetic fields for a propagating wave along the z-direction can be expressed as

$$\vec{E}_{mn}^{TM} = V_{mn}^{TM} \left(\vec{e}_{mn}^{TM} + \frac{\hat{z}(\nabla \cdot \vec{e}_{mn}^{TM})}{jk_{zmn}} \right) \exp(-jk_{zmn}z)$$

$$\vec{H}_{mn}^{TM} = V_{mn}^{TM} Y_{mn}^{TM} \vec{h}_{mn}^{TM} \exp(-jk_{zmn}z)$$

$$(3.22)$$

For propagation along the negative z-direction, the fields are given by

$$\vec{E}_{mn}^{TM} = V_{mn}^{TM} \left(\vec{e}_{mn}^{TM} - \frac{\hat{z}(\nabla \cdot \vec{e}_{mn}^{TM})}{jk_{zmn}} \right) \exp(jk_{zmn}z)$$

$$\vec{H}_{mn}^{TM} = -V_{mn}^{TM} Y_{mn}^{TM} \vec{h}_{mn}^{TM} \exp(jk_{zmn}z)$$

$$(3.23)$$

where V_{mn}^{TM} is the modal voltage or modal amplitude of the TM_{zmn} mode. The complex power per unit cell is

$$P_{mn}^{TM} = |V_{mn}^{TM}|^2 Y_{mn}^{TM*}$$

$$(3.24)$$

3.3 INFINITE ARRAY OF ELECTRIC SURFACE CURRENT ON DIELECTRIC-COATED GROUND PLANE

In this section we demonstrate the application of TE_z and TM_z mode decomposition for obtaining the solution of the electromagnetic fields produced by an infinite array of electric surface current sources. The source elements are situated at the air–dielectric interface of a dielectric coated ground plane. The electric currents are x-directed, as shown in Figure 3.2, and the elements are linearly phased with uniform amplitudes (Floquet excitation). The phase gradients are k_{x0} and k_{y0} along x- and y-directions, respectively. Therefore, the source distribution can be expressed as

$$\vec{I}(x,y) = \hat{x} \sum_{i=-\infty}^{\infty} \sum_{j=-\infty}^{\infty} f(x - x_{ij}, y - y_{ij}) \exp(-jk_{x0}x_{ij} - jk_{y0}y_{ij})$$

$$(3.25)$$

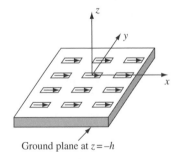

Ground plane at $z = -h$

FIGURE 3.2 Infinite array of x-directed electric current sources.

For the grid structure $[a,b,\gamma]$, x_{ij} and y_{ij} are given by[2]

$$x_{ij} = ia + \frac{jb}{\tan \gamma} \qquad y_{ij} = jb \tag{3.26}$$

We assume a uniform current source of length l and width t, so that $f(x, y)$ can be represented by

$$f(x, y) = \begin{cases} \frac{I_0}{t} & |x| < \frac{l}{2} \quad |y| < \frac{t}{2} \\ 0 & \text{otherwise} \end{cases} \tag{3.27}$$

where I_0 is the total input current. Expanding $\vec{I}(x, y)$ as a Floquet series we obtain

$$\vec{I}(x, y) = \hat{x}\frac{4\pi^2}{ab} \sum_m \sum_n \tilde{f}(k_{xmn}, k_{ymn}) \exp(-jk_{xmn}x - jk_{ymn}y) \tag{3.28}$$

with

$$k_{xmn} = k_{x0} + \frac{2m\pi}{a} \qquad k_{ymn} = k_{y0} - \frac{2m\pi}{a \tan \gamma} + \frac{2n\pi}{b}$$

$$\tag{3.29}$$

$$\tilde{f}(k_{xmn}, k_{ymn}) = \frac{1}{4\pi^2} \int\limits_{-\infty}^{\infty} \int\limits_{-\infty}^{\infty} f(x, y) \exp(jk_{xmn}x + jk_{ymn}y)dx\,dy$$

Substituting the expression for $f(x, y)$ in the above integral and performing the integration, one obtains

$$\tilde{f}(k_{xmn}, k_{ymn}) = I_0 \frac{l}{4\pi^2} \text{sinc}\left(\frac{k_{xmn}l}{2}\right) \text{sinc}\left(\frac{k_{ymn}t}{2}\right) \tag{3.30a}$$

where

$$\text{sinc}(x) = \frac{\sin x}{x} \tag{3.30b}$$

The source function is thus expressed in a Floquet series containing infinite number of terms as given in (3.28). Because the Maxwell equations are linear, the principle of superposition can be employed to obtain the total fields. Accordingly, we will obtain the electromagnetic fields produced by each term of the series in (3.28). To simplify further, we will first decompose the (m, n) term of the series into TE_{zmn} and TM_{zmn} modal sources and then determine the respective fields.

---

[2] The reader must not confuse the index j with the imaginary number, though the same symbol is used.

3.3.1 TE$_{zmn}$ and TM$_{zmn}$ Modal Source Decomposition

The (m, n) current term in (3.28) is given by

$$\vec{I}_{mn}(x, y) = \hat{x} I_{xmn} \exp(-jk_{xmn}x - jk_{ymn}y) \tag{3.31}$$

where

$$I_{xmn} = \frac{4\pi^2}{ab} \hat{f}(k_{xmn}, k_{ymn}) \tag{3.32}$$

We decompose the above current vector into TE$_{zmn}$ and TM$_{zmn}$ modal sources. We use the fact that the TE$_{zmn}$ modal current source vector is parallel to $\hat{z} \times \vec{h}_{mn}^{\text{TE}}$ and the TM$_{zmn}$ modal current source vector is parallel to $\hat{z} \times \vec{h}_{mn}^{\text{TM}}$. Accordingly we set

$$\vec{I}_{mn}(x, y) = I_{mn}^{\text{TE}}(\hat{z} \times \vec{h}_{mn}^{\text{TE}}) + I_{mn}^{\text{TM}}(\hat{z} \times \vec{h}_{mn}^{\text{TM}}) \tag{3.33}$$

where I_{mn}^{TE} and I_{mn}^{TM} are the current amplitudes to be determined. Taking the scalar product with $\vec{e}_{mn}^{\text{TE}*}$ on both sides of (3.33) and then integrating on a unit cell, we obtain

$$\iint_{\text{unit cell}} \vec{I}_{mn}(x, y) \cdot \vec{e}_{mn}^{\text{TE}*} \, dx \, dy = I_{mn}^{\text{TE}} \iint_{\text{unit cell}} (\hat{z} \times \vec{h}_{mn}^{\text{TE}}) \cdot \vec{e}_{mn}^{\text{TE}*} \, dx \, dy$$

$$+ I_{mn}^{\text{TM}} \iint_{\text{unit cell}} (\hat{z} \times \vec{h}_{mn}^{\text{TM}}) \cdot \vec{e}_{mn}^{\text{TE}*} \, dx \, dy \tag{3.34}$$

The second integral on the right hand side vanishes because the TE$_{zmn}$ and TM$_{zmn}$ modal functions are orthogonal. The first integral is equal to -1 according to the relations in (3.16). Therefore we deduce

$$I_{mn}^{\text{TE}} = - \iint_{\text{unit cell}} \vec{I}_{mn}(x, y) \cdot \vec{e}_{mn}^{\text{TE}*} \, dx \, dy \tag{3.35}$$

In a similar fashion, by taking the scalar product with $\vec{e}_{mn}^{\text{TM}*}$, one obtains

$$I_{mn}^{\text{TM}} = - \iint_{\text{unit cell}} \vec{I}_{mn}(x, y) \cdot \vec{e}_{mn}^{\text{TM}*} \, dx \, dy \tag{3.36}$$

Now substituting $\vec{I}_{mn}(x, y)$, $\vec{e}_{mn}^{\text{TE}}$, and $\vec{e}_{mn}^{\text{TM}}$ from (3.31), (3.11), and (3.19a), respectively, we finally obtain

$$I_{mn}^{\text{TE}} = -I_{xmn} \iint_{\text{unit cell}} \hat{x} \cdot (\hat{x} k_{ymn} - \hat{y} k_{xmn}) \frac{1}{\sqrt{ab(k^2 - k_{zmn}^2)}} dx \, dy$$

$$= -I_{xmn} k_{ymn} \sqrt{\frac{ab}{k^2 - k_{zmn}^2}} = -I_{xmn} k_{ymn} \sqrt{\frac{ab}{k_{xmn}^2 + k_{ymn}^2}} \tag{3.37}$$

$$I_{mn}^{TM} = -I_{xmn} \underbrace{\iint}_{\text{unit cell}} \hat{x} \cdot (\hat{x} k_{xmn} + \hat{y} k_{ymn}) \frac{1}{\sqrt{ab(k^2 - k_{zmn}^2)}} dx\, dy$$

$$= -I_{xmn} k_{xmn} \sqrt{\frac{ab}{k^2 - k_{zmn}^2}} = -I_{xmn} k_{xmn} \sqrt{\frac{ab}{k_{xmn}^2 + k_{ymn}^2}} \qquad (3.38)$$

Notice, the TE_{zmn} and TM_{zmn} modal current amplitudes are independent of the media parameter.

3.3.2 TE_{zmn} Fields

The TE_{zmn} fields are produced by the TE_{zmn} modal source current. The TE_{zmn} fields have different expressions in the two regions. For the $z > 0$ region, the fields are purely progressive and can be expressed as

$$\vec{E}_{mn}^{TE+} = V_{mn}^{TE+} \vec{e}_{mn}^{TE} \exp(-jk_{zmn}^+ z)$$

$$\vec{H}_{mn}^{TE+} = V_{mn}^{TE+} Y_{mn}^{TE+} \left(\vec{h}_{mn}^{TE} + \frac{\hat{z}(\nabla \cdot \vec{h}_{mn}^{TE})}{jk_{zmn}^+} \right) \exp(-jk_{zmn}^+ z) \qquad (3.39)$$

where V_{mn}^{TE+} is the modal voltage, Y_{mn}^{TE+} is the modal admittance, k_{zmn}^+ is the wave number in the z-direction, and

$$k_{zmn}^+ = \sqrt{k_0^2 - k_{xmn}^2 - k_{ymn}^2} \qquad Y_{mn}^{TE+} = \frac{k_{zmn}^+}{\omega \mu_0} \qquad (3.40)$$

In the $z < 0$ region, the electric field vector is expressed as a superposition of an incident and a reflected wave. The reflection is caused by the ground plane at $z = -h$. The electric field vector and the modal admittance are given by

$$\vec{E}_{mn}^{TE-} = V_{mn}^{TE-} \vec{e}_{mn}^{TE} [\exp\{jk_{zmn}^-(z+h)\} + R_{mn}^{TE} \exp\{-jk_{zmn}^-(z+h)\}] \qquad (3.41)$$

$$k_{zmn}^- = \sqrt{\varepsilon_r k_0^2 - k_{xmn}^2 - k_{ymn}^2} \qquad Y_{mn}^{TE-} = \frac{k_{zmn}^-}{\omega \mu_0} \qquad (3.42)$$

where ε_r is the dielectric constant of the dielectric layer. The transverse magnetic field vector is given by

$$\vec{H}_{tmn}^{TE-} = V_{mn}^{TE-} Y_{mn}^{TE-} \vec{h}_{mn}^{TE} [-\exp\{jk_{zmn}^-(z+h)\} + R_{mn}^{TE} \exp\{-jk_{zmn}^-(z+h)\}] \qquad (3.43)$$

The transverse electric field must vanish on the ground plane at $z = -h$. Enforcing this condition in (3.41) we obtain

$$R_{mn}^{TE} = -1 \qquad (3.44)$$

Substituting $R_{mn}^{TE} = -1$ in (3.41) and (3.43), we obtain the transverse field vectors as

$$\vec{E}_{mn}^{TE-} = 2jV_{mn}^{TE-}\vec{e}_{mn}^{TE}\sin\{k_{zmn}^-(z+h)\} \tag{3.45}$$

$$\vec{H}_{mn}^{TE-} = -2V_{mn}^{TE-}Y_{mn}^{TE-}\vec{h}_{mn}^{TE}\cos\{k_{zmn}^-(z+h)\} \tag{3.46}$$

To find the unknown modal voltages, we enforce the electric field continuity and magnetic field discontinuity (because of the existence of the electric surface current) conditions at $z = 0$. The continuity of the electric field at $z = 0$ yields

$$2jV_{mn}^{TE-}\vec{e}_{mn}^{TE}\sin\{k_{zmn}^-h\} = V_{mn}^{TE+}\vec{e}_{mn}^{TE} \tag{3.47}$$

The magnetic fields are discontinuous at $z = 0$ and the discontinuity condition is

$$\hat{z}\times(\vec{H}_{mn}^{TE+} - \vec{H}_{mn}^{TE-}) = I_{mn}^{TE}(\hat{z}\times\vec{h}_{mn}^{TE}) \tag{3.48}$$

Substituting the expressions for the magnetic field vectors from (3.39) and (3.46) into (3.48), one obtains

$$V_{mn}^{TE+}Y_{mn}^{TE+} + 2V_{mn}^{TE-}Y_{mn}^{TE-}\cos(k_{zmn}^-h) = I_{mn}^{TE} \tag{3.49}$$

Eliminating V_{mn}^{TE-} from (3.47) and (3.49) we obtain

$$V_{mn}^{TE+} = \frac{I_{mn}^{TE}}{y_{mn}^{TE}} \tag{3.50}$$

with

$$y_{mn}^{TE} = Y_{mn}^{TE+} - jY_{mn}^{TE-}\cot(k_{zmn}^-h) \tag{3.51}$$

The quantity y_{mn}^{TE} represents the equivalent admittance experienced by the TE_{zmn} current source at the air–dielectric interface. This equivalent admittance can be viewed analogously as the input admittance of a shunt source located at the junction of two transmission lines, as shown in Figure 3.3. The transmission lines have characteristic admittances equal to the modal admittances and the lengths equal to the media widths. Since the free-space side is of infinite extent, the associated infinite transmission line is equivalent to a match termination. The transmission line analogy is a very powerful tool for obtaining the equivalent admittance of more

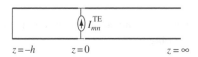

FIGURE 3.3 Transmission line equivalent circuit for the TE_{zmn} Floquet mode.

complex multilayer structures. For a multilayer structure, the equivalent circuit consists of several transmission line sections connected in cascade. Each transmission line section is associated with a homogeneous layer of the dielectric substrate.

For our present structure, the TE_{zmn} electric field vector for the $z > 0$ region is given by

$$\vec{E}_{mn}^{\text{TE}+} = \frac{I_{mn}^{\text{TE}}}{y_{mn}^{\text{TE}}} \vec{e}_{mn}^{\text{TE}} \exp(-jk_{zmn}^{+}z) \tag{3.52a}$$

The magnetic field vector is obtained from (3.39) and is given by

$$\vec{H}_{mn}^{\text{TE}+} = \frac{I_{mn}^{\text{TE}}}{y_{mn}^{\text{TE}}} Y_{mn}^{\text{TE}+} \left(\vec{h}_{mn}^{\text{TE}} + \frac{\hat{z}(\nabla \cdot \vec{h}_{mn}^{\text{TE}})}{jk_{zmn}^{+}} \right) \exp(-jk_{zmn}^{+}z) \tag{3.52b}$$

3.3.3 TM$_{zmn}$ Fields

Following a similar procedure the TM_{zmn} electric and magnetic field vectors for the $z > 0$ region can be obtained as

$$\vec{E}_{mn}^{\text{TM}+} = \frac{I_{mn}^{\text{TM}}}{y_{mn}^{\text{TM}}} \left(\vec{e}_{mn}^{\text{TM}} + \frac{\hat{z}(\nabla \cdot \vec{e}_{mn}^{\text{TM}})}{jk_{zmn}^{+}} \right) \exp(-jk_{zmn}^{+}z) \tag{3.53a}$$

$$\vec{H}_{mn}^{\text{TM}+} = \frac{I_{mn}^{\text{TM}}}{y_{mn}^{\text{TM}}} Y_{mn}^{\text{TM}+} \vec{h}_{mn}^{\text{TM}} \exp(-jk_{zmn}^{+}z) \tag{3.53b}$$

where

$$y_{mn}^{\text{TM}} = Y_{mn}^{\text{TM}+} - jY_{mn}^{\text{TM}-} \cot(k_{zmn}^{-}h) \tag{3.54}$$

$$Y_{mn}^{TM+} = \frac{\omega \varepsilon_0}{k_{zmn}^{+}} \qquad Y_{mn}^{\text{TM}-} = \frac{\omega \varepsilon_0 \varepsilon_r}{k_{zmn}^{-}} \tag{3.55}$$

3.3.4 Floquet Impedance

The Floquet impedance of an array element is defined as the input impedance of the element under the following conditions:

- The element must reside in an infinite array environment.
- The array must have a Floquet excitation, that is, the elements should be excited with uniform amplitudes and linearly progressed phase.

The Floquet impedance is a function of the grid parameters and phase slopes, which are related to the scan location. The Floquet impedance for an element in the array considered here can be obtained from the stationary expression [2, Ch. 7] given by

$$Z^{\text{FL}}(k_{x0}, k_{y0}) = -\frac{<\vec{E}, \vec{I}>}{I_0^2} \tag{3.56}$$

where I_0 is the input current, $\vec{E}$ is the electric field at the current source location, $\vec{I}$ is the current distribution function, and $< \vec{E}, \vec{I} >$ is the self-reaction of the current source under Floquet excitation. The self-reaction is given by

$$< \vec{E}, \vec{I} >= \iint_{\text{unit cell}} \vec{E} \cdot \vec{I} \, dx \, dy \tag{3.57}$$

The electric field at the source current location ($z = 0$) is the superposition of the TE_{zmn} and TM_{zmn} fields where both m and n run from $-\infty$ to $+\infty$. Therefore, the electric field at $z = 0$ becomes

$$\vec{E} = \sum_m \sum_n [\vec{E}_{mn}^{\text{TE}+} + \vec{E}_{mn}^{\text{TM}+}]_{z=0} \tag{3.58}$$

Substituting $z = 0$ in the expressions for the electric fields in (3.52a) and (3.53a) and then substituting in (3.58), we write

$$\vec{E} = \sum_m \sum_n \frac{I_{mn}^{\text{TE}}}{y_{mn}^{\text{TE}}} \vec{e}_{mn}^{\text{TE}} + \frac{I_{mn}^{\text{TM}}}{y_{mn}^{\text{TM}}} \vec{e}_{mn}^{\text{TM}} \tag{3.59}$$

We substitute the above expression for the electric field in (3.56) and then utilize the following integral relations, which directly follow from (3.35) and (3.36):

$$\iint_{\text{unit cell}} \vec{e}_{mn}^{\text{TE}} \cdot \vec{I}^* \, dx \, dy = -I_{mn}^{\text{TE}*} \tag{3.60}$$

$$\iint_{\text{unit cell}} \vec{e}_{mn}^{\text{TM}} \cdot \vec{I}^* \, dx \, dy = -I_{mn}^{\text{TM}*} \tag{3.61}$$

The expression for the Floquet impedance thus becomes

$$Z^{\text{FL}}(k_{x0}, k_{y0}) = \frac{1}{I_0^2} \sum_m \sum_n \left[\frac{|I_{mn}^{\text{TE}}|^2}{y_{mn}^{\text{TE}}} + \frac{|I_{mn}^{\text{TM}}|^2}{y_{mn}^{\text{TM}}} \right] \tag{3.62}$$

After back substitutions of I_{mn}^{TE} and I_{mn}^{TM}, the Floquet impedance is obtained as

$$Z^{\text{FL}}(k_{x0}, k_{y0}) = \frac{l^2}{ab} \sum_m \sum_n \left[\frac{k_{ymn}^2}{y_{mn}^{\text{TE}}} + \frac{k_{xmn}^2}{y_{mn}^{\text{TM}}} \right] \frac{\{\text{sinc}(k_{xmn}l/2)\text{sinc}(k_{ymn}t/2)\}^2}{k_0^2 - k_{zmn}^{+2}} \tag{3.63}$$

The Floquet impedance is computed for various scan angles and plotted in Figures 3.4 a–c for E, H, and diagonal (45°) planes. The cell size was $0.6\lambda_0 \times 0.6\lambda_0$, where λ_0 is the wavelength in free space. The dipole length was $0.4\lambda_0$ and the width of the dipole was $0.1\lambda_0$. The substrate thickness was $0.05\lambda_0$ with a dielectric constant of 2.0. A rectangular lattice structure was considered for these plots. A large capacitive reactance of 136 Ω was observed for the bore-sight scan.

Near 41° scan location on the E-plane, the Floquet impedance has a singularity. This singularity is due to coupling of the TM_0 surface wave mode. The dipole currents have a phase match with the surface wave supported by the grounded dielectric layer. Under such a condition the amplitude of the surface wave field

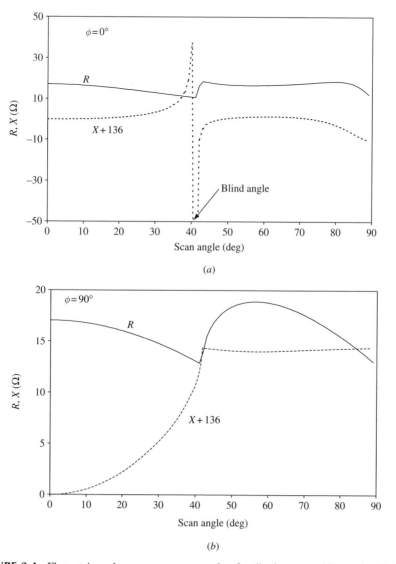

FIGURE 3.4 Floquet impedance versus scan angle of a dipole array with $a = b = 0.6\lambda_0$, $l = 0.4\lambda_0$, $t = 0.1\lambda_0$, $h = 0.05\lambda_0$, $\varepsilon_r = 2$: (a) E-plane scan; (b) H-plane scan; (c) D-plane (45°) scan.

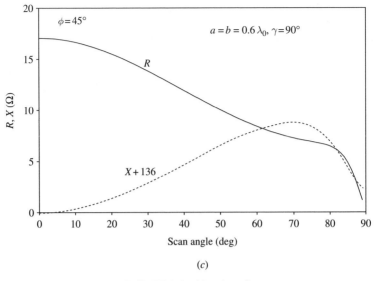

(c)

FIGURE 3.4   (Continued)

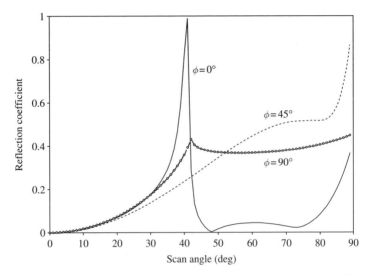

FIGURE 3.5   Floquet reflection coefficient versus scan angle at three scan planes of the dipole array in Figure 3.4.

increases infinitely, resulting in an infinitely large input impedance. Because each source has non zero internal impedance, the input match becomes very poor (almost 100% reflection coefficient), as shown in Figure 3.5. Thus the array does not radiate and the energy does not couple to the surface wave mode. This phenomenon is known as *scan blindness* and the corresponding angle is the *blind angle*. For the reflection coefficient computation in Figure 3.5, the array was assumed to be bore-sight matched, that is, the source internal impedance was assumed equal to the complex conjugate of the bore-sight Floquet impedance. We notice that a blind angle does not occur in the other two planes for the present geometry. However, with increased element spacing blind angles may exist in other planes. A more detailed discussion on array scan blindness is presented in Section 3.4. The discontinuities in the input resistance and reactance near 40° scan angle are attributed to the higher order Floquet modes that begin to propagate.

3.4 DETERMINATION OF BLIND ANGLES

From (3.63) we observe that a zero of either y_{mn}^{TE} or y_{mn}^{TM} causes singularity in the Floquet impedance. A singularity is associated with a surface wave mode. In a dielectric coated ground plane y_{mn}^{TM} has the lowest order zero. For electrically thin dielectric substrate only one zero exists. Therefore, for a thin substrate, the blind angles can be computed from the root of the equation $y_{mn}^{TM} = 0$. Invoking y_{mn}^{TM} from (3.54) we write

$$y_{mn}^{TM} = Y_{mn}^{TM+} - jY_{mn}^{TM-}\cot(k_{zmn}^- h) = 0 \qquad (3.64)$$

Expressions for Y_{mn}^{TM+} and Y_{mn}^{TM-} are given in (3.55). Using (3.64) and after simplification one obtains

$$k_{zmn}^+ + \frac{jk_{zmn}^-\tan(k_{zmn}^- h)}{\varepsilon_r} = 0 \qquad (3.65)$$

with

$$k_{zmn}^+ = \sqrt{k_0^2 - k_{\rho mn}^2} \qquad k_{zmn}^- = \sqrt{\varepsilon_r k_0^2 - k_{\rho mn}^2} \qquad (3.66)$$

In (3.66), $k_{\rho mn}^2 = k_{xmn}^2 + k_{ymn}^2$, where k_{xmn} and k_{ymn} are given in (3.2). Observe that the left-hand side of (3.65) is an explicit function of the single variable $k_{\rho mn}$. We will seek solutions for $k_{\rho mn}$ by solving (3.65). For a small value of $k_0 h\sqrt{\varepsilon_r}$ an approximate solution, denoted by $k_{\rho s}$, can be expressed as [3]

$$k_{\rho s} \approx k_0 \left[1 + \frac{1}{2}\left\{k_0 h \left(1 - \frac{1}{\varepsilon_r}\right)\right\}^2\right] \qquad (3.67)$$

The above solution is obtained with a single iteration using the initial solution as k_0. For $\varepsilon_r = 2.5$, the error is less than 2% if $(k_0 h\sqrt{\varepsilon_r}) < 0.5$. Notice, $k_{\rho s}$ is greater than k_0.

The direction of propagation of the surface wave is necessary to ascertain a surface wave resonance and the subsequent blind spot. In the following development we will articulate a graphical method to identify a surface wave resonance. Suppose the surface wave propagation vector makes an angle α with the x-axis. Then the transverse surface wave numbers would be

$$k_{xs} = k_{\rho s}\cos\alpha \qquad k_{ys} = k_{\rho s}\sin\alpha \qquad (3.68a)$$

Since the surface wave mode is supported by the periodic array, according to the Floquet theorem [4], the following mode numbers are also possible in regard to a rectangular lattice structure of cell size $a \times b$:

$$k_{xs} = k_{\rho s} \cos \alpha + \frac{2m\pi}{a} \qquad k_{ys} = k_{\rho s} \sin \alpha + \frac{2n\pi}{b} \qquad (3.68b)$$

where m and n are integers. Equations (3.68b) represent a family of circles of radius $k_{\rho s}$. The circle with center located at $(2m\pi/a, 2n\pi/b)$ is associated with the (m, n) surface wave mode. If (k_{xs}, k_{ys}) matches with the Floquet mode number pair (k_{xij}, k_{yij}), then the (m, n) surface wave mode strongly couples with the (i, j) Floquet mode, resulting in scan blindness. This is pictorially shown in Figure 3.6.

In Figure 3.6 we show only the (0,0) Floquet mode circle of radius k_0 (to avoid clutter) and five surface wave mode circles. An intersection between a surface wave mode circle and the (0,0) Floquet mode circle indicates the possibility of coupling between the two modes and a subsequent blind spot in the visible region. Each point on the surface wave mode circle that lies inside the (0,0) Floquet mode circle may represent a blind spot. The point P, for instance, is associated with a blind spot. This blind spot occurs on the OP scan plane. The number of surface wave mode circles along a pattern cut is equal to the number of blind spots. The vector joining the center of the surface wave mode circle and the point P represents the propagation direction of the surface wave mode, as shown in Figure 3.6.

For a rectangular lattice, the location of the blind spot on the $\phi = 0°$ plane caused by the (1,0) surface wave mode can be determined from the circle diagram. The blind angle θ_b is given by

$$k_0 \sin \theta_b + k_{\rho s} = \frac{2\pi}{a} \qquad (3.69)$$

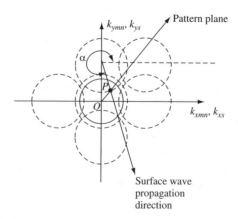

FIGURE 3.6 Circle diagram for surface wave modes and the (0,0) Floquet mode.

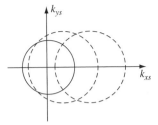

FIGURE 3.7 The (0,0) Floquet mode circle (solid line) and two surface wave mode circles (dashed lines).

In the case of an equilateral triangular grid of element spacing a, the blind angle due to a lowest order surface wave mode coupling is given by

$$k_0 \sin \theta_b + k_{\rho s} = \frac{4\pi}{a\sqrt{3}}$$

(3.70)

As mentioned, (3.69) and (3.70) yield the blind angles corresponding to the possible lowest order surface wave mode that couples with the fundamental Floquet mode. Those blind angles are *not necessarily the closest blind angle to the bore sight*, because for large element spacing, a higher order surface wave mode circle could pass through a point that is closer to the origin than that of a lower order surface wave mode, as shown in Figure 3.7. Notice that the circumference of the (2,0) surface wave mode circle is nearer to the origin than that of the (1,0) surface wave mode circle; thus the nearest blind angle to the bore sight on the $\phi = 0°$ plane will occur due to the (2,0) surface wave mode coupling. Figure 3.8 plots the nearest

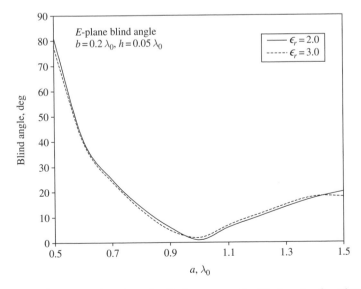

FIGURE 3.8 Element spacing along the E-plane versus the blind angle of a printed dipole array on a dielectric-coated ground plane.

blind angle along the E-plane versus element spacing for two different dielectric substrates. Observe that the blind spot occurs almost near the bore sight if the element spacing is close to one wavelength.

3.5 ACTIVE ELEMENT PATTERN

The active element pattern (AEP) or the embedded element pattern of an array element is defined as the radiation pattern of the element when all other elements are match terminated. The AEP is very effective for precise computations of the gain and radiation patterns of a "finite array," including mutual coupling effects. It turns out that the gain pattern of a finite array, including mutual coupling effects, can be determined using the following simple equation:

$$G_A(\theta, \phi) = G_e(\theta, \phi) \times \mathrm{AF}(\theta, \phi) \tag{3.71}$$

where G_A represents the array gain pattern, G_e represents the active element gain pattern, and AF represents the normalized array factor with the sum of the amplitude squares set to unity. The above formula is exact if all elements of the array are situated in an identical environment. This can only happen if the array size is infinitely large and only a finite number of elements are excited. Therefore, by "finite array" we refer to an *infinite array with a finite number of excited elements*. The remaining elements of the array are not excited but are match terminated. A finite array as defined above does not exist in practice. However, in a large array most of the elements are in similar environments (except the edge elements); therefore, they have similar (if not identical) element patterns. The edge elements, however, have somewhat different element patterns. In a typical phased array the amplitudes of the edge elements are 10–15 dB below the amplitudes of the central elements; therefore (3.71) becomes a good approximation, particularly in the boresight and near-in side-lobe regions.

To derive the AEP we first express the radiation pattern of an infinite array under Floquet excitation. Toward that end we employ superposition of the AEPs with appropriate excitation coefficients. Then we obtain the radiation pattern of the infinite array independently, employing Floquet modal expansion. Equating the two array pattern expressions we identify the unknown AEP.

3.5.1 Array Pattern Using Superposition

Suppose $\vec{E}_a(\theta, \phi) \exp(-jk_0 r)/r$ is the active element far-field pattern. The elements of the array are arranged in a general grid $[a, b, \gamma]$ as shown in Figure 3.1. Suppose the Floquet excitations are set for radiating in the (θ_0, ϕ_0)-direction. Then the phase gradients are

$$k_{x0} = k_0 \sin\theta_0 \cos\phi_0 \qquad k_{y0} = k_0 \sin\theta_0 \sin\phi_0 \tag{3.72}$$

The coordinates of the (m, n) element are

$$x_{mn} = ma + \frac{nb}{\tan \gamma} \qquad y_{mn} = nb \tag{3.73}$$

For uniform excitation the array factor of the infinite array is given by

$$AF(\theta, \phi) = \sum_{m=-\infty}^{\infty} \sum_{n=-\infty}^{\infty} \exp[jx_{mn}(k_x - k_{x0}) + jy_{mn}(k_y - k_{y0})] \tag{3.74}$$

where

$$k_x = k_0 \sin \theta \cos \phi \qquad k_y = k_0 \sin \theta \sin \phi \tag{3.75}$$

We now substitute the expressions for x_{mn} and y_{mn} from (3.73) into (3.74) and after rearranging we obtain

$$AF(\theta, \phi) = \sum_{m=-\infty}^{\infty} \exp[jma(k_x - k_{x0})] \sum_{n=-\infty}^{\infty} \exp\left[jnb\left(\frac{k_x - k_{x0}}{\tan \gamma} + k_y - k_{y0} \right) \right] \tag{3.76}$$

We now use the following identity (see Section A.2) in (3.76):

$$\sum_{n=-\infty}^{\infty} \exp(jnp\sigma) = \frac{2\pi}{\sigma} \sum_{m=-\infty}^{\infty} \delta\left(p - \frac{2m\pi}{\sigma} \right) \tag{3.77}$$

The final expression for the array factor becomes

$$AF(\theta, \phi) = \frac{4\pi^2}{ab} \sum_{m=-\infty}^{\infty} \sum_{n=-\infty}^{\infty} \delta\left(k_x - k_{x0} - \frac{2m\pi}{a} \right) \delta\left(k_y - k_{y0} - \frac{2n\pi}{b} + \frac{2m\pi}{a \tan \gamma} \right) \tag{3.78}$$

The Dirac delta functions in the array factor indicate that the array radiates only at discrete angles. The dominant term ($m = n = 0$) corresponds to the radiation in the intended direction. Therefore, the contribution of the array factor along the intended radiation direction is given by

$$AF_0(\theta, \phi) = \frac{4\pi^2}{ab} \delta(k_x - k_{x0})\delta(k_y - k_{y0}) \tag{3.79}$$

The field intensity of the infinite array along the intended direction, $\vec{E}_\infty(r, \theta, \phi)$, is equal to the array factor times the element pattern, which is

$$\vec{E}_\infty(r, \theta, \phi) = \frac{4\pi^2}{ab} \delta(k_x - k_{x0})\delta(k_y - k_{y0})\vec{E}_a(\theta, \phi)\frac{\exp(-jk_0 r)}{r} \tag{3.80}$$

3.5.2 Array Pattern Using Floquet Modal Expansion

We now derive the infinite array far-field pattern employing Floquet modal expansion theory. The field intensity of the infinite periodic array along the intended radiation direction is equal to the total fields contributed by the TE_{z00} and TM_{z00} modes. The resultant electric field vector contributed by the above two modes is

$$\vec{E}_{00}^{FL}(x, y, z) = \left[V_{00}^{TE}\vec{e}_{00}^{TE} + V_{00}^{TM}\left(\vec{e}_{00}^{TM} + \frac{\hat{z}(\nabla \cdot \vec{e}_{00}^{TM})}{jk_{z00}} \right) \right] \exp(-jk_{z00}z) \qquad (3.81)$$

In short notation, the above electric field vector can be expressed as

$$\vec{E}_{00}^{FL}(x, y, z) = \vec{C} \exp(-jk_{x0}x - jk_{y0}y - jk_{z00}z) \qquad (3.82)$$

where $\vec{C}$ is a constant vector independent of x, y, and z. Now, $\vec{E}_{\infty}(r, \theta, \phi)$ in (3.80) and $\vec{E}_{00}^{FL}(x, y, z)$ in (3.82) must be identical in the far-field region. This equality condition yields the unknown active element pattern $\vec{E}_a(\theta, \phi)$ in terms of the Floquet modal fields. In order to derive an explicit expression for $\vec{E}_a(\theta, \phi)$, we express the right-hand side of (3.82) in spherical coordinates (r, θ, ϕ) and then use its asymptotic form for the far-field representation. To that end we first express the exponential factor of (3.82) in terms of the following integral:

$$\exp(-jk_{x0}x - jk_{y0}y - jk_{z00}z) = \int\limits_{-\infty}^{\infty} \int\limits_{-\infty}^{\infty} \delta(u - k_{x0})\delta(v - k_{y0})$$

$$\times \exp(-jux - jvy - jwz)\, du\, dv \qquad (3.83)$$

where

$$w = \sqrt{k_0^2 - u^2 - v^2}$$

The above relation can be proven almost by inspection. We now substitute

$$x = r\sin\theta\cos\phi \qquad y = r\sin\theta\sin\phi \qquad z = r\cos\theta$$
$$u = k_0\sin\alpha\cos\beta \qquad v = k_0\sin\alpha\sin\beta \qquad w = k_0\cos\alpha \qquad (3.84)$$

and then compute

$$du\, dv = \frac{\partial(u, v)}{\partial(\alpha, \beta)}\, d\alpha\, d\beta = k_0^2 \sin\alpha\cos\alpha\, d\alpha\, d\beta \qquad (3.85)$$

The double integral on the right-hand side of (3.83) becomes

$$I = \iint\limits_{\alpha\beta} \delta(u - k_{x0})\delta(v - k_{y0})\exp[-jk_0 r\{\sin\theta\sin\alpha\cos(\beta - \phi) + \cos\theta\cos\alpha\}]$$

$$\times k_0^2 \sin\alpha\cos\alpha\, d\alpha\, d\beta \qquad (3.86)$$

where $0 \leq \beta \leq 2\pi$. The integration path on the complex α-plane and limits for α are determined from $0 \leq \sin \alpha \leq \infty$. A segment of the path can be considered on the real axis of α joining zero and $\pi/2$.

We integrate with respect to β first. For large r (far field) the saddle point method [5] can be applied to obtain the asymptotic form of the integral. The result is

$$I = k_0^2 \int_\alpha \delta(k_0 \sin \alpha \cos \phi - k_{x0}) \delta(k_0 \sin \alpha \sin \phi - k_{y0}) \sqrt{\frac{2\pi}{rk_0 \sin \theta \sin \alpha}}$$

$$\times \exp[-jk_0 r \cos(\alpha - \theta) + \frac{1}{4} j\pi] \sin \alpha \cos \alpha \, d\alpha \qquad (3.87)$$

Once again we apply the saddle point method of integration with respect to the variable α. A saddle point exists at $\alpha = \theta$ for $0 \leq \theta \leq \pi/2$. The final expression of I becomes

$$I = j2\pi k_0 \delta(k_x - k_{x0}) \delta(k_y - k_{y0}) \frac{\exp(-jk_0 r)}{r} \cos \theta \qquad (3.88)$$

In (3.82) we now replace the exponential term by the quantity on the right-hand side of (3.88) to obtain the far field $\vec{E}_{00}^{\mathrm{FL}}$ in the spherical coordinate system. Then we equate $\vec{E}_{00}^{\mathrm{FL}}$ with $\vec{E}_\infty$ in (3.80) and obtain the active element pattern. The result is

$$\vec{E}_a(\theta_0, \phi_0) = j\vec{C} \frac{k_0}{2\pi} ab \cos \theta_0 \qquad (3.89)$$

Comparing (3.81) and (3.82) we obtain the following expression for the vector $\vec{C}$:

$$\vec{C} = V_{00}^{\mathrm{TE}} \frac{\hat{x}k_{y0} - \hat{y}k_{x0}}{\sqrt{ab(k_{x0}^2 + k_{y0}^2)}} + V_{00}^{\mathrm{TM}} \frac{\hat{x}k_{x0} + \hat{y}k_{y0} - \hat{z}(k_{x0}^2 + k_{y0}^2)/k_{z00}}{\sqrt{ab(k_{x0}^2 + k_{y0}^2)}} \qquad (3.90)$$

Substituting the expressions for k_{x0} and k_{y0} from (3.72) and then expressing the vector $\vec{C}$ in terms of spherical basis vectors, we obtain

$$\vec{C} = -\hat{\phi}_0 \frac{V_{00}^{\mathrm{TE}}}{\sqrt{ab}} + \hat{\theta}_0 \frac{V_{00}^{\mathrm{TM}}}{\cos \theta_0 \sqrt{ab}} \qquad (3.91)$$

Using (3.91) in (3.89) we obtain the active element pattern as

$$\vec{E}_a(\theta_0, \phi_0) = \hat{\theta}_0 \left\{ jV_{00}^{\mathrm{TM}} \sqrt{\frac{ab}{\lambda_0^2}} \right\} - \hat{\phi}_0 \left\{ jV_{00}^{\mathrm{TE}} \sqrt{\frac{ab}{\lambda_0^2}} \cos \theta_0 \right\} \qquad (3.92)$$

Equation (3.92) yields the relation between Floquet modal voltages and the active element pattern. It is interesting to observe that the TM_{z00} mode is responsible for the θ-component of the electric field, while the TE_{z00} mode is responsible for the ϕ-component of the electric field in the far field region.

3.5.3 Active Element Gain Pattern

The active element pattern expressed in (3.92) may not be very convenient for array
pattern computation because, in general, the modal voltages vary with the scan angle.
The active element pattern expressed with respect to the incident current (or voltage)
is preferable. This is primarily due to the fact that the incident quantities (current,
voltage, or power) remain unchanged with the scan angle assuming that the source
electromotive force (EMF) and its internal *impedance* do not vary with the antenna
input impedance. *Thus the principle of superposition is directly applicable only if
the active element pattern is normalized with respect to the incident quantities.* We
will first express the active element pattern in terms of the incident current and then
obtain the active element gain.

The incident current I_{inc} and the antenna current (dipole current for the present
case) I_0 are related through the input impedance of the dipole and the internal
impedance Z_s of the source[3]. The incident current is given by

$$I_{inc} = \frac{I_0(Z^{FL} + Z_s)}{2R_s} \tag{3.93}$$

In (3.93) Z^{FL} is the dipole input impedance (Floquet impedance) and R_s is the real
part of the source impedance Z_s. The incident current is defined here as the load
current if the maximum power is delivered to the load by the source. This happens
when the load impedance becomes equal to the complex conjugate of the source
impedance.

Now let us find V_{00}^{TE} in terms of I_0. Substituting $m = n = 0$ in (3.50) and (3.51),
we write

$$V_{00}^{TE} = \frac{I_{00}^{TE}}{Y_{00}^{TE+} - jY_{00}^{TE-}\cot(k_{z00}^- h)} \tag{3.94}$$

By virtue of (3.30a), (3.32), (3.37), and (3.72) we obtain

$$I_{00}^{TE} = -I_0 \frac{l}{\sqrt{ab}} \sin\phi_0 \, \text{sinc}\left(k_0\frac{l}{2}\sin\theta_0\cos\phi_0\right) \text{sinc}\left(k_0\frac{t}{2}\sin\theta_0\sin\phi_0\right) \tag{3.95}$$

Using (3.72) in (3.40) and (3.42) we obtain

$$Y_{00}^{TE+} = \frac{\cos\theta_0}{\eta} \qquad Y_{00}^{TE-} = \frac{\sqrt{\varepsilon_r - \sin^2\theta_0}}{\eta} \tag{3.96}$$

---

[3] The source is assumed to be an ideal voltage (or current) source with internal impedance connected in
series (or parallel).

with $\eta = 377\Omega$. Using (3.94), (3.95), and (3.96) we express V_{00}^{TE} in terms of I_0, θ_0, and ϕ_0. We then use (3.93) to express I_0 in terms of I_{inc}. The final expression for V_{00}^{TE} in terms of I_{inc} becomes

$$
V_{00}^{\mathrm{TE}} = -I_{\mathrm{inc}}\eta \left[\frac{2R_s}{Z^{\mathrm{FL}}+Z_s} \right]
$$
$$
\frac{(l/\sqrt{ab})\sin\phi_0\,\mathrm{sinc}[k_0(l/2)\sin\theta_0\cos\phi_0]\,\mathrm{sinc}[k_0(t/2)\sin\theta_0\sin\phi_0]}{\cos\theta_0 - j\sqrt{\varepsilon_r - \sin^2\theta_0}\,\cot(k_0 h\sqrt{\varepsilon_r - \sin^2\theta_0})} \tag{3.97}
$$

From (3.92) the ϕ-component of the far field is given by

$$
E_{a\phi} = -jV_{00}^{\mathrm{TE}}\sqrt{\frac{ab}{\lambda_0^2}}\cos\theta_0 \tag{3.98}
$$

Substituting V_{00}^{TE} from (3.97) into (3.98) we obtain

$$
E_{a\phi} = jI_{\mathrm{inc}}\eta \left(\frac{l}{\lambda_0}\right)\left[\frac{2R_s}{Z^{\mathrm{FL}}+Z_s} \right]
$$
$$
\frac{\sin\phi_0\,\mathrm{sinc}[k_0(l/2)\sin\theta_0\cos\phi_0]\,\mathrm{sinc}[k_0(t/2)\sin\theta_0\sin\phi_0]}{1 - j\sqrt{\varepsilon_r - \sin^2\theta_0}\,\cot(k_0 h\sqrt{\varepsilon_r - \sin^2\theta_0})/\cos\theta_0} \tag{3.99}
$$

In a similar fashion we find the θ-component of the active element pattern as

$$
E_{a\theta} = -jI_{\mathrm{inc}}\eta \left(\frac{l}{\lambda_0}\right)\left[\frac{2R_s}{Z^{\mathrm{FL}}+Z_s} \right]
$$
$$
\frac{\cos\theta_0\cos\phi_0\,\mathrm{sinc}[k_0(l/2)\sin\theta_0\cos\phi_0]\,\mathrm{sinc}[k_0(t/2)\sin\theta_0\sin\phi_0]}{1 - j[\varepsilon_r\cos\theta_0/\sqrt{\varepsilon_r - \sin^2\theta_0}]\cot(k_0 h\sqrt{\varepsilon_r - \sin^2\theta_0})} \tag{3.100}
$$

The normalized far-field pattern is obtained by setting the total incident power equal to 4π watts. The incident current in that case will be given by

$$
|I_{\mathrm{inc}}| = \sqrt{\frac{4\pi}{R_s}} \tag{3.101}
$$

Substituting the above expression of the incident current in (3.99) and (3.100), we directly obtain far-field patterns, normalized with respect to the incident power. The active element gain G_e is given by

$$
G_e(\theta_0, \phi_0) = \sqrt{\{|E_{a\theta}|^2 + |E_{a\phi}|^2\}/\eta} \tag{3.102}
$$

The above expression for G_e should be used in (3.71) for the array gain computation. Figure 3.9 shows the active element gain pattern of the array with the physical

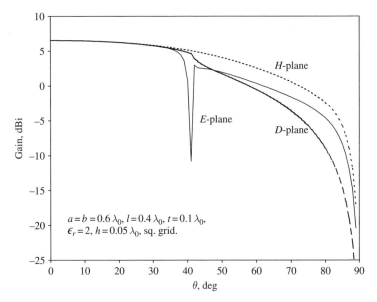

FIGURE 3.9 Embedded element gain versus scan angle of the dipole array of Figure 3.4.

parameters as in Figure 3.4. As before, the array is assumed to be bore-sight matched. A blind spot is observed near the 41° scan angle on the *E*-plane pattern. This observation is consistent with the active reflection coefficient plot in Figure 3.5.

3.6   ARRAY OF RECTANGULAR HORN APERTURES

In this section we perform Floquet modal analysis of an infinite array of rectangular horn apertures. The geometry of the array structure is shown in Figure 3.10. We assume that a horn aperture supports several waveguide modes. We will derive the input reflection matrix of a horn aperture in the array environment. This would

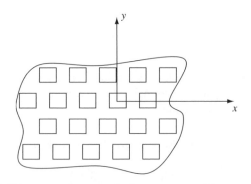

FIGURE 3.10   Infinite array of rectangular horn apertures. Horn aperture size $a' \times b'$.

be followed by the derivation of active element patterns with respect to the TE_{10} incident mode. In the following section waveguide modal fields will be defined.

3.6.1 Waveguide Modes

The modal field vectors are obtained following the procedure in Section 3.2. The modal vector functions (with z-components suppressed) for a rectangular waveguide of aperture size $a' \times b'$ satisfying appropriate boundary conditions are

$$\vec{e}_{mn}^{g\text{TE}}(x, y) = \frac{\hat{x}k_y \cos k_x \left(x - \frac{a'}{2}\right) \sin k_y \left(y - \frac{b'}{2}\right) - \hat{y}k_x \sin k_x \left(x - \frac{a'}{2}\right) \cos k_y \left(y - \frac{b'}{2}\right)}{\sqrt{\frac{a'b'}{4}\{k_y^2 \sigma_x (2 - \sigma_y) + k_x^2 \sigma_y (2 - \sigma_x)\}}}$$

(3.103a)

$$\vec{h}_{mn}^{g\text{TE}}(x, y) = \frac{\hat{x}k_x \sin k_x \left(x - \frac{a'}{2}\right) \cos k_y \left(y - \frac{b'}{2}\right) + \hat{y}k_y \cos k_x \left(x - \frac{a'}{2}\right) \sin k_y \left(y - \frac{b'}{2}\right)}{\sqrt{\frac{a'b'}{4}\{k_y^2 \sigma_x (2 - \sigma_y) + k_x^2 \sigma_y (2 - \sigma_x)\}}}$$

(3.103b)

$$\vec{e}_{mn}^{g\text{TM}}(x, y) = -\frac{\hat{x}k_x \cos k_x \left(x - \frac{a'}{2}\right) \sin k_y \left(y - \frac{b'}{2}\right) + \hat{y}k_y \sin k_x \left(x - \frac{a'}{2}\right) \cos k_y \left(y - \frac{b'}{2}\right)}{\sqrt{\frac{a'b'}{4}\{k_x^2 + k_y^2\}}}$$

(3.103c)

$$\vec{h}_{mn}^{g\text{TM}}(x, y) = \frac{\hat{x}k_y \sin k_x \left(x - \frac{a'}{2}\right) \cos k_y \left(y - \frac{b'}{2}\right) - \hat{y}k_x \cos k_x \left(x - \frac{a'}{2}\right) \sin k_y \left(y - \frac{b'}{2}\right)}{\sqrt{\frac{a'b'}{4}\{k_x^2 + k_y^2\}}}$$

(3.103d)

with

$$k_x = \frac{m\pi}{a'} \qquad k_y = \frac{n\pi}{b'}$$

$$\sigma_x = \begin{cases} 2 & \text{if } k_x = 0 \\ 1 & \text{if } k_x \neq 0 \end{cases} \qquad \sigma_y = \begin{cases} 2 & \text{if } k_y = 0 \\ 1 & \text{if } k_y \neq 0 \end{cases}$$

(3.104)

The reference coordinate system is shown in Figure 3.10. For the TE_{zmn} modes integers m and n are not zero simultaneously. For the TM_{zmn} modes m and n are non zero. The modal admittances are

$$Y_{mn}^{g\text{TE}} = \frac{k_{zmn}}{\omega\mu_0} \qquad Y_{mn}^{g\text{TM}} = \frac{\omega\epsilon_0}{k_{zmn}}$$

$$k_{zmn} = \sqrt{k_0^2 - k_x^2 - k_y^2}$$

(3.105)

The above modal functions satisfy normalization condition, that is,

$$\int\limits_{-a'/2}^{a'/2} \int\limits_{-b'/2}^{b'/2} \vec{e}_{mn}^{g\mathrm{TE}} \times \vec{h}_{ij}^{g\mathrm{TE}} \cdot \hat{z}\, dx\, dy = \delta_{mi}\delta_{nj} \qquad (3.106)$$

where δ_{pq} is the Kronecker delta. The TM_z modes also satisfy similar relations. Further, the TE_z and the TM_z modal vectors are mutually orthogonal.

3.6.2 Waveguide Modes to Floquet Modes

Suppose the cell size is $a \times b$ and γ is the lattice angle of the array of Figure 3.10. We assume that the apertures are excited uniformly with linear phase progression (i.e., Floquet excitation). The field of the aperture located at the origin can be expressed as a superposition of several waveguide modes as follows:

$$\vec{E}^g = \sum_{m=1}^{M} A_m \vec{e}_m^{g\mathrm{TE}} + \sum_{n=1}^{N} B_n \vec{e}_n^{g\mathrm{TM}} \qquad (3.107)$$

In the above, A_m and B_n are the modal voltages with respect to the TE_z and TM_z waveguide modes and $\vec{e}_m^{g\mathrm{TE}}$ and $\vec{e}_n^{g\mathrm{TM}}$ are the corresponding modal vectors. Notice that for simplicity of presentation we use only one index, instead of two, to represent a waveguide mode.

Under Floquet excitations, the fields in the $z > 0$ region can be expressed as

$$\vec{E}^f = \sum_{i=1}^{\infty} C_i \vec{e}_i^{f\mathrm{TE}} \exp(-jk_{zi}z) + \sum_{i=1}^{\infty} D_i \vec{e}_i^{f\mathrm{TM}} \exp(-jk_{zi}z) \qquad (3.108)$$

where $\vec{e}_i^{f\mathrm{TE}}$ and $\vec{e}_i^{f\mathrm{TM}}$ are the vector modal functions for the TE_z and TM_z Floquet modes, respectively. Here also we use a single index to represent a double-index Floquet mode. The vector modal functions are deduced in Section 3.2. The coefficients C_i and D_i are the Floquet modal voltages. We enforce the continuity condition at $z = 0$ to obtain the relation between the waveguide modal voltages and the Floquet modal voltages. This yields

$$\sum_{i=1}^{\infty} C_i \vec{e}_i^{f\mathrm{TE}} + \sum_{i=1}^{\infty} D_i \vec{e}_i^{f\mathrm{TM}} = \sum_{m=1}^{M} A_m \vec{e}_m^{g\mathrm{TE}} + \sum_{n=1}^{N} B_n \vec{e}_n^{g\mathrm{TM}} \qquad (3.109)$$

We now take the inner product on both sides of (3.109) with $\vec{e}_j^{f\mathrm{TE}*}$ and integrate over a unit cell. Using orthogonality of the Floquet modes shown in Section 3.2, we obtain

$$C_j = \sum_{m=1}^{M} A_m \tau_{ee}(j, m) + \sum_{n=1}^{N} B_n \tau_{em}(j, n) \qquad (3.110)$$

where

$$\tau_{ee}(j, m) = \iint\limits_{\text{unit cell}} \vec{e}_j^{f\text{TE}*} \cdot \vec{e}_m^{g\text{TE}} dx \, dy$$

$$\tau_{em}(j, n) = \iint\limits_{\text{unit cell}} \vec{e}_j^{f\text{TE}*} \cdot \vec{e}_n^{g\text{TM}} dx \, dy$$

$$(3.111)$$

It is assumed that the regions between the apertures are filled with perfect conducting planes. Therefore, the integrals in (3.111) should be performed over a waveguide aperture only, because the transverse components of the waveguide electric fields must vanish on the ground-plane region.

Taking the inner product of (3.109) with $\vec{e}_j^{f\text{TM}*}$ and then integrating, we obtain

$$D_j = \sum_{m=1}^{M} A_m \tau_{me}(j, m) + \sum_{n=1}^{N} B_n \tau_{mm}(j, n) \qquad (3.112)$$

with

$$\tau_{me}(j, m) = \iint\limits_{\text{unit cell}} \vec{e}_j^{f\text{TM}*} \cdot \vec{e}_m^{g\text{TE}} dx \, dy$$

$$\tau_{mm}(j, n) = \iint\limits_{\text{unit cell}} \vec{e}_j^{f\text{TM}*} \cdot \vec{e}_n^{g\text{TM}} dx \, dy$$

$$(3.113)$$

Equations (3.110) and (3.112) yield the Floquet modal voltages in terms of the waveguide modal voltages on an aperture.

3.6.3 Reflection and Transmission Matrices

We now proceed to determine the reflection and transmission matrices of a waveguide aperture in an infinite array environment. The reflection matrix provides complex reflection coefficients of various waveguide modes on the aperture, radiating in an infinite array environment. The reflection matrix is related to the input match of an element. The transmission matrix yields the radiating Floquet mode amplitudes with respect to the incident waveguide modes. The active element pattern of the array is determined directly from the transmission matrix elements.

In order to determine the reflection matrix, we determine the aperture admittance matrix first. The aperture admittance matrix relates the modal current (magnetic field) vector with the modal voltage (electric field) vector. The magnetic field at $z = 0-$ can be expressed in terms of magnetic modal vector functions of the waveguide modes. This yields

$$\vec{H}^g = \sum_{m=1}^{M} P_m Y_m^{g\text{TE}} \vec{h}_m^{g\text{TE}} + \sum_{n=1}^{N} Q_n Y_n^{g\text{TM}} \vec{h}_n^{g\text{TM}} \qquad (3.114)$$

where P_m and Q_n, respectively, are the modal currents for the TE_m and TM_n modes normalized with respect to the corresponding modal admittances $Y_m^{g\text{TE}}$ and $Y_n^{g\text{TM}}$. The magnetic field at $z = 0+$ can be expressed in terms of Floquet modes. Since waves are progressive in the $z > 0$ region, the magnetic field is given by

$$\vec{H}^f = \sum_{i=1}^{\infty} C_i Y_i^{f\text{TE}} \vec{h}_i^{f\text{TE}} \exp(-jk_{zi}z) + \sum_{i=1}^{\infty} D_i Y_i^{f\text{TM}} \vec{h}_i^{f\text{TM}} \exp(-jk_{zi}z) \qquad (3.115)$$

Continuity of the magnetic fields at $z = 0$ yields

$$\sum_{m=1}^{M} P_m Y_m^{g\text{TE}} \vec{h}_m^{g\text{TE}} + \sum_{n=1}^{N} Q_n Y_n^{g\text{TM}} \vec{h}_n^{g\text{TM}} = \sum_{i=1}^{\infty} C_i Y_i^{f\text{TE}} \vec{h}_i^{f\text{TE}} + \sum_{i=1}^{\infty} D_i Y_i^{f\text{TM}} \vec{h}_i^{f\text{TM}} \qquad (3.116)$$

We now take the scalar product with $\vec{h}_k^{g\text{TE}}$ on both sides and then integrate over the waveguide aperture. Using orthogonality between the waveguide modes once again, we obtain

$$P_k = \sum_{i=1}^{\infty} C_i \kappa_{ee}(k, i) + \sum_{i=1}^{\infty} D_i \kappa_{em}(k, i) \qquad (3.117)$$

where

$$\kappa_{ee}(k, i) = \frac{Y_i^{f\text{TE}}}{Y_k^{g\text{TE}}} \iint\limits_{\text{waveguide}} \vec{h}_k^{g\text{TE}} \cdot \vec{h}_i^{f\text{TE}} \, dx \, dy$$

$$\kappa_{em}(k, i) = \frac{Y_i^{f\text{TM}}}{Y_k^{g\text{TE}}} \iint\limits_{\text{waveguide}} \vec{h}_k^{g\text{TE}} \cdot \vec{h}_i^{f\text{TM}} \, dx \, dy \qquad (3.118)$$

Similarly, Q_k can be obtained as

$$Q_k = \sum_{i=1}^{\infty} C_i \kappa_{me}(k, i) + \sum_{i=1}^{\infty} D_i \kappa_{mm}(k, i) \qquad (3.119)$$

with

$$\kappa_{me}(k, i) = \frac{Y_i^{f\text{TE}}}{Y_k^{g\text{TM}}} \iint\limits_{\text{waveguide}} \vec{h}_k^{g\text{TM}} \cdot \vec{h}_i^{f\text{TE}} \, dx \, dy$$

$$\kappa_{mm}(k, i) = \frac{Y_i^{f\text{TM}}}{Y_k^{g\text{TM}}} \iint\limits_{\text{waveguide}} \vec{h}_k^{g\text{TM}} \cdot \vec{h}_i^{f\text{TM}} \, dx \, dy \qquad (3.120)$$

We now substitute C_i and D_i from (3.110) and (3.112) into (3.117) and (3.119). After straightforward algebraic manipulation, we obtain

$$P_k = \sum_{m=1}^{M} A_m y_{ee}(k, m) + \sum_{n=1}^{N} B_n y_{em}(k, n) \qquad k = 1, 2, \ldots, M \qquad (3.121a)$$

$$Q_k = \sum_{m=1}^{M} A_m y_{me}(k, m) + \sum_{n=1}^{N} B_n y_{mm}(k, n) \qquad k = 1, 2, \ldots, N \qquad (3.121b)$$

where

$$y_{ee}(k, m) = \sum_{i=1}^{\infty} [\tau_{ee}(i, m)\kappa_{ee}(k, i) + \tau_{me}(i, m)\kappa_{em}(k, i)]$$

$$y_{em}(k, m) = \sum_{i=1}^{\infty} [\tau_{em}(i, m)\kappa_{ee}(k, i) + \tau_{mm}(i, m)\kappa_{em}(k, i)]$$

$$\qquad (3.122)$$

$$y_{me}(k, m) = \sum_{i=1}^{\infty} [\tau_{ee}(i, m)\kappa_{me}(k, i) + \tau_{me}(i, m)\kappa_{mm}(k, i)]$$

$$y_{mm}(k, m) = \sum_{i=1}^{\infty} [\tau_{em}(i, m)\kappa_{me}(k, i) + \tau_{mm}(i, m)\kappa_{mm}(k, i)]$$

Equations (3.121a) and (3.121b) together represent $M + N$ linear equations. Using matrix notation the above set of equations can be expressed as:

$$\begin{bmatrix} P_1 \\ \vdots \\ P_M \\ Q_1 \\ \vdots \\ Q_N \end{bmatrix} = \begin{bmatrix} y_{ee}(1, 1) & \cdots & y_{ee}(1, M), & y_{em}(1, 1) & \cdots & y_{em}(1, N) \\ \vdots & & \vdots & \vdots & & \vdots \\ y_{ee}(M, 1) & \cdots & y_{ee}(M, M), & y_{em}(1, 1) & \cdots & y_{em}(M, N) \\ y_{me}(1, 1) & \cdots & y_{me}(1, M), & y_{mm}(1, 1) & \cdots & y_{mm}(1, N) \\ \vdots & & \vdots & \vdots & & \vdots \\ y_{me}(N, 1) & \cdots & y_{me}(N, M), & y_{mm}(N, 1) & \cdots & y_{mm}(N, N) \end{bmatrix} \begin{bmatrix} A_1 \\ \vdots \\ A_M \\ B_1 \\ \vdots \\ B_N \end{bmatrix} \qquad (3.123)$$

The above equation can be written simply as

$$[I^g] = [y][V^g] \qquad (3.124)$$

The $[y]$ matrix on the right-hand side is a square matrix of order $(M + N) \times (M + N)$. It is known as the aperture admittance matrix of a waveguide aperture under Floquet excitation. Elements of the matrix represent coupling between two waveguide modes through the aperture. The current vector $[I^g]$ and voltage vector $[V^g]$ can be represented in terms of the incident voltage vector $[V^{g+}]$ and the aperture reflection matrix $[\Gamma]$ as follows:

$$[V^g] = [I + \Gamma][V^{g+}]$$

$$[I^g] = [I - \Gamma][V^{g+}]$$

$$\qquad (3.125)$$

where $[I]$ is an identity matrix of the same order as $[\Gamma]$. The notation $[I \pm \Gamma]$ should be understood as $[[I] \pm [\Gamma]]$. Similar notations are used at other places as deemed appropriate. Substituting $[V^g]$ and $[I^g]$ from (3.125) into (3.124) we obtain the aperture reflection matrix as

$$[\Gamma] = [I + y]^{-1}[I - y] \tag{3.126}$$

The reflection matrix relates the amplitudes of the incident modes with the amplitudes of the reflected modes. It essentially provides information about impedance mismatch under Floquet excitation, which is very useful information for an array design. Another useful information necessary for array performance is the Floquet modal contents of the array. The radiating Floquet modal voltages can be found from the aperture voltages using (3.110) and (3.112). In matrix format, the relation is

$$\begin{bmatrix} C_1 \\ C_2 \\ \vdots \\ D_1 \\ D_2 \\ \vdots \end{bmatrix} = \begin{bmatrix} \tau_{ee}(1,1) & \cdots & \tau_{ee}(1,M), & \tau_{em}(1,1) & \cdots & \tau_{em}(1,N) \\ \tau_{ee}(2,1) & \cdots & \tau_{ee}(2,M), & \tau_{em}(2,1) & \cdots & \tau_{em}(2,N) \\ \vdots & & \vdots & \vdots & & \vdots \\ \tau_{me}(1,1) & \cdots & \tau_{me}(1,M), & \tau_{mm}(1,1) & \cdots & \tau_{mm}(1,N) \\ \tau_{me}(2,1) & \cdots & \tau_{me}(2,M), & \tau_{mm}(2,1) & \cdots & \tau_{mm}(2,N) \\ \vdots & & \vdots & \vdots & & \vdots \end{bmatrix} \begin{bmatrix} A_1 \\ \vdots \\ A_M \\ B_1 \\ \vdots \\ B_N \end{bmatrix} \tag{3.127}$$

In abbreviated notation, we write

$$[V^{f+}] = [\tau][V^g] \tag{3.128}$$

Using (3.125) we have

$$[V^{f+}] = [\tau][I + \Gamma][V^{g+}] \tag{3.129}$$

The transmission matrix, $[T]$, relating the incident waveguide modal voltage and the radiating Floquet modal voltage is

$$[T] = [\tau][I + \Gamma] \tag{3.130}$$

For the sake of completeness we determine the reflection and transmission matrix with respect to a finite number of Floquet modes incident upon the waveguide array aperture. For M' TE$_z$ and N' TM$_z$ incident Floquet modes, the electric fields at $z = 0+$ can be expressed as

$$\vec{E}'^f = \sum_{i=1}^{M'} C'_i \vec{e}_i^{fTE} + \sum_{i=1}^{N'} D'_i \vec{e}_i^{fTM} + \sum_{i=M'+1}^{\infty} C'_i \vec{e}_i^{fTE} + \sum_{i=N'+1}^{\infty} D'_i \vec{e}_i^{fTM} \tag{3.131}$$

The first two terms on the right-hand side of (3.131) represent incident and reflected Floquet modal fields and the last two terms represent additional Floquet modes produced by the discontinuity. At $z = 0-$, the electric field can be expressed in terms of waveguide modes as

$$\vec{E}'^g = \sum_{m=1}^{M} A'_m \vec{e}_m^{g\text{TE}} + \sum_{n=1}^{N} B'_n \vec{e}_n^{g\text{TM}} \tag{3.132}$$

Applying the continuity condition of the electric field at $z = 0$ and using mode orthogonality, as was done in (3.110) and (3.112), we obtain

$$C'_j = \sum_{m=1}^{M} A'_m \tau_{ee}(j, m) + \sum_{n=1}^{N} B'_n \tau_{em}(j, n) \tag{3.133}$$

$$D'_j = \sum_{m=1}^{M} A'_m \tau_{me}(j, m) + \sum_{n=1}^{N} B'_n \tau_{mm}(j, n) \tag{3.134}$$

The magnetic fields at the two regions ($z = 0+$ and $z = 0-$) are

$$\vec{H}'^f = -\sum_{i=1}^{M'} G'_i Y_i^{f\text{TE}} \vec{h}_i^{f\text{TE}} - \sum_{i=1}^{N'} H'_i Y_i^{f\text{TM}} \vec{h}_i^{f\text{TM}} + \sum_{i=M'+1}^{\infty} C'_i Y_i^{f\text{TE}} \vec{h}_i^{f\text{TE}}$$

$$+ \sum_{i=N'+1}^{\infty} D'_i Y_i^{f\text{TM}} \vec{h}_i^{f\text{TM}} \tag{3.135}$$

$$\vec{H}'^g = -\sum_{m=1}^{M} A'_m Y_m^{g\text{TE}} \vec{h}_m^{g\text{TE}} - \sum_{n=1}^{N} B'_n Y_n^{g\text{TM}} \vec{h}_n^{g\text{TM}} \tag{3.136}$$

Note that G'_i and H'_i are the normalized modal currents for the Floquet modes at $z = 0$. The last two terms in (3.135) correspond to the higher order Floquet modes that are created at the discontinuity, propagating along z. Further, since the waveguides are match terminated, the normalized modal currents become identical with the modal voltages with negative signs. The negative sign is due to the propagation along the $-z$-direction. Applying the continuity of the magnetic fields at $z = 0$ and then using orthogonality, we obtain

$$-A'_k = -\sum_{i=1}^{M'} G'_i \kappa_{ee}(k, i) - \sum_{i=1}^{N'} H'_i \kappa_{em}(k, i) + \sum_{i=M'+1}^{\infty} C'_i \kappa_{ee}(k, i) + \sum_{i=N'+1}^{\infty} D'_i \kappa_{em}(k, i) \tag{3.137}$$

Substituting C'_i and D'_i from (3.133) and (3.134) we obtain

$$-A'_k = -\sum_{i=1}^{M'} G'_i \kappa_{ee}(k, i) - \sum_{i=1}^{N'} H'_i \kappa_{em}(k, i) + \sum_{m=1}^{M} A'_m \bar{y}_{ee}(k, m) + \sum_{n=1}^{N} B'_n \bar{y}_{em}(k, n) \tag{3.138}$$

Similarly

$$-B'_k = -\sum_{i=1}^{M'} G'_i \kappa_{me}(k, i) - \sum_{i=1}^{N'} H'_i \kappa_{mm}(k, i) + \sum_{m=1}^{M} A'_m \bar{y}_{me}(k, m) + \sum_{n=1}^{N} B'_n \bar{y}_{mm}(k, n) \tag{3.139}$$

with

$$\bar{y}_{ee}(k,m) = \sum_{i=M'+1}^{\infty} \tau_{ee}(i,m)\kappa_{ee}(k,i) + \sum_{i=N'+1}^{\infty} \tau_{me}(i,m)\kappa_{em}(k,i)$$

$$\bar{y}_{em}(k,m) = \sum_{i=M'+1}^{\infty} \tau_{em}(i,m)\kappa_{ee}(k,i) + \sum_{i=N'+1}^{\infty} \tau_{mm}(i,m)\kappa_{em}(k,i)$$

$$\bar{y}_{me}(k,m) = \sum_{i=M'+1}^{\infty} \tau_{ee}(i,m)\kappa_{me}(k,i) + \sum_{i=N'+1}^{\infty} \tau_{me}(i,m)\kappa_{mm}(k,i) \qquad (3.140)$$

$$\bar{y}_{mm}(k,m) = \sum_{i=M'+1}^{\infty} \tau_{em}(i,m)\kappa_{me}(k,i) + \sum_{i=N'+1}^{\infty} \tau_{mm}(i,m)\kappa_{mm}(k,i)$$

Equations (3.138) and (3.139) can be written in matrix format as

$$-\begin{bmatrix} A' \\ B' \end{bmatrix} = -[\kappa]\begin{bmatrix} G' \\ H' \end{bmatrix} + [\bar{y}]\begin{bmatrix} A' \\ B' \end{bmatrix} \qquad (3.141)$$

From (3.133) and (3.134) we write

$$\begin{bmatrix} C' \\ D' \end{bmatrix} = [\tau]\begin{bmatrix} A' \\ B' \end{bmatrix} \qquad (3.142)$$

From (3.141) and (3.142) we deduce

$$\begin{bmatrix} C' \\ D' \end{bmatrix} = [\tau][I + \bar{y}]^{-1}[\kappa]\begin{bmatrix} G' \\ H' \end{bmatrix} \qquad (3.143)$$

Equation (3.143) relates the Floquet modal voltage with the Floquet modal current. Thus the impedance matrix seen by the Floquet modes is

$$[z'] = [\tau][I + \bar{y}]^{-1}[\kappa] \qquad (3.144)$$

The reflection matrix of the Floquet modes is then

$$[\Gamma'] = [z' + I']^{-1}[z' - I'] \qquad (3.145)$$

The identity matrix $[I']$ in (3.145) is of order $(M' + N') \times (M' + N')$. The transmission matrix $[T']$ relates the incident Floquet modes to the waveguide modes. This can be deduced as

$$[T'] = [I + \bar{y}]^{-1}[\kappa][I' - \Gamma'] \qquad (3.146)$$

In the GSM terminology defined in Chapter 6 $[\Gamma]$, $[T]$, $[\Gamma']$, and $[T']$ are $[S_{11}]$, $[S_{21}]$, $[S_{22}]$, and $[S_{12}]$, respectively, where port 1 is located inside the waveguide at $z = 0-$ and port 2 is on the aperture at $z = 0+$.

3.6.4 TE₁₀ Mode Incidence

In a waveguide array, the incident mode generally is the dominant TE_{10} mode. Therefore, the input reflection coefficient and the active element pattern with respect to the TE_{10} mode is of importance. The input reflection coefficient of a waveguide with respect to the TE_{10} mode incidence is one of the diagonal elements of the $[\Gamma]$ matrix in (3.126). If the TE_{10} mode is designated as mode 1 in (3.107), then the (1,1) element of $[\Gamma]$ would be the desired reflection coefficient. Similarly, the k th radiating Floquet modal voltage with respect to the TE_{10} incident waveguide mode will be equal to the $(k,1)$ element of the $[T]$ matrix in (3.130).

Figure 3.11 shows the input reflection coefficient versus scan angle of an infinite array ($\gamma = 90°$) under Floquet excitation. The reflection coefficient varies significantly with the scan angle. For the E-plane scan, a large reflection occurs near the 23° scan location. This reflection represents array blindness. This is primarily due to the appearance of the grating lobe along the grazing angle that causes strong E-plane coupling between the waveguides. The scan blindness does not occur for the H-plane scan because of low mutual coupling along the H-plane. Interpretations of array blindness in light of surface and leaky wave propagation are presented in Chapter 8.

The active element patterns of the array are obtained using (3.92). For the waveguide array V_{00}^{TE} and V_{00}^{TM} should be replaced by the $(j,1)$ and $(k,1)$ elements, respectively, of the $[T]$ matrix in (3.130), where the index j is associated with the

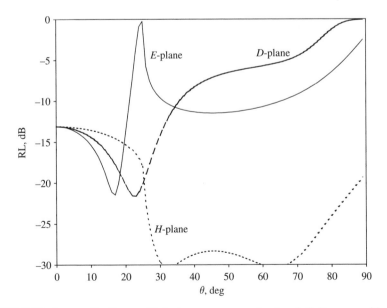

FIGURE 3.11 Input reflection coefficient versus scan angle of an infinite waveguide aperture array (square grid) under Floquet excitation of cell size $0.7\lambda_0 \times 0.7\lambda_0$, waveguide size $0.6\lambda_0 \times 0.6\lambda_0$.

TE_{00} Floquet mode and the index k is associated with the TM_{00} Floquet mode. A normalization factor should be multiplied in order to make the total incident power equal to 4π. The incident power for unity incident voltage with respect to the TE_{10} mode is equal to its modal admittance, Y_{10}^{gTE}; therefore, the normalization constant should be

$$C_{norm} = \sqrt{\frac{4\pi}{Y_{10}^{gTE}}} \tag{3.147}$$

The normalized electric field thus becomes

$$\vec{E}_a(\theta_0, \phi_0) = \sqrt{\frac{4\pi}{Y_{10}^{gTE}}} \left[\hat{\theta}_0 \left\{ jT(k,1)\sqrt{\frac{ab}{\lambda_0^2}} \right\} - \hat{\phi}_0 \left\{ jT(j,1)\sqrt{\frac{ab}{\lambda_0^2}} \cos\theta_0 \right\} \right] \tag{3.148}$$

Figure 3.12 shows the active element pattern cuts along the E, H, and diagonal planes. The E-plane cut has a blind spot near 23°. This blind location was also predicted from the return loss results plotted in Figure 3.11. It is interesting to observe that the gain of the active E-plane pattern remains flat and then reduces abruptly after the blind angle. As mentioned before, the blind angle is due to the appearance of a grating lobe. A finite amount of power is lost in the grating lobe, causing such a gain reduction in the main beam. A more detailed study of waveguide aperture and horn arrays is undertaken in Chapter 8.

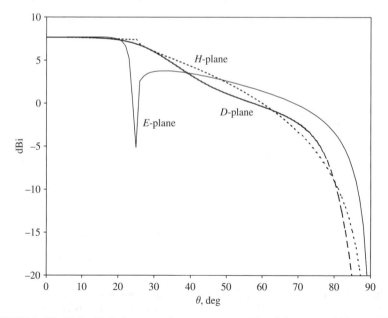

FIGURE 3.12 Embedded element gain versus scan angle of the array of Figure 3.11.

REFERENCES

[1] N. Marcuvitz, in *Waveguide Handbook*, Dover, New York, 1951, "Transmission Line Models," Chapter 2.

[2] R. F. Harrington, "Transmission Line Models," *Time Harmonic Electromagnetic Fields*, McGraw-Hill, New York, 1961.

[3] W. C. Chew, J. A. Kong, and L. C. Shen, "Radiation Characteristics of Circular Microstrip Antennas," *J. Appl. Phys.*, Vol. 51, No. 7, pp. 3907–3915, July 1980.

[4] R. E. Collin, in Field *Theory of Guided Waves*, IEEE, New York, 1991, "Periodic Structures," Chapter 9.

[5] A. K. Bhatacharyya, in *Electromagnetic Fields in Multilayered Structures*, Artech House, Norwood, MA, pp.155–160, 1994.

BIBLIOGRAPHY

PRINTED DIPOLE ELEMENTS AND ARRAYS

Delaveaud, C., and C. Brocheton, "Dual-Band Behavior of Printed Dipoles," *Electron. Lett.*, Vol. 36, No. 14, pp. 1175–1177, July 2000.

Dey, S., C. K. Anandan, P. Mohanan, and K. G. Nair, "Analysis of Cavity Backed Printed Dipoles," *Electron. Lett.*, Vol. 30, No. 3, pp. 173–174, Feb. 1994.

Evtioushkine, G. A., J. W. Kim, and K. S. Han, "Very Wideband Printed Dipole Antenna Array," *Electron. Lett.*, Vol. 34, No. 24, pp. 2292–2293, Nov. 1998.

Hansen, V., "Finite Array of Printed Dipoles with a Dielectric Cover," *IEE Proc.*, Vol. 134, Pt. H, No. 3, pp. 261–269, 1987.

Oltman, H. G., and D. A. Huebner, "Electromagnetically Coupled Microstrip Dipoles," *IEEE Trans. Antennas Propagat.*, Vol. AP-29, No. 1, pp. 151–157, Jan. 1981.

Paul, A., and I. Gupta, "An Analysis of Log Periodic Antenna with Printed Dipoles," *IEEE Trans. Microwave Theory Tech.*, Vol. MTT-50, No. 2, pp. 114–117, Feb. 1981.

Pozar, D. M., and D. H. Schaubert, "Scan Blindness in Infinite Phased Array of Printed Dipoles," *IEEE Trans. Antennas Propagat.*, Vol. AP-32, pp. 602–610, 1984.

Rana, I. E., and N. G. Alexopoulos, "Current Distribution and Input Impedance of Printed Dipoles," *IEEE Trans. Antennas Propagat.*, Vol. AP-29, No. 1, pp. 99–105, Jan. 1981.

Suh, Y-H., and K. Chang, "Low Cost Microstrip-Fed Dual Frequency Printed Dipole Antenna for Wireless Communications," *Electron. Lett.* Vol. 36, No. 14, pp. 1177–1179, July 2000.

WAVEGUIDE ARRAYS

Amitay, N., and V. Galindo, "Characteristics of Dielectric Loaded and Covered Circular Waveguide Phased Arrays," *IEEE Trans. Antennas Propagat.*, Vol. AP-17, No. 6, pp. 722–729, Nov. 1969.

Amitay, N., V. Galindo, and C. P. Wu, in *Theory and Analysis of Phased Array Antennas*, Wiley-Interscience, New York, 1972, "Arrays of Rectangular Waveguides," Chapter 5.

Diamond, B. L., "A Generalized Approach to the Analysis of Infinite Planar Array Antennas," *Proc. IEEE*, Vol. 56, No. 11, pp. 1837–1851, Nov. 1968.

Galindo, V., and C. P. Wu, "Numerical Solutions for an Infinite Phased Array of Rectangular Waveguide with Thick Walls," *IEEE Trans. Antennas Propagat.*, Vol. AP-14, No. 2, pp. 149–158, Mar. 1966.

Farrell, G. F., and D. H. Kuhn, "Mutual Coupling in Infinite Planar Arrays of Rectangular Waveguide Horns," *IEEE Trans. Antennas Propagat.*, Vol. AP-16, No. 4, pp. 405–414, July 1968.

Knittel, G. H., A. Hessel, and A. A. Oliner, "Element Pattern Nulls in Phased Arrays and Their Relation to Guided Waves," *Proc. IEEE*, Vol. 56, No. 11, pp. 1822–1836, Nov. 1968.

Lee, S. W., "Aperture Matching for an Infinite Circular Polarized Array of Rectangular Waveguides," *IEEE Trans. Antennas Propagat*, Vol. AP-19, No. 3, pp. 332–342, May 1971.

Wu, C. P., and V. Galindo, "Properties of Phased Array of Rectangular Waveguides with Thin Walls," *IEEE Trans. Antennas Propagat.*, Vol. AP-14, No. 2, pp. 163–173, Mar. 1966.

PROBLEMS

3.1 A two-dimensional infinite array of planar current sources is radiating into free space. The current distribution of the array can be expressed as

$$\vec{I}(x, y) = \hat{x} \sum_{m=-\infty}^{\infty} \sum_{n=-\infty}^{\infty} f(x - x_{mn}, y - y_{mn}) \exp(-jk_{x0} x_{mn} - jk_{y0} y_{mn})$$

with

$$f(x, y) = \begin{cases} \cos(2\pi x/a) & |x| \le a/4, |y| \le b/4 \\ 0 & \text{otherwise} \end{cases}$$

where $a \times b$ is the cell size of the array. The grid angle is 90°.

(a) Express the current source in Floquet series.

(b) Obtain the current amplitudes for the TE_{z00} and TM_{z00} Floquet modes.

(c) Compute the total power radiated from a unit cell into the above two modes if $a = b = 0.75\lambda_0$, where λ_0 is the free-space wavelength. Intended radiation direction is $\theta = 30°$, $\phi = 45°$.

3.2 An infinite array of z-directed current sources are radiating in free space. The source distribution is given by

$$\vec{I}(x, y, z) = \hat{z} \sum_{m=-\infty}^{\infty} \sum_{n=-\infty}^{\infty} \delta(x - ma)\delta(y - nb)\delta(z) \exp(-jk_{x0}x - jk_{y0}y)$$

Show that the above source will produce the TM_z Floquet modes only. Obtain an expression of the electric field in the $z > 0$ region produced by the array. [*Hint*: Find the magnetic vector potential and from that obtain the field components.]

3.3 An infinite array of current sources is located at the interface of two dielectric layers of thicknesses h_1 and h_2 with dielectric constants ε_{r1} and ε_{r2}, respectively. The cell size of the array is $a \times b$ and the grid angle is γ.

    (a) Obtain the expressions for the equivalent admittances seen by the TE_{zmn} and TM_{zmn} modes for an intended radiation angle (θ_s, ϕ_s).

    (b) Find the amplitudes of the TE_{z00} and TM_{z00} Floquet modes in the region $z > h_1$ if the source current is y-directed and uniformly distributed within a unit cell.

3.4 Show that for an infinite array the real part of the Floquet impedance is contributed by the TE_{z00} and TM_{z00} Floquet modes only if the cell dimensions a and b are both less than one-half of the free-space wavelength.

3.5 Deduce an expression for the Floquet impedance for an infinite array of electric surface current elements immersed in vacuum. The current distribution of an element is given by

$$\vec{I}(x, y) = \begin{cases} \hat{x} \cos(\pi x / l) & |x| \leq l/2, |y| \leq t/2 \\ 0 & \text{otherwise} \end{cases}$$

The cell size is $a \times b$ and the grid angle is γ. The intended radiation direction is (θ_s, ϕ_s).

3.6 Find the blind angles nearest to the bore-sight directions of the following infinite arrays with different grid angles as given below:

    (a) $\gamma = 90°$,
    (b) $\gamma = 60°$,
    (c) $\gamma = 45°$.

The cell dimensions are: $a = b = 2\lambda_0/3$, where λ_0 is the free-space wavelength. The array is located over a grounded dielectric layer with thickness $0.05\lambda_0$ and with a dielectric constant of 2.5. Use circle diagrams to verify your results.

3.7 Suppose $\vec{E}_a(\theta, \phi)$ is the active element pattern of an infinite array. Then obtain the array pattern of the infinite array, $\vec{E}_\infty(\theta, \phi)$, under Floquet excitation. The power radiated by an element along the desired radiation direction is given by

$$P_a = \frac{1}{\eta} \iint_\Omega \vec{E}_a \cdot \vec{E}_\infty^* \, d\Omega$$

where Ω is a small solid angle that encompasses only the desired beam of the array. It is also known that the power radiated by an element in the desired direction is given by

$$P_a = |V_{00}^{\text{TE}}|^2 \; Y_{00}^{\text{TE}} + |V_{00}^{\text{TM}}|^2 \; Y_{00}^{\text{TM}}$$

where V_{00}^{TE} and V_{00}^{TM} are the modal voltages for the dominant Floquet modes radiated by the array under Floquet excitation. Equating the above two expressions for P_a, obtain the magnitude of the active element pattern in terms of the Floquet modal voltages.

3.8 The transverse electric field distribution on an infinite aperture at $z = 0$ is given by

$$\vec{E} = V_{00}^{\text{TE}} \vec{e}_{00}^{\text{TE}} + V_{00}^{\text{TM}} \vec{e}_{00}^{\text{TM}}$$

where V_{00}^{TE} and V_{00}^{TM} are the modal voltages for the TE_{z00} and TM_{z00} modes, respectively, and $\vec{e}_{00}^{\text{TE}}$ and $\vec{e}_{00}^{\text{TM}}$ are the modal vectors defined in (3.11) and (3.19a), respectively.

(a) Find the equivalent magnetic surface current.

(b) Obtain the far-field pattern of the infinite aperture.

(c) Equate the far-field pattern with $\vec{E}_\infty(\theta, \phi)$ in (3.80) to deduce $\vec{E}_a(\theta, \phi)$ in terms of the modal voltages.

3.9 Suppose $\Gamma(\theta, \phi)$ is the Floquet reflection coefficient when the intended radiation angle is (θ, ϕ). Suppose the cell dimensions a and b are both less than $\lambda_0/2$. Then prove that the active element pattern is given by

$$|\vec{E}_a(\theta, \phi)| = C\sqrt{\cos\theta[1 - |\Gamma(\theta, \phi)|^2]}$$

where C is independent of θ and ϕ. Hence prove that the gain of an active element pattern cannot exceed $4\pi ab/\lambda_0^2$.

3.10 Only four selected elements of an infinite array are excited with the following incident power and phase:
Find the bore-sight gain of the beam produced by the four elements. Assume that the active element gain of the infinite array is 15 dBi.

Element Number	Incident Power (W)	Phase (deg)
1	0.5	0
2	1.5	45
3	1.0	60
4	1.2	30

3.11 The aperture efficiency of an aperture antenna is defined as

$$\eta_{apr} = \frac{G\lambda_0^2}{4\pi A}$$

where G is the gain of the antenna, A is the area of the aperture, and λ_0 is the free-space wavelength. For aperture efficiency of an array element we use G as the active element gain and A as the unit cell area. Then show that the aperture efficiency of an element in an infinite array is given by

$$\eta_{apr} = \frac{[|V_{00}^{TE}|^2 + |V_{00}^{TM}|^2]}{377 P_{inc}}$$

where P_{inc} is the incident power to a unit cell and V_{00}^{TE} and V_{00}^{TM} are the Floquet modal voltages for the TE_{z00} and TM_{z00} Floquet modes, respectively. The array is excited for bore-sight radiation.

3.12 Show that for an infinitely large cell size (a and b are infinitely large) the Floquet impedance of a current element on a grounded dielectric layer given in (3.63) becomes independent of the scan angle and approaches the following infinite integral:

$$\lim_{a,b\to\infty} Z^{FL} = \frac{l^2}{4\pi^2} \int_{-\infty}^{\infty}\int_{-\infty}^{\infty} \left[\frac{k_y^2}{y^{TE}} + \frac{k_x^2}{y^{TM}} \right] \frac{\text{sinc}(k_x l/2)\,\text{sinc}(k_y t/2)}{k_x^2 + k_y^2} \, dk_x \, dk_y$$

where y^{TE} and y^{TM} are obtained from (3.51) and (3.54), respectively, replacing k_{xmn} and k_{ymn} by k_x and k_y, respectively.

3.13 The self-impedance of a current element in an infinite array is defined as the input impedance of the current element when all other elements are not excited (zero current). It can be shown that for a rectangular grid of cell size $a \times b$, the self-impedance, Z_{11}, can be computed from the Floquet impedance using the relation (see Chapter 4)

$$Z_{11} = \frac{ab}{4\pi^2} \int_0^{2\pi/a}\int_0^{2\pi/b} Z^{FL}(k_{x0}, k_{y0})\, dk_{x0}\, dk_{y0}$$

Use the above relation to prove that for a current element on a grounded dielectric substrate, the self-impedance is given by

$$Z_{11} = \frac{l^2}{4\pi^2} \int_{-\infty}^{\infty}\int_{-\infty}^{\infty} \left[\frac{k_y^2}{y^{TE}} + \frac{k_x^2}{y^{TM}} \right] \frac{\text{sinc}(k_x l/2)\text{sinc}(k_y t/2)}{k_x^2 + k_y^2} \, dk_x \, dk_y$$

(*Note*: The self-impedance and the isolated impedance becomes identical if a and b are infinitely large.)

3.14 For small scan angles, the aperture field distribution of a waveguide can be assumed as the field distribution of the TE_{01} waveguide mode as

$$\vec{E}^{apr} = \hat{x}E_0 \cos\left(\frac{\pi y}{b_1}\right) \qquad |x| \leq \frac{1}{2}a_1 \qquad |y| \leq \frac{1}{2}b_1$$

where $a_1 \times b_1$ is the waveguide aperture size. Assume $a \times b$ is the cell size of the array. Use rectangular grid.

  (a) Deduce the Floquet modal fields emanating from the infinite array in terms of the waveguide aperture field amplitude E_0. Assume (θ_s, ϕ_s) as the scan direction.
  (b) We define the Floquet aperture admittance as

$$Y_{apr}^{FL}(\theta_s, \phi_s) = \frac{\iint\limits_{apr} \vec{E}^{apr} \times \vec{H}^{apr} \cdot ds}{\iint\limits_{apr} \vec{E}^{apr} \cdot \vec{E}^{apr} ds}$$

  Deduce the expression for the Floquet aperture admittance.
  (c) Compute $Y_{apr}^{FL}(0,0)$ and find the input reflection coefficient of a waveguide under Floquet excitation. *Use $a_1 = b_1 = 0.6\lambda_0, a = b = 0.62\lambda_0$.*
  (d) Find the active element gain in the bore-sight direction.
  (e) The input reflection of the array elements can be eliminated invoking a stub matching procedure. Under such a "matched" condition, find the active element gain of the array.

3.15 The aperture field distribution of a multi mode horn antenna is given by

$$\vec{E}^{apr} = \hat{y}E_0 \left\{\cos\left(\frac{\pi x}{a'}\right) - 0.3\cos\left(\frac{3\pi x}{a'}\right)\right\} \qquad |x| \leq \frac{1}{2}a' \qquad |y| \leq \frac{1}{2}b'$$

An array is made with the above horns. The cell size of the array is $1.01a' \times 1.01b'$ with $a' \times b'$ as the horn aperture size. The intermediate space between horn apertures is perfectly conducting. The horn is designed to have a perfect input match in the infinite array environment for the bore-sight beam. Assume that the aperture distribution does not change in the array environment.

  (a) Determine the active element gain of the array of horns.
  (b) Find the locations of the grating lobes if the array is excited for an intended bore-sight beam.
  (c) Determine the percentage of the total radiated power that goes to the main beam.

Use $a' = 2\lambda_0, b' = 2\lambda_0$ for the computations. The grid angle of the array is 63.43°.

CHAPTER FOUR

Finite Array Analysis Using Infinite Array Results: Mutual Coupling Formulation

4.1 INTRODUCTION

In Chapter 3 we presented analyses of infinite arrays under Floquet excitations. In this chapter we demonstrate that the results of an infinite array can be utilized to predict the performances of a finite array with arbitrary excitations. We emphasize here that the predicted result would be *exact* if we use a very special definition of a finite array, which is somewhat contradictory to our notion. We define a finite array as *a physically infinite array with a finite number of excited elements*. The remaining elements are nonexcited, though they *must be physically present*. Such a finite array[1] exists only in theory. A real finite array, however, has only a finite number of physical elements. In many cases, radiation characteristics of a real finite array can be approximated as that of a finite array as defined above, because the nonexcited elements generally do not contribute significantly to the radiated fields, particularly in the main-lobe region.

An accurate analysis of a finite array with an arbitrary excitation necessitates an accurate estimation of the mutual coupling parameters (such as mutual impedance or admittance) between the array elements. In this chapter we establish analytically that the mutual coupling parameters between the elements can be estimated by

---

[1] Throughout the chapter, we will use the word *finite array* to refer to an infinite array with a finite number of excited elements. We use the word *real finite array* to refer to a practical finite array.

Phased Array Antennas. By Arun K. Bhattacharyya
© 2006 John Wiley & Sons, Inc.

invoking the Floquet modal analysis of an infinite array. Then we demonstrate that the mutual coupling data can be utilized to determine the active impedance (driving point impedance) and radiation patterns for an arbitrary amplitude distribution. Numerical results for active return loss and radiation patterns of a finite array of open-ended waveguides are presented. The pertinence of the finite array results to a real finite array is discussed in light of prediction uncertainty. Finally, an alternative formulation for finite array analysis, based on a convolution integral involving the infinite array characteristics and the finite array excitation function, is presented.

4.2 SYMMETRY PROPERTY OF FLOQUET IMPEDANCE

Before dealing with the mutual coupling aspect between the array elements, we first establish the symmetry property of the Floquet impedance. This important property is relevant to prove the symmetry of the $[Z]$ matrix of a finite array, which is a required condition for a reciprocal network.

Consider the one-dimensional infinite array of aperture sources shown in Figure 4.1. It is apparent that an array with symmetrical shaped elements will have symmetrical Floquet impedance with respect to the scan angle. In the following development we will prove a theorem that the symmetry property is valid not only for symmetrically shaped elements but also for arbitrarily shaped elements. This property is contingent on a primary source distribution that can be represented by a real function. For analytical simplicity, we place an infinite array of arbitrarily shaped, perfectly conducting scattering objects in front of an array of ideal aperture sources. The array of the scatters is in vacuum. A source aperture combined with the front scatterer makes an antenna element that can be considered arbitrarily shaped. We will prove that under this scenario the Floquet impedance remains symmetrical with respect to the scan angle, that is, $Z^{FL}(\theta) = Z^{FL}(-\theta)$, where $Z^{FL}(\theta)$ is the Floquet impedance seen by a source with respect to a scan angle θ from the normal direction. The aperture source array is located at $z = 0$. At $z = 0$ the source distribution can be expressed as

$$\vec{E}_s = V_s \sum_{n=-\infty}^{\infty} \vec{e}^s(x - na) \exp(-jn\psi) \qquad (4.1)$$

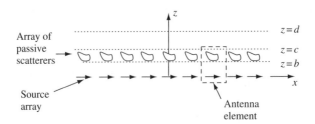

FIGURE 4.1   Infinite array of sources with arbitrarily shaped scatterers.

where a is the element spacing of the array, V_s represents the modal voltage of the aperture source, and $\vec{e}_s(x)$ is the modal vector of the source field, which is assumed to be a real function of x. For instance, if the source is an array of waveguide apertures, then $\vec{e}_s(x)$ will be equal to the modal electric field vector [1] of a waveguide. The above array of apertures will produce an incident wave to the scatterers at an angle θ with the z-axis, where θ is given by

$$\psi = k_0 a \sin \theta \tag{4.2}$$

where k_0 is the wave number in free space. The source field can be expanded in terms of the Floquet modal fields as follows:

$$\vec{E}_s(x) = \sum_{n=-\infty}^{\infty} V_n^{FL}(\psi)\, \vec{e}_n(x) \tag{4.3}$$

where $V_n^{FL}(\psi)$ is the Floquet modal voltage for the nth Floquet mode and $\vec{e}_n(x)$ is the corresponding normalized modal electric field vector (transverse components only). The modal field vectors are mutually orthogonal such that

$$\int_{-a/2}^{a/2} \vec{e}_m(x) \cdot \vec{e}_n^*(x)\, dx = \delta_{mn} \tag{4.4}$$

where δ_{mn} is the Kronecker delta. For the TE$_z$ fields, $\vec{e}_n(x)$ is given by

$$\vec{e}_n(x) = \hat{y}\frac{1}{\sqrt{a}} \exp(-jk_{xn}x) \tag{4.5}$$

with

$$k_{xn} = (2n\pi + \psi)/a \tag{4.6}$$

The TM$_z$ field vector (transverse components only) will have a similar expression as in (4.5) except that the $\hat{y}$ vector is replaced by the $\hat{x}$ vector. Using orthogonality the Floquet modal voltage can be obtained as

$$V_n^{FL}(\psi) = V_s \int_{-a/2}^{a/2} \vec{e}_s(x) \cdot \vec{e}_n^*(x)\, dx \tag{4.7}$$

Equation (4.3) essentially expresses the aperture source field in terms of Floquet modal source fields. Our next task is to find the input admittance seen by each Floquet modal source. Then we apply the principle of superposition to obtain the input admittance seen by the aperture source.

4.2.1 Admittance Seen by Floquet Modal Source

Consider the nth Floquet modal source with unity modal voltage. Inside the region $0 < z < b$ (see Figure 4.1), the electromagnetic fields produced by this mode can be expressed in terms of incident and reflected fields as below:

$$\vec{E}(x, z) = \frac{\vec{e}_n(x)[\exp(-jk_{zn}z) + \Gamma_n \exp(jk_{zn}z)]}{1 + \Gamma_n} + \sum_{i \neq n} A_i \vec{e}_i \sin(k_{zi}z) \qquad (4.8)$$

Notice, at $z = 0$, the total field (incident + reflected) is made equal to $\vec{e}_n(x)$. The transverse components of the magnetic field can be expressed as

$$\vec{H}(x, z) = \frac{Y_{0n}\vec{h}_n(x)[\exp(-jk_{zn}z) - \Gamma_n \exp(jk_{zn}z)]}{1 + \Gamma_n} + j \sum_{i \neq n} A_i Y_{0i} \vec{h}_i \cos(k_{zi}z) \quad (4.9)$$

where Y_{0i} is the modal admittance and $\vec{h}_i(x)$ is the modal magnetic field vector, given by

$$\vec{h}_i(x) = \hat{z} \times \vec{e}_i(x) \qquad (4.10)$$

It is important to realize that the electric field at $z = 0$ in (4.8) should not contain other Floquet harmonics than the source harmonic, because the total electric field distribution is *enforced* as $\vec{e}_n(x)$. On the other hand, the magnetic field may have all possible harmonics as represented in (4.9).

The fields in the $z > c$ region are purely progressive and can be expressed as

$$\begin{aligned} \vec{E}(x, z) &= \sum_i B_i \vec{e}_i(x) \exp(-jk_{zi}z) \\ \vec{H}(x, z) &= \sum_i B_i Y_{0i} \vec{h}_i(x) \exp(-jk_{zi}z) \end{aligned} \qquad (4.11)$$

Now consider an auxiliary Floquet modal source vector $\vec{e}_n'(x)$, which is equal to the complex conjugate of $\vec{e}_n(x)$. The fields of the new source in the $0 < z < b$ region are

$$\vec{E}'(x, z) = \frac{\vec{e}_n'(x)[\exp(-jk_{zn}z) + \Gamma_n' \exp(jk_{zn}z)]}{1 + \Gamma_n'} + \sum_{i \neq n} C_i \vec{e}_i' \sin(k_{zi}z) \qquad (4.12a)$$

$$\vec{H}'(x, z) = \frac{Y_{0n}\vec{h}_n'(x)[\exp(-jk_{zn}z) - \Gamma_n' \exp(jk_{zn}z)]}{1 + \Gamma_n'} + j \sum_{i \neq n} C_i Y_{0i} \vec{h}_i' \cos(k_{zi}z)$$

$$(4.12b)$$

Notice that no prime sign is used for the characteristic admittances of the auxiliary source, as they are identical to that of the primary source modes. In the $z > c$ region, the fields are assumed as

$$\begin{aligned} \vec{E}'(x, z) &= \sum_i B_i' \vec{e}_i'(x) \exp(-jk_{zi}z) \\ \vec{H}'(x, z) &= \sum_i B_i' Y_{0i} \vec{h}_i'(x) \exp(-jk_{zi}z) \end{aligned} \qquad (4.13)$$

As mentioned before, the modal electric vector of the new source is given by

$$\vec{e}'_n(x) = \vec{e}^*_n(x) = \hat{y}\frac{1}{\sqrt{a}} \exp(jk_{xn}x) \tag{4.14}$$

The corresponding modal magnetic vector would be

$$\vec{h}'_i(x) = \hat{z} \times \vec{e}'_i(x) \tag{4.15}$$

The modal fields denoted by primes are produced by a source array that has reversed phase distribution as compared to that in (4.1). The corresponding scan angle is $-\theta$. The purpose of considering the auxiliary source is to establish the symmetry property of the admittance seen by a Floquet modal source. Toward that end we use the reciprocity relation given by

$$\oiint_S (\vec{E} \times \vec{H}' - \vec{E}' \times \vec{H}) \cdot ds = \iiint_V (\vec{E}' \cdot \vec{J} - \vec{E} \cdot \vec{J}') \, dv \tag{4.16}$$

where E, H, E', and H' are given in (4.8)–(4.13); J and J' are the induced surface currents on the perfect conducting scatterers. Consider the surface S as the surface of a unit cell with the boundaries at $z = 0, z = d, x = -a/2$ and $x = a/2$ (see Figure 4.1). The right-hand side of (4.16) vanishes because the scatterer is a perfect electric conductor and the tangential electric field must vanish. Furthermore, in the case of a one-dimensional array, the field quantities are independent of y; therefore the closed surface integral becomes a closed contour integral multiplied by a constant factor. The contour integral is decomposed into four line integrals as below:

$$\oint_c \vec{P}(x, z) \cdot \hat{n} \, dl = - \int_{x=-a/2}^{a/2} \vec{P}(x, 0) \cdot \hat{z} \, dx + \int_{x=-a/2}^{a/2} \vec{P}(x, d) \cdot \hat{z} \, dx$$
$$- \int_{z=0}^{d} \vec{P}\left(-\frac{1}{2}a, z\right) \cdot \hat{x} \, dz + \int_{z=0}^{d} \vec{P}\left(\frac{1}{2}a, z\right) \cdot \hat{x} \, dz \tag{4.17}$$

In (4.17) $\hat{n}$ is the unit vector directed outward to the closed contour c. For brevity we use the symbol $P(x, z)$, which is defined as

$$\vec{P}(x, z) = \vec{E} \times \vec{H}' - \vec{E}' \times \vec{H} \tag{4.18}$$

Since $P(x, z)$ is a periodic function of x with periodicity a, $P(-a/2, z)$ and $P(a/2, z)$ are identical for all z. Thus the last two integrals on the right-hand side of (4.17) cancel each other. Now consider the first integral. From (4.4), (4.14), and (4.15) we observe

$$\int_{-a/2}^{a/2} \vec{e}_n(x) \times \vec{h}'_i \cdot \hat{z} \, dx = \delta_{ni} \tag{4.19}$$

and

$$\int_{-a/2}^{a/2} \vec{e}_i'(x) \times \vec{h}_n \cdot \hat{z} \, dx = \delta_{ni} \tag{4.20}$$

Expressing $\vec{P}(x, 0)$ in terms of $\vec{E}, \vec{H}, \vec{E}', \vec{H}'$ and then utilizing (4.19) and (4.20), we deduce

$$\int_{-a/2}^{a/2} \vec{P}(x, 0) \cdot \hat{z} \, dx = \frac{Y_{0n}(1 - \Gamma_n')}{1 + \Gamma_n'} - \frac{Y_{0n}(1 - \Gamma_n)}{1 + \Gamma_n} \tag{4.21}$$

Further, from (4.11) and (4.13) it is straightforward to show that

$$\int_{-a/2}^{a/2} \vec{P}(x, d) \cdot \hat{z} \, dx = 0 \tag{4.22}$$

Therefore (4.16) becomes

$$\frac{Y_{0n}(1 - \Gamma_n')}{1 + \Gamma_n'} - \frac{Y_{0n}(1 - \Gamma_n)}{1 + \Gamma_n} = 0 \tag{4.23}$$

yielding

$$\frac{Y_{0n}(1 - \Gamma_n')}{1 + \Gamma_n'} = \frac{Y_{0n}(1 - \Gamma_n)}{1 + \Gamma_n} \tag{4.24}$$

One must recognize the left-hand side of (4.24) as the input admittance seen by the Floquet modal voltage source $e_n'(x)$, which may be denoted as $Y_n'(\psi)$, and the right-hand side of (4.24) as the input admittance seen by $e_n(x)$, which may be denoted as $Y_n(\psi)$. Thus we write

$$Y_n'(\psi) = Y_n(\psi) \tag{4.25}$$

The Floquet modal vector $e_n'(x)$ can be generated from $e_n(x)$ by replacing n and ψ by $-n$ and $-\psi$, respectively. Therefore $Y_n'(\psi)$ can alternatively be represented as

$$Y_n'(\psi) = Y_{-n}(-\psi) \tag{4.26}$$

From (4.24) and (4.26) we write

$$Y_n(\psi) = Y_{-n}(-\psi) \tag{4.27}$$

Observe that the Floquet modal input admittances are identical if the mode number and the phase slope both reverse their signs. We will utilize this property to establish symmetry in the aperture admittance with respect to the incident angle.

4.2.2 Aperture Admittance

Now consider the source in (4.3), which is expressed in terms of Floquet modal voltages. Using superposition, the magnetic field at $z = 0$ can be expressed as

$$\vec{H}_s = \sum_n V_n^{FL} Y_n \vec{h}_n(x) \tag{4.28}$$

The input admittance seen by the aperture source is

$$Y^{FL} = \frac{1}{V_s^2} \int_{-a/2}^{a/2} \vec{E}_s \times \vec{H}_s \cdot \hat{z} \, dx \tag{4.29}$$

The above input admittance expression is *permissible* only if the aperture field vector $\vec{e}_s(x)$ is a real function of x. Using (4.1) we get

$$Y^{FL} = \frac{1}{V_s} \int_{-a/2}^{a/2} \vec{e}_s \times \sum_n V_n^{FL} Y_n \vec{h}_n \cdot \hat{z} \, dx \tag{4.30}$$

Using the expression for $\vec{h}_n$ from (4.10) we write

$$Y^{FL} = \frac{1}{V_s} \sum_n V_n^{FL} Y_n \int_{-a/2}^{a/2} \vec{e}_s \cdot \vec{e}_n \, dx \tag{4.31}$$

Recalling (4.7), the integral in (4.31) becomes

$$\int_{-a/2}^{a/2} \vec{e}_s \cdot \vec{e}_n \, dx = \frac{V_{-n}^{FL}(-\psi)}{V_s} \tag{4.32}$$

Thus, the input admittance becomes

$$Y^{FL}(\psi) = \frac{1}{V_s^2} \sum_n V_n^{FL}(\psi) Y_n(\psi) V_{-n}^{FL}(-\psi) \tag{4.33}$$

In the summation, n varies from $-\infty$ to ∞. Using (4.27) it is straightforward to show that

$$Y^{FL}(\psi) = Y^{FL}(-\psi) \tag{4.34}$$

In terms of the scan angle, we will have

$$Y^{FL}(\theta) = Y^{FL}(-\theta) \qquad Z^{FL}(\theta) = Z^{FL}(-\theta) \tag{4.35a}$$

Equation (4.35a) indicates that the Floquet admittance and impedance are symmetrical with respect to the incident angle.

For two-dimensional arrays, if ψ_x and ψ_y are the phase differences between the adjacent elements along the x- and y-directions, respectively, then the following symmetry property holds:

$$Y^{\mathrm{FL}}(\psi_x, \psi_y) = Y^{\mathrm{FL}}(-\psi_x, -\psi_y) \tag{4.35b}$$

In a spherical coordinate system, the above relation is equivalent to

$$Y^{\mathrm{FL}}(\theta, \phi) = Y^{\mathrm{FL}}(\theta, \phi + \pi) \tag{4.35c}$$

A similar symmetry property holds for input impedance.

It is worth noting that the symmetry theorem is proven for perfect electric conducting scatteres. The same symmetry property can be established for dielectric or dielectric-coated conducting scatterers as long as the dielectric material is isotropic (see problem 4.1).

4.3 MUTUAL COUPLING

A finite array can be analyzed rigorously if the mutual coupling between the elements in the array environment is known. The mutual coupling between the elements is generally quantified in terms of the following three measurable quantities:

(a) mutual impedance,
(b) mutual admittance, and
(c) scattering parameters.

The above three measurable quantities are related to each other by simple algebraic relations. In this section we will first derive the mutual impedance from Floquet impedance of an infinite array [2]. For simplicity we will first consider a one-dimensional array. The more general array structure will be considered later.

4.3.1 Mutual Impedance

Consider the infinite array shown in Figure 4.1. The elements are arranged along the x-direction with element spacing a. Suppose the elements are excited uniformly with linearly progressed phase (Floquet excitation). Suppose ψ is the phase difference between two adjacent elements. Then following the definition of mutual impedance, the input voltage for the zeroth element can be obtained as

$$V_0(\psi) = \sum_{n=-\infty}^{\infty} I_n Z_{0n} \tag{4.36}$$

In the above V_0 is the input voltage for the element located at $x = 0$; I_n is the input current of the nth element and Z_{0n} is the mutual impedance between the two elements that are located at $x = 0$ and at $x = na$, respectively. For Floquet excitations, the input currents can be expressed as

$$I_n = I_0 \exp(-jn\psi) \tag{4.37}$$

where I_0 is the input current for the element at $x = 0$. The input impedance seen by the $n = 0$ element is

$$Z_0(\psi) = \frac{V_0(\psi)}{I_0} \tag{4.38}$$

Substituting (4.36) and (4.37) into (4.38) we obtain

$$Z_0(\psi) = \sum_{n=-\infty}^{\infty} Z_{0n} \exp(-jn\psi) \tag{4.39}$$

For Floquet excitation, the above input impedance must be equal to the Floquet impedance, $Z^{\text{FL}}(\psi)$. Therefore we write

$$Z^{\text{FL}}(\psi) = \sum_{n=-\infty}^{\infty} Z_{0n} \exp(-jn\psi) \tag{4.40}$$

The right-hand side of (4.40) is the Fourier series expansion of the Floquet impedance where the Fourier coefficients are equal to the mutual impedances. Thus, the mutual impedance Z_{0n} is readily obtained in terms of the Fourier integral as follows:

$$Z_{0n} = \frac{1}{2\pi} \int_{-\pi}^{\pi} Z^{\text{FL}}(\psi) \exp(jn\psi)\, d\psi \tag{4.41}$$

If the two elements are located at $x = ma$ and $x = na$, respectively, then the mutual impedance between these two elements can be expressed as

$$Z_{mn} = \frac{1}{2\pi} \int_{-\pi}^{\pi} Z^{\text{FL}}(\psi) \exp\{-j(m-n)\psi\}\, d\psi \tag{4.42a}$$

Equation (4.42a) establishes the relation between the Floquet impedance and the mutual impedance between the elements. It is important to observe that (4.42a) yields the mutual impedance in the array environment. Also observe that Z_{mn} and Z_{nm} are identical because $Z^{\text{FL}}(\psi) = Z^{\text{FL}}(-\psi)$. The symmetry property of $Z^{\text{FL}}(\psi)$ can be utilized to express Z_{mn} in a convenient form from a computational point of view as follows:

$$Z_{mn} = \frac{1}{\pi} \int_{0}^{\pi} Z^{\text{FL}}(\psi) \cos\{(m-n)\psi\}\, d\psi \tag{4.42b}$$

The mutual impedance deduced in (4.42b) includes the effects of scattering from the intermediate and surrounding elements that are open circuited. The element-by-element approach [3] typically ignores the scattering effects; therefore the present formulation is generally more accurate than the element-by-element approach.

It is worth pointing out that for some arrays the Floquet impedance Z^{FL} may have a finite number of singularities due to resonances of selective Floquet modes with the guided wave modes supported by the array structures. A typical example of this kind is a patch array on a dielectric substrate that supports surface wave modes. Under such a situation, a singularity extraction technique [4] must be employed to compute the integral near a singular point.

4.3.2 Mutual Admittance

Following a similar procedure, the mutual admittance between the two elements in an array environment can be obtained as

$$Y_{mn} = \frac{1}{\pi} \int_0^\pi Y^{FL}(\psi) \cos\{(n-m)\psi\} \, d\psi \qquad (4.43)$$

where $Y^{FL}(\psi)$ is the Floquet admittance, reciprocal to the Floquet impedance $Z^{FL}(\psi)$; Y_{mn} is the mutual admittance between the mth and nth elements. The distance between the two elements is $|(m-n)a|$. The relation in (4.43) can also be proven from (4.42a) directly. Also note that $Y_{mn} = Y_{nm}$.

4.3.3 Scattering Matrix Elements

The scattering parameters between the elements also follow the similar relation as the mutual admittance/impedance. If S_{mn} represents the scattering parameter defined as the voltage received by the mth element when the nth element is excited with all other elements including the mth element match terminated, then

$$S_{mn} = \frac{1}{\pi} \int_0^\pi \Gamma^{FL}(\psi) \cos\{(m-n)\psi\} \, d\psi \qquad (4.44)$$

where $\Gamma^{FL}(\psi)$ is the reflection coefficient of an array element under Floquet excitation. Since $\Gamma^{FL}(\psi) = \Gamma^{FL}(-\psi)$, which follows from (4.34), one can see that $S_{mn} = S_{nm}$.

4.4   ARRAY OF MULTIMODAL SOURCES

For many applications the array element may be excited with more than one mode at the input level. For instance, if a square horn array has to radiate a circularly polarized wave, the input waveguide must contain the TE_{10} and TE_{01} modes simultaneously. For symmetrical horns, these two modes may couple with each other through the aperture. Therefore, the Floquet input impedance essentially becomes an impedance matrix (in this case 2×2 matrix because of only two input modes). To accommodate such multimodal input into our formulation (4.42b) should be expressed in a more general form as

$$[Z_{mn}] = \frac{1}{\pi} \int\limits_0^\pi [Z^{\text{FL}}(\psi)] \cos\{(m-n)\psi\} \, d\psi \qquad (4.45)$$

In (4.45) $[Z_{mn}]$ represents the mutual impedance matrix[2] between the mth and nth elements of the array. The diagonal elements represent the mutual impedance between the same type of modes (say between TE_{10} modes of the waveguides), and the off-diagonal elements represent the mutual impedance between different types of modes (say the TE_{10} and TE_{01} modes). A similar relation holds for mutual admittance and scattering parameters.

The relation in (4.45) can be utilized to obtain the impedance matrix with respect to any reference plane of the array element. If $[Z^{\text{FL}}(\psi)]$ represents the Floquet impedance matrix with reference to the aperture plane, then $[Z_{mn}]$ would represent the mutual impedance matrix between the mth and nth apertures. On the other hand, if $[Z^{\text{FL}}(\psi)]$ represents the Floquet impedance matrix at the input location, then $[Z_{mn}]$ would be equal to the mutual impedance matrix transformed to the input level. It is often convenient to deal with the input level impedance than the aperture level impedance. To illustrate this point, consider an array of rectangular horns where the input waveguide supports only the TE_{10} mode and the other modes are much below cut-off. The input waveguide is flared to a large aperture that supports several waveguide modes. The aperture level impedance matrix would be much larger in size than that of the input level impedance matrix because the former contains explicit mutual coupling data between multiple modes on the apertures. Thus, it is advantageous computationally to deal with the input level impedance matrix than the aperture level impedance matrix to incorporate the mutual coupling effects in a finite array.

4.5 MUTUAL COUPLING IN TWO-DIMENSIONAL ARRAYS

The mutual coupling formulation can be extended for a two-dimensional planar array. First we will present the relation between Floquet impedance and mutual

---

[2] The "mutual impedance matrix" should not be confused with the impedance matrix of a finite array. A mutual impedance matrix essentially describes coupling between two particular elements of the array. The impedance matrix consists of $N \times N$ mutual impedance matrices, where N represents the number of elements in the array.

impedance for a rectangular lattice, which is a simple two-dimensional extension of (4.42a). Then we will present the relation for an arbitrary lattice.

4.5.1 Rectangular Lattice

For a rectangular lattice, the mutual impedance between two elements can be expressed as

$$Z_{mn} = \frac{1}{4\pi^2} \int_{-\pi}^{\pi} \int_{-\pi}^{\pi} Z^{FL}(\psi_x, \psi_y) \exp\left[j\left(\frac{d_x \psi_x}{a} + \frac{d_y \psi_y}{b}\right)\right] d\psi_x \, d\psi_y \qquad (4.46a)$$

where a and b are the element spacings along the x- and y-directions; ψ_x, ψ_y are the phase differences between the adjacent elements along the x- and y-directions; and (d_x, d_y) is the relative coordinate of the mth element with respect to the nth element, as shown in Figure 4.2. The quantities d_x and d_y are integral multiples of a and b, respectively. If the symmetry property of the Floquet impedance is exploited, then the integral of (4.46a) reduces to

$$Z_{mn} = \frac{1}{\pi^2} \int_0^{\pi} \int_0^{\pi} Z^{FL}(\psi_x, \psi_y) \cos(p\psi_x) \cos(q\psi_y) \, d\psi_x \, d\psi_y \qquad (4.46b)$$

where

$$p = \frac{d_x}{a} \qquad q = \frac{d_y}{b}$$

The expressions for mutual admittance and scattering parameter will have similar relations.

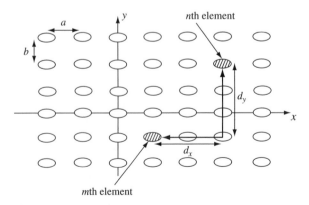

FIGURE 4.2 Two-dimensional array of rectangular lattice.

4.5.2 Arbitrary Lattice

For an arbitrary lattice represented by $[a, b, \gamma]$, the mutual impedance between the mth and nth elements is given by

$$Z_{mn} = \frac{1}{4\pi^2} \int\limits_{-\pi}^{\pi} \int\limits_{-\pi}^{\pi} Z^{FL}(\psi_x, \psi_y') \exp[j(p\psi_x + q\psi_y')] \, d\psi_x \, d\psi_y' \qquad (4.47)$$

In the above equation $\{p, q\}$ is a pair of integers that symbolically represents the relative cell location between two elements as shown in Figure 4.3. The quantity ψ_x represents the phase difference between two adjacent elements along x. The quantity ψ_y' represents the phase difference between the adjacent elements along the oblique direction that is inclined at an angle γ with respect to the x-axis. Notice that, to maintain uniformity of symbols, we use a prime to distinguish from ψ_y. The expression of ψ_y' is given by

$$\psi_y' = \psi_y + \psi_x \frac{b}{a \tan \gamma} \qquad (4.48)$$

Recall from Chapter 3 that in terms of scan location, ψ_x and ψ_y are given by

$$\psi_x = k_0 a \sin\theta \cos\phi \qquad \psi_y = k_0 b \sin\theta \sin\phi \qquad (4.49)$$

The relation in (4.47) is the direct consequence of the Fourier expansion of the Floquet impedance, as derived in (4.41) for a linear array. It should be pointed out that Z^{FL} is not symmetric with respect to ψ_x and ψ_y'. Therefore for a nonrectangular lattice, Z_{mn} cannot be simplified as in (4.46b). This is also supported by the fact that the mutual impedance between the elements with relative cell location $\{p, q\}$ should not be identical to that with relative cell locations $\{p, -q\}$ and $\{-p, q\}$ because of different distances between the coupling cells. The mutual impedance would be identical to that with relative cell location $\{-p, -q\}$, however.

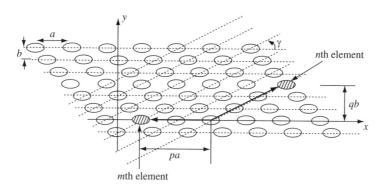

FIGURE 4.3 Two-dimensional array of arbitrary lattice $[a, b, \gamma]$.

The mutual admittance and the scattering matrix elements in an infinite array environment are

$$Y_{mn} = \frac{1}{4\pi^2} \int\limits_{-\pi}^{\pi} \int\limits_{-\pi}^{\pi} Y^{FL}(\psi_x, \psi_y') \exp[j(p\psi_x + q\psi_y')] \, d\psi_x \, d\psi_y' \qquad (4.50)$$

$$S_{mn} = \frac{1}{4\pi^2} \int\limits_{-\pi}^{\pi} \int\limits_{-\pi}^{\pi} S^{FL}(\psi_x, \psi_y') \exp[j(p\psi_x + q\psi_y')] \, d\psi_x \, d\psi_y' \qquad (4.51)$$

The mutual coupling (S_{mn}) data between waveguide elements are shown in Figures 4.4 and 4.5 for two different arrays. The waveguide dimensions for the two arrays were $0.6\lambda_0 \times 0.6\lambda_0$ and $1.2\lambda_0 \times 1.2\lambda_0$, respectively. Expectedly, the

−34.1	−41.3	−36.7	−41.3	−34.1
−32.8	−37.6	−26.0	−37.6	−32.8
−24.9	−21.0	−18.1	−21.0	−24.9
−32.8	−37.6	−26.0	−37.6	−32.8
−34.1	−41.3	−36.7	−41.3	−34.1

FIGURE 4.4 Mutual coupling (S_{ij}) between the center element and other elements of an open-ended waveguide array in infinite array environment. The numbers indicate mutual coupling in decibels of the respective element with the center (shaded) element. Waveguide aperture size $0.6\lambda_0 \times 0.6\lambda_0$, cell size $0.7\lambda_0 \times 0.7\lambda_0$. E-plane along the horizontal direction.

−65.1	−63.6	−62.4	−63.6	−65.1
−58.4	−56.2	−63.0	−56.2	−58.4
−44.0	−32.0	−26.9	−32.0	−44.0
−58.4	−56.2	−63.0	−56.2	−58.4
−65.1	−63.6	−62.4	−63.6	−65.1

FIGURE 4.5 Mutual coupling (S_{ij}) between the center element and other elements of an open-ended waveguide array in infinite array environment. The numbers indicate mutual coupling in decibels of the respective element with the center (shaded) element. Waveguide aperture size $1.2\lambda_0 \times 1.2\lambda_0$, cell size $1.4\lambda_0 \times 1.4\lambda_0$. E-plane along horizontal direction.

coupling reduces as the element spacing increases. The coupling between the E-plane elements is stronger than that between the H-plane elements.

4.6   ACTIVE INPUT IMPEDANCE OF FINITE ARRAY

The mutual coupling information between the elements is utilized to determine the active impedance or return loss of an element with respect to given amplitude and phase distributions. The active impedance of an element of a finite array changes with the amplitude and phase distributions as well as the input conditions of the nonexcited elements. To illustrate we consider three different input conditions in the following sections.

4.6.1   Nonexcited Elements Open Circuited

Suppose N elements are excited and the nonexcited elements are open circuited, as shown in Figure 4.6. Therefore, the magnitude of the input current is zero for a nonexcited element. It follows that the elements of the impedance matrix ($[Z]$ matrix) of the finite array should be identical to the mutual impedances in the infinite array environment. The mutual impedance in this case can be computed employing (4.47) for a general lattice structure. However, if the nonexcited elements have other load terminations (e.g., match terminations), then (4.47) cannot be employed for computing the $[Z]$ matrix elements. Therefore, the impedance matrix formulation is the simplest way to handle a finite array if the nonexcited elements are open circuited.

Suppose $[I]$ represents the current excitation vector of the array of N excited elements. The input voltage vector is then obtained from

$$[V] = [Z][I] \tag{4.52}$$

Expanding (4.52) the active impedance of the ith element is obtained as

$$Z_i^a = \frac{V_i}{I_i} = \sum_{j=1}^{N} Z_{ij} \frac{I_j}{I_i} \tag{4.53}$$

where V_i and I_i are the ith elements of the voltage and current vectors, respectively, and Z_{ij} represents the mutual impedance between the ith and jth elements of the array.

FIGURE 4.6   Array with few excited elements with remaining elements open circuited.

4.6.2 Nonexcited Elements Short Circuited

If the nonexcited elements are short circuited (as opposed to open circuited, as shown in Figure 4.6), then (4.53) will not yield the correct solution for active impedances of the elements, because the short-circuited elements will have induced currents at the input ports. In this situation the admittance matrix formulation would simplify the finite array analysis. The admittance matrix elements would be the same as the mutual admittance in an infinite array environment, given in (4.50). The relation between the input voltage vector and input current vector is

$$[I] = [Y][V] \tag{4.54}$$

where $[Y]$ is the $N \times N$ admittance matrix. The above equation can be rearranged to yield

$$[V] = [Y]^{-1}[I] = [Z^{\text{sh}}][I] \tag{4.55}$$

where $[Y]^{-1}$ is denoted as $[Z^{\text{sh}}]$. The superscript "sh" indicates that the nonexcited elements are short circuited. It should be mentioned that for a finite array $[Z]$ and $[Z^{\text{sh}}]$ are not identical. Equation (4.55) should be used for active impedance computations for this type of finite array of N excited elements with other elements terminated by shorts.

4.6.3 Nonexcited Elements Match Terminated

If the nonexcited elements are match terminated, then the scattering matrix relation will yield the exact active input impedance solution. The relation in this situation is

$$[V^-] = [S][V^+] \tag{4.56}$$

where V^+ and V^- are the incident and reflected voltage vectors. Elements of $[S]$ are obtained using (4.51). Equation (4.56) can be used to find the impedance matrix of the array. The impedance matrix $[Z^{\text{mat}}]$ is given by

$$[Z^{\text{mat}}] = Z_0[I + S][I - S]^{-1} \tag{4.57}$$

where Z_0 is the internal impedance of the sources. The superscript "mat" stands for match termination of the nonexcited elements. The active input impedance can be found from the impedance matrix if the impressed current or voltage is known.

We presented methods to determine the impedance matrices of a finite array when the nonexcited elements have three different load conditions, namely open circuit, short circuit, and match termination. The question arises: Which would be most appropriate model in the case of a real finite array that does not have any nonexcited element as such. Precisely speaking, none of the models would be exact for a real finite array. However, depending on the practical situation, one model may be more appropriate than others. To illustrate this point, consider a real finite patch array that is printed on a large ground plane. For this case, short-circuited nonexcited elements would closely simulate the ground plane. On the other hand, in the case of a real finite horn array that has no extra ground plane, match-terminated elements surrounding the array perhaps would be a good approximation, because not much scattering is expected from the outside region. Furthermore, in the case of a large array, the three models yield similar results for most of the elements except the edge elements. In the following section we present numerical results for an array of open-ended waveguides.

4.7   ACTIVE RETURN LOSS OF OPEN-ENDED WAVEGUIDE ARRAY

We compute the active return loss of a finite array of open-ended waveguides applying Floquet modal theory of mutual coupling. The geometry of the array and the numbering scheme for the excited elements are shown in Figure 4.7a. The array consists of 9×9 excited elements situated in an infinite array environment. We compute the active return loss applying the theory developed in the previous sections for three distinct input termination types for the nonexcited elements, namely (a) short, (b) open, and (c) match terminations. In principle it is not possible to derive the active return loss data for a real finite array of 9×9 elements from the results of the above three distinct finite arrays. However, we can conjecture on the bounds of the active return loss data for the elements of the real array, looking at the three sets of finite array data, as we explain shortly.

The element aperture size was $0.6\lambda_0 \times 0.6\lambda_0$, and the element spacing was $0.7\lambda_0$ in both planes. The small element size was chosen deliberately to examine the effects of strong mutual coupling on three different finite arrays as described before. Square grids were used for this study. The reference plane for active impedance computation was considered at the aperture plane. The impedance matrix with respect to each model is converted to the 81×81 scattering matrix. The active reflection coefficients are then obtained by multiplying the scattering matrix with the excitation vector. For uniform incident voltage distribution the active reflection coefficients of an array element turns out to be the sum of the corresponding row elements of the scattering matrix.

Figure 4.7b shows the numerical data for the active return loss of the three finite arrays. The three arrays differ from each other only in terminating loads of the nonexcited elements. For each array we observe that the active return loss varies from about -11 dB to about -19 dB between the elements, signifying the effects of mutual coupling between the elements. More importantly, the three models yield very similar results for most of the array elements. The difference in numerical results is found for the edge elements, particularly near the E-plane edges, due to strong coupling effects on the E-plane. The three sets of results lie within 1.5 dB at about -17 dB return loss level. Because the three load types drastically differ from each other in terms of reflected field amplitude and phase, one speculates that the results of the real finite array (with no nonexcited elements) lie within the bounds of the three sets of results. The array structure under consideration represents a

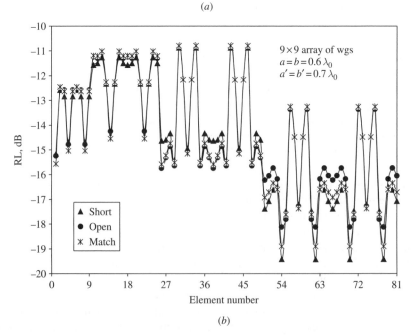

FIGURE 4.7 (*a*) Aperture of a finite array of open-ended waveguides and the numbering scheme to identify the 9 × 9 excited elements. Only few nonexcited elements are shown. (*b*) Active return loss (RL) with uniform amplitude and phase distributions of the excited elements with three types of terminating loads for the nonexcited elements (●, open; ▲, short; ✖, match). Waveguide aperture size $0.6\lambda_0 \times 0.6\lambda_0$, cell size $0.7\lambda_0 \times 0.7\lambda_0$.

somewhat extreme example where the mutual coupling is a dominating factor in deciding the active return loss of the elements. For larger element size, the mutual coupling is much weaker (see Figure 4.5); thus all three models would show very similar results.

4.8 RADIATION PATTERNS OF FINITE ARRAY

In this section we derive the radiation patterns of finite arrays with three types of load terminations for the nonexcited elements. The numerical results with respect to the three different load conditions will allow estimating the uncertainty in the computed radiation pattern of a real finite array.

4.8.1 Nonexcited Elements Open Circuited

The radiation pattern of the finite array in Figure 4.6 can be determined from the active element pattern of the associated infinite array. The active element pattern is defined as the radiation pattern with one element excited and the rest matched terminated (see Section 3.5). Suppose V_n^+ is the incident voltage of the nth element. Then, the electromagnetic far field radiated by the element is given by

$$\vec{E}(\theta, \phi) = V_n^+ E^a(\theta, \phi) \tag{4.58}$$

where E^a is the active element pattern with respect to the incident voltage. For the present situation, it is convenient to find the radiation pattern of an infinite array when only one element is excited by an ideal current source and the rest of the elements are open circuited. To simplify the presentation, we assume a linear array. We assume that the excited element has an input current I_0. If we find the radiation pattern of this arrangement, then it is a matter of superposition to find the radiation pattern of the finite array under consideration.

Suppose the excited element is denoted by the index $k = 0$. The induced voltages for the elements of the entire infinite array can be determined from the mutual impedance between the elements as

$$V_k = Z_{k0} I_0 \qquad -\infty < k < \infty \tag{4.59}$$

where V_k is the total induced voltage. To be consistent with the definition of the active element pattern, the forward-moving part V_k^+ of the port voltage is of interest for computing the array pattern. Now,

$$
\begin{aligned}
V_k^+ + V_k^- &= V_k \\
V_k^+ - V_k^- &= Z_0 I_k = 0 \qquad k \neq 0
\end{aligned}
\tag{4.60}
$$

In the above equations, the plus and minus signs represent incident and reflected voltages and Z_0 is the characteristic impedance of the input transmission line or the source internal impedance (Thevenin's impedance). The second equation of (4.60) represents the open-circuit condition of the kth element. The above two equations yield

$$V_k^+ = \frac{1}{2}V_k = \frac{1}{2}Z_{k0}I_0 \qquad k \neq 0. \tag{4.61}$$

For the zeroth element (excited element) the following relations hold:

$$V_0^+ + V_0^- = Z_{00}I_0 \qquad V_0^+ - V_0^- = Z_0 I_0 \tag{4.62}$$

In the above, Z_{00} is the self-impedance. From the above we obtain

$$V_0^+ = \frac{1}{2}(Z_0 + Z_{00})I_0 \tag{4.63}$$

The total radiated far field is given by

$$E_0^{\mathrm{cur}} = E^a \sum_k V_k^+ \exp(jkak_0 \sin\theta) \tag{4.64}$$

The superscript "cur" represents radiation pattern with respect to the input current. Substituting expressions for V_k^+ from (4.61) and (4.63) we obtain

$$E_0^{\mathrm{cur}} = \left\{ \frac{1}{2}Z_0 I_0 + \frac{1}{2}I_0 \sum_k Z_{k0} \exp(jkak_0 \sin\theta) \right\} E^a \tag{4.65}$$

Substituting the integral expression for Z_{k0} from (4.42a) we obtain

$$E_0^{\mathrm{cur}} = \left\{ \frac{1}{2}Z_0 I_0 + \frac{1}{2}I_0 \sum_k \frac{1}{2\pi} \int_{-\pi}^{\pi} Z^{\mathrm{FL}}(\psi) \exp(jk\psi)d\psi \exp(jkak_0 \sin\theta) \right\} E^a \tag{4.66}$$

Interchanging the summation and integration sign, we write

$$E_0^{\mathrm{cur}} = I_0 E^a \left\{ \frac{1}{2}Z_0 + \frac{1}{4\pi} \int_{-\pi}^{\pi} Z^{\mathrm{FL}}(\psi) \sum_{k=-\infty}^{\infty} \exp[jk(\psi + \psi')]d\psi \right\} \tag{4.67}$$

In (4.67) $\psi' = k_0 a \sin\theta$. The double infinite summation becomes a series of Delta functions because of the following identity:

$$\sum_{k=-\infty}^{\infty} \exp(jkx) = 2\pi \sum_{n=-\infty}^{\infty} \delta(x - 2n\pi) \tag{4.68}$$

Using the above identity we deduce

$$E_0^{\text{cur}} = I_0 E^a \left\{ \frac{1}{2} Z_0 + \frac{1}{2} \int_{-\pi}^{\pi} Z^{\text{FL}}(\psi) \sum_{k=-\infty}^{\infty} \delta(\psi + \psi' - 2k\pi) d\psi \right\}$$

$$= \frac{1}{2} I_0 E^a \{ Z_0 + Z^{\text{FL}}(2i\pi - \psi') \} \qquad |2i\pi - \psi'| < \pi \qquad (4.69)$$

Now, $Z^{\text{FL}}(\psi')$ is a periodic function of ψ' with period 2π, because the Floquet impedance remains unaffected when the element phase is increased or decreased by an integral multiple of 2π. Furthermore, $Z^{\text{FL}}(\psi')$ is a symmetrical function of ψ'. This yields

$$E_0^{\text{cur}} = \frac{1}{2} I_0 E^a [Z_0 + Z^{\text{FL}}(\psi')] \qquad (4.70)$$

Notice, the effective radiation pattern of an element in terms of its input current (with other elements open circuited) differs from that of the active element pattern.

It should be pointed out that (4.70) can also be derived following the procedure in Section 3.5. Under Floquet excitation, the incident voltage V_k^+ is equal to $I_k[Z_0 + Z^{\text{FL}}(\psi')]/2$, where I_k represents the input current. The incident voltage V_k^+ and the input current I_k essentially represent the same source in regard to Floquet modal amplitudes. It follows that a factor $[Z_0 + Z^{\text{FL}}(\psi')]/2$ must be multiplied with the active element pattern in order to represent the element pattern in terms of the input current.

The radiation pattern of the array in Figure 4.6 can be obtained using superposition. The radiation pattern becomes

$$E^{\text{array}} = \frac{E^a}{2} [Z_0 + Z^{\text{FL}}(\psi')] \sum_{n=1}^{N} I_n \exp\{j(n-1)\psi'\} \qquad (4.71)$$

The summation term in (4.71) represents the array factor. We need to express the array pattern in terms of the incident voltage because the incident voltage generally represents the intended excitation. The relation between the input current and the incident voltage can be found through the impedance matrix. As explained in Section 4.6.1, the impedance matrix for the structure of Figure 4.6 consists of the mutual impedances in an infinite array environment. The relation between the input current and the incident voltage is

$$[I] = 2\{[Z] + [Z_0]\}^{-1} [V^+] \qquad (4.72)$$

where, $[I]$ is the input current vector, $[Z]$ is the impedance matrix, and $[Z_0]$ is the characteristic impedance matrix (diagonal) or the internal impedance matrix of the source and $[V^+]$ is the incident voltage vector. Substituting (4.72) in (4.71) we obtain

$$E^{\text{array}} = E^a [Z_0 + Z^{\text{FL}}(\psi')][P]\{[Z] + [Z_0]\}^{-1} [V^+] \qquad (4.73)$$

where $[P]$ represents a row vector comprising exponential terms that appear in the array factor of (4.71). For a one-dimensional array

$$[P] = [1, \exp(j\psi'), \exp(2j\psi'), \ldots, \exp\{(N-1)j\psi'\}] \qquad (4.74)$$

It is straightforward to generalize (4.73) for a two-dimensional array. In that case, Z^{FL} and the elements of $[P]$ would be functions of two variables, ψ_x and ψ_y. For the array gain computation directly from (4.73), the incident voltage vector $[V^+]$ should be normalized with respect to the total incident power, and E_a should represent the active element gain pattern.

4.8.2 Nonexcited Elements Short Circuited

If the nonexcited elements are short circuited, then the array pattern of the finite array becomes

$$E^{\text{array}} = E^a[Y_0 + Y^{\text{FL}}(\psi')][P]\{[Y] + [Y_0]\}^{-1}[V^+] \qquad (4.75)$$

where Y^{FL} is the Floquet admittance, Y_0 the source admittance, $[Y]$ the array admittance matrix, and the remaining terms are as defined earlier.

4.8.3 Nonexcited Elements Match Terminated

The radiation pattern of the finite array for this particular situation can be obtained directly from superposition. The result becomes

$$E^{\text{array}} = E^a[P][V^+] \qquad (4.76)$$

In the following section we will compare the numerical results of a finite array.

4.9 RADIATION PATTERNS OF OPEN-ENDED WAVEGUIDE ARRAY

The radiation patterns of an open-ended waveguide array of 81 elements considered in Section 4.7 are computed using three different types of terminating loads at the inputs of the nonexcited elements. Equations (4.73), (4.75), and (4.76) are employed for open, short, and match terminations, respectively. The E-plane patterns are plotted in Figure 4.8a and the H-plane patterns are plotted in Figure 4.8b. A uniform distribution for the incident voltage was considered for this numerical exercise. The peak gains were 26.64, 26.76, and 26.86 dBi for the open-circuit, match, and short-circuit conditions, respectively. For the E-plane case, the patterns differ near the far-out side-lobe region while matching closely in the near-in side-lobe and main-lobe regions. The uncharacteristic dips near 25° are due to a blind spot. The H-plane patterns are almost indistinguishable from each other, because the edge elements have minimal effects due to weak mutual coupling. In general, the patterns

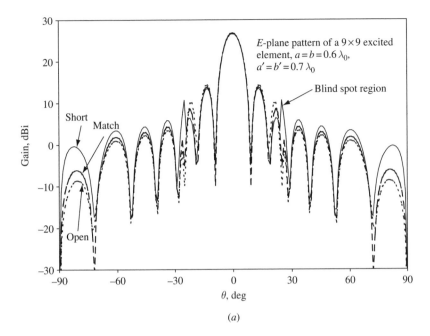

(a)

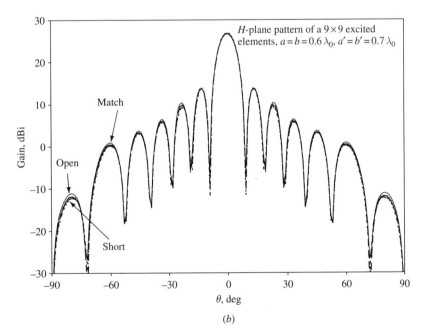

(b)

FIGURE 4.8 Radiation patterns of the array described in Figure 4.7b for three types of terminating loads for the nonexcited elements: (a) E-plane patterns; (b) H-plane patterns.

are less sensitive to the type of terminating loads on the nonexcited elements, particularly near the main-lobe and near-in side-lobe regions. These results lead us to believe that the radiation pattern of the real finite array (without nonexcited elements) would reside within the bounds of the above three patterns.

4.10   ARRAY WITH NONUNIFORM SPACING

In principle, an array of open-ended waveguides or horn apertures with nonuniform spacing opened onto an infinite ground plane can be analyzed rigorously using Floquet modal approach. The mutual admittance between two waveguide apertures with respect to various waveguide modes can be determined using the mutual admittance formulation derived in (4.50). The element spacing for computing the Floquet aperture admittances Y^{FL} (or aperture admittance matrix for large apertures) under the integral sign of (4.50) should correspond to the relative location of the apertures under consideration. Because of nonuniform element spacing, the mutual admittance between the elements should be computed one pair at a time (the coupling between the uniformly spaced elements can be obtained in one sort, however). Once the admittance matrix is filled in, the array can be characterized accurately. However, this procedure may not be very practical because of extensive numerical computations involved in the analysis.

4.11   FINITE ARRAY ANALYSIS USING CONVOLUTION

Ishimaru et al. introduced an alternative formulation for obtaining the results of a finite array[3] utilizing infinite array results [5]. It involves computation of a convolution integral involving the "finite array excitation function" and the "infinite array response." The underlying principle has been successfully applied [6–13] for analyzing finite arrays. Mathematically, the convolution method and the method that has been articulated in this chapter (and also in Chapter 3) are equivalent. We will present the mathematical foundation of the convolution method for the aperture field computation of a finite array. Also, we will deduce the mutual impedance between array elements to show the concurrence of the two formulations.

4.11.1   Convolution Relation for Aperture Field

Suppose $\tilde{T}_{\infty}(\psi)\exp(-jx\psi/a)$ is the transmitted field intensity emanating from an infinite array under Floquet excitation on the aperture plane situated at $z = 0$. We assume a one-dimensional array with element spacing a where $\psi = k_0 a \sin\theta$, θ

---

[3] The finite array must have the identical definition as defined in the introduction.

being the scan angle. Suppose the corresponding Floquet excitation coefficients in terms of the input current is given by

$$I_\infty(x, \psi) = \sum_{m=-\infty}^{\infty} \delta(x - ma) \exp(-jm\psi) \qquad (4.77)$$

We will show that a finite array excitation can be expressed as a continuum of Floquet excitations. Suppose the finite excitation is given by

$$I_N(x) = \sum_{n=1}^{N} A_n \delta(x - na) \qquad (4.78)$$

where A_n is the complex excitation coefficient of the nth element of the finite array of N excited elements. For m and n integers we can write

$$\delta(x - na) = \frac{1}{2\pi} \sum_{m=-\infty}^{\infty} \delta(x - ma) \int_{-\pi}^{\pi} \exp(jn\psi) \exp(-jm\psi) d\psi \qquad (4.79)$$

The validity of the above identity relies on the fact that the integral inside the summation sign is 2π only when $m = n$ and is zero when $m \neq n$. Interchanging the integration and summation signs, we have

$$\delta(x - na) = \frac{1}{2\pi} \int_{-\pi}^{\pi} \exp(jn\psi) \sum_{m=-\infty}^{\infty} \delta(x - ma) \exp(-jm\psi) \, d\psi$$

$$= \frac{1}{2\pi} \int_{-\pi}^{\pi} \exp(jn\psi) I_\infty(x, \psi) d\psi \qquad (4.80)$$

Thus a delta function is represented as a continuum of Floquet excitations. Utilizing the above expression in (4.78), we obtain

$$I_N(x) = \frac{1}{2\pi} \int_{-\pi}^{\pi} \left[\sum_{n=1}^{N} A_n \exp(jn\psi) \right] I_\infty(x, \psi) \, d\psi \qquad (4.81)$$

Since $\tilde{T}_\infty(\psi) \exp(-jx\psi/a)$ is the response of the Floquet excitation $I_\infty(x, \psi)$, applying superposition we can write the response of $I_N(x)$ as

$$T_N(x) = \int_{-\pi}^{\pi} \left[\frac{1}{2\pi} \sum_{n=1}^{N} A_n \exp(jn\psi) \right] \tilde{T}_\infty(\psi) \exp\left(-\frac{jx\psi}{a} \right) d\psi \qquad (4.82)$$

The term inside the square bracket is the FT of the finite array source $I_N(x)$. We now define a space domain function $T_\infty(x)$ as the inverse Fourier transform of $\tilde{T}_\infty(\psi)$. Taking the FT of (4.82) or directly from (4.82) we observe that the FT of

$T_N(x)$ is equal to the product of the FTs of $I_N(x)$ and $\tilde{T}_\infty(\psi)$. From the convolution theorem [14] it follows that in the space domain, $T_N(x)$ can be expressed in terms of the following convolution integral:

$$T_N(x) = \frac{1}{2\pi} \int_{-\infty}^{\infty} I_N(x') T_\infty(x - x') \, dx' \qquad (4.83)$$

Equation (4.83) implies that the aperture field response of a finite array source is obtained by convolving the finite source distribution and the infinite array response.

4.11.2 Mutual Impedance

We will now demonstrate that the mutual impedance between two elements deduced from the convolution relation has an identical expression with that in (4.42a). Toward that end, we define $T_N(x)$ as the electric field at the source input (instead of the aperture plane), which is proportional to the input voltage. Ignoring the proportionality constant, the voltage $T_N(ma)$ at the mth input location can be found from (4.82), which is given by

$$T_N(ma) = \int_{-\pi}^{\pi} \left[\frac{1}{2\pi} \sum_{n=1}^{N} A_n \exp(jn\psi) \right] \tilde{T}_\infty(\psi) \exp(-jm\psi) \, d\psi \qquad (4.84)$$

The above equation can be expressed as

$$T_N(ma) = \sum_{n=1}^{N} A_n Z_{mn} \qquad (4.85)$$

with

$$Z_{mn} = \frac{1}{2\pi} \int_{-\pi}^{\pi} \tilde{T}_\infty(\psi) \exp[-j(m-n)\psi] d\psi \qquad (4.86)$$

Because $T_N(ma)$ is considered as the voltage at mth-element's input location associated with the finite array excitation, $\tilde{T}_\infty(\psi) \exp(-jn\psi)$ must be the voltage at nth-element's input location associated with the Floquet excitation. Furthermore, in (4.77), if we consider the Floquet excitation as the input current excitation, then (4.85) would represent the voltage–current relationship of the finite array with Z_{mn} as the mutual impedance. Moreover, $\tilde{T}_\infty(\psi)$ would play the role of the Floquet impedance $Z^{FL}(\psi)$ because $\tilde{T}_\infty(\psi)$, in this case, turns out to be the voltage response with respect to a Floquet unity current excitation [see (4.77)]. The mutual impedance thus becomes

$$Z_{mn} = \frac{1}{2\pi} \int_{-\pi}^{\pi} Z^{FL}(\psi) \exp[-j(m-n)\psi] d\psi \qquad (4.87)$$

Thus the mutual impedance expression becomes identical to that derived in (4.42a). Following a similar procedure, the active element pattern and other array attributes can be shown to concur with the previously derived expressions.

REFERENCES

[1] R. F. Harrington, in *Time Harmonic Electromagnetic Fields*, McGraw-Hill, New York, 1961, "Microwave Networks," Chapter 8.

[2] A. K. Bhattacharyya, "Floquet-Modal-Based Analysis for Mutual Coupling Between Elements in an Array Environment," *IEE Proc.*, Vol. MAP-144, No. 6, pp. 491–497, Dec. 1997.

[3] D. M. Pozar, "Input Impedance and Mutual Coupling of Rectangular Microstrip Antennas," *IEEE Trans.*, Vol. AP-30, pp. 1191–11196, 1982.

[4] A. K. Bhattacharyya, in *Electromagnetic Fields in Multilayered Structures: Theory and Applications*, Artech House, Norwood, 1994, pp. 161–164.

[5] A. Ishimaru, R. J. Coe, G. E. Miller, and W. P. Green, "Finite Periodic Structure Approach to Large Scanning Array Problems," *IEEE Trans.*, Vol. AP-33, No. 11, pp. 1213–1220, Nov. 1985.

[6] A. K. Skrivervik and J. R. Mosig, "Finite Phased Array of Microstrip Patch Antennas," *IEEE Trans.*, Vol. AP-40, No. 5, pp. 579–582, May 1992.

[7] A. K. Skrivervik and J. R. Mosig, "Analysis of Finite Phased Arrays of Microstrip Patches," *IEEE Trans.*, AP-41, pp. 1105–1114, 1993.

[8] A. J. Roscoe and R. A. Perrott, "Large Finite Array Analysis Using Infinite Array Data," *IEEE Trans.*, Vol. AP-42, No. 7, pp. 983–992, July 1994.

[9] S. K. N. Yeo and J. A. Parfitt, "Finite Array Analysis Using Iterative Spatial Fourier Windowing of the Generalized Periodic Green's Function," *IEEE APS Symp. Dig.*, Vol.1, pp. 392–395, 1996.

[10] K. K. Chan and K. Chadwick, "Accurate Prediction of Finite Waveguide Array Performance Based on Infinite Array Theory," *IEEE APS Symp. Phased Array Syst. Tech. Dig.*, pp. 150–154, Oct. 1996.

[11] B. Tomasic and A. Hessel, "Analysis of Finite Arrays: A New Approach," IEEE Trans., Vol. AP-47, No. 3, pp. 555–565, Mar. 1999.

[12] A. Nato, S. Maci, G. Vecchi, and M. Sabbadini, "A Truncated Floquet Wave Diffraction Method for the Full Wave Analysis of Large Phased Arrays— Part I: Basic Principles and 2-D Cases," *IEEE Trans.*, AP-48, No. 3, pp. 594–600, Mar. 2000.

[13] A. Cucini, M. Albani, and S. Maci, "Truncated Floquet Wave Full-Wave $(T(FW)^2)$ Analysis of Large Periodic Arrays of Rectangular Waveguides," *IEEE Trans.*, Vol. AP-51, No. 6, pp. 1373–1384, June 2003.

[14] J. Mathews and R. L. Walker, *Mathematical Methods of Physics*, 2nd ed., W. A. Benjamin, Reading, MA, 1970.

BIBLIOGRAPHY

Bhattacharyya, A. K., and L. Shafai, "Effect of Mutual Coupling on the Radiation Pattern of Phased Array Antennas," *IEEE APS Symp. Dig.*, pp. 891–893, 1986.

Bird, T. S., "Analysis of Mutual Coupling in Finite Arrays of Different-Sized Rectangular Waveguides," *IEEE Trans.*, Vol. AP-38, pp. 166–172, Feb. 1990.

Civi, O. A., P. H. Pathak, and H-T Chou, "On the Poisson Sum Formula for the Analysis of Wave Radiation and Scattering from Large Finite Arrays," *IEEE Trans.*, Vol. AP-47, No. 5, pp. 968–959, May 1999.

Galindo, A. W., and C. P. Wu, *Theory and Analysis of Phased Array Antennas*, Wiley, New York, 1972.

Hansen, R. C., and D. Gammon, "A Gibbsian Model for Finite Scanned Arrays," *IEEE Trans.*, Vol. AP-44, No. 2, pp. 243–248, Feb. 1996.

Jackson, D. R., W. F. Richard, and A. Ali-Khan, "Series Expansions for Mutual Coupling in Microstrip Patch Arrays," *IEEE Trans.*, Vol. AP-37, pp. 269–274, 1989.

Katehi, P. B., "Mutual Coupling Between Microstrip Dipoles in Multielement Arrays," *IEEE Trans.*, Vol. AP-37, pp. 275–280, 1989.

Skinner, J. P., and P. J. Collins, "A One-Sided Version of the Poisson Sum Formula for Semi-Infinite Array Green's Function," *IEEE Trans.*, Vol. AP-45, No. 4, pp. 601–607, Apr. 1997

Mailloux, R. J., "Radiation and Near-Field Coupling Between Two Collinear Open-Ended Waveguides," *IEEE Trans.*, Vol. AP-17, pp. 49–55, Jan. 1969.

Mohammadian, A. H., N. W., Martin and D. W. Griffin, "Mutual Impedance Between Microstrip Patch Antennas", *IEEE APS Symp. Dig.*, Vol.3, pp. 932–935, 1988.

Pan, S. G., and I. Wolff, "Computation of Mutual Coupling Between Slot-Coupled Microstrip Patches in a Finite Array," *IEEE Trans.*, AP-40, pp. 1047–1053, 1992.

PROBLEMS

4.1 Consider an infinite array of conducting scattering objects coated with a dielectric layer. Using the procedure in Section 4.2 (or otherwise) show that for this structure the Floquet modal admittances are symmetric, that is the relation in (4.25) is still valid. *Hint:* Use the volume equivalence principle in the dielectric region for the equivalent source to be used in the reciprocity relation in (4.16).

4.2 Show that for a linear array the Floquet impedance Z^{FL} is a periodic function of ψ where $\psi = k_0 a \sin \theta$, θ is the scan angle, a is the element spacing, and k_0 is the wave number. Explain the physical significance of this periodic property.

4.3 The Floquet impedance of an infinite linear array is given by

$$Z^{FL}(\theta) = \frac{1}{4}\{3 + \cos^2(4\pi \sin \theta) + j \sin^2(4\pi \sin \theta)\}$$

where θ is the scan angle. The element spacing is $2\lambda_0$, where λ_0 is the free-space wavelength. First, examine whether the impedance function obeys the periodic and symmetry properties. Then determine (a) the self impedance of an element in the array, (b) the mutual impedance between the two adjacent elements, and (c) the mutual impedance between two elements that are three elements away.

4.4 If in problem 4.3 the three consecutive elements are excited uniformly with a current source and the other elements are open circuited, find the active input impedances of the elements.

4.5 Repeat problem 4.4 assuming that the other elements of the finite array are (a) short circuited, and (b) match terminated.

4.6 Using the relation between the mutual impedance and the Floquet impedance of an infinite array as

$$Z_{mn} = \frac{1}{2\pi} \int_{-\pi}^{\pi} Z^{FL}(\psi) \exp\{j(m-n)\psi\}d\psi$$

show that the mutual admittance between two elements can be deduced from the above relation as

$$Y_{mn} = \frac{1}{2\pi} \int_{-\pi}^{\pi} \frac{1}{Z^{FL}(\psi)} \exp\{j(m-n)\psi\}d\psi$$

Hint: Use $[Y] = [Z]^{-1}$, where both matrices are of infinite order. Notice $Z^{FL}(\psi)$ is the eigenvalue of the $[Z]$ matrix with eigenvector $(d\psi/2\pi)[\ldots, \exp(-jn\psi), \exp\{-j(n-1)\psi\}, \ldots, 1, \ldots, \exp(jn\psi), \ldots]$. Use the $[S][\Lambda][S]^{-1}$ format to invert $[Z]$.

4.7 Prove the relation in (4.47) for an arbitrary lattice structure represented by $[a, b, \gamma]$.

4.8 In an infinite array, only one element is excited and the remaining elements are short circuited. Prove that the radiation pattern of the element under such an array environment is given by

$$E_0^{volt} = \frac{1}{2}V_0 E^a[Y_0 + Y^{FL}(\psi')]/Y_0$$

where the symbols are defined in Section 4.8.2. Show that the finite array pattern is given by

$$E^{array} = E^a[Y_0 + Y^{FL}(\psi')][P]\{[Y] + [Y_0]\}^{-1}[V^+]$$

4.9 The input impedance of a single-mode, open-ended waveguide a quarter wavelength long is related to the aperture impedance of the horn by the following relation:

$$Z_{inp}Z_{apr} = Z_0^2$$

where Z_{inp} is the input impedance, Z_{apr} is the aperture impedance, and Z_0 is the characteristic impedance of the waveguide. Assume $Z_{apr}^{FL}(\psi_x, \psi_y)$ is the Floquet impedance of the waveguide array.

(a) Find the Floquet impedance at the input level.

(b) Find the mutual impedance at the aperture level.

(c) Obtain the mutual impedance at the input level.

(d) Establish a mathematical relation between the mutual impedances at the aperture level and mutual admittances at the input level.

4.10 In a multimode horn, the aperture supports N propagating modes, while the input section of the horn supports only one mode. The incident and reflected voltages at the two ports (input and aperture) are related via the S-matrix of the horn as

$$V_{inp}^- = [S_{11}]V_{inp}^+ + [S_{12}]V_{out}^- \qquad V_{out}^+ = [S_{21}]V_{inp}^+ + [S_{22}]V_{out}^-$$

where $[S_{11}]$ is a 1×1 matrix, $[S_{12}]$ is an $1 \times N$ matrix, $[S_{21}]$ is an $N \times 1$ matrix, and $[S_{22}]$ is an $N \times N$ matrix. The Floquet reflection (matrix) at the aperture plane for an infinite array of such horns is given by

$$V_{out}^- = [\Gamma^{FL}(\psi_x, \psi_y)]V_{out}^+$$

Obtain an expression for the Floquet reflection coefficient at the input location and then obtain the scattering parameters (mutual coupling) between the elements at the input level. (Notice, for a finite array of M elements, the input level S-matrix is of order $M \times M$. Contrary to this, the S-matrix would be of order $MN \times MN$ if the coupling is analyzed at the aperture level, because the aperture is equivalent to N ports due to N modes).

4.11 A real finite array of open-ended rectangular waveguides opening onto an infinite ground plane is excited uniformly. The waveguides support only the dominant mode (TE_{01} mode). Use the most appropriate formulation and derive the active return loss of each waveguide. Assume that the Floquet aperture impedances for an infinite array of such waveguides are known. The input ports are at a distance of quarter wavelength behind the aperture plane.

CHAPTER FIVE

Array of Subarrays

5.1 INTRODUCTION

In an array antenna, it is often beneficial to excite a group of radiating elements by a single source instead of exciting each element individually. Such a group of radiating elements is called a subarray. The subarray concept is very advantageous, particularly for an active array antenna because of the following reasons:

(a) It requires a smaller number of excitation sources than that of a normal array of equal number of radiating elements.

(b) For active array applications, the number of phase shifters and amplifiers are reduced by the same factor as the number of elements in a subarray.

(c) The interelement spacing in a subarray can be kept small; thus the aperture efficiency can be very close to 100%.

Some of the disadvantages are as follows:

(a) Each subarray requires a power divider circuit in order to excite the elements of the subarray with desired taper.

(b) The scan loss (gain loss due to off bore-sight scan) is higher as compared to an array of individually excited elements.

(c) A large cell size limits the scanning range because unwanted grating lobe and scan blindness problems may occur.

In this chapter we present a simple procedure for analyzing an infinite array of subarrays [1]. Employing Floquet modal theory and matrix algebra, we deduce the

Phased Array Antennas. By Arun K. Bhattacharyya
© 2006 John Wiley & Sons, Inc.

impedance matrix of the elements in a subarray. This impedance matrix corresponds to a Floquet excitation of the infinite array of subarrays. We first illustrate the procedure for a 2×2-element subarray on a rectangular grid and then generalize for more complex subarray structures. It is well known that the impedance matrix of a multiport network can be expressed in terms of eigenvalues and eigenvectors of the impedance matrix [2]. In the case of a subarray, the number of eigenvectors is equal to the number of elements[1] in a subarray. Interestingly, it is found that these eigenvectors can be constructed from Floquet excitation coefficients associated with the entire array of individually fed elements. This implies that the subarray impedance matrix can be determined from the characteristics of the array of individually fed elements. It is found that each eigenvector excitation yields a set of grating lobes. Thus an array of subarrays produces multiple sets of grating lobes associated with multiple eigenvectors. As a result, the grating lobes are closely spaced as compared with that of the array of individually fed elements. However, unlike the array of individually fed elements, the grating lobe intensities are nonuniform.[2] The grating lobe locations and grating lobe intensities of an array of subarrays are analyzed from the eigenexcitation viewpoint. The active subarray pattern, active return loss, and scan blindness phenomena are studied. At the end, a procedure for analyzing a finite array of subarrays is outlined.

5.2 SUBARRAY ANALYSIS

Figure 5.1 shows a generic diagram of an infinite array of subarrays. Each subarray is fed by a power divider network (PDN) which is excited by a single source. The number of output ports of the power divider is equal to the number of elements in a subarray. We will restrict our analysis for an infinite array of subarrays under Floquet excitation, that is, uniform and linearly phased voltage distribution with respect to the input ports of the PDNs is assumed. The PDN primarily controls

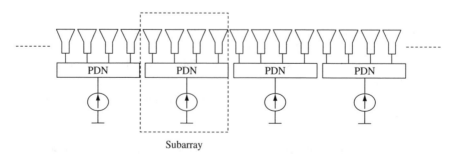

FIGURE 5.1   Infinite array of subarrays. Subarray elements are fed with a PDN.

---

[1] We will use the word *element* to represent a radiating element in a subarray.

[2] Under the assumption that the elements are isotropic.

the excitation coefficients (voltage distribution) of the elements within a subarray. In order to estimate the voltage distribution of the elements within a subarray, the equivalent load impedances experienced by the output ports of a PDN must be determined. To account for mutual coupling between the radiating elements, the terminating loads are collectively represented by an impedance matrix (equivalently, by admittance or by scattering matrices). Two different methods exist for obtaining the above impedance matrix. The first method considers a subarray as a unit cell for the infinite array of subarrays and then performs the Floquet modal analysis of the infinite periodic structure of such large unit cells [3]. This method requires excessive analytical and computational efforts, because a large number of unknowns need to be solved simultaneously. The large number of unknowns is primarily due to multiple elements inside a large cell, requiring several basis functions to represent unknown aperture fields or currents for the elements. In this section, we will articulate an alternative approach to determine the impedance matrix of a subarray. The method is elegant and accurate and computationally simpler but applicable for elements that are equally spaced. Using matrix theory and Floquet modal analysis, we prove that the subarray characteristics can be estimated from the characteristics of an individually fed element array. We present the derivation of a subarray impedance matrix under Floquet excitation of the array of subarrays. The Floquet impedance at the subarray input can then be determined from the subarray impedance matrix and the PDN impedance matrix. Treating each subarray as an element and then employing the theory developed in Chapter 4, one can determine the active impedances of a finite array of subarrays.

5.2.1 Subarray Impedance Matrix: Eigenvector Approach

The subarray impedance matrix yields the relation between the input voltages and input currents of the radiating elements within a subarray. The impedance matrix we develop in this section can be termed as the Floquet impedance matrix, because a Floquet excitation of the array of subarrays is assumed. For simplicity of presentation we consider rectangular lattice structures for the elements as well as for the subarrays, as shown in Figure 5.2. Also, we assume that a subarray consists of 2×2 elements. The subarray cell size is $a \times b$. The element cell size is therefore $a/2 \times b/2$. Suppose the infinite array of subarrays is excited to radiate in the (θ, ϕ)-direction with respect to a spherical coordinate system. The phase difference between two adjacent subarrays along the x- and y-directions should be

$$\psi_x = k_0 a \sin\theta \cos\phi \tag{5.1a}$$

$$\psi_y = k_0 b \sin\theta \sin\phi \tag{5.1b}$$

where $k_0 = 2\pi/\lambda_0$, λ_0 is the wavelength in free space. The amplitude and the phase distributions of the elements within a subarray depend on the PDN circuitry and the impedance matrix of the subarray.

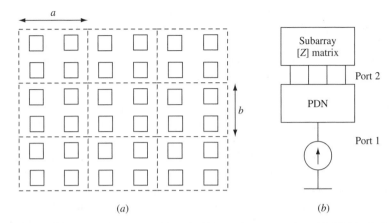

FIGURE 5.2 (*a*) Array of subarrays of 2×2 elements in rectangular grid. (*b*) Block diagram for a subarray impedance analysis.

In order to determine the impedance matrix, we apply the following property of a matrix. It is well known that a matrix (in this case, the subarray impedance matrix $[Z]$) can be constructed from its eigenvectors and eigenvalues [4] due to the following identity:

$$[Z] = [S][\Lambda][S]^{-1} \qquad (5.2)$$

where $[S]$ represents the eigenvector matrix[3] of $[Z]$ with the eigenvectors as columns and $[\Lambda]$ is a diagonal matrix with the eigenvalues as diagonal elements. The above identity can be proven directly from the definition of the eigenvector of $[Z]$. Thus, we need to determine the eigenvectors and the eigenvalues of the impedance matrix. Toward that end we choose the phase difference between the adjacent elements as $\psi_x/2$ and $\psi_y/2$, respectively, along x- and y-directions with uniform amplitudes. This particular selection of phase and amplitude distributions among the elements will allow us to treat the entire array as an infinite periodic structure of cell size $a/2 \times b/2$ radiating in the (θ, ϕ)-direction under Floquet excitation. In other words, the "subarray grouping" would not have any influence on the field solutions. For such an excitation all the elements inside a subarray (for that matter, all the elements of the entire array) have identical impedances, equal to the Floquet impedance corresponding to the scan angle (θ, ϕ). This type of subarray excitation can be regarded as an eigenvector excitation, because the input voltage vector (of four dimensions because of the four elements in a subarray) and the input current vector are parallel. This voltage (or current) vector is one of the eigenvectors

---

[3] Not to be confused with the scattering matrix.

of the subarray impedance matrix. The Floquet impedance is the corresponding eigenvalue. Mathematically, we can write

$$
\begin{bmatrix}
Z_{11} & Z_{12} & Z_{13} & Z_{14} \\
Z_{21} & Z_{22} & Z_{23} & Z_{24} \\
Z_{31} & Z_{32} & Z_{33} & Z_{34} \\
Z_{41} & Z_{42} & Z_{43} & Z_{44}
\end{bmatrix}
\begin{bmatrix}
s_1^{(1)} \\
s_2^{(1)} \\
s_3^{(1)} \\
s_4^{(1)}
\end{bmatrix}
= Z_1^{\mathrm{FL}}
\begin{bmatrix}
s_1^{(1)} \\
s_2^{(1)} \\
s_3^{(1)} \\
s_4^{(1)}
\end{bmatrix}
\tag{5.3}
$$

In the above equation, the 4×4 matrix is the subarray impedance matrix ($[Z]$ matrix), the column vector is the input current vector, and Z_1^{FL} is the Floquet impedance. The current vector constitutes a column of the $[S]$ matrix in (5.2). The elements of the current vector of a reference subarray may be considered as

$$
s_1^{(1)} = \frac{1}{2}
$$

$$
s_2^{(1)} = \frac{1}{2} \exp\left(-\frac{1}{2} j\psi_x\right)
$$

$$
s_3^{(1)} = \frac{1}{2} \exp\left(-\frac{1}{2} j\psi_y\right)
\tag{5.4}
$$

$$
s_4^{(1)} = \frac{1}{2} \exp\left(-\frac{1}{2} j\psi_x\right) \exp\left(-\frac{1}{2} j\psi_y\right)
$$

Notice that the input currents are associated with the Floquet excitation. The current vector is normalized to make the magnitude unity. The advantage of this normalization would be found later.

In order to construct the $[Z]$ matrix as per (5.2), we need three other eigenvectors and the associated eigenvalues. It is apparent that if the phase difference between the adjacent subarrays changes by an integral multiple of 2π, then the overall Floquet excitation with respect to the array of subarrays remains unchanged, because $\exp(j2\pi) = 1$. On the other hand, if this phase change between the subarrays is distributed linearly between the elements, then we obtain a new Floquet excitation for the infinite array of elements of cell size $a/2 \times b/2$. The excitation coefficients of the elements inside a subarray constitute the second eigenvector. A 2π-phase change between two adjacent subarrays along x yields the following eigenvector elements for the subarray impedance matrix:

$$
s_1^{(2)} = \frac{1}{2}
$$

$$
s_2^{(2)} = \frac{1}{2} \exp\left(-\frac{1}{2} j\psi_x - j\pi\right)
$$

$$
s_3^{(2)} = \frac{1}{2} \exp\left(-\frac{1}{2} j\psi_y\right)
\tag{5.5}
$$

$$
s_4^{(2)} = \frac{1}{2} \exp\left(-\frac{1}{2} j\psi_x - j\pi\right) \exp\left(-\frac{1}{2} j\psi_y\right)
$$

For this particular excitation, the phase difference between the adjacent elements along x becomes $\pi + \psi_x/2$ and the phase difference between the elements along y remains unchanged as $\psi_y/2$. The above excitation produces a beam in the (θ', ϕ')-direction, where $\sin \theta' \cos \phi' = \sin \theta \cos \phi + \lambda_0/a$, $\sin \theta' \sin \phi' = \sin \theta \sin \phi$. We assume Z_2^{FL} as the Floquet impedance, which becomes the second eigenvalue of $[Z]$.

The remaining two eigenvectors are selected as

$$s_1^{(3)} = \frac{1}{2}$$

$$s_2^{(3)} = \frac{1}{2} \exp\left(-\frac{1}{2} j\psi_x\right)$$

$$s_3^{(3)} = \frac{1}{2} \exp\left(-\frac{1}{2} j\psi_y - j\pi\right) \qquad (5.6)$$

$$s_4^{(3)} = \frac{1}{2} \exp\left(-\frac{1}{2} j\psi_x\right) \exp\left(-\frac{1}{2} j\psi_y - j\pi\right)$$

and

$$s_1^{(4)} = \frac{1}{2}$$

$$s_2^{(4)} = \frac{1}{2} \exp\left(-\frac{1}{2} j\psi_x - j\pi\right)$$

$$s_3^{(4)} = \frac{1}{2} \exp\left(-\frac{1}{2} j\psi_y - j\pi\right) \qquad (5.7)$$

$$s_4^{(4)} = \frac{1}{2} \exp\left(-\frac{1}{2} j\psi_x - j\pi\right) \exp\left(-\frac{1}{2} j\psi_y - j\pi\right)$$

As before, the above two eigenvectors are associated with different Floquet excitations of the infinite array of elements of cell size $a/2 \times b/2$. We assume that the corresponding eigenvalues are Z_3^{FL} and Z_4^{FL}, respectively. Notice the eigenvectors are orthonormal. Therefore, $[S]$ becomes a unitary matrix that can be inverted simply by complex conjugating its transpose [5]. Therefore from (5.2) we write

$$[Z] = [S][\Lambda][S^*]^T \qquad (5.8)$$

where $[S]$ and $[\Lambda]$ are given by

$$[S] = \begin{bmatrix} s_1^{(1)} & s_1^{(2)} & s_1^{(3)} & s_1^{(4)} \\ s_2^{(1)} & s_2^{(2)} & s_2^{(3)} & s_2^{(4)} \\ s_3^{(1)} & s_3^{(2)} & s_3^{(3)} & s_3^{(4)} \\ s_4^{(1)} & s_4^{(2)} & s_4^{(3)} & s_4^{(4)} \end{bmatrix} \qquad (5.9)$$

$$[\Lambda] = \begin{bmatrix} Z_1^{\mathrm{FL}} & 0 & 0 & 0 \\ 0 & Z_2^{\mathrm{FL}} & 0 & 0 \\ 0 & 0 & Z_3^{\mathrm{FL}} & 0 \\ 0 & 0 & 0 & Z_4^{\mathrm{FL}} \end{bmatrix} \qquad (5.10)$$

The Floquet impedance being complex, the impedance matrix is non-Hermitian[4] in general [5]. It can also be verified that $[Z]$ is not a symmetrical matrix in general. In fact, it can be shown directly that the individual subarray elements do not satisfy the reciprocity relation under the constraint of Floquet excitation; hence $[Z]$ is nonsymmetric. This is not a violation of the reciprocity theorem, however, because the reciprocity relation holds for the source of the entire array (see problem 5.2).

5.3 SUBARRAY WITH ARBITRARY NUMBER OF ELEMENTS

The above procedure can be generalized for a subarray of $M \times N$ elements. In this case, the $[S]$ matrix in (5.2) should consist of MN number of eigenvectors. An eigenvector can be constructed in the following manner. Consider the phase difference between the two adjacent subarrays along x as $\psi_x + 2\pi i$ and along y as $\psi_y + 2\pi k$, where $0 \leq i \leq (M-1)$, $0 \leq k \leq (N-1)$. The above phase increments between the adjacent subarrays are linearly distributed between the elements inside a subarray. Thus the phase differences between adjacent elements become $(\psi_x + 2\pi i)/M$ and $(\psi_y + 2\pi k)/N$ for the elements along x- and y-directions, respectively. Using these incremental phase values, the components of the eigenverctor are constructed as

$$s_{mn} = \frac{\exp(-j\chi_{mn})}{\sqrt{MN}} \tag{5.11}$$

with

$$\chi_{mn} = \frac{m(\psi_x + 2\pi i)}{M} + \frac{n(\psi_y + 2\pi k)}{N} \tag{5.12}$$

where $(ma/M, nb/N)$ is the relative coordinate of the element associated with the excitation coefficient s_{mn}. Since m runs from 0 to $M-1$ and n runs from 0 to $N-1$, we would have MN number of s_{mn}'s that constitute an eigenvector. The eigenvalue corresponding to this eigenvector is the Floquet impedance of the array of cell size $a/M \times b/N$. Since i and k in (5.12) run from 0 to $M-1$ and 0 to $N-1$, respectively, we should have MN number of independent eigenvectors required to construct the subarray impedance matrix. Further, these eigenvectors being orthonormal, (5.8) can be applied directly to obtain the impedance matrix of the subarray.

It may be observed that each eigenvector excitation makes the array radiate at different scan angles. Only the first eigenvector ($i = k = 0$) produces a beam at the

---

[4] Every Hermitian matrix has real eigenvalues.

desired scan angle. The scan angle (θ_{ik}, ϕ_{ik}) for a general eigenvector excitation corresponding to (5.12) can be computed from the following equations:

$$\frac{\psi_x + 2\pi i}{M} = k_0 \left(\frac{a}{M}\right) \sin \theta_{ik} \cos \phi_{ik}$$

$$\frac{\psi_y + 2\pi k}{N} = k_0 \left(\frac{b}{N}\right) \sin \theta_{ik} \sin \phi_{ik}$$

(5.13)

Eliminating the $1/M$ factor from the first equation and $1/N$ factor from the second, we obtain the identical equations for finding grating lobe locations for an array of cell size $a \times b$. Thus the scan direction associated with an eigenvector is along a grating lobe of the array of subarrays.

5.4 SUBARRAYS WITH ARBITRARY GRIDS

The procedure can be generalized for obtaining the $[Z]$ matrix of a general array of subarrays with arbitrary lattice structures. It can be visualized that the elements in a sub-array (at least the driving point locations) can be accommodated inside a parallelogram that does not intercept with the neighboring subarray elements (or the driving point locations). In order to construct an eigenvector excitation, the phase difference of two consecutive subarrays along the two sides of the parallelogram should be considered.

Consider the array of subarrays in Figure 5.3. The subarray lattice is characterized by $[a, b, \gamma]$, and the element lattice is characterized by $[a_1, b_1, \gamma_1]$. The number of elements inside a subarray unit cell is $N = ab/a_1b_1$, which must be an integer. In order to obtain an eigenvector, we consider that the adjacent subarrays have phase differences of $\psi'_x = \psi_x + i2\pi$ and $\psi'_y = \psi_y + k2\pi$, respectively along the two oblique directions inclined at an angle γ. In the above ψ_x and ψ_y (both less than 2π) are the actual phase differences between the adjacent subarrays for scanning

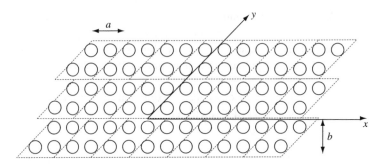

FIGURE 5.3 Array of subarrays with fictitious boundaries of subarrays.

at the desired angle (θ, ϕ) and are given by

$$\psi_x = k_0 a \sin\theta \cos\phi$$

$$\psi_y = k_0 b \sin\theta \sin\phi + \frac{k_0 b \sin\theta \cos\phi}{\tan\gamma} \tag{5.14}$$

Now the phase differences ψ_x' and ψ_y' are linearly distributed between the elements inside the subarray. This can be accomplished using the phase gradients along the two oblique directions. The phase gradient along x- and y-axes (Figure 5.3) are

$$\frac{\partial\psi_x'}{\partial x} = \frac{\psi_x + 2\pi i}{a} \qquad \frac{\partial\psi_y'}{\partial y} = \frac{(\psi_y + 2\pi k)\sin\gamma}{b} \tag{5.15}$$

Suppose the nth element in the subarray has the coordinates (x_n, y_n) with respect to the oblique coordinate system (x, y). Then the phase difference of the nth element with respect to a reference point becomes

$$\chi_n = x_n \frac{\partial\psi_x'}{\partial x} + y_n \frac{\partial\psi_y'}{\partial y} \tag{5.16}$$

Therefore, the nth component of the desired eigenvector can be considered as

$$s_n = \frac{\exp(-j\chi_n)}{\sqrt{N}} \tag{5.17}$$

where $N = ab/a_1 b_1$ is the number of elements in a subarray. Since n varies from 1 to N, we can have all components of the eigenvector. Selecting different sets of (i, k) we can have N distinct sets of eigenvectors. The eigenvectors are orthogonal (hence orthonormal because of the factor $\sqrt{N}$ in the denominator) because the differential phase between two eigenvectors summing over the elements is an integral multiple of 2π. The Floquet impedances associated with each eigenvector excitation are the eigenvalues of the $[Z]$ matrix. The $[Z]$ matrix can be constructed from the eigenvectors and the eigenvalues using (5.8).

5.5 SUBARRAY AND GRATING LOBES

One disadvantage of subarraying is the appearance of grating lobes at the close proximity to the main beam. This is primarily due to the effective large element spacing of the array of subarrays. The scanning range thus becomes narrower for a scanned beam antenna. The grating lobe locations of a subarray can be computed from the Floquet wave numbers of an infinite array of subarrays.

Suppose $[a, b, \gamma]$ represents the lattice structure of an array of subarrays, where $a \times b$ is the subarray size and γ is the grid angle. The Floquet wave numbers are (see Section 2.2)

$$k_{xmn} = \frac{2m\pi + \psi_x}{a} \qquad k_{ymn} = \frac{2n\pi + \psi_y}{b} - \frac{2m\pi}{a\tan\gamma} \tag{5.18}$$

where ψ_x and ψ_y are given by

$$\psi_x = k_{x0}a = k_0 a \sin\theta_s \cos\phi_s, \qquad \psi_y = k_{y0}b = k_0 b \sin\theta_s \sin\phi_s \qquad (5.19)$$

where (θ_s, ϕ_s) is the desired scan direction. The grating lobe direction (θ_{mn}, ϕ_{mn}) associated with the Floquet wave numbers in (5.18) is given by

$$k_{xmn} = k_0 \sin\theta_{mn} \cos\phi_{mn} \qquad k_{ymn} = k_0 \sin\theta_{mn} \sin\phi_{mn} \qquad (5.20)$$

Combining (5.18), (5.19), and (5.20) we obtain the relations between the desired beam location and grating lobe locations given by

$$
\begin{aligned}
\sin\theta_{mn}\cos\phi_{mn} &= \frac{m\lambda_0}{a} + \sin\theta_s\cos\phi_s \\
\sin\theta_{mn}\sin\phi_{mn} &= \frac{n\lambda_0}{b} + \sin\theta_s\sin\phi_s - \frac{m\lambda_0}{a\tan\gamma}
\end{aligned}
\qquad (5.21)
$$

For a given scan direction, we obtain finite sets of real solutions for (θ_{mn}, ϕ_{mn}). These solutions correspond to the grating lobe locations. It can be verified that the grating lobe locations correspond to the radiation directions of the N eigenvector excitations of the array, where N is the number of elements in a subarray.

Figure 5.4 shows the grating lobe locations in the $k_x k_y$-plane for two subarray grids (rectangular and triangular). Each subarray has 2×2 elements, as shown in the figures. Expectedly, "subarraying" brings in additional grating lobes closer to the desired radiation direction, thus limiting the scanning range. For a given subarray size a triangular grid has advantages with regard to the beam scanning range.

Each eigenvector is responsible for a set of grating lobes. Thus the grating lobe intensities are directly related to the amplitudes of the eigenvectors. To illustrate, consider that V_i, $i = 1, 2, 3, 4$, are the input voltages at the input port of the four radiating elements. The input voltage vector can be expressed as a superposition of the subarray eigenvectors given in (5.4)–(5.7) for the 2×2 element subarray in a rectangular grid. Thus we have

$$[V] = c_1[s^{(1)}] + c_2[s^{(2)}] + c_3[s^{(3)}] + c_4[s^{(4)}]$$

Using orthogonality, we obtain the amplitudes of the eigenvector excitations as

$$c_i = [s^{(i)*}]^T[V] \qquad i = 1, 2, 3, 4$$

Figure 5.5 shows the grating lobe intensities along the E-plane of an array of subarrays with four radiating elements shown in Figure 5.4a. Only two sets of grating lobes exist for the E-plane scan. The elements are assumed to be isotropic. The intended radiation direction is $10°$ off bore sight and along the E-plane. The subarray dimensions are $2\lambda_0 \times 2\lambda_0$. The intensities of the grating lobes that resulted from the subarray grouping are smaller in this case than that of the main lobe and the affiliated grating lobes.

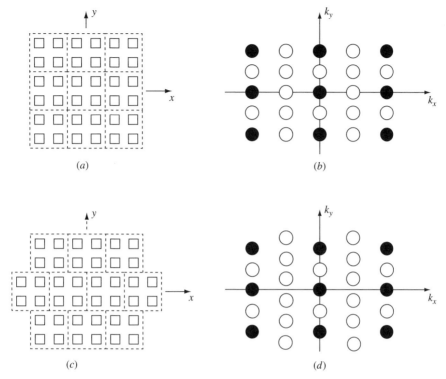

FIGURE 5.4   Array of subarrays and the location of grating lobes. (*a*) Subarrays in rectangular grid. (*b*) Grating lobe locations for (*a*). (*c*) Subarrays in triangular grid. (*d*) Grating lobe locations for (*c*). The dark circles represents grating lobe locations of the array of individually fed elements. The unfilled circles are the additional grating lobe locations due to subarray groupings.

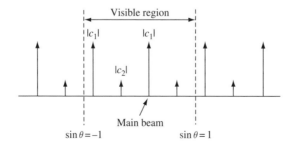

FIGURE 5.5   Main-beam and grating lobe intensities of an infinite array of subarrays scanning along the *E*-plane. Each subarray has 2×2 elements. Subarray size $2\lambda_0 \times 2\lambda_0$. Intended beam direction 10° off bore sight. A uniform source distribution (amplitude and phase) for the subarray elements is assumed. Isotropic radiation pattern for the elements.

5.6 ACTIVE SUBARRAY PATTERNS

In order to determine the radiation patterns of a finite array of subarrays, the active subarray pattern must be determined. Following the definition of the active element pattern, we define the active subarray pattern as the radiation pattern of an infinite array of subarrays when only one subarray is excited and the other subarrays are match terminated at the PDN input ends. The active subarray patterns yield the array gain, scan loss, cross-polar power, and relative power level of the grating lobes with respect to the main beam. The active subarray pattern can be computed from the active element pattern, as illustrated in this section.

Suppose ab is the subarray cell area and $a_1 b_1$ is the element cell area. The number of elements within a subarray is thus $N = ab/(a_1 b_1)$. To obtain the active subarray pattern we invoke the formulation in Section 3.5. To that end, we apply a Floquet excitation to the infinite array of subarrays. The active subarray pattern can be determined from the dominant Floquet modal voltages on the aperture, as deduced in (3.92).

To find the Floquet modal voltages on the aperture of the infinite array of subarrays, we need to determine the incident voltages at the input ends of the elements of a subarray. For that, the electrical characteristics of the feed network (the PDN) must be known. The feed network can be characterized in terms of its impedance, admittance, or scattering matrix. We will apply the impedance matrix formulation for the present development. Suppose the subarray feed is an $(N+1)$-port network with only one input port (the analysis can be easily extended if the feed network has more than one input ports). The number of output ports is N, which is equal to the number of elements in a subarray. Suppose $[Z_{\text{feed}}]$ is the $(N+1) \times (N+1)$ impedance matrix of the feed. We also assume $[Z_{\text{sub}}]$ as the $N \times N$ impedance matrix of the subarray for a given scan angle, obtained in Section 5.2. The subarray and its feed network are shown schematically in Figure 5.2b. The port voltage and port current of the feed network are related through the feed impedance matrix as

$$
\begin{bmatrix} v_1 \\ \hline v_2 \\ v_3 \\ \vdots \\ v_{N+1} \end{bmatrix} = \begin{bmatrix} z_{11}^f & z_{12}^f & z_{13}^f & \cdots \\ \hline z_{21}^f & z_{22}^f & z_{23}^f & \cdots \\ z_{31}^f & z_{32}^f & z_{33}^f & \cdots \\ \vdots & \vdots & \vdots & \ddots \end{bmatrix} \begin{bmatrix} i_1 \\ \hline -i_2 \\ -i_3 \\ \vdots \\ -i_{N+1} \end{bmatrix}
\tag{5.22}
$$

where v_1 and i_1 are the voltage and current, respectively, at the input port (port 1) of the PDN and $v_2, v_3, \ldots, v_{N+1}$ are the voltages and $i_2, i_3, \ldots, i_{N+1}$ are the currents at the output ports of the PDN. The output ports are collectively designated as port 2. The output ports are directly connected to the radiating elements. The negative signs for the output port currents are due to the usual conventions of the port currents, which are opposite to the input current directions of the elements. We now separate out the input and output ports by means of partitioning the impedance matrix in (5.22). This yields

$$[V_1] = [Z_{11}^f][I_1] - [Z_{12}^f][I_2] \qquad [V_2] = [Z_{21}^f][I_1] - [Z_{22}^f][I_2] \tag{5.23}$$

with

$$[V_1] = [v_1] \qquad [I_1] = [i_1]$$

$$[V_2] = \begin{bmatrix} v_2 \\ v_3 \\ \vdots \\ v_{N+1} \end{bmatrix} \qquad [I_2] = \begin{bmatrix} i_2 \\ i_3 \\ \vdots \\ i_{N+1} \end{bmatrix} \tag{5.24}$$

The $[Z^f]$ matrices are the submatrices of $[Z_{feed}]$ as partitioned in (5.22). The input voltages and currents at port 2 are related through the subarray impedance matrix as

$$[V_2] = [Z_{sub}][I_2] \tag{5.25}$$

From (5.23) and (5.25) we can deduce the following relations:

$$[I_2] = [Z_{22}^f + Z_{sub}]^{-1}[Z_{21}^f][I_1]$$
$$[V_2] = [Z_{sub}][Z_{22}^f + Z_{sub}]^{-1}[Z_{21}^f][I_1] \tag{5.26}$$
$$[V_1] = \{[Z_{11}^f] - [Z_{12}^f][Z_{22}^f + Z_{sub}]^{-1}[Z_{21}^f]\}[I_1]$$

The objective here is to find the incident voltage at the element ports with respect to the incident voltage at the input port of the feed network. From the last equation of (5.26), we can write the input impedance matrix seen by the feed input as

$$[Z_{in}] = [Z_{11}^f] - [Z_{12}^f][Z_{22}^f + Z_{sub}]^{-1}[Z_{21}^f] \tag{5.27}$$

Since the input has only one port, $[Z_{in}]$ is a 1×1 matrix. Suppose z_{in} is the matrix element, which is equal to the input impedance seen by port 1 when the output ports are loaded with the elements. For normalization with respect to the incident input power, we assume the incident voltage at port 1 is $v_1^+ = \sqrt{z_{01}}$, where z_{01} is the characteristic impedance of the input transmission line (or the source internal impedance). The input current at the input port is then

$$i_1 = i_1^+ + i_1^- = \frac{v_1^+}{z_{01}}[1 - \Gamma_{in}] = \frac{2v_1^+}{z_{in} + z_{01}} = \frac{2\sqrt{z_{01}}}{z_{in} + z_{01}} \tag{5.28}$$

In the above equation, Γ_{in} is the input reflection coefficient, which is equal to $(z_{in} - z_{01})/(z_{in} - z_{01})$. Substituting the expression for the input current in (5.26), we obtain $[V_2]$ and $[I_2]$ vectors that are normalized with respect to the incident power.

We now use the following relations to obtain the incident voltage vector at the subarray elements ports:

$$[V_2] = [V_2^+] + [V_2^-] \qquad [I_2] = \frac{\{[V_2^+] - [V_2^-]\}}{z_{02}} \tag{5.29}$$

We obtain

$$[V_2^+] = [Z_{02} + Z_{\text{sub}}][Z_{22}^f + Z_{\text{sub}}]^{-1}[Z_{21}^f]\frac{\sqrt{z_{01}}}{z_{\text{in}} + z_{01}} \tag{5.30}$$

In the above equations, z_{02} is the characteristic impedance of the transmission lines at the output port of the feed network and $[Z_{02}]$ is a diagonal matrix with z_{02} as diagonal elements. The normalized incident voltage vector for the subarray elements then becomes

$$[\hat{V}_2^+] = \sqrt{\frac{1}{z_{02}}}[V_2^+] \tag{5.31}$$

It has been established in Chapter 3 that the far-field components of the active element pattern are proportional to the TE_{00} and TM_{00} Floquet modal voltages. Our next task therefore is obtaining the Floquet modal voltages produced by the excitation vector $[\hat{V}_2^+]$. Toward that end, we expand $[\hat{V}_2^+]$ in terms of subarray eigenvectors. The amplitude of the eigenvector that corresponds to the desired look angle yields the fundamental Floquet modal voltages.

The process of obtaining the eigenvectors has been discussed in the previous section. Suppose $[s^{(1)}], [s^{(2)}], \ldots, [s^{(N)}]$ are the orthonormal eigenvectors for the subarray of N elements. Then we can write

$$[\hat{V}_2^+] = c_1[s^{(1)}] + c_2[s^{(2)}] + \cdots + c_N[s^{(N)}] \tag{5.32}$$

with

$$c_i = [\hat{V}_2^+]^T[s^{(i)}]^* \qquad i = 1, 2, \ldots, N \tag{5.33}$$

Each eigenvector excitation produces an infinite number of Floquet modes. However, the fundamental Floquet modes that radiate along the desired direction are only produced by the eigenvector $[s^{(1)}]$. Suppose V_{00}^{TE} and V_{00}^{TM} are the Floquet modal voltages for the TE_{00} and TM_{00} modes, respectively, normalized with respect to the incident power at the element input. For $[s^{(1)}]$ as the incident voltage vector for the subarray, the modal voltages become $V_{00}^{TE}/\sqrt{N}$ and $V_{00}^{TM}/\sqrt{N}$, respectively. The factor $1/\sqrt{N}$ appears because the magnitude of the incident voltage at an element port corresponding to the eigenvector $[s^{(1)}]$ is $1/\sqrt{N}$. Further, to be consistent with the definition of modal functions in a subarray cell, the modal voltages, denoted with overbar, should be modified as

$$\bar{V}_{00}^{\text{TE}} = \left[\frac{V_{00}^{\text{TE}}}{\sqrt{N}}\right]\sqrt{\frac{ab}{a_1 b_1}} = V_{00}^{\text{TE}} \qquad (5.34)$$

$$\bar{V}_{00}^{\text{TM}} = \left[\frac{V_{00}^{\text{TM}}}{\sqrt{N}}\right]\sqrt{\frac{ab}{a_1 b_1}} = V_{00}^{\text{TM}} \qquad (5.35)$$

which turn out to be identical with the element cell modal voltages. The multiplication factor $\sqrt{ab/(a_1 b_1)}$ is due to the difference in the modal function definitions in two different cells (see Section 3.2). Now from equation (3.92), we write the active radiated field in the desired (θ_0, ϕ_0) direction as

$$\vec{E}_{\text{sub}}(\theta_0, \phi_0) = c_1\left[\hat{\theta}_0\left\{jV_{00}^{\text{TM}}\sqrt{\frac{ab}{\lambda_0^2}}\right\} - \hat{\phi}_0\left\{jV_{00}^{\text{TE}}\sqrt{\frac{ab}{\lambda_0^2}}\cos\theta_0\right\}\right] \qquad (5.36)$$

We now rearrange the above equation to write

$$\vec{E}_{\text{sub}}(\theta_0, \phi_0) = c_1\sqrt{\frac{ab}{a_1 b_1}}\left[\hat{\theta}_0\left\{jV_{00}^{\text{TM}}\sqrt{\frac{a_1 b_1}{\lambda_0^2}}\right\} - \hat{\phi}_0\left\{jV_{00}^{\text{TE}}\sqrt{\frac{a_1 b_1}{\lambda_0^2}}\cos\theta_0\right\}\right] \qquad (5.37)$$

The term inside the square bracket is the active element pattern. Therefore we obtain the active subarray pattern as

$$\vec{E}_{\text{sub}}(\theta_0, \phi_0) = c_1\sqrt{\frac{ab}{a_1 b_1}}\vec{E}_{\text{ele}}(\theta_0, \phi_0) = c_1\sqrt{N}\ \vec{E}_{\text{ele}}(\theta_0, \phi_0) \qquad (5.38)$$

where $\vec{E}_{\text{ele}}$ is the active element pattern. The left-hand side of (5.38) would directly represent the subarray gain pattern if the active element pattern were normalized with respect to the input power. Notice, the active subarray pattern differs from the active element pattern because c_1 is a function of (θ_0, ϕ_0). Also, for an equal-split power divider, $c_1 = 1$ for the bore-sight scan; therefore the active subarray gain is N times the active element gain. This is not true for other scan angles, because the values of c_1 differ from unity as the array scans off bore sight. One may observe that c_1 effectively becomes the array factor for the elements in a subarray corresponding to the excitation vector $[\hat{V}_2^+]$.

5.7 FOUR-ELEMENT SUBARRAY FED BY POWER DIVIDER

A four-element subarray of open-ended waveguide elements is studied in this section. A one-to-four equal-split power divider excites the subarray. The schematic of the power divider is shown in Figure 5.6. Such a power divider can be realized

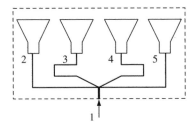

FIGURE 5.6 Schematic of a four-element subarray fed by an equal-split power divider of equal arm lengths.

by waveguide T-junctions or by a multifurcated waveguide or by using transmission line circuits. In order to examine the interaction between the subarray and the power divider network, we use a simplified model of the power divider which consists of an input transmission line that branches out to four transmission line sections (output lines) of equal lengths, as shown in Figure 5.6. For the best performance, the characteristic impedance of the input transmission line is one-quarter of the output transmission lines. The length of an open-ended waveguide element is chosen such that the impedance is purely resistive at the input end when the array radiates at the bore-sight direction under Floquet excitation. The characteristic impedances of the output transmission lines of the power divider are assumed to be equal to the input impedance of the open-ended waveguide array for the bore-sight radiation. This condition ensures a perfect impedance match for the power-divider-fed subarray when the array of subarrays radiates in the bore-sight direction.

Following a standard procedure [6] the impedance matrix elements of the five port power divider network can be deduced as

$$
\begin{aligned}
z_{11}^{f} &= -j \cot \beta l \\
z_{mm}^{f} &= j[3 \tan \beta l - \cot \beta l] \qquad m = 2, 3, 4, 5 \\
z_{1m}^{f} &= z_{m1}^{f} = -j \mathrm{cosec} \beta l \qquad m \neq 1 \\
z_{mn}^{f} &= -2j \mathrm{cosec}(2\beta l) \qquad m, n \neq 1 \qquad m \neq n
\end{aligned}
\tag{5.39}
$$

The normalized characteristic impedances (normalized with respect to the input transmission line) of the output lines (ports 2, 3, 4, and 5) are assumed to be 4. The line length of an output line is l and the propagation constant is β. The input transmission line has zero length. In this study we consider the following two linear subarrays: (a) a subarray along the E-plane and (b) a subarray along the H-plane of the waveguides.

5.7.1 *E*-Plane Subarray

The subarray impedance matrix $[Z_{\mathrm{sub}}]$ is obtained for a given scan angle using the procedure of Section 5.2. The input return loss is computed from the input impedance at (5.27) for various scan angles along the E-plane and plotted in Figure 5.7a. This input return loss corresponds to the Floquet excitation of the infinite array of subarrays. The waveguide inner dimensions were $0.6\lambda_0 \times 0.6\lambda_0$ and the subarray dimensions were $2.8\lambda_0 \times 0.7\lambda_0$. The waveguide heights were $0.75\lambda_0$, where λ_0 is the wavelength in free space. The plot shows better than 35 dB return loss for the bore-sight scan. However, several resonant spikes are observed in the return loss plot, implying poor impedance match for certain scan angles. Resonant spikes nearest to the bore sight occur at $\pm 3.4°$ scan angles. The poor return loss results in a significant reduction of the subarray gain, causing scan blindness. These spikes are primarily due to grazing lobes that cause strong mutual coupling between the elements, leading to scan blindness. The blind angles move with the transmission line length.

To compute the active subarray pattern we utilize (5.38). For computing c_1 in (5.38), the elements of the eigenvector $[s^{(1)}]$ for the subarray with 4×1 elements are considered as

$$s_1^{(1)} = \frac{1}{2}$$

$$s_2^{(1)} = \frac{1}{2} \exp\left(-\frac{1}{4} j \psi_x\right)$$

$$s_3^{(1)} = \frac{1}{2} \exp\left(-\frac{1}{2} j \psi_x\right) \tag{5.40}$$

$$s_4^{(1)} = \frac{1}{2} \exp\left(-\frac{1}{4} j 3 \psi_x\right)$$

with

$$\psi_x = k_0 a \sin \theta_0 \cos \phi_0 \tag{5.41}$$

where a is the subarray dimension along x. Figure 5.7b shows the active subarray pattern. Gain loss is observed near the $3.4°$ scan angle, which is consistent with the return loss data.

5.7.2 H-Plane Subarray

The return loss and the active subarray pattern of a four-element H-plane subarray are computed and plotted in Figure 5.8. Unlike the E-plane subarray, no resonant spike is observed. This is due to the fact that no grazing lobe exists on this plane, because of the $\cos \theta_0$ factor in the H-plane pattern function [see (5.37)].

It is found that the subarray gain is about 0.8 dB higher than that of a single linearly flared horn of the same aperture dimension as that of the subarray. The aperture field distribution of the single horn has a cosine taper along the H-plane, resulting in low aperture efficiency. However, multimode step horns can be designed to have almost uniform field distributions in both planes [7]. For a narrow-band application (typically 5–10% bandwidth), a multimode horn element (with large aperture size) is a better option than using a subarray of small horns, because the

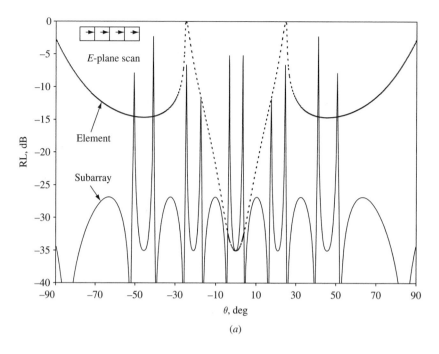

(a)

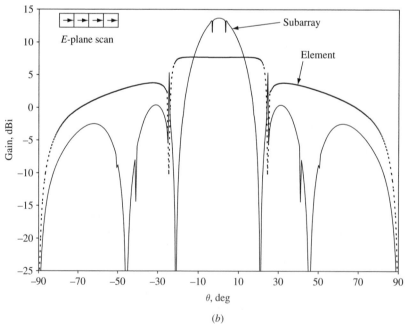

(b)

FIGURE 5.7 Return loss and active E-plane pattern of an E-plane subarray. Element aperture size $0.6\lambda_0 \times 0.6\lambda_0$, element cell size $0.7\lambda_0 \times 0.7\lambda_0$. (a) Return loss versus scan angle. (b) Active subarray pattern versus scan angle.

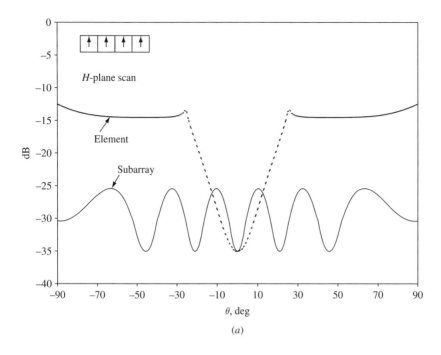

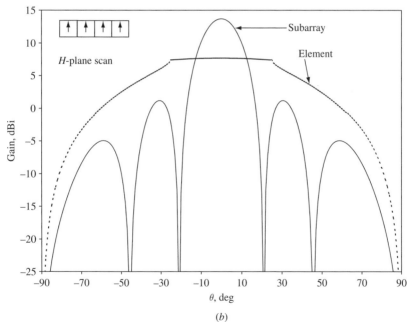

FIGURE 5.8 Return loss and active H-plane pattern of an H-plane subarray. Element aperture size $0.6\lambda_0 \times 0.6\lambda_0$, element cell size $0.7\lambda_0 \times 0.7\lambda_0$. ($a$) Return loss versus scan angle. (b) Active subarray pattern versus scan angle.

former does not require a power divider network. However, for resonant radiating elements, such as patch radiators, subarraying may be the only viable option to minimize the number of phase shifters while maintaining high aperture efficiency, because unlike a horn element, the patch element size cannot be increased arbitrarily.

5.8 SUBARRAY BLINDNESS

The scan blindness in the subarray observed in Figure 5.7b can be explained from the eigenexcitation point of view. The aperture voltage vector (assuming that the waveguide aperture supports only one mode, TE_{10} mode, for instance) for the waveguide elements of a four-element subarray, for instance, can be expressed as a linear combination of four eigenvectors of the subarray as follows:

$$[V_{in}] = c_1 [s^{(1)}] + c_2 [s^{(2)}] + c_3 [s^{(3)}] + c_4 [s^{(4)}] \qquad (5.42)$$

where $[s^{(i)}]$, $i = 1, 2, 3, 4$, are the eigenvectors and c_1, c_2, c_3, and c_4 are the voltage excitation coefficients that are determined using (5.33). The input current can be obtained by multiplying respective Floquet impedances with the eigenvectors:

$$[I_{in}] = c_1 Y_1^{FL} [s^{(1)}] + c_2 Y_2^{FL} [s^{(2)}] + c_3 Y_3^{FL} [s^{(3)}] + c_4 Y_4^{FL} [s^{(4)}] \qquad (5.43)$$

where Y_i^{FL}, $i = 1, 2, 3, 4$, are the Floquet admittances at the aperture plane. The complex power radiated by the eigenmodes is given by

$$P = [V_{in}]^T [I_{in}^*] \qquad (5.44)$$

Since the eigenvectors are orthonormal, we deduce

$$P = |c_1|^2 Y_1^{FL*} + |c_2|^2 Y_2^{FL*} + |c_3|^2 Y_3^{FL*} + |c_4|^2 Y_4^{FL*} \qquad (5.45)$$

A grazing grating lobe is associated with a cut-off Floquet mode with $k_z = 0$. This indicates that the corresponding TM Floquet modal admittance would be infinitely large at the radiating aperture plane (see Section 3.2.2). Such a large Floquet admittance corresponds to a large P. Since the input impedance is related to P, an impedance mismatch results, leading to scan blindness. The sharpness of the resonant spike in the return loss or gain curve depends on the relative amplitude of the cut-off mode. The lesser the relative amplitude, the shaper is the spike.

It must be pointed out that a complete blindness (with 0 dB return loss) does not happen always, particularly for large apertures, because the PDN adjusts the voltage distribution between the ports to compensate for high aperture admittance. A similar observation can be made with a large-aperture horn array. In that case multiple waveguide modes on the horn aperture adjust the voltage distribution to minimize the array blindness effect. A detailed study is undertaken in Chapter 8.

A similar type of resonance may occur when one of the eigenvectors excites a guided wave mode supported by the infinite array structure. The microstrip array structure is a typical example that supports guided wave modes, called surface wave modes. In that case coupling between a Floquet mode and the guided wave mode causes scan blindness.

The subarray blind angle can also be estimated employing the circle diagram method detailed in Section 3.4. In that case, the array unit cell size should be equal to the subarray size and the mode circles should be spaced accordingly.

5.9 CONCLUDING REMARKS

The purpose of this chapter was to demonstrate the eigenvector approach to analyze an array of subarrays. We primarily focused on an infinite array of subarrays under Floquet excitations because the results would be necessary for analyzing a finite array of subarrays as established in Chapter 4. To analyze a finite array of subarrays, the Floquet impedance at the input port of the PDN needs to be determined employing (5.27). This should be followed by mutual impedance (or admittance or scattering parameter) computation between subarrays using the procedure of Section 4.5. The impedance matrix (or scattering matrix) of a finite array of subarrays can be used to estimate the active return loss with respect to an amplitude distribution. The radiation pattern of the array can be determined by multiplying the active subarray pattern and the array factor associated with the amplitude distribution. Since the subarray cells are electrically large in size, one of the three models presented in Section 4.8 would be fairly accurate to estimate the active return loss and the radiation patterns.

REFERENCES

[1] A. K. Bhattacharyya, "Analysis of Multilayer Infinite Periodic Array Structures with Different Periodicities and Axes Orientations," *IEEE Trans.*, Vol. AP-48, No. 3, pp. 357–369, Mar. 2000.

[2] C. G. Montgomery, R. H. Dicke, and E. M. Purcell, in *Principles of Microwave Circuits* Dover, New York, 1965, "The Symmetry of Waveguide Junctions", Chapter 12.

[3] D. M. Pozar, "Scanning Characteristics of Infinite Array of Printed Antenna Subarrays," *IEEE Trans.*, Vol. AP-40, pp. 666–674, June 1992.

[4] G. Strang, *Linear Algebra and Its Applications*, Academic, New York, 1976.

[5] J. Mathews and R. L. Walker, *Mathematical Methods of Physics*, W. A. Benjamin, Reading, MA, 1979.

[6] R. E. Collin, *Foundation for Microwave Engineering*, McGraw-Hill, New York, 1966.

[7] A. K. Bhattacharyya and G. Goyette, "A Novel Horn Radiator with High Aperture Efficiency and Low Cross-Polarization and Applications in Arrays and Multi-Beam Reflector Antennas," *IEEE Trans*, Vol. AP-52, No. 11, pp. 2850–2859, Nov. 2004.

BIBLIOGRAPHY

Au, T. M., K. F. Tong, and K. M. Luk, "Characteristics of Aperture-Coupled Coplanar Microstrip Subarrays," *IEE Proc.*, Vol. MAP-144, pp. 137–140, Apr. 1997.

Coetzee, J. C., J. Joubert, and D. A. McNamara, "Off-Center-Frequency Analysis of a Complete Planar Slotted-Waveguide Array Consisting of Subarrays," *IEEE Trans.*, Vol. AP-48, No. 11, pp. 1746–1755, Nov. 2000.

Demir, S., "Efficiency Calculation of Feed Structures and Optimum Number of Antenna Elements in a Subarray for Highest G/T," *IEEE Trans.*, Vol. AP-52, No. 4, pp. 1024–1029, Apr. 2004.

Duffy, S. M., and D.M. Pozar, "Aperture Coupled Microstrip Subarrays," *Electron. Lett.*, Vol. 30, No. 23, pp. 1901–1902, Nov. 1994.

Fan, Z., Y. M. M. Antar, A. Ittipiboon, and A. Petosa, "Parasitic Coplanar Three-Element Dielectric Resonator Antenna Subarray," *Electron. Lett.*, Vol. 32, No. 9, pp. 789–790, Apr. 1996.

Fante, R. L., "System Study of Overlapped Subarrayed Scanning Antennas," *IEEE Trans.*, Vol. AP-28, No. 9, pp. 668–679, Sept. 1980.

Hansen, R. C., and G. G. Charlton, "Subarray Quantization Lobe Decollimation," *IEEE Trans.*, Vol. AP-47, No. 8, pp. 1237–1239, Aug. 1999.

Haupt, R. L., "Reducing Grating Lobes Due to Subarray Amplitude Tapering," *IEEE Trans.*, Vol. AP-33, No. 8, pp. 846–850, Aug. 1985.

Lee, K. F., R. Q. Lee, and T. Talty, "Microstrip Subarray with Coplanar and Stacked Parasitic Elements," *Electron. Lett.*, Vol. 26, No. 10, pp. 668–669, May 1990.

Lee, T-S., and T-K. Tseng, "Subarray-Synthesized Low-Side-Lobe Sum and Difference Patterns with Partial Common Weights," *IEEE Trans.*, Vol. AP-41, No. 6, pp. 791–800, June 1993.

Legay, H., and L. Shafai, "New Stacked Microstrip Antenna with Large Bandwidth and High Gain," *IEE Proc.*, Vol. MAP-141, pp. 199–204, June 1994.

Legay, H., and L. Shafai, "Analysis and Design of Circularly Polarized Series Fed Planar Subarray," *IEE Proc.*, Vol. MAP-142, pp. 173–177, Apr. 1995.

Lopez, P., J. A. Rodriguez, and E. Moreno, "Subarray Weighting for the Difference Patterns of Monopulse Antennas: Joint Optimization of Subarray Configurations and Weights," *IEEE Trans.*, Vol. AP-49, No. 11, pp. 1606–1608, Nov. 2001.

Mailloux, R. J., *Phased Array Antenna Handbook*, Artech House, Boston, 1994.

Mailloux, R. J., "Space-Fed Subarrays Using Displaced Feed," *Electron. Lett.*, Vol. 38, No. 21, pp. 1241–1243, Oct. 2002.

Mailloux, R. J., "A Low-Sidelobe Partially Overlapped Constrained Feed Network for Time-Delayed Subarrays," *IEEE Trans.*, Vol. AP-49, No. 2, pp. 280–291, Feb. 2001.

Motta, E. C., M. Hamidi, and J. P. Daniel, "Broadside Printed Antenna Arrays Built with Dissymmetrical Subarrays," *Electron. Lett.*, Vol. 27, No. 5, pp. 425–426, Feb. 1991.

Reddy, C. A., K. V. Janardhanan, K. K. Mukundan, and K. S. V. Shenoy, "Concept of an Interlaced Phased Array for Beam Switching," *IEEE Trans.*, Vol. AP-38, No. 4, pp. 573–575, Apr. 1990.

Sanzgiri, S., D. Bostrom, W. Pottenger, and R. Q. Lee, "A Hybrid Tile Approach for Ka Band Subarray Modules," *IEEE Trans.*, Vol. AP-43, No. 9, pp. 953–959, Sept. 1995.

Simons, R. N., R. Q. Lee, and G. R. Lindamood, "New Coplanar Waveguide/Stripline Feed Network for Seven Patch Hexagonal CP Subarray," *Electron. Lett.*, Vol. 27, No. 6, pp. 533–535, Mar. 1991.

Tamijani, A. A., and K. Sarabandi, "An Affordable Millimeter-Wave Beam Steerable Antenna Using Interleaved Planar Subarrays," *IEEE Trans.*, Vol. AP-51, No. 11, pp. 2193–2202, Nov. 2003.

PROBLEMS

5.1 An infinite array of subarrays is excited to radiate in the (θ, ϕ)-direction. Each subarray consists of 3×1 elements, and the subarray elements are aligned along the E-plane. The cell size of the elements within a subarray is $1.0\lambda_0 \times 1.0\lambda_0$, where λ_0 is the free-space wavelength.

   (a) Obtain the eigenvectors for the subarray excitation.

   (b) Assume that σ_1, σ_2, and σ_3 are three complex impedances associated with the eigencurrent excitations; obtain the impedance matrix of the subarray.

   (c) Show that the impedance matrix of the subarray is not symmetrical in general.

   (d) Recall that σ_1, σ_2, and σ_3 are the Floquet impedances with respect to three different scan angles of an infinite array of elements with cell size $1.0\lambda_0 \times 1.0\lambda_0$. If $\theta = 10°$, $\phi = 45°$, then compute the scan angles associated with σ_1, σ_2, and σ_3. Assume a square lattice of the subarrays.

5.2 An infinite array of magnetic line current sources is shown in Figure P5.2. The magnetic currents are uniform along y and lying on an infinite ground plane. The magnitudes of the two consecutive line currents are V_1 and V_2, respectively, and this "magnitude pattern" repeats (as in an array of two-element subarrays) with linear phase progression, as shown in the figure. Suppose ψ $(0 \leq \psi < 2\pi)$ is the phase difference between two sources of identical magnitudes. Then show that the magnetic field produced by all the sources of magnitude V_1 is given by

$$\vec{H}^{(1)} = -\hat{y}V_1 \frac{k_0}{\eta a} \sum_{m=-\infty}^{\infty} \frac{1}{\sqrt{k_0^2 - k_{xm}^2}} \exp[-jk_{xm}(x + \tfrac{1}{4}a) - j\sqrt{k_0^2 - k_{xm}^2}\, z]$$

The magnetic field produced by all the sources of magnitude V_2 is

$$\vec{H}^{(2)} = -\hat{y}V_2 \frac{k_0}{\eta a} \sum_{m=-\infty}^{\infty} \frac{1}{\sqrt{k_0^2 - k_{xm}^2}} \exp[-jk_{xm}(x - \tfrac{1}{4}a) - j\sqrt{k_0^2 - k_{xm}^2}\, z]$$

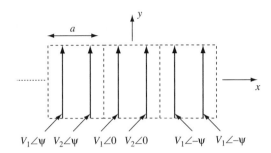

FIGURE P5.2 Array of subarrays of magnetic line currents.

with $k_{xm} = (\psi + 2m\pi)/a$. Hence show that

$$\iint_{xz} \vec{M}_{10} \cdot \vec{H}^{(2)} dx\, dz \neq \iint_{xz} \vec{M}_{20} \cdot \vec{H}^{(1)} dx dz \qquad \text{if } \psi \neq 0$$

where

$$\vec{M}_{1n} = \hat{y} V_1 \delta(x + \tfrac{1}{4}a + na) \exp(-jn\psi)\delta(z)$$

$$\vec{M}_{2n} = \hat{y} V_2 \delta(x - \tfrac{1}{4}a + na) \exp(-jn\psi)\delta(z)$$

are the magnetic, current densities. The above relation indicates that the $[Z]$ matrix of a subarray is nonsymmetrical. Also show that

$$\sum_n \iint_{xz} \vec{M}_{1n} \cdot \vec{H}^{(2)} dx\, dz = \sum_n \iint_{xz} \vec{M}_{2n} \cdot \vec{H}^{(1)} dx\, dz = 0$$

for $\psi \neq 0$. (The reciprocity relation is satisfied if the entire source of the array is considered. The integrals diverge for $\psi = 0$.)

5.3 With reference to problem 5.1, obtain the eigenvalues of the subarray scattering matrix in terms of σ_1, σ_2, and σ_3. Then obtain the scattering matrix of the subarray.

5.4 Show that the diagonal elements of a subarray $[Z]$ matrix are identical. Explain the physical significance of this property of the $[Z]$ matrix.

5.5 An infinite array of subarrays is under Floquet excitation for a scan angle of (θ_s, ϕ_s). Suppose $f(\theta, \phi)$ is the Floquet impedance of the elements in a subarray. Each subarray has 3×1 elements. The element cell size is $0.7\lambda_0 \times 0.7\lambda_0$. The subarrays are arranged in rectangular grids. The elements of a subarray are excited by ideal voltage sources with zero internal impedances.

The magnitudes of the source voltages within a subarray are in the ratio $1:2:1$ with equal phase angles.

(a) Obtain the eigenvectors of the subarray

(b) Expand the source voltage in terms of the orthogonal eigenvectors

(c) Using superposition of the eigenexcitations, determine the input impedance seen by each element within a subarray.

5.6 The Floquet impedance of an infinite array of elements is given by the function $Z^{FL}(\theta, \phi) = f(\theta, \phi)$, where f is a complex function of θ and ϕ, (θ, ϕ) being the scan direction. The elements are arranged in a triangular grid of grid angle $60°$ and a as the side of the triangle. Every four elements are grouped together to make subarrays as shown in Figure P5.6.

(a) Find the subarray cell dimensions and the grid angle.

(b) Obtain the subarray eigenvectors and the eigenvalues for a scan angle (θ, ϕ).

(c) Obtain the subarray $[Z]$ matrix elements in terms of the function f.

5.7 Let the $[Z]$ matrix of a subarray with 2×1 elements be given by

$$[Z] = \begin{bmatrix} z_{11} & z_{12} \\ z_{21} & z_{22} \end{bmatrix}$$

when the array of subarrays scan in the (θ_s, ϕ_s)-direction. The elements and the subarrays have rectangular grids. The subarray cell size is $a \times b$.

(a) Show that $z_{11} = z_{22}$, $|z_{12}| = |z_{21}|$.

(b) Show that $z_{12}/z_{21} = \exp(jk_0 a \sin\theta_s \cos\phi_s)$.

(c) Explain with reasoning that for a subarray of $M \times 1$ elements the mutual impedance between two elements within a subarray satisfies the relation

$$z_{ij} = z_{i+1, j+1}, \quad i, j < M.$$

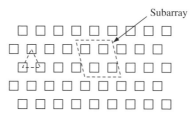

FIGURE P5.6 Array of elements in triangular grid. A subarray consists of four elements as shown.

5.8   For a subarray of $N \times 1$ elements, show that the mutual impedance between the first and nth elements is given by

$$z_{n1} = \frac{1}{N} \sum_{i=0}^{N-1} Z^{FL}\left(\frac{\psi_x + 2\pi i}{N}, \psi_y\right) \exp\left\{-j(n-1)\frac{\psi_x + 2\pi i}{N}\right\}$$

where $\psi_x = k_0 a \sin\theta\cos\phi$, $\psi_y = k_0 b \sin\theta\sin\phi$, (θ, ϕ) is the direction of radiation, and $a \times b$ is the subarray cell size; Z^{FL} is the Floquet impedance of the infinite array of elements of unit cell size $a/N \times b$. Show that when N approaches infinity, the above summation approaches

$$z_{n1} = \frac{1}{2\pi} \int_0^{2\pi} Z^{FL}(\alpha, \psi_y) \exp\{-j(n-1)\alpha\} d\alpha$$

which is equal to the mutual impedance between two elements in an infinite array environment as deduced in Chapter 3.

5.9   Deduce the relations in (5.39). Obtain the reflection coefficient at the input port of the power divider in terms of the scattering matrices of the power divider network and the subarray.

5.10   Obtain the 3×3 $[Z]$ matrix of a 1:2 binary-split power divider. The characteristic impedance of the input transmission line is z_{01} and that of the output transmission lines is z_{02}. The input transmission line length is zero and the output transmission line length is l with β as the propagation constant of the output transmission lines.

5.11   Suppose $Z^{FL}(\theta)$ is the Floquet impedance for the elements of a one-dimensional array for a scan angle θ. Every two elements are fed by an equal-split power divider. The element spacing is a. Use $\lambda_0 = 1$. Obtain the input impedance seen at the power divider input as a function of θ. Assume z_{01} and z_{02} as the characteristic impedances of the input and output transmission lines, respectively. If $f(\theta)$ is the active element pattern, then find the expression for the active subarray pattern.

5.12   A subarray with 4×1 elements is excited by a 1:4 equal-split power divider as in Figure 5.6. The length of an output transmission line section is $\lambda_0/2$. All transmission line sections have equal lengths. Using superposition of the eigenvector excitations, show that the admittance at the power divider input port is given by

$$y_{in} = \sum_{i=1}^4 y_{in}^{(i)}$$

where

$$y_{in}^{(i)} = \sum_{k=1}^4 c_k Y_k^{FL} \zeta_i^{(k)} \qquad c_k = \sum_{n=1}^4 \zeta_n^{(k)*}$$

$\zeta_n^{(k)}$ being the nth element of the kth eigenvector and Y_k^{FL} the Floquet admittance at the element input for the kth eigenvector excitation. Also, deduce that

$$y_{\text{in}} = \sum_{i=1}^{4} |c_i|^2 Y_i^{\text{FL}}$$

5.13 It is found that a blind spot occurs at a scan angle of $\theta = 10°$ on the E-plane of an infinite array of patch elements of element cell size $0.8\lambda_0 \times 0.8\lambda_0$, where λ_0 is the free-space wavelength. The elements are in square grids. If every group of four elements is excited by a single power divider network to form a subarray, then determine the nearest blind angle that may occur for such an array of subarrays for the following four cases:

  (a) Subarray made with 4×1 elements, subarrays are in rectangular grids.
  (b) Subarray made with 1×4 elements, subarrays are in rectangular grids.
  (c) Subarray made with 2×2 elements, subarrays are in rectangular grids.
  (d) Subarray made with 2×2 elements, subarrays are in triangular grids of grid angle $63.4°$.

5.14 A subarray consists of 2×2 horn elements. The horn elements are in square grids of cell size $1.2\lambda_0 \times 1.2\lambda_0$. Using circle diagrams or otherwise determine the scan direction nearest to the boresight for which array blindness may occur for the following two cases:

  (a) subarrays in square grids and
  (b) subarrays in triangular grids of grid angle $63.4°$.

5.15 Show that the maximum scan angle, θ_m, for a linear array of subarrays within which no array blindness occurs is given by $\sin\theta_m = |1 - n\lambda_0/a|$, where n is the integer nearest to a/λ_0, and a is the subarray dimension (notice, if a/λ_0 is an integer, then the resonance may occur in the bore-sight scan angle).

5.16 Show that the self-impedance matrix for the elements of a subarray, when all other elements outside the subarray are open circuited, is given by

$$[Z_{\text{self}}] = \frac{1}{4\pi^2} \int\limits_{-\pi}^{\pi} \int\limits_{-\pi}^{\pi} [Z]\, d\psi_x d\psi_y$$

where $[Z]$ represents the impedance matrix when the array of subarrays is under Floquet excitation with the phase difference pair (ψ_x, ψ_y) between adjacent subarrays, deduced in Section 5.2.1. The subarrays are in rectangular grids.

CHAPTER SIX

GSM Approach for Multilayer Array Structures

6.1 INTRODUCTION

In this chapter we present the generalized scattering matrix (GSM) approach to analyze multilayer array structures [1]. Multilayer structures have applications in printed antennas to enhance the bandwidth performance [2]. Multilayer arrays are also used as frequency-selective surfaces [3], as screen polarizers [4], and for realizing photonic bandgap materials [5]. It will be seen that the GSM approach is very convenient for analyzing such structures. The GSM approach primarily evolved from the well-established mode matching approach [6] that became prevalent for analyzing passive microwave components such as step and corrugated horns, iris-coupled filters, and polarizers. The GSM approach essentially is a modular approach, where each layer of a multilayer structure is analyzed independently and then characterized in terms of a multidimensional matrix. The multidimensional matrix is called the GSM of the layer, because the reflection and transmission characteristics of the layer with respect to several incident modes are embedded within the matrix. The complete characterization of a multilayer structure is obtained by cascading the individual GSMs of the layers.

The chapter begins with a conceptual illustration of the GSM approach followed by a mathematical definition of the GSM. Then the GSM cascading rules associated with two consecutive layers are elaborated in great detail. Advantages of the GSM approach over a transmission matrix approach are discussed from a numerical stability standpoint. The GSMs of basic building blocks, relevant for a multilayer array analysis, are derived. In particular, the GSM of a patch layer is obtained invoking the two-dimensional Galerkin periodic method-of-moment (MoM) procedure. The analysis is limited to a thin patch layer, though, in principle, the GSM approach can be applied for electrically thick patch arrays also, which require a three-dimensional MoM procedure. Stationary expressions for the GSM elements are derived. Such expressions are found to match with Galerkin's MoM solution, implying that the MoM-based GSM is stationary. Convergence criteria and limitations of Galerkin's MoM analysis with respect to printed arrays are investigated. Important features of the GSM approach are outlined and other numerical methods are discussed at the end.

6.2 GSM APPROACH

The essence of the GSM approach lies in its modularity. Each individual layer of a multilayer structure is considered as a module, which is analyzed independently. Then, a matrix is constructed which represents the input–output characteristics of the module with respect to multiple Floquet modes. This matrix is called the GSM of the layer or module. The GSMs of all modules associated with the entire structure are then combined to obtain an overall GSM of the structure. The overall GSM essentially provides the input–output characteristics with respect to the Floquet modes.

To illustrate the GSM approach pictorially, consider a three-layer periodic array structure (patch–dielectric–patch), as shown in Figure 6.1*a*. Figure 6.1*b* is the

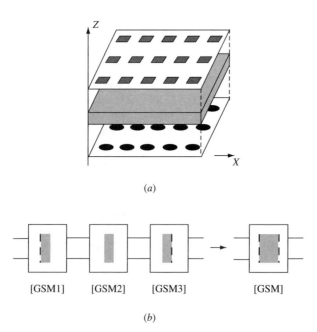

(*a*)

[GSM1]   [GSM2]   [GSM3]   [GSM]

(*b*)

FIGURE 6.1   (*a*) Exploded view of a three-layer array structure. (*b*) Two-port representation of each module and of the entire structure.

two-port representation of each module and of the entire structure. Identical cell sizes and cell orientations for the periodic arrays are assumed. Also, the structure is assumed to be of infinite extent along x- and y-directions and is under Floquet excitation.

The GSM of a module is defined using the relation between incident and reflected voltages as

$$\begin{bmatrix} a_1^- \\ a_2^- \end{bmatrix} = \begin{bmatrix} S_{11} & S_{12} \\ S_{21} & S_{22} \end{bmatrix} \begin{bmatrix} a_1^+ \\ a_2^+ \end{bmatrix} \tag{6.1}$$

where $[a_1^+]$ and $[a_2^+]$ are the incident voltage vectors with respect to the Floquet modes at the two sides (or ports) of the module and $[a_1^-]$ and $[a_2^-]$ are the corresponding reflected voltage vectors. The $[S]$ matrix on the right-hand side of (6.1) is called the GSM of the layer. It consists of four submatrices, namely $[S_{11}]$, $[S_{12}]$, $[S_{21}]$, and $[S_{22}]$, respectively. For equal number of modes considered at the two ports, the submatrices are square matrices of order $n \times n$, where n, is the number of modes at each port. However, for many applications, different numbers of modes at the two ports are used. In that case, $[S_{12}]$ and $[S_{21}]$ are not square matrices while $[S_{11}]$ and $[S_{22}]$ still remain square matrices of different orders. Determination of the matrix elements is considered later.

6.3 GSM CASCADING RULE

Cascading of multiple GSMs is necessary to obtain the overall GSM of several consecutive layers of periodic arrays. Toward that goal, we first derive the expression for the combined GSM of two consecutive layers in terms of the GSMs of individual layers. This formula can be applied repeatedly to obtain the overall GSM of a multilayer array. The GSM cascading rule is applicable only if the layers have identical periodicities and have identical cell orientations that can be uniquely denoted by $[a, b, \gamma]$. This ensures that a Floquet modal vector function has an identical expression for all the layers. Suppose the GSMs of layer A and layer B in Figure 6.2 are given by

$$\begin{bmatrix} a_1^- \\ a_2^- \end{bmatrix} = \begin{bmatrix} S_{11}^A & S_{12}^A \\ S_{21}^A & S_{22}^A \end{bmatrix} \begin{bmatrix} a_1^+ \\ a_2^+ \end{bmatrix} \tag{6.2}$$

$$\begin{bmatrix} b_1^- \\ b_2^- \end{bmatrix} = \begin{bmatrix} S_{11}^B & S_{12}^B \\ S_{21}^B & S_{22}^B \end{bmatrix} \begin{bmatrix} b_1^+ \\ b_2^+ \end{bmatrix} \tag{6.3}$$

Suppose m is the number of modes in port 1 of layer A, n is the number of modes in port 2 of layer A, and p is the number of modes in port 2 of layer B. The number of modes in port 1 of layer B must be equal to the number of modes in port 2 of layer A. Therefore, $[a_1^+]$ and $[a_1^-]$ are of order $m \times 1$, $[a_2^+]$, $[a_2^-]$, $[b_1^+]$ and $[b_1^-]$ are of order $n \times 1$ and $[b_2^+]$ and $[b_2^-]$ are of order $p \times 1$. Now (6.2) and (6.3) can be written as

$$[a_1^-] = [S_{11}^A][a_1^+] + [S_{12}^A][a_2^+] \tag{6.4}$$

$$[a_2^-] = [S_{21}^A][a_1^+] + [S_{22}^A][a_2^+] \tag{6.5}$$

$$[b_1^-] = [S_{11}^B][b_1^+] + [S_{12}^B][b_2^+] \tag{6.6}$$

$$[b_2^-] = [S_{21}^B][b_1^+] + [S_{22}^B][b_2^+] \tag{6.7}$$

For the cascade connection shown in Figure 6.2, the following two conditions must be satisfied:

$$[a_2^-] = [b_1^+] \tag{6.8}$$

$$[a_2^+] = [b_1^-] \tag{6.9}$$

Our objective is to find the combined GSM defined by the following equation:

$$\begin{bmatrix} a_1^- \\ b_2^- \end{bmatrix} = \begin{bmatrix} S_{11}^{AB} & S_{12}^{AB} \\ S_{21}^{AB} & S_{22}^{AB} \end{bmatrix} \begin{bmatrix} a_1^+ \\ b_2^+ \end{bmatrix} \tag{6.10}$$

Toward that goal, we use (6.5), (6.6), (6.8), and (6.9) to express $[a_2^+]$ in terms of $[a_1^+]$ and $[b_2^+]$. This yields

$$[a_2^+] = [I - S_{11}^B S_{22}^A]^{-1}[S_{11}^B][S_{21}^A][a_1^+] + [I - S_{11}^B S_{22}^A]^{-1}[S_{12}^B][b_2^+] \tag{6.11}$$

In the above $[I]$ represents an identity matrix of order $n \times n$. Next, we eliminate $[a_2^+]$ from (6.4) using (6.11) and obtain the following equation:

$$[a_1^-] = \{[S_{11}^A] + [S_{12}^A][I - S_{11}^B S_{22}^A]^{-1}[S_{11}^B][S_{21}^A]\}[a_1^+] + [S_{12}^A][I - S_{11}^B S_{22}^A]^{-1}[S_{12}^B][b_2^+] \tag{6.12}$$

Observe that $[a_1^+]$ and $[b_2^+]$ become the incident voltage vectors for the combination of two layers. Therefore (6.12) yields the two submatrices of the combined GSM as

$$[S_{11}^{AB}] = [S_{11}^A] + [S_{12}^A][I - S_{11}^B S_{22}^A]^{-1}[S_{11}^B][S_{21}^A] \tag{6.13a}$$

$$[S_{12}^{AB}] = [S_{12}^A][I - S_{11}^B S_{22}^A]^{-1}[S_{12}^B] \tag{6.13b}$$

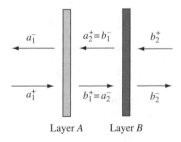

Layer A   Layer B

FIGURE 6.2 Incident, reflected, and transmitted voltage vectors of two layers in cascade.

The remaining two submatrices of the combined GSM can be obtained in a similar fashion. They are derived as

$$[S_{21}^{AB}] = [S_{21}^{B}][I - S_{22}^{A}S_{11}^{B}]^{-1}[S_{21}^{A}] \tag{6.13c}$$

$$[S_{22}^{AB}] = [S_{22}^{B}] + [S_{21}^{B}][I - S_{22}^{A}S_{11}^{B}]^{-1}[S_{22}^{A}][S_{12}^{B}] \tag{6.13d}$$

By inspection one can also write the expressions for $[S_{21}^{AB}]$ and $[S_{22}^{AB}]$ directly from the right-hand sides of (6.13b) and (6.13a), respectively, by swapping A and B and 1 and 2.

Equations (6.13) can also be established by taking proper account of all reflected and transmitted voltage vectors. The multiple reflections between two layers are illustrated in Figure 6.3. Suppose the voltage vector $[a_1^+]$ is incident on layer A (denoted by the line AA). The transmitted voltage vector that emerged from AA becomes $[S_{21}^{A}][a_1^+]$, which now is incident on layer B (denoted by the line BB). This voltage vector is partly reflected and partly transmitted by BB. The output port of B is assumed to be match terminated. To obtain the reflected and transmitted voltage vectors from BB, the incident voltage vector should be premultiplied by $[S_{11}^{B}]$ and $[S_{21}^{B}]$, respectively. The transmitted voltage vector along the direct path therefore becomes equal to $[S_{21}^{B}S_{21}^{A}][a_1^+]$. The reflected voltage vector from BB rereflects from AA, which again re-transmits through BB. The transmitted voltage vector after two reflections becomes equal to $[S_{21}^{B}S_{22}^{A}S_{11}^{B}S_{21}^{A}][a_1^+]$. Similarly, after four reflections, the transmitted voltage vector would be $[S_{21}^{B}S_{22}^{A}S_{11}^{B}S_{22}^{A}S_{11}^{B}S_{21}^{A}][a_1^+]$ and so on. The total transmitted voltage vector after an infinite number of reflections becomes

$$[b_2^-] = \{[S_{21}^{B}][S_{21}^{A}] + [S_{21}^{B}][S_{22}^{A}S_{11}^{B}][S_{21}^{A}] + [S_{21}^{B}][S_{22}^{A}S_{11}^{B}]^2[S_{21}^{A}] + \cdots \infty\}[a_1^+] \tag{6.14}$$

The above infinite summation can be written as

$$[b_2^-] = [S_{21}^{B}]\{[I + S_{22}^{A}S_{11}^{B} + [S_{22}^{A}S_{11}^{B}]^2 + \cdots \infty]\}[S_{21}^{A}][a_1^+] \tag{6.15}$$

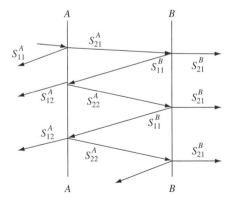

FIGURE 6.3 Sketch of multiple reflection phenomena between two modules.

If the above infinite sum of matrices converges (which happens if magnitudes of the eigenvalues of the matrix $[S_{22}^A S_{11}^B]$ are less than unity), then the sum will have a simple expression as follows:

$$[I + S_{22}^A S_{11}^B + [S_{22}^A S_{11}^B]^2 + \cdots \infty] = [I - S_{22}^A S_{11}^B]^{-1} \tag{6.16}$$

Substituting (6.16) into (6.15) we obtain

$$[b_2^-] = [S_{21}^B][I - S_{22}^A S_{11}^B]^{-1}[S_{21}^A][a_1^+] \tag{6.17}$$

Since the output port of B is match terminated, $[b_2^+] = [0]$ (a null vector). Equations (6.10) and (6.17) lead to

$$[S_{21}^{AB}] = [S_{21}^B][I - S_{22}^A S_{11}^B]^{-1}[S_{21}^A] \tag{6.18}$$

which is same as given in (6.13c). The other submatrices for the two layers can be determined similarly.

6.4 TRANSMISSION MATRIX REPRESENTATION

A periodic array structure can also be characterized by a transmission matrix. The transmission matrix, $[T]$, of a layer relates the input port voltage vectors (both reflected and transmitted) to the output port voltage vectors. With reference to Figure 6.2, the following relation defines the transmission matrix:

$$\begin{bmatrix} a_2^- \\ a_2^+ \end{bmatrix} = \begin{bmatrix} T_{11} & T_{12} \\ T_{21} & T_{22} \end{bmatrix} \begin{bmatrix} a_1^+ \\ a_1^- \end{bmatrix} \tag{6.19}$$

For some cases, the transmission matrix method could be advantageous for analyzing multiple layers. This is primarily due to the fact that the overall transmission matrix of two consecutive layers can be obtained by simple multiplication of the individual transmission matrices. Unlike the GSM cascading process, no matrix inversion is required here. This can be proven by direct substitution. However, in many situations, the $[T]$ matrix formulation suffers from a numerical instability problem that will be explained shortly.

Let us represent the $[T]$ matrix of a layer in terms of its GSM. From (6.1) we can write

$$[a_1^-] = [S_{11}][a_1^+] + [S_{12}][a_2^+] \tag{6.20}$$

$$[a_2^-] = [S_{21}][a_1^+] + [S_{22}][a_2^+] \tag{6.21}$$

From (6.19) we write

$$[a_2^+] = [T_{21}][a_1^+] + [T_{22}][a_1^-] \tag{6.22}$$

Equations (6.22) and (6.20) can be compared to determine $[T_{21}]$ and $[T_{22}]$. For that we premultiply (6.20) by $[S_{21}]$ and then with minor algebraic manipulation we obtain

$$[a_2^+] = -[PS_{21}S_{11}][a_1^+] + [PS_{21}][a_1^-] \tag{6.23}$$

where

$$[P] = [S_{21}S_{12}]^{-1} \tag{6.24}$$

Comparing (6.22) and (6.23) we readily obtain

$$[T_{21}] = -[PS_{21}S_{11}] = -[S_{21}S_{12}]^{-1}[S_{21}S_{11}] \tag{6.25}$$

$$[T_{22}] = [PS_{21}] = [S_{21}S_{12}]^{-1}[S_{21}] \tag{6.26}$$

For obtaining $[T_{11}]$ and $[T_{12}]$ we substitute $[a_2^+]$ from (6.23) into (6.21). This yields

$$[a_2^-] = [S_{21} - S_{22}PS_{21}S_{11}][a_1^+] + [S_{22}PS_{21}][a_1^-] \tag{6.27}$$

Comparing (6.19) and (6.27) we have

$$[T_{11}] = [S_{21} - S_{22}PS_{21}S_{11}] = [S_{21}] - [S_{22}][S_{21}S_{12}]^{-1}[S_{21}S_{11}] \tag{6.28}$$

$$[T_{12}] = [S_{22}PS_{21}] = [S_{22}][S_{21}S_{12}]^{-1}[S_{21}] \tag{6.29}$$

It is important to note that the above $[T]$ matrix becomes a square matrix only if the numbers of modes at the two sides of a layer are equal. Furthermore, $[S_{21}S_{12}]$ in (6.24) is a square matrix; however, it is invertible only if the number of modes in port 2 is not greater than the number of modes in port 1. Otherwise the rank of $[S_{21}S_{12}]$ will be lower than its dimension. Thus, conversion from a GSM to a $[T]$ matrix fails if the number of modes in port 2 is greater than the number of modes in port 1.

As mentioned earlier, the $[T]$ matrix formulation for a multilayer array structure is not recommended because of the following reason. Typically, to obtain the $[T]$ matrix, one obtains the GSM first and then converts, which requires inversion of the $[S_{21}S_{12}]$ matrix. If a large number of modes are considered in the analysis, which may be required for convergence of the end results, then $[S_{21}]$ and $[S_{12}]$ may become ill-conditioned matrices. As a result, the numerical computation may suffer from instability problems (see problem 6.5).

6.5 BUILDING BLOCKS FOR GSM ANALYSIS

Most useful multilayer array structures generally consist of several dielectric layers, dielectric interfaces, and layers of conducting patch arrays. Therefore, in order to analyze such structures one needs to know the GSMs of these basic "building blocks".

6.5.1 Dielectric Layer

Consider an isotropic and homogeneous dielectric layer of thickness d and dielectric constant ε_r. Suppose M Floquet modes are incident on port 1 of the layer, as depicted in Figure 6.4. Suppose β_i is the propagation constant of the ith mode. For a propagating mode β_i is real and for an evanescent mode β_i is imaginary in a lossless dielectric medium.

Let us first obtain the $[S_{11}]$ submatrix of the GSM. From (6.1) we can write

$$[a_1^-] = [S_{11}][a_1^+] \qquad \text{if } [a_2^+] = [0] \qquad (6.30)$$

Now, $[a_2^+] = [0]$ indicates that nothing reflects back from the output port, which can be simulated by a match termination. Further, the incident waves do not undergo any change of the medium; thus no reflection occurs, indicating $[a_1^-] = [0]$. This is true for an arbitrary incident voltage vector $[a_1^+]$. In order to satisfy this condition, the submatrix $[S_{11}]$ must be a null matrix. Therefore, for a dielectric layer we obtain

$$[S_{11}] = [0] \qquad (6.31)$$

From symmetry we can also write

$$[S_{22}] = [0] \qquad (6.32)$$

The $[S_{21}]$ matrix is defined by

$$[S_{21}][a_1^+] = [a_2^-] \qquad \text{if } [a_2^+] = [0] \qquad (6.33)$$

If $[a_2^+] = 0$, that is, port 2 is match terminated for all the modes, then the incident modes remain purely progressive in nature. The modal voltages between the input port and the output port differ by a phase factor only. Since the modes are mutually orthogonal, we can write

$$a_2^-(i) = a_1^+(i)\exp(-j\beta_i d) \qquad (6.34)$$

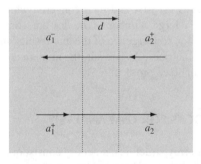

FIGURE 6.4 Incident and reflected voltage vectors in a dielectric layer of thickness d. The incident voltage vectors are launched from the same dielectric medium (no change of medium occurs). The GSM is contributed by the propagation delay only.

where $a_2^-(i)$ and $a_1^+(i)$ are the modal voltages at the output and input ports, respectively, for the ith mode. Equation (6.34) yields the $[S_{21}]$ matrix as

$$[S_{21}] = \begin{bmatrix} t_1 & 0 & 0 & 0 & \cdots & 0 \\ 0 & t_2 & 0 & 0 & \cdots & 0 \\ 0 & 0 & t_3 & 0 & \cdots & 0 \\ \cdots\cdots\cdots\cdots & \cdots & \cdots \\ 0 & 0 & 0 & \cdots & \cdots & \cdots \end{bmatrix} \tag{6.35}$$

which is a diagonal matrix of order $N \times M$, with N as the number of modes considered in port 2. The diagonal elements t_i are given by

$$t_i = \exp(-j\beta_i d) \tag{6.36}$$

Similarly, $[S_{12}]$ is also a diagonal matrix of order $M \times N$ with t_i as diagonal elements. If $M = N$, then $[S_{12}] = [S_{21}]$.

6.5.2 Dielectric Interface

The interface of two dielectric media is shown in Figure 6.5. The dielectric constants of the media are ε_{r1} and ε_{r2}, respectively. Suppose $Z_{01}^{FL}(i)$ and $Z_{02}^{FL}(i)$ are the characteristic impedances of the ith Floquet mode at the two dielectric media. Then the reflection coefficient of the mode, incident from medium 1 to medium 2 is given by

$$\Gamma_{21}(i) = \frac{Z_{02}^{FL}(i) - Z_{01}^{FL}(i)}{Z_{02}^{FL}(i) + Z_{01}^{FL}(i)} \tag{6.37}$$

The subscript 21 in Γ indicates that the wave travels toward medium 2 from medium 1. The above reflection coefficient is valid only if medium 2 is of infinite extent or equivalently the output port is match terminated. Therefore, using the definition of $[S_{11}]$ in (6.30), we can write $[S_{11}]$ for the interface as

$$[S_{11}] = \begin{bmatrix} \Gamma_{21}(1) & 0 & 0 & 0 & \cdots & 0 \\ 0 & \Gamma_{21}(2) & 0 & 0 & \cdots & 0 \\ \cdots & \cdots & \cdots\cdots\cdots & \cdots & \cdots \\ 0 & 0 & 0 & \cdots\cdots & \cdots & \Gamma_{21}(M) \end{bmatrix} \tag{6.38}$$

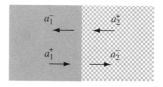

Medium 1   Medium 2

FIGURE 6.5 Incident and reflected voltage vectors at the interface of two dielectric media. The incident voltage vectors are launched from one medium to the other medium.

which is a diagonal matrix of order $M \times M$. The off-diagonal terms are zero because an interface does not cause any coupling between the Floquet modes assuming that TE_z and TM_z types of modes are used, where the z-axis is normal to the interface.

To obtain $[S_{21}]$, we need to find the transmitted voltage with respect to the incident voltage when the output port is match terminated. Form the continuity condition we know that the total voltage before the interface and after the interface must remain unchanged. The total voltage before the interface for the ith mode is given by

$$v(i) = a_1^+(i) + a_1^-(i) = [1 + \Gamma_{21}(i)]a_1^+(i) \tag{6.39}$$

Now $v(i)$ must be the voltage after the interface because of the continuity condition. Further, this voltage must be equal to $a_2^-(i)$ because the output port is match terminated. Thus we write

$$a_2^-(i) = [1 + \Gamma_{21}(i)]a_1^+(i) \tag{6.40}$$

According to (6.33) we obtain

$$[S_{21}] = \begin{bmatrix} \{1 + \Gamma_{21}(1)\} & 0 & 0 & 0 & \cdots & 0 \\ 0 & \{1 + \Gamma_{21}(2)\} & 0 & 0 & \cdots & 0 \\ \cdots & \cdots & \cdots & \cdots & \cdots & \cdots \\ 0 & 0 & 0 & \cdots & \cdots & \cdots \end{bmatrix} \tag{6.41}$$

which is another diagonal matrix of order $N \times M$. The submatrices $[S_{22}]$ and $[S_{12}]$ can be determined from (6.38) and (6.41), respectively, replacing Γ_{21} by Γ_{12}. From (6.31) we obtain

$$\Gamma_{12}(i) = -\Gamma_{21}(i) \tag{6.42}$$

Therefore we finally obtain $[S_{22}]$ and $[S_{12}]$ as

$$[S_{22}] = \begin{bmatrix} -\Gamma_{21}(1) & 0 & 0 & 0 & \cdots & 0 \\ 0 & -\Gamma_{21}(2) & 0 & 0 & \cdots & 0 \\ \cdots & \cdots & \cdots & \cdots & \cdots & \cdots \\ 0 & 0 & 0 & \cdots & \cdots & -\Gamma_{21}(N) \end{bmatrix} \tag{6.43}$$

$$[S_{12}] = \begin{bmatrix} \{1 - \Gamma_{21}(1)\} & 0 & 0 & 0 & \cdots & 0 \\ 0 & \{1 - \Gamma_{21}(2)\} & 0 & 0 & \cdots & 0 \\ \cdots & \cdots & \cdots & \cdots & \cdots & \cdots \\ 0 & 0 & 0 & \cdots & \cdots & \cdots \end{bmatrix} \tag{6.44}$$

6.5.3 Array of Patches

In this section we will obtain the GSM of a layer consisting of an infinite array of thin conducting patch elements [1]. For the purpose of generality, we assume that the patch elements are lying at the interface of two dielectric media, as shown in

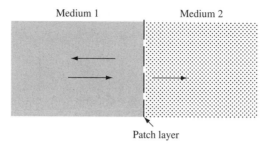

FIGURE 6.6   Infinite array of patches at the interface of two dielectric media.

Figure 6.6. The GSM will include the effects of the media interface in addition to the scattering from the conducting patches.

The lattice structure is represented by $[a, b, \gamma]$. Suppose $[a_1^+]$ is the Floquet modal voltage vector of order M that is incident on the patch surface from the left. This incident modal packet will induce surface current on the patch surface. The scattered fields essentially are the fields radiated by the induced current. In order to determine the induced current on the patch surface, we set the integral equation formulation and solve the equation employing Galerkin's periodic MoM. The integral equation to be satisfied on the patch surface is given by

$$\vec{E}_{\text{inc}}(x, y) + \iint_{\text{Patch}} \overline{\overline{G}}(x, y/x_0, y_0) \cdot \vec{I}_p(x_0, y_0) dx_0 dy_0 = 0 \qquad (6.45)$$

where $\vec{E}_{\text{inc}}(x, y)$ is the incident electric field (transverse components only) on the patch surface, $\vec{I}_p(x, y)$ is the induced electric surface current on a patch element, and $\overline{\overline{G}}$ is the dyadic Green's function. The integral in (6.45) represents the tangential electric field, $\vec{E}_p(x, y)$, produced by the patch surface current, $\vec{I}_p(x, y)$. The incident electric field can be expressed in terms of the incident modal voltages and the modal functions as follows:

$$\vec{E}_{\text{inc}}(x, y) = \sum_{i=1}^{M} a_1^+(i)\vec{e}_i(x, y)(1 + \Gamma_i) \qquad (6.46)$$

where $a_1^+(i)$ is the modal voltage for the ith Floquet mode, $\vec{e}_i(x, y)$ is the corresponding normalized modal function, and Γ_i is the reflection coefficient of the ith mode due to the dielectric media interface. The expression for Γ_i is given by

$$\Gamma_i = \frac{Z_{02}^{\text{FL}}(i) - Z_{01}^{\text{FL}}(i)}{Z_{02}^{\text{FL}}(i) + Z_{01}^{\text{FL}}(i)} \qquad (6.47)$$

where $Z_{01}^{\text{FL}}(i)$ and $Z_{02}^{\text{FL}}(i)$ are the characteristic impedances of the ith Floquet mode in the two dielectric media. Expressions for these impedances can be found in Chapter 3 for the TE_z and TM_z modes.

In order to obtain the unknown induced current in terms of the incident fields, we employ Galerkin's MoM procedure to solve (6.45). To that end we express the induced current in terms of a linear combination of K known vector basis functions as

$$\vec{I}_p(x, y) = \sum_{k=1}^{K} A_k \vec{f}_k(x, y) \tag{6.48}$$

where $\vec{f}_k(x, y)$ represents the kth vector basis function and A_k is its coefficient to be determined. For a good convergence, it is desirable that each basis function satisfies the edge conditions at the edge of the patch. For regular geometrical shaped patches, like circular or rectangular shapes, entire domain basis functions are preferable. For arbitrary shaped patches, subdomain basis functions are used. We now substitute $\vec{I}_p(x, y)$ from (6.48) into (6.45) to yield

$$\vec{E}_{inc}(x, y) + \sum_{k=1}^{K} A_k \iint_{Patch} \overline{\overline{G}}(x, y/x_0, y_0) \cdot \vec{f}_k(x_0, y_0) \, dx_0 \, dy_0 = 0 \tag{6.49}$$

Following Galerkin's test procedure, we take the scalar product of $\vec{f}_l(x, y)$ on both sides of (6.49) and then integrate over the patch surface. This yields

$$\iint_{Patch} \vec{f}_l(x, y) \cdot \vec{E}_{inc}(x, y) \, dx \, dy + \sum_{k=1}^{K} A_k \iint_{Patch} \vec{f}_l(x, y)$$

$$\cdot \iint_{Patch} \overline{\overline{G}}(x, y/x_0, y_0) \cdot \vec{f}_k(x_0, y_0) \, dx_0 \, dy_0 \, dx \, dy = 0 \tag{6.50}$$

We now substitute $\vec{E}_{inc}(x, y)$ from (6.46) into (6.50) and obtain

$$\sum_{i=1}^{M} a_1^+(i) fe(l, i)(1 + \Gamma_i) = \sum_{k=1}^{K} A_k ff(l, k) \qquad l = 1, 2, \ldots, K \tag{6.51}$$

where

$$fe(l, i) = \iint_{Patch} \vec{f}_l(x, y) \cdot \vec{e}_i(x, y) \, dx \, dy \tag{6.52}$$

$$ff(l, k) = -\iint_{Patch} \vec{f}_l(x, y) \cdot \iint_{Patch} \overline{\overline{G}}(x, y/x_0, y_0) \cdot \vec{f}_k(x_0, y_0) \, dx_0 \, dy_0 \, dx \, dy \tag{6.53}$$

Notice, $ff(l,k)$ represents the mutual reaction between two current basis functions $\vec{f}_l(x, y)$ and $\vec{f}_k(x, y)$. Equation (6.51) can be written in matrix format as

$$[fe][I + \Gamma][a_1^+] = [ff][A] \tag{6.54}$$

where $[fe]$ is a rectangular matrix of order $K \times M$, $[ff]$ is a square matrix of order $K \times K$, and $[I + \Gamma]$ is a diagonal matrix of $M \times M$. The elements of these matrices can be easily identified from (6.51). The unknown current amplitude vector $[A]$ can be obtained from (6.54) as

$$[A] = [ff]^{-1}[fe][I + \Gamma][a_1^+] \tag{6.55}$$

The GSM is constructed from the reflected and the transmitted fields. The total reflected field is the sum of the field produced by the patch current and the field reflected due to the change of the medium in the absence of the patch. The electric field produced by the patch current is

$$\vec{E}_p(x, y) = \sum_{k=1}^{K} A_k \iint_{\text{Patch}} \overline{\overline{G}}(x, y/x_0, y_0) \cdot \vec{f}_k(x_0, y_0) dx_0\, dy_0 \tag{6.56}$$

The modal voltages can be determined by expanding the electric field in terms of Floquet modes. This yields

$$a_{1p}^-(i) = \iint_{\text{Cell}} \vec{E}_p(x, y) \cdot \vec{e}_i^*(x, y)\, dx\, dy \qquad i = 1, 2, \ldots, M \tag{6.57}$$

Combining (6.56) and (6.57) we obtain

$$a_{1p}^-(i) = \sum_{k=1}^{K} A_k aef(i, k) \tag{6.58}$$

where

$$aef(i, k) = \iint_{\text{Cell}} \vec{e}_i^*(x, y) \cdot \iint_{\text{Patch}} \overline{\overline{G}}(x, y/x_0, y_0) \cdot \vec{f}_k(x_0, y_0)\, dx_0\, dy_0\, dx\, dy \tag{6.59}$$

Equation (6.58) can be written in matrix format as

$$[a_{1p}^-] = [aef][A] \tag{6.60}$$

Now substituting $[A]$ from (6.55) we obtain

$$[a_{1p}^-] = [aef][ff]^{-1}[fe][I + \Gamma][a_1^+] \tag{6.61}$$

The reflected voltage vector due to the change of medium is

$$[a_{1m}^-] = [\Gamma][a_1^+] \tag{6.62}$$

where $[\Gamma]$ is a diagonal matrix with Γ_i as the ith diagonal element given in (6.47). The total reflected voltage vector thus becomes

$$[a_1^-] = [a_{1p}^- + a_{1m}^-] = \{[aef][ff]^{-1}[fe][I + \Gamma] + [\Gamma]\}[a_1^+] \tag{6.63}$$

The reflected field is obtained assuming that medium 2 is of infinite extent, which is equivalent to a match termination. Therefore, the $[S_{11}]$ submatrix is derived as

$$[S_{11}] = [\Gamma] + [aef][ff]^{-1}[fe][I + \Gamma] \tag{6.64}$$

The total voltage vector at the patch interface is given by

$$[v'] = [a_1^+] + [a_1^-] = [I + S_{11}][a_1^+] \tag{6.65}$$

The total voltage at the interface must be equal to the transmitted voltage because the output port is matched. It follows that

$$[S_{21}] = [I + S_{11}] \tag{6.66}$$

From (6.64) and (6.66) we thus obtain

$$[S_{21}] = \{[I] + [aef][ff]^{-1}[fe]\}[I + \Gamma] \tag{6.67}$$

The above expression for $[S_{21}]$ is valid if the numbers of modes are equal in two ports, because $[S_{21}]$ in (6.67) is a square matrix of order $M \times M$. If, however, the number of modes N in port 2 is assumed different from the number of modes M in port 1, then (6.67) needs a modification. For $N < M$, one can still use (6.67) after removing the $M - N$ rows associated with the modes that are not considered in port 2. For $N > M$, the situation is more complex. In this case $N - M$ rows must be added to be consistent with the definition of $[S_{21}]$. Now, going back to (6.65), the additional transmitted modal voltages are $a_1^-(i)$, $M < i \leq N$. The induced current on the patch generates these transmitted modes. Therefore,

$$a_2^-(i) = a_1^-(i) = a_{1p}^-(i) \qquad M < i \leq N \tag{6.68}$$

For $a_{1p}^-(i)$ we use (6.61) and accordingly modify (6.67). The final result is

$$[S_{21}] = \{[I'] + [aef'][ff]^{-1}[fe]\}[I + \Gamma] \tag{6.69}$$

where $[I']$ is a diagonal matrix of order $N \times M$ with diagonal elements as unity, $[aef']$ is a rectangular matrix of order $N \times K$, and the elements are given in (6.59). Now the order of $[S_{21}]$ becomes $N \times M$, as expected.

In order to obtain $[S_{12}]$ and $[S_{22}]$ submatrices, one can follow the same method as for $[S_{21}]$ and $[S_{11}]$. From symmetry we can also write the expressions as

$$[S_{22}] = [\Gamma'] + [aef'][ff]^{-1}[fe'][I + \Gamma'] \tag{6.70}$$

$$[S_{12}] = \{[I''] + [aef''][ff]^{-1}[fe']\}[I + \Gamma'] \tag{6.71}$$

where

$$\Gamma_i' = -\Gamma_i \tag{6.72}$$

$[I'']$ is a rectangular diagonal matrix with diagonal elements as unity, and $[aef'']$ is a rectangular matrix whose elements are given in (6.59). The matrix $[fe']$ basically evolved from $[fe]$ with N columns. The dimensions of the above matrices are set to make $[S_{12}]$ and $[S_{22}]$ of proper orders.

Simplification of Integrals The integrals in (6.53) and (6.59) can be simplified after deducing a closed-form expression for the electric field. The transverse electric field produced by a surface current $\vec{f}_i(x, y)$ at the interface of two dielectric media is given by

$$\vec{E}_i(x, y) = \iint_{\text{Patch}} \overline{\overline{G}}(x, y/x_0, y_0) \cdot \vec{f}_i(x_0, y_0)\, dx_0\, dy_0 \tag{6.73}$$

where $\overline{\overline{G}}$ is Green's function. Instead of finding Green's function explicitly, we will obtain $\vec{E}_i(x, y)$ using Floquet modal expansion. To that end let us expand the current distribution $\vec{f}_i(x, y)$ in terms of the following Floquet series:

$$\vec{f}_i(x, y) = \sum_n B(i, n)\{\hat{z} \times \vec{h}_n(x, y)\} \tag{6.74}$$

where $\vec{h}_n(x, y)$ is the normalized magnetic modal function with respect to the array lattice. Theoretically, the number of terms in the series should be infinitely large because an infinite number of modes are produced by each current basis. The expression for $\vec{h}_n(x, y)$ is derived in Section 3.2 for the TE$_z$ and TM$_z$ modes. To determine the modal coefficients, $B(i, n)$, we scalar multiply on both sides of (6.74) by the complex conjugate of the normalized electric modal function. Using orthogonality of Floquet modes and the normalization condition, we obtain

$$B(i, m) = -\iint_{\text{Patch}} \vec{f}_i(x, y) \cdot \vec{e}_m^*(x, y)\, dx\, dy \tag{6.75}$$

If $\vec{f}_i(x, y)$ are real functions, then from (6.52) we can write

$$B(i, m) = -fe^*(i, m) \tag{6.76}$$

Now let us determine the electromagnetic fields produced by the surface current $\vec{I}_n(x, y) = \hat{z} \times \vec{h}_n(x, y)$ at $z = 0$, the interface of two media. The transverse electric field vector at $z = 0$ must be continuous; therefore we write

$$\vec{E}_n(x, y, z) = \begin{cases} C_n \vec{e}_n(x, y)\exp(-j\beta_n^{(2)}z) & \text{for } z > 0 \\ C_n \vec{e}_n(x, y)\exp(j\beta_n^{(1)}z) & \text{for } z < 0 \end{cases} \tag{6.77}$$

From inspection, the transverse magnetic field vector would be

$$\vec{H}_n(x, y, z) = \begin{cases} C_n Y_{02}^{\text{FL}}(n)\,\vec{h}_n(x, y)\exp(-j\beta_n^{(2)}z) & \text{for } z > 0 \\ -C_n Y_{01}^{\text{FL}}(n)\vec{h}_n(x, y)\exp(j\beta_n^{(1)}z) & \text{for } z < 0 \end{cases} \tag{6.78}$$

where $\beta_n^{(1)}$ and $\beta_n^{(2)}$ are propagation constants. Discontinuity of the magnetic fields at $z = 0$ yields the surface current. Therefore

$$\hat{z} \times \{\vec{H}_n(x, y, 0+) - \vec{H}_n(x, y, 0-)\} = \vec{I}_n = \hat{z} \times \vec{h}_n(x, y) \tag{6.79}$$

From (6.78) and (6.79) we solve

$$C_n = Z^{\text{FL}}(n) \tag{6.80a}$$

with

$$Z^{\text{FL}}(n) = \frac{1}{Y_{01}^{\text{FL}}(n) + Y_{02}^{\text{FL}}(n)} \tag{6.80b}$$

Using superposition, the total electric field at $z = 0$ produced by $\vec{f}_i(x, y)$ in (6.74) is obtained as

$$\vec{E}_i(x, y) = \sum_n B(i, n) Z^{\text{FL}}(n) \vec{e}_n(x, y) \tag{6.81}$$

From (6.76) and (6.81) we obtain

$$\vec{E}_i(x, y) = -\sum_n fe^*(i, n) Z^{\text{FL}}(n) \vec{e}_n(x, y) \tag{6.82}$$

From (6.53) and (6.73) we write

$$ff(l, k) = -\iint_{\text{Patch}} \vec{f}_l(x, y) \cdot \vec{E}_k(x, y) \, dx \, dy$$

$$= \sum_n fe^*(k, n) Z^{\text{FL}}(n) \iint_{\text{Patch}} \vec{f}_l(x, y) \cdot \vec{e}_n(x, y) \, dx \, dy$$

$$= \sum_n fe(l, n) Z^{\text{FL}}(n) fe^*(k, n) \tag{6.83}$$

From (6.59) we also obtain

$$aef(i, k) = \iint_{\text{Cell}} \vec{e}_i^*(x, y) \cdot \vec{E}_k(x, y) \, dx \, dy$$

$$= -\sum_n fe^*(k, n) Z^{\text{FL}}(n) \iint_{\text{Cell}} \vec{e}_i^*(x, y) \cdot \vec{e}_n(x, y) \, dx \, dy$$

$$= -fe^*(k, i) Z^{\text{FL}}(i) \tag{6.84}$$

The above simplified expressions can be used to compute the $[ff]$, $[aef]$, $[aef']$, and $[aef'']$ matrices that appear in the expressions for $[S_{11}]$, $[S_{21}]$, $[S_{12}]$, and $[S_{22}]$. Notice, the fundamental quantity $fe(m, n)$ needs to be computed in order to obtain the GSM of a patch. For a fast computation it is preferable to have closed-form expressions for $fe(m, n)$, which is possible if the basis functions have closed-form Fourier transforms.

As derived in (6.80b), $Z^{FL}(i)$ turns out to be the equivalent impedance experienced by the ith modal current source offered by the two adjacent layers that are assumed to be infinitely wide. This assumption is rigorously valid for the coupling modes only ($i \leq M$, assume $M = N$ for simplicity). For a noncoupling mode ($i > M$) that is produced locally by the patch current, (6.80b) could be a reasonable approximation for the modal impedance if the modal fields decay so rapidly that they practically die down within the adjacent dielectric layers. If this condition is violated, which may happen for a very thin adjacent layer, then one can either increase the value of M until the condition is fulfilled to a desired limit or consider multilayer equivalent impedance for the noncoupling modes. The first option increases the computation time, while the second option becomes more complex to implement as the GSM of the patch layer depends not only on the adjacent layers but also on the other layers of the structure.

Basis Functions To obtain the GSM of a patch array, a proper selection of basis functions is vital from a convergence point of view. For regular geometrical shaped patches (rectangular, circular, etc.) entire domain basis functions are preferable. For a nongeometrical shaped patch, a suitable set of entire domain basis functions may not be available. In such situations, subdomain basis functions are used. The patch is divided into many rectangular or triangular segments. A subdomain basis function is defined on a small segment only [7]. Therefore, the number of basis functions K in this case would be equal to the number of segments.

For regular geometrical patch shapes, an orthogonal set of basis functions is preferred because the error function becomes orthogonal to the estimated current distribution, minimizing the root-mean-square (rms) error. Typically, for a geometrical shaped patch, the basis functions are considered as the modal electric field vectors associated with a waveguide of magnetic walls (see problem 6.10). Subdomain basis functions also satisfy orthogonality [8].

6.6 EQUIVALENT IMPEDANCE MATRIX OF PATCH LAYER

A multilayer array can be modeled as multiple transmission line sections connected in cascade. Each transmission line section represents a homogeneous layer, allowing propagation of multiple Floquet modes, such as an oversized waveguide section. A thin patch array layer is equivalent to a shunt impedance, because the voltage vectors before and after the patch layer are identical. For a multimodal analysis, the shunt impedance effectively is an impedance matrix, $[Z]$, which relates the voltage vector with the current vector. In the following development we will derive the expression for the $[Z]$ matrix. For simplicity, we assume that the number of incident modes M and the number of transmitted modes N are equal. The total modal voltage vector (incident plus reflected) at the patch location with respect to an incident modal voltage vector $[a_1^+]$ is given by (assuming that the transmitted port is match terminated)

$$[v] = [S_{21}][a_1^+] \tag{6.85}$$

Substituting $[S_{21}]$ from (6.67) we obtain

$$[v] = [I + \Gamma][a_1^+] + [aef][ff]^{-1}[fe][I + \Gamma][a_1^+] \qquad (6.86)$$

Using (6.55) in (6.86) we obtain

$$[v] = [I + \Gamma][a_1^+] + [aef][A] \qquad (6.87)$$

Now, from (6.48) and (6.74) the patch current can be expressed as

$$\vec{I}_p(x, y) = \sum_n \{\hat{z} \times \vec{h}_n\} \left[\sum_{k=1}^{K} B(k, n) A_k \right] \qquad (6.88)$$

The term inside the square bracket is the amplitude of the nth modal current, I_n. By virtue of (6.76) we express the modal current as

$$I_n = \sum_{k=1}^{K} B(k, n) A_k = - \sum_{k=1}^{K} fe^*(k, n) A_k \qquad (6.89)$$

In matrix notation, the above relation is equivalent to

$$[I^p] = -[fe^*]^T [A] \qquad (6.90)$$

where $[I^p]$ represents the patch modal current vector of order M. Now the voltage vector at the patch location can be expressed in terms of the patch current vector via the shunt impedance, $[Z]$, as

$$[v] = -[Z][I^p] \qquad (6.91)$$

The negative sign is due to the fact that the patch current is considered as the load current (not a source current). Equations (6.87), (6.90) and (6.91) lead to

$$[Z][fe^*]^T [A] = [I + \Gamma][a_1^+] + [aef][A] \qquad (6.92)$$

To derive an explicit expression for $[Z]$ we replace $[A]$ in terms of $[a_1^+]$ using (6.55). This yields

$$[Z][fe^*]^T [ff]^{-1}[fe][I + \Gamma][a_1^+] = [I + \Gamma][a_1^+] + [aef][ff]^{-1}[fe][I + \Gamma][a_1^+] \quad (6.93)$$

The above relation must be true for any arbitrary vector $[a_1^+]$. Therefore, we have

$$[Z][fe^*]^T [ff]^{-1}[fe] = [I] + [aef][ff]^{-1}[fe] \qquad (6.94)$$

Now, in matrix format (6.84) is equivalent to the following equation:

$$[aef] = -[Z^{FL}][fe^*]^T \qquad (6.95)$$

with $[Z^{FL}]$ a diagonal matrix of order $M \times M$. The nth diagonal element is $Z^{FL}(n)$ given in (6.80b). Equations (6.94) and (6.95) yield

$$[Z] = [[fe^*]^T[ff]^{-1}[fe]]^{-1} - [Z^{FL}] \tag{6.96}$$

An alternative relation involving $[Z]$ can be found by combining (6.84), (6.92) and (6.54). The relation is given by

$$[fe][Z][fe^*]^T = [ff] - [fe][Z^{FL}][fe^*]^T \tag{6.97}$$

Equation (6.97) helps in understanding the origin of the equivalent shunt impedance matrix. Writing the expression of $[ff]$ in terms of $fe(m, n)$ from (6.83), we obtain

$$[ff] = [f\bar{e}][\bar{Z}^{FL}][f\bar{e}^*]^T \tag{6.98}$$

where $[f\bar{e}]$ is a matrix of order $K \times \infty$. The first M columns of $[f\bar{e}]$ constitute $[fe]$. The matrix $[\bar{Z}^{FL}]$ is diagonal and has order $\infty \times \infty$. The first $M \times M$ elements of $[\bar{Z}^{FL}]$ constitute the $[Z^{FL}]$ matrix. Using (6.98) in (6.97) with minor manipulation we derive

$$[fe][Z][fe^*]^T = [Z_f] \tag{6.99}$$

with

$$[Z_f] = [f\tilde{e}][\tilde{Z}^{FL}][f\tilde{e}^*]^T \tag{6.100}$$

The $[f\tilde{e}]$ matrix is of order $K \times \infty$; $[\tilde{Z}^{FL}]$ is a diagonal matrix of order $\infty \times \infty$ and their elements are given by

$$f\tilde{e}(i, j) = fe(i, M+j) = \iint\limits_{\text{Patch}} \vec{f}_i(x, y) \cdot \vec{e}_{M+j}(x, y) \, dx \, dy \tag{6.101}$$

$$\tilde{Z}^{FL}(i, i) = Z^{FL}(i+M) \tag{6.102}$$

Equation (6.99) indicates that $[Z_f]$, of order $K \times K$, is related to the shunt impedance matrix $[Z]$, of order $M \times M$, via two multiplier matrices $[fe]$ and $[fe^*]^T$. The above relation permits one to construct an equivalent circuit, shown in Figure 6.7a. Impedance transformers are introduced to account for the multiplier matrices. The matrix $[fe]$ acts as the turn ratio of the transformers. The two transmission lines are associated with two dielectric media at the two sides of the thin layer of the patch array. Each transmission line carries M Floquet modes. Figure 6.7b is a simplified version of the equivalent circuit where the transformers are removed. This equivalent circuit model can also be used to construct the GSM of a patch layer.

We established that a thin patch array layer is equivalent to a shunt impedance matrix $[Z]$ that can be computed using (6.99) and (6.100). Furthermore, $[Z_f]$ (hence $[Z]$) in (6.100) is contributed only by the *higher order Floquet modes* with

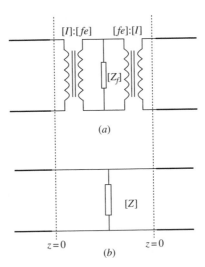

FIGURE 6.7   Equivalent circuit of an infinite array of patch elements: (a) with transformers; (b) simplified version.

modal indices greater than M. To compute $[Z]$, one needs computation of $[Z_f]$, which involves multiplication of infinite-order matrices. Recall that the infinite-order resulted from the expression of $ff(l, k)$ in (6.83), which contains summation of a series. Ideally, the number of terms in the series should be infinitely large. In practice, the series is truncated with a finite number of terms, N_{ff}, which is equal to the number of expansion modes used. This makes $[f\tilde{e}]$ and $[\tilde{Z}^{\mathrm{FL}}]$ finite matrices of order $K \times (N_{ff} - M)$ and $(N_{ff} - M) \times (N_{ff} - M)$, respectively, assuming $N_{ff} \geq M$. Furthermore, the $[Z]$ matrix elements can be determined unambiguously if $K \geq M$. If $K < M$, $[Z]$ cannot be determined as the number of unknowns in $[Z]$ exceeds the number of equations [also, note that the triple-product matrix in (6.96) becomes singular, hence noninvertible (see problem 6.16)]. Thus, $K \geq M$ is a required condition for obtaining an equivalent circuit for a patch array.

An interesting shortcoming of the equivalent impedance can be found if the number of basis functions K and the number of expansion modes N_{ff} are set equal. The impedance matrix becomes a null matrix. To prove that, we first combine (6.97) and (6.98) to write

$$[fe]\{[Z]+[Z^{\mathrm{FL}}]\}[fe^*]^T = [f\bar{e}][\bar{Z}^{\mathrm{FL}}][f\bar{e}^*]^T \tag{6.103}$$

For $K = N_{ff}$, $[f\bar{e}]$ and $[f\bar{e}^*]^T$ are of order $K \times K$ that are invertible. Thus we obtain

$$[f\bar{e}]^{-1}[fe]\{[Z]+[Z^{\mathrm{FL}}]\}[fe^*]^T\{[f\bar{e}^*]^T\}^{-1} = [\bar{Z}^{\mathrm{FL}}] \tag{6.104}$$

We assume $M \leq K$. Since the first M columns of $[f\bar{e}]$ constitute $[fe]$, the product of $[f\bar{e}]^{-1}$ and $[fe]$ becomes a rectangular identity matrix of order $K \times M$. Similarly, the product of $[fe^*]^T$ and $\{[f\bar{e}^*]^T\}^{-1}$ is also a rectangular identity matrix of order $M \times K$. Thus from (6.104) it is straightforward to show that the elements of the first M rows and M columns of $[\bar{Z}^{FL}]$ are identical with that of $[Z] + [Z^{FL}]$, implying $[Z] + [Z^{FL}] = [Z^{FL}]$. Thus for $K = N_{ff}$, we have

$$[Z] = [0] \tag{6.105}$$

that is, the shunt impedance matrix becomes a null matrix, which is intuitively incorrect. Thus the analysis fails for $K = N_{ff}$. This implies that the number of basis functions must not be equal to the number of expansion modes. In Section 6.8 we reconsider this limitation of MoM analysis from a different viewpoint.

6.7 STATIONARY CHARACTER OF MoM SOLUTIONS

We have seen that the MoM solution is not an exact solution because the current distribution is estimated by a finite number of basis functions. It is desirable that the final solution be stationary because that would eliminate the first order error in the solution with respect to the error in the estimated current [9]. Fortunately, Galerkin's MoM solution is stationary. It will be shown that the GSM of a patch array that resulted from Galerkin's MoM analysis is stationary. The patch layer is assumed to be located at the interface of two dielectric layers, as shown in Figure 6.6. The stationary characteristic is demonstrated indirectly in two steps. First, we derive stationary expressions for the elements of the GSM. Then we demonstrate that the GSM derived from the stationary expressions coincides with the GSM derived employing Galerkin's MoM analysis.

6.7.1 Stationary Expression

To deduce the stationary expression of an element of the GSM, say $S_{11}(2,1)$, we assume that only one Floquet mode is incident on the patch surface. Suppose

$$\vec{E}_a^+ = V_1^+ \vec{e}_1(x, y) \tag{6.106}$$

is the electric field vector (with the z-component suppressed) of the incident mode with V_1^+ as the modal voltage and $\vec{e}_1(x, y)$ as the normalized modal function. The incident mode will induce electric current on the patch. We assume that $\vec{g}_1(x, y)$ is the *actual* induced current on the patch. To derive $S_{11}(2,1)$, we need to determine the reflected field. The reflected field is produced by the induced current. The reflected electric field (transverse components only) is given by

$$\vec{E}^{\text{ref}} = \iint_{\text{Patch}} \vec{g}_1(x_0, y_0) \cdot \overline{\overline{G}}(x, y/x_0, y_0) \, dx_0 \, dy_0 \tag{6.107}$$

where $\overline{\overline{G}}$ represents the dyadic Green's function. The total tangential field on the patch surface must vanish. This condition yields

$$V_1^+ \vec{e}_1(1+\Gamma_1) = -\iint_{\text{Patch}} \vec{g}_1(x_0, y_0) \cdot \overline{\overline{G}}(x, y/x_0, y_0) \, dx_0 \, dy_0 \qquad (6.108)$$

where Γ_1 is the reflection coefficient of the incident mode due to the change in medium only. The induced current will produce all possible modes. Following the results in (6.82) we express the voltage associated with the modal function $\vec{e}_2$ produced by $\vec{g}_1(x, y)$ as

$$V_2^- = -\frac{1}{Y(2)} \iint_{\text{Patch}} \vec{g}_1 \cdot \vec{e}_2^* \, dx \, dy \qquad (6.109)$$

where $Y(2) = Y_{01}^{\text{FL}}(2) + Y_{02}^{\text{FL}}(2)$ and $Y_{01}^{\text{FL}}(n)$, $Y_{02}^{\text{FL}}(n)$ are nth Floquet modal admittances of media 1 and 2, respectively. From (6.108) and (6.109) we obtain

$$S_{11}(2, 1) = \frac{V_2^-}{V_1^+} = \frac{1+\Gamma_1}{Y(2)} \frac{\vec{e}_1 \iint_{\text{Patch}} \vec{g}_1 \cdot \vec{e}_2^* \, dx \, dy}{\iint_{\text{Patch}} \vec{g}_1(x_0, y_0) \cdot \overline{\overline{G}}(x, y/x_0, y_0) \, dx_0 \, dy_0} \qquad (6.110)$$

Observe that $S_{11}(2, 1)$ is expressed as a ratio of two vector functions. The ratio of two vectors is meaningful only if the vectors are parallel and that happens for an exact solution only. Thus, (6.110) is rigorously valid and error free if $\vec{g}_1(x, y)$ represents the exact current distribution. On the other hand, if $\vec{g}_1(x, y)$ represents an estimated current distribution, then (6.110) may not be usable as is, because the vectors in the numerator and denominator may not be parallel. Besides, the resultant expression may not be independent of x and y even if the parallel condition is satisfied by some means.

To circumvent this situation, a preferable option is to scalar multiply both numerator and denominator with a suitable vector function and integrate on the patch surface. This will mitigate the ratio-of-two-vectors problem as well as make $S_{11}(2, 1)$ independent of x and y. It is preferable that the vector function be such that the modified expression becomes stationary, that is, insensitive to the error in the current distribution. Hence we will pursue for a vector function $\vec{g}_2(x, y)$ that would make our modified expression stationary. As stated, the modified expression would look like

$$S_{11}(2, 1) = \frac{(1+\Gamma_1) \iint_{\text{Patch}} \vec{g}_2 \cdot \vec{e}_1 \, dx \, dy \iint_{\text{Patch}} \vec{g}_1 \cdot \vec{e}_2^* \, dx \, dy}{Y(2) \iint_{\text{Patch}} \iint_{\text{Patch}} \vec{g}_1(x_0, y_0) \cdot \overline{\overline{G}}(x, y/x_0, y_0) \cdot \vec{g}_2 \, dx_0 \, dy_0 \, dx \, dy} \qquad (6.111)$$

Now let us first compute the deviation in $S_{11}(2, 1)$ with respect to an error in $\vec{g}_1(x, y)$ by introducing an error term $\partial g_1(x, y)$. Cross-multiplying and then taking differentials, we deduce

$$S_{11}(2, 1)Y(2) \iint_{\text{Patch}} \vec{\partial g_1}(x, y) \cdot \iint_{\text{Patch}} \vec{g}_2 \cdot \overline{\overline{G}} \, dx_0 \, dy_0 \, dx \, dy$$

$$+ \partial S_{11}(2, 1)Y(2) \iint_{\text{Patch}} \vec{g}_1(x, y) \cdot \iint_{\text{Patch}} \vec{g}_2 \cdot \overline{\overline{G}} \, dx_0 \, dy_0 \, dx \, dy$$

$$= (1 + \Gamma_1) \iint_{\text{Patch}} \vec{g}_2 \cdot \vec{e}_1 \, dx \, dy \iint_{\text{Patch}} \vec{\partial g_1} \cdot \vec{e}_2^* \, dx \, dy \qquad (6.112)$$

Note, $\overline{\overline{G}}$ is a symmetric dyadic so that $\overline{\overline{G}} \cdot \vec{g}_2 = \vec{g}_2 \cdot \overline{\overline{G}}$. Setting $\partial S_{11}(2, 1) = 0$ we obtain the stationary condition. This becomes

$$\iint_{\text{Patch}} \vec{\partial g_1}(x, y) \cdot \left[S_{11}(2, 1)Y(2) \iint_{\text{Patch}} \vec{g}_2 \cdot \overline{\overline{G}} \, dx_0 \, dy_0 - \vec{e}_2^*(1 + \Gamma_1) \right.$$

$$\left. \iint_{\text{Patch}} \vec{g}_2 \cdot \vec{e}_1 \, dx \, dy \right] dx \, dy = 0 \qquad (6.113)$$

Since $\vec{\partial g_1}(x, y)$ is an arbitrary function, the term inside the square bracket must vanish. After rearrangement we get

$$S_{11}(2, 1) = \frac{1 + \Gamma_1}{Y(2)} \frac{\vec{e}_2^* \iint\limits_{\text{Patch}} \vec{g}_2 \cdot \vec{e}_1 \, dx \, dy}{\iint\limits_{\text{Patch}} \vec{g}_2 \cdot \overline{\overline{G}} \, dx_0 \, dy_0} \qquad (6.114)$$

Now let us investigate $\vec{g}_2(x, y)$ that would satisfy (6.114). We postulate that $\vec{g}_2(x, y)$ is the exact current distribution with respect to an incident Floquet mode with the following electric field:

$$\vec{E}_b^+ = \tilde{V}_2^+ \vec{e}_2^*(x, y) \qquad (6.115)$$

(anticipated from Galerkin's test procedure). Then following the same procedure used for deriving (6.110) we obtain,

$$\tilde{S}_{11}(1, 2) = \frac{1 + \Gamma_2}{Y(1)} \frac{\vec{e}_2^* \iint\limits_{\text{Patch}} \vec{g}_2 \cdot \vec{e}_1 dx \, dy}{\iint\limits_{\text{Patch}} \vec{g}_2(x_0, y_0) \cdot \overline{\overline{G}}(x, y/x_0, y_0) \, dx_0 \, dy_0} \qquad (6.116)$$

where $\tilde{S}_{11}(1,2) = \tilde{V}_1^-/\tilde{V}_2^+$, $Y(1) = Y_{01}^{\text{FL}}(1) + Y_{02}^{\text{FL}}(1)$, and $\Gamma_2 = [Y_{01}^{\text{FL}}(2) - Y_{02}^{\text{FL}}(2)]/Y(2)$. Equations (6.114) and (6.116) would be consistent if the following condition is satisfied:

$$S_{11}(2,1)Y(2)(1+\Gamma_2) = \tilde{S}_{11}(1,2)Y(1)(1+\Gamma_1) \tag{6.117}$$

Substituting expressions for $S_{11}(2,1)$, $\tilde{S}_{11}(1,2)$, Γ_1, and Γ_2, we obtain

$$[V_1^+ Y_{01}^{\text{FL}}(1)]\tilde{V}_1^- = [\tilde{V}_2^+ Y_{01}^{\text{FL}}(2)]V_2^- \tag{6.118}$$

The terms inside the square brackets are the incident modal currents for the two sources. Thus (6.118) is equivalently expressed as

$$I_1^+ \tilde{V}_1^- = \tilde{I}_2^+ V_2^- \tag{6.119}$$

Equation (6.119) can be recognized as a special form of the reciprocity relation between two voltage sources associated with the incident fields $\vec{E}_a^+$ and $\vec{E}_b^+$, given in (6.106) and (6.115), respectively (see problem 6.13). This implies that our postulate is true and the expression of $S_{11}(2,1)$ in (6.111) is stationary provided $\vec{g}_2(x,y)$ represents the induced current distribution on the patch with respect to the incident field $\vec{E}_b^+$. From symmetry, one also observes that the same expression would be stationary with respect to $\vec{g}_2(x,y)$. Thus, from the chain rule of partial differentiation one concludes that the expression for $S_{11}(2,1)$ in (6.111) is stationary with respect to both current distributions.

Equation (6.111) can be generalized for the other elements of the $[S_{11}]$ submatrix by inspection. The general expression for $m \neq n$ would be

$$S_{11}(m,n) = \frac{(1+\Gamma_n) \underset{\text{Patch}}{\iint} \vec{g}_m \cdot \vec{e}_n \, dx \, dy \underset{\text{Patch}}{\iint} \vec{g}_n \cdot \vec{e}_m^* \, dx \, dy}{Y(m) \underset{\text{Patch}}{\iint} \underset{\text{Patch}}{\iint} \vec{g}_n(x_0,y_0) \cdot \overline{\overline{G}}(x,y/x_0,y_0) \cdot \vec{g}_m \, dx_0 \, dy_0 \, dx \, dy} \tag{6.120}$$

For $m = n$, the reflection term from the dielectric interface must be added. The general stationary expression for all m and n should be

$$S_{11}(m,n) = \Gamma_m \delta_{mn} + \frac{(1+\Gamma_n) \underset{\text{Patch}}{\iint} \vec{g}_m \cdot \vec{e}_n \, dx \, dy \underset{\text{Patch}}{\iint} \vec{g}_n \cdot \vec{e}_m^* \, dx \, dy}{Y(m) \underset{\text{Patch}}{\iint} \underset{\text{Patch}}{\iint} \vec{g}_n(x_0,y_0) \cdot \overline{\overline{G}}(x,y/x_0,y_0) \cdot \vec{g}_m \, dx_0 \, dy_0 \, dx \, dy} \tag{6.121}$$

where δ_{mn} is the Kronecker delta. Similarly, the stationary expression for $S_{21}(m,n)$ would be

$$S_{21}(m,n) = (1+\Gamma_m)\delta_{mn}$$

$$+ \frac{(1+\Gamma_n) \underset{\text{Patch}}{\iint} \vec{g}_m \cdot \vec{e}_n \, dx \, dy \underset{\text{Patch}}{\iint} \vec{g}_n \cdot \vec{e}_m^* \, dx \, dy}{Y(m) \underset{\text{Patch}}{\iint} \underset{\text{Patch}}{\iint} \vec{g}_n(x_0,y_0) \cdot \overline{\overline{G}}(x,y/x_0,y_0) \cdot \vec{g}_m \, dx_0 \, dy_0 \, dx \, dy} \tag{6.122}$$

6.7.2 GSM from Stationary Expression

Thus far we have obtained stationary expressions for the elements of the GSM of a patch array. However, we are yet to prove that the GSM derived using MoM analysis is stationary. Toward that pursuit, we will demonstrate that the GSM constructed using the above stationary expressions for the elements would be identical with that derived using the MoM formulation. This is an indirect way to prove the stationary character of the MoM solution. To derive the submatrix $[S_{11}]$ of the GSM using (6.121), we express the induced current functions in terms of known basis functions as

$$\vec{g}_n(x, y) = \sum_{i=1}^{K} A_i^{(n)} \vec{f}_i(x, y) \tag{6.123}$$

$$\vec{g}_m(x, y) = \sum_{j=1}^{K} B_j^{(m)} \vec{f}_j(x, y) \tag{6.124}$$

Substituting (6.123) and (6.124) into (6.121) we obtain after rearrangement

$$\{S_{11}(m, n) - \Gamma_m \delta_{mn}\} \sum_i \sum_j A_i^{(n)} B_j^{(m)} \iint_{\text{Patch}} \iint_{\text{Patch}} \vec{f}_i \cdot \overline{\overline{G}} \cdot \vec{f}_j \, dx_0 \, dy_0 \, dx \, dy$$

$$= (1 + \Gamma_n) \sum_i A_i^{(n)} \frac{1}{Y(m)} \iint_{\text{Patch}} \vec{f}_i \cdot \vec{e}_m^* \, dx \, dy \sum_j B_j^{(m)} \iint_{\text{Patch}} \vec{f}_j \cdot \vec{e}_n \, dx \, dy \quad (6.125)$$

Using the symbols defined in (6.52), (6.53), and (6.84) we express (6.125) as

$$\{S_{11}(m, n) - \Gamma_m \delta_{mn}\} \sum_i \sum_j A_i^{(n)} B_j^{(m)} ff(i, j)$$

$$= (1 + \Gamma_n) \sum_i A_i^{(n)} aef(m, i) \sum_j B_j^{(m)} fe(j, n) \tag{6.126}$$

In matrix notation, (6.126) can be expressed as (considering [ff] symmetric)

$$\{S_{11}(m, n) - \Gamma_m \delta_{mn}\}[B^{(m)}]^T [ff][A^{(n)}] = (1 + \Gamma_n)[B^{(m)}]^T \begin{bmatrix} fe(1, n) \\ fe(2, n) \\ \vdots \\ fe(K, n) \end{bmatrix}$$

$$\times [aef(m, 1) \; aef(m, 2) \; \cdots \; aef(m, K)] [A^{(n)}] \tag{6.127}$$

Typically, in a stationary formulation, the induced current distribution is guessed. This is contrary to the MoM analysis, where the induced current is estimated by enforcing boundary conditions. However, for the present situation we will consider the MoM current as the guessed current, because our ultimate goal is to prove that the MoM solution satisfies the stationary condition. The current distribution used in

MoM formulation can be obtained from (6.48) and (6.55) with respect to an incident modal voltage vector. Recall that $V_n^+ \vec{e}_n$ is the only incident field in the present case and $\sum_i A_i^{(n)} \vec{f}_i$ is the estimated induced current on the patch corresponding to the incident field. Therefore from (6.55) or (6.54) we write

$$[ff][A^{(n)}] = [fe][I + \Gamma][V^+] \tag{6.128}$$

where $[V^+]$ is a column matrix that has only one nonzero element, V_n^+. Also, $[I+\Gamma]$ is a diagonal matrix. Therefore (6.128) reduces to

$$[ff][A^{(n)}] = \begin{bmatrix} fe(1,n) \\ fe(2,n) \\ \vdots \\ fe(K,n) \end{bmatrix} (1+\Gamma_n)V_n^+ \tag{6.129}$$

If we eliminate $[A^{(n)}]$ from (6.127), using (6.129) we get

$$\{S_{11}(m,n) - \Gamma_m \delta_{mn}\}[B^{(m)}]^T \begin{bmatrix} fe(1,n) \\ fe(2,n) \\ \vdots \\ fe(K,n) \end{bmatrix} V_n^+$$

$$= [B^{(m)}]^T \begin{bmatrix} fe(1,n) \\ fe(2,n) \\ \vdots \\ fe(K,n) \end{bmatrix} [aef(m,1) \ aef(m,2) \ \cdots \ aef(m,K)]$$

$$\times [ff]^{-1} \begin{bmatrix} fe(1,n) \\ fe(2,n) \\ \vdots \\ fe(K,n) \end{bmatrix} (1+\Gamma_n)V_n^+ \tag{6.130}$$

Rearranging the above equation we obtain

$$[B^{(m)}]^T \begin{bmatrix} fe(1,n) \\ fe(2,n) \\ \vdots \\ fe(K,n) \end{bmatrix} \Big[\{S_{11}(m,n) - \Gamma_m \delta_{mn}\}$$

$$- [aef(m,1) \ aef(m,2) \ \cdots \ aef(m,K)]$$

$$\times [ff]^{-1} \begin{bmatrix} fe(1,n) \\ fe(2,n) \\ \vdots \\ fe(K,n) \end{bmatrix} (1+\Gamma_n) \Big] V_n^+ = 0 \tag{6.131}$$

If (6.131) is valid, then the term inside the large square bracket must vanish. Thus we obtain

$$S_{11}(m, n) = \Gamma_m \delta_{mn} + [aef(m, 1) \ aef(m, 2) \quad \cdots \quad aef(m, K)]$$

$$\times [ff]^{-1} \begin{bmatrix} fe(1, n) \\ fe(2, n) \\ \vdots \\ fe(K, n) \end{bmatrix} (1 + \Gamma_n) \qquad (6.132)$$

Equation (6.132) is valid for all m and n. Thus, for all m and n the consolidated format for (6.132) becomes

$$[S_{11}] = [\Gamma] + [aef][ff]^{-1}[fe][I + \Gamma] \qquad (6.133)$$

Equation (6.133) is identical with (6.64), obtained using Galerkin's MoM.

The above development signifies that the stationary expression and the MoM solution yield identical values for $S_{11}(m, n)$ if identical patch current distribution is assumed in both formulations. We emphasize here that the current distribution does not represent the exact current distribution; it represents an estimated one (may be very close to the exact distribution if the number of basis functions K is sufficiently large). Hence, the solutions for $S_{11}(m, n)$ are estimated solutions. Nevertheless, both formulations yield identical estimates for $S_{11}(m, n)$. This result leads us to conclude that the MoM-based GSM is also stationary. In a similar fashion it is very straightforward to prove that $[S_{21}]$, $[S_{12}]$, and $[S_{22}]$ in (6.67), (6.71), and (6.70) are stationary. The GSM of a patch array deduced from Galerkin's MoM is thus stationary.

6.8 CONVERGENCE OF MoM SOLUTIONS

To construct the GSM of a patch layer, it is very important that the MoM solution be converged. Four parameters decide the convergence of a solution: the number of basis functions K, the number of incident modes (or coupling modes) M, the number of transmitted modes N, and the number of expansion modes N_{ff}. The computation time increases rapidly with increasing K, M, and N because the matrix sizes increase. On the other hand, smaller values of these parameters may not yield an acceptable accuracy. The parameter K has a direct influence on the accuracy in the patch current distribution, while M and N are related to the coupling between adjacent layers. For a smaller patch size, K could be smaller. For a one-layered array structure (one-layered frequency selective surface, for example), smaller M and N are permissible because no coupling layer exists. For a multilayer structure, selections of M and N are critical. The smaller is the layer thickness (between two patch layers), the larger the values of M and N should be. This is primarily due to the fact that a large number of modal fields arrive at the following layer;

hence their interactions must be taken into consideration for accurate end results. Further, M and N should be large for large cells, because the number of coupling modes increases with the cell size. We will assume $M = N$, which will not alter our general conclusion of the present study.

The number of expansion modes plays a very important role for convergence. The contributions of the higher order modes must be included in the field expansion; therefore N_{ff} must be sufficiently large. In the following sections we will develop guidelines for selecting these parameters with respect to an acceptable error.

6.8.1 Selection of Number of Coupling Modes

As a rule of thumb, the following criteria can be applied in deciding the number of coupling modes (M). We assume that the highest order coupling evanescent mode must decay to its 5% initial value when it reaches the following patch or dielectric layer. Suppose the highest order mode index is M_1. Therefore for a one-dimensional array problem, we write

$$\exp(-jk_z d) = 0.05 \tag{6.134}$$

where $k_z = \sqrt{\varepsilon_r k_0^2 - k_x^2}$, $k_x = (\psi_x + 2M_1 \pi)/a$, a is the cell dimension along x, d is spacer thickness, and ψ_x is the phase difference between the elements along x. For large M_1 we have

$$\exp\left(\frac{-2M_1 \pi d}{a}\right) \approx 0.05 \tag{6.135}$$

which yields $M_1 = 0.48a/d$. Now considering positive and negative indices and TE_z and TM_z modes, the total number of coupling modes becomes

$$M = 4M_1 = \frac{1.92a}{d} \approx \frac{2a}{d} \tag{6.136}$$

The total number of coupling modes for a two-dimensional array, therefore, may be considered as (considering indices along x, y, positive, and negative values) equal to $4 \times 0.48 \times (a/d) \times 0.48 \times (b/d) = 0.92(ab/d^2)$. Now including both types of modes (TE_z and TM_z) this final number becomes

$$M = 2 \times 0.92\left(\frac{ab}{d^2}\right) \approx \frac{2ab}{d^2} \tag{6.137}$$

Interestingly, M is independent of the wavelength and the dielectric constant of the spacer.

The numerical example in Table 6.1 shows M versus percentage of error in the final result. In this example we consider a two-layer patch array of cell size $0.6\lambda_0 \times 0.6\lambda_0$ and patch size $0.3\lambda_0 \times 0.3\lambda_0$. The thickness of the spacer (space between the patch layers) is $0.1\lambda_0$ with a dielectric constant of 1.1. We consider

TABLE 6.1 Convergence Test for Two-Layer Patch Array

M	Reflection Coefficient	Error, %
10	0.3416	13.19
20	0.3134	3.84
40	0.3029	0.44
60	0.3025	0.30
80	0.3020	0.13
100	0.3018	0.07

Note: Cell size $0.6\lambda_0 \times 0.6\lambda_0$, patch size (in both layers) $0.3\lambda_0 \times 0.3\lambda_0$, spacer thickness $0.1\lambda_0$, dielectric constant of spacer 1.1, $N_{ff} = 1088$, $K = 50$. The reflection coefficient corresponds to the TE_{z00} Floquet mode.

normal incidence. The convergence is tested on the reflection coefficient of the TE_{z00} Floquet mode. For our example the estimated value of M from (6.137) should be about 72. We notice that the result converges (within 0.3% error) for $M > 60$.

6.8.2 Failure of MoM Analysis

Interestingly, the MoM analysis fails if the number of basis functions, K, and the number of expansion modes, N_{ff}, are of the same order, even if they are sufficiently large individually. In fact, it can be proven theoretically that for $K = N_{ff}$, the computed reflection coefficient of a single patch layer becomes unity, regardless of patch dimensions, which is obviously incorrect. We will formally prove this important theorem and then seek for a proper value of N_{ff} for a reasonable end result.

For simplicity we assume only one incident mode to a patch scatterer given by

$$\vec{E}^{\text{inc}} = \vec{e}_1(x, y)\exp(-jk_{1z}z) \tag{6.138}$$

The patch is located at $z = 0$. According to the MoM formulation, the patch current is expanded in terms of K basis functions as

$$\vec{I}_p(x, y) = \sum_{i=1}^{K} A_i \vec{f}_i(x, y) \tag{6.139}$$

From (6.82), the electric field produced by the current source $\vec{f}_i(x, y)$ is given by

$$\vec{E}_i = -\sum_n fe^*(i, n)Z^{\text{FL}}(n)\vec{e}_n(x, y) \tag{6.140}$$

with $Z^{\text{FL}}(n) = 1/[Y_{01}^{\text{FL}}(n) + Y_{02}^{\text{FL}}(n)]$. In principle, the above summation should include all of the modes generated by the induced current; hence an infinite number of terms should be summed over. However, for computation purpose, the series must be truncated with a finite number of terms, N_{ff}. As a special case we assume $N_{ff} = K$ (= number of basis functions) and obtain

$$\vec{E}_i = -\sum_{n=1}^{K} fe^*(i, n) Z^{FL}(n) \vec{e}_n(x, y) \qquad (6.141)$$

The total field produced by $\vec{I}_p(x, y)$ is obtained using superposition as

$$\vec{E}^p(x, y) = \sum_{i=1}^{K} A_i \vec{E}_i = -\sum_{n=1}^{K} B_n \vec{e}_n(x, y) \qquad (6.142)$$

with

$$B_n = \sum_{i=1}^{K} fe^*(i, n) Z^{FL}(n) A_i \qquad (6.143)$$

Notice, B_n's are a new set of unknowns that are related linearly with A_i's. Therefore, if B_n's are found, A_i's can be uniquely determined from (6.143). Furthermore, in the conventional MoM procedure usually one does not determine B_n's explicitly. However, we will determine B_n's here because we observe from (6.142) that $-B_1$ turns out to be the reflection coefficient of the incident mode. To obtain B_n's we set the boundary condition on the patch surface at $z = 0$ as

$$\vec{E}^{inc} = -\vec{E}^p \qquad (6.144)$$

which yields

$$\vec{e}_1(x, y) = B_1 \vec{e}_1(x, y) + B_2 \vec{e}_2(x, y) + \cdots + B_K \vec{e}_K(x, y) \qquad (6.145)$$

If we test the above equation in Galerkin's way (or any other way, for that matter) we have K linear equations as

$$fe(m, 1) = fe(m, 1) B_1 + fe(m, 2) B_2 + \cdots + fe(m, K) B_K \qquad m = 1, 2, \ldots, K \qquad (6.146)$$

Solving (6.146) using Cramer's rule (or any other method) one obtains the solutions for B_n's as

$$B_1 = 1 \qquad B_2 = B_3 = \cdots = B_K = 0 \qquad (6.147)$$

This indicates that the incident mode completely reflects back irrespective of the patch size. If the incident field consists of several modes, then the reflection coefficient matrix would be obtained as

$$[S_{11}] = -[I] \qquad (6.148)$$

where $[I]$ is the identity matrix. This proves that when the number of expansion modes becomes equal to the number of basis functions, the reflection coefficient matrix always becomes a negative identity matrix irrespective of the patch dimensions. This cannot be true in reality. Thus the MoM fails for $K = N_{ff}$.

The above result derived in (6.148) is consistent with the equivalent impedance matrix formulation developed in section 6.6. It was shown that for $K = N_{ff}$, the impedance matrix becomes a null matrix, which implies $[S_{11}] = -[I]$.

6.8.3 Lower Limit for Number of Expansion Modes

Because the MoM fails for $K = N_{ff}$, it is logical to contemplate that the end results would be inaccurate if K and N_{ff} are close to each other, even if they are not equal. In fact, it will be seen that N_{ff} must be greater than K and for a given value of K there exists a lower limit of N_{ff} for a converged solution [8]. Conversely, for a given value of N_{ff}, K should not exceed beyond a certain limit.

To establish the above-mentioned facts, let us examine carefully the MoM analysis procedure. In the MoM analysis, the induced current on the patch surface is expressed in terms of known basis functions with unknown coefficients. The scattered electric field produced by the induced current is obtained in terms of several Floquet modes (theoretically infinite, computationally N_{ff} number of modes). The unknown coefficients for the basis functions are the solutions of a set of linear equations that is generated by enforcing the boundary condition that the scattered electric field on the patch surface is negative of the incident electric field (tangential components only). What is missing in this procedure is the enforcement of the continuity condition of the tangential magnetic field in the open region coplanar to the patch. In principle, a planar electric current does not produce any tangential magnetic field at a coplanar point outside the source region [9]. Thus the continuity of the magnetic field is automatically satisfied, because the patch current produces a null tangential magnetic field in the open region. However, to ensure such a null field in the open region, one needs to use an infinite number of Floquet modes in the series expansion. A truncation of the infinite series causes a nonzero scattered magnetic field in that region, introducing inaccuracy to the final solution. (This is somewhat analogous to a Fourier series expansion of a periodic pulse train. One needs an infinite number of Fourier terms to satisfy zero intensity in the off region.)

For a quantitative analysis, consider a one-dimensional problem of printed strips as shown in Figure 6.8. A y-polarized plane wave is incident on the strip array. Assuming a normal incidence, the induced strip current, which has y-component only, can be expressed in terms of K basis functions as

$$I_y^{\text{strip}} = \sum_{n=1}^{K} A_n \cos\left(\frac{(n-1)\pi x}{a_1}\right) \tag{6.149}$$

In (6.149), a_1 is the strip width. Now let us determine the magnetic field produced by the above strip current. The x-component magnetic field can be expressed in terms of the following infinite series:

$$H_x^S = \pm \sum_{m=0}^{\infty} B_m \exp\left(\frac{-2jm\pi x}{a}\right) \exp(\mp jk_{mz}z) \qquad \text{for} \quad z > \text{ or } < 0 \tag{6.150}$$

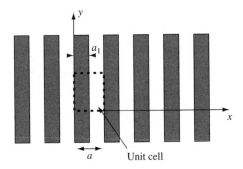

FIGURE 6.8 Infinite array of strips.

We assume that no dielectric layer exists in either side of the strip array. The discontinuity of the scattered magnetic field yields the strip current; therefore at $z = 0$, we write

$$I_y^{\text{strip}} = 2 \sum_{m=-\infty}^{\infty} B_m \exp\left(\frac{-2jm\pi x}{a}\right) \tag{6.151}$$

The right-hand side of (6.151) must be equal to the right-hand side of (6.149) for $0 < x < a_1$ and zero for $x > a_1$. Now expanding (6.149) in Fourier series and comparing with (6.151), we obtain

$$B_m = \frac{j\pi}{a^2} \sum_{n=1}^{K} A_n \left[(-1)^{n-1} \exp\left(\frac{2jm\pi a_1}{a}\right) - 1\right] \left[\frac{m}{\{(n-1)\pi/a_1\}^2 - (2m\pi/a)^2}\right] \tag{6.152}$$

Let us look for the dominating coefficient of A_n by varying m. The coefficient of A_n becomes maximum if

$$m = \pm(n-1)\frac{a}{2a_1} \tag{6.153}$$

The above result is interesting. It indicates that the nth basis function will have the most significant contributions on the mth expansion mode, where $m = \pm(n-1)a/(2a_1)$. This phenomenon is shown in Figure 6.9. Therefore, to capture the largest contributions of the nth basis function to the magnetic field, the minimum number of terms in the modal expansion should be about $(n-1)a/a_1$. This result is apparent because the nth basis function and the $[(n-1)a/2a_1]$th expansion mode have identical periodicity along x; hence they interact strongly. It follows that for K basis functions the minimum number of expansion modes should be about $(K-1)a/a_1$. However, we need few additional expansion modes to capture all the modes within, say, -13.4 dB amplitude with respect to the peak value. The minimum number of expansion modes should be

$$N_{ff}(\text{min}) \approx (K+3)\frac{a}{a_1} \tag{6.154}$$

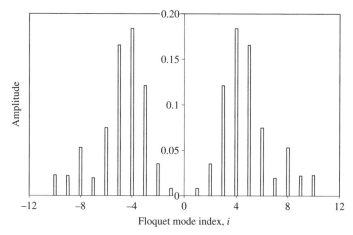

FIGURE 6.9 Floquet mode amplitudes versus mode index for the fourth-order basis function ($n = 4$) with $a/a_1 = 2.7$.

The above lower limit for N_{ff} is true for sinusoidal basis functions. For other types of basis functions the limit may vary slightly, but the general order remains unchanged. In principle, no upper limit of N_{ff} exists. However, for a given K, the end results converge very rapidly after N_{ff} crosses the lower limit, as seen in Figure 6.10. It is worth mentioning that a converged result does not necessarily mean an accurate result. The accuracy depends on the selection of K. The higher the value of K, the better is the accuracy at the cost of computation time.

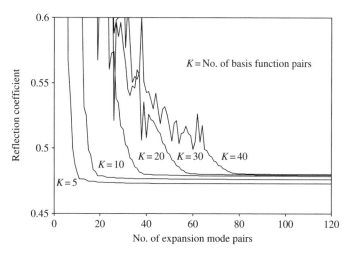

FIGURE 6.10 Computed reflection coefficient versus number of expansion modes in an infinite array of strips with normal incidence: $a = 1.6\lambda_0$, $a_1 = 0.8\lambda_0$; number of incident mode pairs is 15.

In this development, we use the entire domain basis functions to establish the lower limit of the expansion modes. However, the same relation would be equally valid if the subdomain basis functions are used, like rooftop or pulse basis functions. A train of K pulse basis functions of arbitrary heights (representing the patch current) can equivalently be represented by K entire-domain orthogonal square basis functions [8]. The spectrum of a square wave is very similar to that of a cosine wave function of identical periodicity. Thus the required number of expansion modes would be of the same order.

The theory can be extended for a two-dimensional patch array. The minimum number of expansion modes should be approximately given by

$$N_{ff}(\min) \sim K \frac{ab}{a_1 b_1} \tag{6.155}$$

where ab is the cell size and $a_1 b_1$ is the patch size.

The above results can also be understood from a point-matching point of view. As mentioned earlier, the electromagentic fields produced by the strip current must satisfy the following boundary conditions: (i) The tangential electric field must be negative of the incident electric field on the strip and (ii) the tangential magnetic field must vanish in the open region coplanar to the strip. If the above electromagnetic fields are expressed by N_{ff} expansion functions, then the unknown amplitudes of the expansion functions need to be solved using the above two conditions. If the point-matching method is used and if the points are evenly spaced, then the number of points should be N_{ff}, of which $N_{ff}(1 - a_1/a)$ points belong to the open region and the remaining points belong to the strip region. This allows eliminating $N_{ff}(1 - a_1/a)$ unknowns by virtue of condition (or constraint) (ii). Thus the number of unknowns to be solved using condition (i) becomes $N_{ff} a_1/a$, which is equivalent to using $N_{ff} a_1/a$ basis functions for the strip current.

The orthogonal (Floquet) expansion method, typically invoked in the periodic MoM analysis, implicitly satisfies condition (ii). If K basis functions are used and expanded in Floquet series with $N_{ff} = Ka/a_1$ terms, then condition (ii) is automatically satisfied at $K(a - a_1)/a_1$ sampling points in the open region. In fact, a larger number of Floquet terms may be allowed even with K basis functions (that is, $N_{ff} > Ka/a_1$) if suitable basis functions are found to express the patch current distribution (for instance, satisfying edge conditions). It means that the number of independent equations, K, employed toward satisfying condition (i) could be fewer than the due proportion. The appropriate shapes of the basis functions provide additional reinforcements for obtaining an accurate current distribution on the patch.

6.8.4 Number of Basis Functions

The number of basis functions K controls the accuracy of the solution. There is no hard and fast rule that can be established theoretically. It depends on the shape of the obstacle (in this case patch) and the fitness of the basis functions in regard to the shape of the patch and physical dimensions. However, it is reasonable (but not strictly required) that K exceeds the number of coupling modes M for the following reasons: (a) The basis vector space should have M degree of freedoms (M dimensions) to account for the current variation in regard to the arbitrary incident voltage vector and (b) for $K \geq M$, the equivalent impedance is well defined, which ensures computational stability [assuming N_{ff} satisfies (6.155)]. Evidently, we notice from Figure 6.10 that convergence to the actual solution is reached rapidly as K becomes closer to the number of incident modes.

6.9 ADVANTAGES OF GSM APPROACH

The GSM approach for multilayer array analysis is very convenient both analytically and computationally due to the following reasons:

- The GSM approach is a modular approach. Each module is analyzed independently with a limited number of unknowns as opposed to an integrated approach [10].
- The required Green's function has a simple form because a module at most has two dielectric media.
- The computation time is proportional to the number of layers. In an integrated approach, the computation time increases exponentially with the number of layers.
- It is easy to develop a computer program that is flexible to handle any number of layers.

The GSM approach will be applied to analyze multilayer arrays in the chapter that follows.

6.10 OTHER NUMERICAL METHODS

While the MoM-based GSM approach has been well established and accurate, other numerical approaches also exist for analyzing multilayer arrays. In particular, the finite element frequency-domain (FEFD) approach [11] and the finite difference time-domain (FDTD) approach [12] are gaining popularity for analyzing infinite periodic arrays. These approaches utilize periodic radiation conditions and periodic side-wall conditions to model infinite arrays under Floquet excitations. A unit cell is filled with a number of node points. The FEFD and FDTD methods express Maxwell's equations in terms of difference equations and solve the electromagnetic field components on these nodes inside a unit cell. These methods have natural abilities to handle dielectric inhomogeneity and anisotropy in all directions within a unit cell. However, both methods demand extensive computer capacity in regard to electromagnetic simulation of a structure.

REFERENCES

[1] A. K. Bhattacharyya, "A Numerical Model for Multilayered Microstrip Phased-Array Antennas," *IEEE Trans.*, Vol. AP-44, No. 10, pp. 1386–1393, Oct. 1996.

[2] A. K. Bhattacharyya, "A Modular Approach for Probe-Fed and Capacitively Coupled Multilayered Patch Arrays," *IEEE Trans.*, Vol. AP-45, No. 2, pp. 193–202, Feb. 1997.

[3] R. Mittra, C. H. Chan, and T. Cwik, "Techniques for Analyzing Frequency Selective Surfaces—A Review," *Proc. IEEE*, Vol. 76, pp. 1593–1615, Dec. 1988.

[4] R-S. Chu and K-M. Lee, "Analytical Model of a Multilayered Meander-Line Polarizer Plate with Normal and Oblique Plane-Wave Incidence," *IEEE Trans.*, Vol. AP-35, No. 6, pp. 652–661, June 1987.

[5] H-Y. D. Yang, N. G. Alexopoulos, and E. Yablonovitch, "Photonic Band-Gap Materials for High-Gain Printed Antennas," *IEEE Trans.*, Vol. AP-45, pp. 185–187, Jan. 1997.

[6] P. H. Masterman and P. J. B. Clarricoats, "Computer Field-Matching Solution of Waveguide Transverse Discontinuities," *Proc. IEE*, Vol. 118, pp. 51–63, 1971.

[7] C. H. Chan and R. Mittra, "On the Analysis of Frequency-Selective Surfaces Using Subdomain Basis Functions," *IEEE Trans.*, Vol. AP-38, No. 1, pp. 40–50, Jan. 1990.

[8] A. K. Bhattacharyya, "On the Convergence of MoM and Mode Matching Solutions for Infinite Array and Waveguide Problems," *IEEE Trans.*, Vol. AP-51, No. 7, pp. 1599–1606, July 2003.

[9] R. F. Harrington, in *Time-Harmonic Electromagnetic Fields*, McGraw-Hill, New York, 1961, "Perturbational and Variational Techniques," Chapter 7.

[10] Y. Lubin and A. Hessel, "Wide-Band, Wide-Angle Microstrip Stacked-Patch-Element Phased Arrays," *IEEE Trans.*, Vol. AP-39, No. 8, pp. 1062–1070, Aug. 1991.

[11] D. T. McGrath and V. P. Pyati, "Phased Array Antenna Analysis with the Hybrid Finite Element Method," *IEEE Trans.*, Vol. AP-42, No. 12, pp. 1625–1630, Dec. 1994.

[12] P. Harms, R. Mittra, and W. Ko, "Implementation of the Periodic Boundary Condition in the Finite-Difference Time-Domain Algorithm for FSS Structures," *IEEE Trans.*, Vol. AP-42, No. 9, pp. 1317–1324, Sept. 1994.

BIBLIOGRAPHY

Anantha, V., and A. Taflove, "Efficient Modeling of Infinite Scatterers Using a Generalized Total-Field/Scattered-Field FDTD Boundary Partially Embedded Within PML," *IEEE Trans.*, Vol. AP-50, No. 10, pp. 1337–1349, Oct. 2002.

Changhua, W., and J. A. Encinar, "Efficient Computation of Generalized Scattering Matrix for Analyzing Multilayered Periodic Structures," *IEEE Trans.*, Vol. AP-43, No. 11, pp. 1233–1242, Nov. 1995.

Gay-Balmaz, P., J. A. Encinar, and J. R. Mosig, "Analysis of Multilayer Printed Arrays by a Modular Approach Based on the Generalized Scattering Matrix," *IEEE Trans.*, Vol. AP-48, No. 1, pp. 26–34, Jan. 2000.

Hall, R. C., R. Mittra, and K. M. Mitzner, "Analysis of Multilayered Periodic Structures Using Generalized Scattering Matrix Theory," *IEEE Trans.*, Vol. AP-36, No. 4, pp. 511–517, Apr. 1988.

Lee, S. W., W. R. Jones, and J. J. Campbell, "Convergence of Numerical Solutions of Iris-Type Discontinuity Problems," *IEEE Trans. Microwave Theory Tech.*, Vol. MTT-19, No. 6, pp. 528–536, June 1971.

Leroy, M., "On the Convergence of Numerical Results in Modal Analysis," *IEEE Trans. Antennas Propagat.*, Vol. AP-31, No. 7, pp. 655–659, July 1983.

Mautz, J. R., "Variational Aspects of the Reaction in the Method of Moments," *IEEE Trans.*, Vol. AP-42, No. 12, pp. 1631–1638, Dec. 1994.

Mittra, R., T. Itoh, and T-S. Li, "Analytical and Numerical Studies of the Relative Convergence Phenomenon Arising in the Solution of an Integral Equation by Moment Method," *IEEE Trans. Microwave Theory Tech.*, Vol. MTT-20, No. 2, pp. 96–104, Feb. 1972.

Peterson, A. F., D. R. Wilton, and R. E. Jorgenson, "Variational Nature of Galerkin and Non-Galerkin Moment Method Solutions," *IEEE Trans.*, Vol. AP-44, No. 4, pp. 500–503, Apr. 1996.

Richmond, J. H., "On the Variational Aspects of the Moment Method," *IEEE Trans.*, Vol. AP-39, No. 4, pp. 473–479, Apr. 1991.

Shih, Y. C., "The Mode-Matching Method," in T. Itoh (Ed.), Numerical Techniques for Microwave and Millimeter-Wave Passive Structures, Wiley, New York, 1989, Chapter 9.

Veysoglu, M. E., R. T. Shin, and J. A. Kong, "A Finite-Difference Time-Domain Analysis of Wave Scattering from Periodic Surfaces: Oblique Incidence Case," *J. Electron. Wave Appl.*, Vol. 7, No. 12, pp. 1595–1607, 1993.

Webb, K. J., P. W. Grounds, and R. Mittra, "Convergence in the Spectral Domain Formulation of Waveguide and Scattering Problems," *IEEE Trans. Antennas Propagat.*, Vol. AP-38, No. 6, pp. 511–517, June 1990.

PROBLEMS

6.1 Derive expressions for $[S_{21}^{AB}]$ and $[S_{22}^{AB}]$ submatrices of the combined GSM of two infinite arrays A and B placed one above the other. The individual GSMs are given in (6.2) and (6.3).

6.2 Derive the combined $[S_{21}^{AB}]$ of two layers in cascade using the multiple reflections between two modules as shown in Figure 6.3.

6.3 Show that a nonsingular matrix $[A]$ can be expressed as

$$[A] = [S][\Lambda][S]^{-1}$$

where $[S]$ is the eigenvector matrix of $[A]$, that is, the columns of $[S]$ are the eigenvectors of $[A]$; $[\Lambda]$ is a diagonal matrix with the diagonal elements as the eigenvalues of $[A]$. Hence show that $[A]^N$ approaches a null matrix as N approaches infinity if the absolute values for all eigenvalues are less than unity.

6.4 For a lossy passive device, prove that the eigenvalues of $[S_{11}]$ have magnitudes less than unity. Hence prove that the infinite series of (6.16) is convergent and the sum of the series converges to the matrix on the right-hand side of (6.16). (*Hint*: Use eigenvector excitation for the incident field and then use the condition that the reflected power must be less than the incident power.)

6.5   Obtain the $[S_{21}]$ submatrix of a dielectric layer of thickness h for two incident modes with propagation constants β and $j\gamma$, respectively, where β and γ are both real numbers (one propagating mode and one evanescent mode). Verify that if $\gamma h \gg 1$, then $[S_{21}]$ becomes an ill-conditioned matrix for inversion [hence the $[P]$ matrix in (6.24) is unstable, implying that the $[T]$ matrix formulation is not robust for numerical manipulations].

6.6   Suppose the TM_{00} Floquet mode is incident on an interface of two dielectric media of relative dielectric constants ε_1 and ε_2, respectively. The wave is incident from the first medium at an angle θ from the interface normal. Obtain the condition for the total reflection of the incident mode. (*Hint:* Obtain the Floquet impedances for the TE_{00} mode in both media and then obtain the reflection coefficient. The magnitude of the reflection coefficient should be unity for a total reflection.)

6.7   Repeat problem 6.6 for the TE_{00} mode of incidence.

6.8   A surface current source is placed inside a parallel-plate waveguide of dimension $a \times b$. The surface current is y-directed and placed at $z = 0$. The source plane is perpendicular to z. The current distribution is a linear combination of two known functions f_1 and f_2, which are given by

$$f_1(x, y) = u\left(x + \frac{w}{2}\right) - u\left(x - \frac{w}{2}\right)$$

$$f_2(x, y) = \cos\left(\frac{\pi x}{w}\right)\left[u\left(x + \frac{w}{2}\right) - u\left(x - \frac{w}{2}\right)\right]$$

where $u(x)$ is the unit step function of x and $w < a$. The origin of the coordinate system is at the center of the waveguide. The mutual reaction between the two source functions is defined as

$$< f_1, f_2 > = -\iint_{\text{Source}} \vec{E}_1 \cdot \hat{y} f_2 \, dx \, dy$$

where $\vec{E}_1$ is the electric field produced by the current function f_1. Obtain the mutual reaction (infinite series form) between the two source functions. Also obtain the self-reaction for the source function f_1 and show that the real part of the self-reaction is equal to the power radiated by the source f_1.

6.9   Derive (6.97). Use this relation to obtain the equivalent shunt impedance matrix with respect to the TE_{z00} Floquet mode incidence for an infinite patch array. Patch dimensions are $a_1 \times b_1$ and the unit cell dimensions are $a \times b$. Use rectangular grids. Scan direction is $\theta = 30°$ and $\phi = 0°$. Use only one basis function for the patch current as $\vec{f}_1(x, y) = \hat{y}\cos(\pi y/b_1)$. The origin of the coordinate system is at the patch center.

6.10   The basis functions for the patch current on a rectangular patch of dimension $a \times b$ is taken as the transverse electric field modal functions of a fictitious rectangular waveguide of identical cross-sectional area with four magnetic walls. Obtain the TE_{zmn} and TM_{zmn} modal functions for the transverse components of the electric field with respect to the equivalent waveguide. The propagation takes place along the z-direction and m and n are the modal indices along x and y, respectively.

6.11 Repeat problem 6.10 for a circular patch element of radius a.

6.12 Show that the first-order error in $S_{11}(2, 1)$ due to an error in the assumed current distribution $\vec{g}_1(x, y)$ is nonzero if the following expression is used for computing $S_{11}(2, 1)$ [hence the expression for $S_{11}(2, 1)$ is not stationary]:

$$S_{11}(2, 1) = \frac{(1+\Gamma_1) \iint\limits_{\text{Patch}} \vec{g}_1 \cdot \vec{e}_2^* \, dx \, dy}{Y(2) \iint\limits_{\text{Patch}} \left[\iint\limits_{\text{Patch}} \vec{g}_1(x_0, y_0) \cdot \overline{\overline{G}}(x, y/x_0, y_0) \, dx_0 \, dy_0 \right] \cdot \vec{e}_1^*(x, y) \, dx \, dy}$$

6.13 Two sources A and B produce the following electromagnetic fields on an aperture through a periodic conducting scatterer:

$$\vec{E}_a = (V_1^+ + V_1^-)\vec{e}_1(x, y) + \sum_{n\neq 1} V_n^- \vec{e}_n(x, y)$$

$$\vec{H}_a = Y_{01}^{\text{FL}}(1)(V_1^+ - V_1^-)\vec{h}_1(x, y) - \sum_{n\neq 1} Y_{01}^{\text{FL}}(n)V_n^- \vec{h}_n(x, y)$$

$$\vec{E}_b = (\tilde{V}_2^+ + \tilde{V}_2^-)\vec{e}_2^*(x, y) + \sum_{n\neq 2} \tilde{V}_n^- \vec{e}_n^*(x, y)$$

$$\vec{H}_b = Y_{01}^{\text{FL}}(2)(\tilde{V}_2^+ - \tilde{V}_2^-)\vec{h}_2^*(x, y) - \sum_{n\neq 2} Y_{01}^{\text{FL}}(n)\tilde{V}_n^- \vec{h}_n^*(x, y)$$

where V_i^+, V_i^-, $\vec{e}_i(x, y)$, $\vec{h}_i(x, y)$ are the transmitted and reflected voltages and normalized electric and magnetic field vectors, respectively, for the ith Floquet mode associated with source A and $\tilde{V}_i^+$, $\tilde{V}_i^-$, $\vec{e}_i^*(x, y)$, $\vec{h}_i^*(x, y)$ are the same quantities for source B; $Y_{01}^{\text{FL}}(i)$ is the modal admittance of the ith Floquet mode. The asterisk indicates complex conjugate. Using reciprocity on a unit cell, show that

$$[V_1^+ Y_{01}^{\text{FL}}(1)]\tilde{V}_1^- = [\tilde{V}_2^+ Y_{01}^{\text{FL}}(2)]V_2^-$$

(*Hint*: For perfect-conducting scattering objects, the reciprocity relation finally boils down to integration on the source plane only.)

6.14 Consider M incident modes, K basis functions, and K expansion modes. Then following the procedure in Section 6.8.2, prove that $[S_{11}] = -[I]$ irrespective of the patch size. Hence the MoM fails if the number of basis functions equals the number of expansion modes.

6.15 Explain why the MoM does not fail in the case of a waveguide step junction analysis even if equal numbers of modes are used in two sides of the junction (see the explanation in [8]).

6.16 If $[A] = [B] \times [C]$, where $[B]$ is of order $M \times N$, $[C]$ is of order $N \times M$, and $M > N$, then show that the rank of $[A]$ is less than M, hence $[A]$ is a singular matrix.

(*Hint*: By direct multiplication, show that each column of $[A]$ can be expressed as a linear combination of the columns of $[B]$. Hence, the column space of $[A]$ has only N independent vectors.)

CHAPTER SEVEN

Analysis of Microstrip Patch Arrays

7.1 INTRODUCTION

In this chapter we apply the GSM approach developed in Chapter 6 for analyzing probe-fed [1] and slot-fed [2] multilayer patch array antennas. Each layer of the array is analyzed independently and then characterized by its GSM with respect to a finite number of Floquet modes. The overall GSM of the array is obtained by combining the individual GSMs. The GSM of a patch layer has been derived in the previous chapter. We develop the analyses for the probe layer and the slot–feedline transition layer, pertinent to the probe-fed and slot-coupled arrays, respectively. For the probe layer, the generalized impedance matrix (GIM) is derived directly from the electromagnetic fields produced by the probe current and then the GIM is converted to a GSM. For the slot–feedline transition an equivalent circuit is obtained employing reciprocity, which is then used to construct the GSM of the transition. The impedance and bandwidth characteristics of one- and two-layer patch arrays are studied. The active element patterns are derived and numerical results are presented. An electromagnetically coupled (EMC) patch array is treated as a special case of a two-layer array with a finite offset between the layers. It is found that the radiation efficiency of a slot-coupled array fed by a microstripline is somewhat lower than that of a probe-fed array because of the undesired feed-side radiation. Characteristics of a stripline-fed patch array with mode suppressing vias are discussed and the effects of vias on the input impedance are shown. The chapter ends with a study of the active return loss of a finite array with respect to uniform and nonuniform amplitude distributions.

Phased Array Antennas.   By Arun K. Bhattacharyya
© 2006 John Wiley & Sons, Inc.

7.2 PROBE-FED PATCH ARRAY

Figure 7.1*a* shows the geometry of a multilayer (two-layer in this case) probe-fed patch array. Figure 7.1*b* is the block diagram of the array, where each block is associated with a layer represented mathematically by a GSM. However, for the probe layer, it is convenient analytically to construct the GIM instead of the GSM. The GSM and the GIM are combined in an appropriate fashion to yield the input impedance seen by the probe. The procedure of obtaining the GSM of a patch layer has been detailed in Chapter 6 and thus will not be repeated here. In the following development we will determine the GIM of the probe layer.

7.2.1 Generalized Impedance Matrix of Probe Layer

Suppose a and b, respectively, represent the x and y dimensions of a unit cell of an array of probe-fed patch elements. As stated before, for the probe layer we will determine the GIM instead of the GSM. The GIM naturally incorporates the effect of the ground plane, which is advantageous both analytically and computationally. However, it is possible to convert the GIM to a GSM (which will be required later for active element pattern computation). To obtain the GIM of the probe layer, we assume that the probe has a finite diameter, d. We define port 1 at the ground plane end of the probe (with a fictitious delta gap between the probe and the ground plane) and port 2 at the other end. Suppose I_1 is the probe current, which we assume to be uniform along z. This is a reasonable assumption if the probe is directly connected or very close to the patch surface. If the probe end is far from the patch surface, a sine function may be used for representing the current distribution on the probe. However, the analysis would be very similar to that of a uniform current distribution.

The probe current will excite several Floquet modes. It is convenient to represent the electromagnetic fields in terms of the Floquet modal voltages and currents.

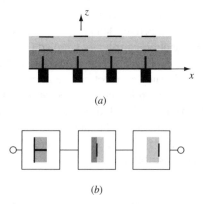

(a)

(b)

FIGURE 7.1 (*a*) Side view of a two-layer patch array fed by probes. (*b*) Modular representation of the array.

Suppose φ_x and φ_y are the phase differences between the adjacent elements along x and y, respectively. Then for propagating waves along z, the Floquet wave numbers (for rectangular grids) are given by

$$k_{xm} = \frac{2m\pi + \varphi_x}{a} \qquad -\infty < m < \infty$$

$$k_{yn} = \frac{2n\pi + \varphi_y}{b} \qquad -\infty < n < \infty \tag{7.1}$$

$$\beta_{mn}^2 = k_0^2\varepsilon_r - k_{xm}^2 - k_{yn}^2 \tag{7.2}$$

where k_0 is the wave number in free space and ε_r is the dielectric constant of the layer. The GIM of a two-port module is defined as

$$\begin{bmatrix} V_1 \\ V_2 \end{bmatrix} = \begin{bmatrix} Z_{11} & Z_{12} \\ Z_{21} & Z_{22} \end{bmatrix} \begin{bmatrix} I_1 \\ I_2 \end{bmatrix} \tag{7.3}$$

where $[V_1]$ and $[V_2]$ are the voltage vectors and $[I_1]$ and $[I_2]$ are the current vectors at the two ports. The vectors $[V_1]$ and $[I_1]$ have only one element each[1] because the current distribution on a probe can be approximated by a single function (for example, uniform, sine). On the other hand, the electromagnetic fields in port 2 are expressed in terms of several Floquet modes, implying that $[V_2]$ and $[I_2]$ vectors are multidimensional. To obtain the submatrices $[Z_{11}]$ and $[Z_{21}]$ we excite port 1 by a current source and terminate port 2 by an open circuit (magnetic wall). Under this open-circuit condition, $[I_2]$ becomes a null vector; thus the above two submatrices are directly obtained from the voltages in port 1 and port 2, respectively. The Floquet modal fields produced by the probe current are obtained next under the open-circuit condition.

Suppose (x_p, y_p) is the coordinate of the probe location in a unit cell. The probe current can be expressed in terms of Dirac delta functions as (to simplify the analysis we temporarily assume an infinitely thin probe; the effect of finite diameter will be incorporated later)

$$J_z(x, y, z) = I_1\delta(x - x_p)\delta(y - y_p) \tag{7.4}$$

The probe current excites the TM_z modes only because the TE_z modes do not react with the z-directed electric current. The TM_z mode fields are determined from the magnetic vector potential, A_z [3]. In the source region, the magnetic vector potential satisfies the following Helmholtz equation [3]:

$$\nabla^2 A_z + k_1^2 A_z = -J_z \tag{7.5}$$

---

[1] For a two-probe-fed patch, these vectors consist of two elements each.

where $k_1 = k_0\sqrt{\varepsilon_r}$. To find the solution for A_z, we express A_z as

$$A_z(x, y, z) = \sum_{m,n} \sum_p A_{mnp} \exp(-jk_{xm}x - jk_{yn}y) \cos\left\{\frac{p\pi}{2h}z\right\} \qquad p = 1, 3, 5, \ldots$$

(7.6)

where h represents the probe height. The cosine functions essentially satisfy the electric and magnetic wall boundary conditions at $z = 0$ and $z = h$, respectively. Substituting (7.6) into (7.5) and expanding J_z in Fourier series, one obtains

$$A_{mnp} = (-1)^{(p+1)/2} I_1 \frac{4\exp(jk_{xm}x_p + jk_{yn}y_p)}{\pi abp[k_1^2 - k_\rho^2 - \{p\pi/2h\}^2]}$$

(7.7)

with $k_\rho^2 = k_{xm}^2 + k_{yn}^2$. The electric field produced by the probe current is obtained from the magnetic vector potential using the relation

$$\vec{E} = \frac{1}{j\omega\varepsilon}\nabla \times \nabla \times \vec{A}$$

(7.8)

with ε as the permittivity of the medium. The magnetic vector potential, $\vec{A}$, has only one component A_z given in (7.6). Using (7.8) the z-component of the electric field is obtained as

$$E_z(x, y, z) = \frac{1}{j\omega\varepsilon}\sum_{m,n}\sum_p A_{mnp}(k_1^2 - k_{zp}^2)\exp(-jk_{xm}x - jk_{yn}y)\cos\{k_{zp}z\}$$

(7.9)

with $k_{zp} = p\pi/2h$. We now use the following stationary expression [3] for the input impedance Z_{11}:

$$Z_{11} = -\frac{\int_V \vec{E}\cdot\vec{J}\,dV}{I_1^2}$$

(7.10)

Using (7.4) and (7.9) in (7.10) and then using (7.7), we obtain

$$Z_{11} = \frac{8h^2}{j\omega\varepsilon abh\pi^2}\sum_{mnp}\frac{k_1^2 - k_{zp}^2}{[\beta_{mn}^2 - k_{zp}^2]p^2}$$

(7.11)

where $\beta_{mn}^2 = k_1^2 - k_\rho^2$. The above equation can be rearranged to yield

$$Z_{11} = \frac{8h^2}{j\omega\varepsilon abh\pi^2}\sum_{mn}\sum_p\left[\frac{1}{p^2} + \frac{k_\rho^2}{(\beta_{mn}^2 - k_{zp}^2)p^2}\right]$$

(7.12)

We now take the summation over p and use the following formula [4]:

$$\sum_{n=1,3,5,\ldots}^{\infty}\frac{\cos nx}{n^2(n^2 - a^2)} = \frac{\pi}{4a^2}\left[\frac{\sin a(\pi/2 - x)}{a\cos(\pi a/2)} - x \pm \frac{\pi}{2}\right]$$

The positive sign applies for $0 < x < \pi$ and the negative sign applies for $x = 0$.[2] Thus (7.12) simplifies to

$$Z_{11} = \frac{h}{j\omega\varepsilon ab} \sum_{mn} \frac{1}{\beta_{mn}^2} \left(k_1^2 - \frac{k_\rho^2 \tan(\beta_{mn}h)}{\beta_{mn}h} \right) \tag{7.13}$$

The right-hand side of the above equation is a divergent series, which is expected for an infinitely thin current source. To incorporate the effect of finite probe diameter, d, each term should be multiplied by the factor $[J_0(k_\rho d/2)]^2$, because the two-dimensional Fourier transform of a uniform circular source is a Bessel function of order zero. Therefore the self-impedance of the probe of diameter d is given by

$$Z_{11} = \frac{120\pi h}{jk_0\varepsilon_r ab} \sum_{m,n} \frac{1}{\beta_{mn}^2} \left(k_1^2 - \frac{k_\rho^2 \tan(\beta_{mn}h)}{\beta_{mn}h} \right) J_0^2 \left(\frac{k_\rho d}{2} \right) \tag{7.14}$$

Expressions for the elements in $[Z_{21}]$ are obtained directly from the Floquet modal voltages associated with the electric field solution in (7.8). The element corresponding to the ith mode becomes

$$Z_{21}(i, 1) = \frac{120\pi k_\rho}{k_0\varepsilon_r \sqrt{ab}} \exp(jk_{xm}x_p + jk_{yn}y_p) \frac{\tan(\beta_{mn}h)}{\beta_{mn}} J_0 \left(\frac{k_\rho d}{2} \right) \tag{7.15}$$

where i represents the sequence number of the TM_{zmn} mode.

To calculate $[Z_{12}]$ and $[Z_{22}]$ we excite port 2 by a modal current source vector $[I_2]$ and set $[I_1] = 0$, that is, the probe is eliminated. Under this condition, the electromagnetic field solution yields the elements of the matrices. Because the Floquet modes are mutually orthogonal, the matrix $[Z_{22}]$ is diagonal. The diagonal elements are the input impedances experienced by the Floquet modal sources placed at $z = h$. The ith diagonal element is given by

$$Z_{22}(i, i) = jZ_{0i} \tan(\beta_{ii}h) \tag{7.16}$$

where Z_{0i} is the characteristic impedance of the ith mode. To obtain the elements of $[Z_{12}]$, we first find E_z in the region $0 < z < h$. Then we use the following stationary expression for mutual impedance.

$$Z_{12}(1, i) = \frac{-\int_V E_{zi}J_z dV}{I_1 I_{2i}} \tag{7.17}$$

---

[2] Reference [4] does not include the result for $x = 0$.

where E_{zi} is the z-component of the electric field produced by the ith modal current I_{2i}, J_z is the probe current density, and I_1 is the input probe current at $z = 0$. The final expression for the ith element is derived as

$$Z_{12}(1, i) = -\frac{120\,\pi\,k_\rho}{k_0\varepsilon_r\sqrt{ab}}\exp(-jk_{xm}x_p - jk_{yn}y_p)\frac{\tan(\beta_{mn}h)}{\beta_{mn}}J_0\left(\frac{k_\rho d}{2}\right) \qquad (7.18)$$

The mutual impedances for the TE$_z$ modes are zero because E_z is zero for a TE$_z$ mode. It is interesting to observe that Z_{12} and Z_{21} do not have identical expressions, that is, the GIM is a nonsymmetrical matrix (see problem 7.5 for an explanation).

7.2.2 Input Impedance

The input impedance seen by a probe feed can be calculated form the GIM of the probe layer and the combined GSM of the layers above the probe layer, which include dielectric layers, dielectric interfaces, and patch layers. With regard to the input impedance analysis the layers above the probe layer may be considered as a load terminated at port 2 of the probe layer. The load impedance, $[Z_L]$, can be calculated from the combined GSM. Suppose $[S^p]$ is a combined GSM of all the layers above the probe layer, expressed as

$$[S^p] = \begin{bmatrix} S^p_{11} & S^p_{12} \\ S^p_{21} & S^p_{22} \end{bmatrix} \qquad (7.19)$$

Port 1 of the above GSM is located at $z = h$. The load impedance matrix seen by the probe layer is then

$$[Z_L] = [I + S^p_{11}][I - S^p_{11}]^{-1}[Z_0] \qquad (7.20)$$

where $[Z_0]$ is the characteristic impedance matrix (diagonal) of the Floquet modes. The input impedance (Floquet impedance) seen by the probe can be derived as

$$Z_{\mathrm{in}} = Z_{11} - [Z_{12}][Z_L + Z_{22}]^{-1}[Z_{21}] \qquad (7.21)$$

7.2.3 Impedance Characteristics

The input impedance in (7.21) is computed to examine the characteristics of one- and two-layer patch arrays. To generate numerical results, we use about 1000 Floquet expansion modes and about 100 coupling modes, which show reasonable numerical convergence. About 50 basis functions are used for the patch current.

The input impedances of one- and two-layer patch arrays are presented in Figures 7.2 and 7.3, respectively. From Figure 7.2 we observe that the input reactance curve shows a positive offset when the probe touches the patch. This offset is primarily due to the inductive self-reactance of the probe. Making the probe somewhat shorter than the dielectric layer thickness can reduce this reactance. In such

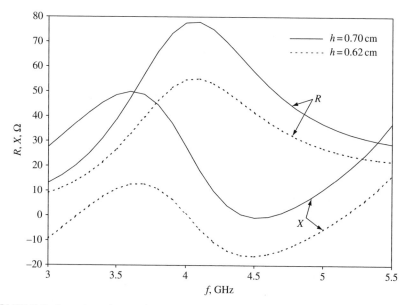

FIGURE 7.2   Input impedance of a one-layer probe-fed patch array under Floquet excitation (bore-sight beam): square lattice, cell size $4 \times 4\,\mathrm{cm}^2$, patch size $2 \times 2\,\mathrm{cm}^2$, substrate thickness $0.7\,\mathrm{cm}$, dielectric constant 2.33, $x_p = -0.8\,\mathrm{cm}$, $y_p = 0$, probe diameter $= 0.7\,\mathrm{cm}$, patch center at $(0, 0)$.

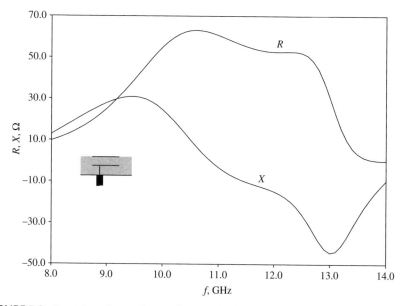

FIGURE 7.3   Input impedance of a two-layer probe-fed patch array under Floquet excitation (bore-sight beam): cell size $1.1 \times 1.1\,\mathrm{cm}^2$; dielectric constant of both layers 2.33; thickness of each layer $0.115\,\mathrm{cm}$; patch size $0.75 \times 0.75\,\mathrm{cm}$ and $0.66 \times 0.66\,\mathrm{cm}^2$ for lower and upper layers, respectively; $x_p = -0.32$; $y_p = 0$; probe diameter $0.1\,\mathrm{cm}$; probe height $0.115\,\mathrm{cm}$.

an arrangement, the gap between the tip of the probe and the patch acts as a series capacitor, which compensates for the positive offset reactance. Notice that the input resistance also decreases if the probe height is shortened. However, by adjusting the probe height and location, it is possible to achieve a desired value of input resistance at resonance with zero reactance.

Figure 7.3 shows the input impedance versus frequency for a two-layer patch array. The input resistance curve is almost flat in the 9.6–12.6-GHz band, caused by the dual resonance behavior of the structure. This flatness results in about 30% bandwidth with respect to -10 dB return loss. The bandwidth is larger than that of a one-layer array of equal dielectric thickness.

In Figure 7.4, the Floquet reflection coefficient versus scan angle is presented for the one-layer patch array considered in Figure 7.2. The surface wave resonance (SWR) occurs near 55° for the E-plane scan. The first grating lobe starts appearing near the 61° scan angle. The appearance of the grating lobe effectively improves the match; hence the reflection coefficient curve has a sharp negative slope near the 61° scan angle (note that the sharp peak on the H-plane scan curve is due to the grating lobe). The grating lobe is more pronounced in the E-plane than in the H-plane. A similar scan characteristic of a two-layer patch array can also be observed [1].

The -15-dB-return-loss bandwidths of one- and two-layer patch arrays are presented in Table 7.1. The dielectric thickness in both cases is about $0.1\lambda_0$,

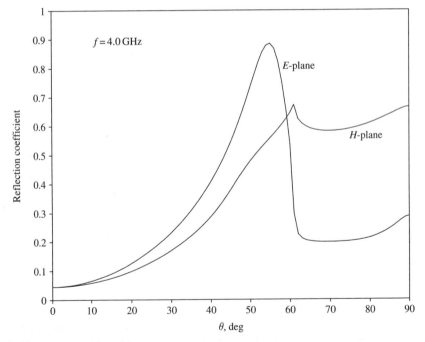

FIGURE 7.4 Reflection coefficient of a one-layer patch array with the scan angle under Floquet excitation. The parameters are the same as in Figure 7.2.

TABLE 7.1 Bandwidth Observed in One- and Two-Layer Patch Arrays as Function of Element Spacing for Bore-Sight Beam

Element Spacing, λ_0	Percent Bandwidth	
	One-Layer Patch array	Two-Layer Patch Array
0.35	24.0	—
0.47	22.0	25.0
0.58	11.0	19.0
0.70	5.0	14.0
0.82	3.0	8.0
0.94	1.5	4.0
1.06	5.0	—

Note: Square grid is assumed. Total substrate thickness $\lambda_0/10$, $\varepsilon_r = 2.33$.

where λ_0 is the wavelength in free space for the center frequency of the band. For both structures the bandwidth decreases as the element spacing increases. This can be explained from a mutual coupling point of view. For small element spacing, the mutual conductance between the adjacent patches is large and positive, resulting in a large value of active conductance; hence a large bandwidth results. The bandwidth becomes very small when the element spacing is about one wavelength. Interestingly, the bandwidth at this point suddenly jumps to a higher value due to the appearance of the first grating lobe. For two-layer structures, the patch sizes and layer thickness (keeping the total thickness fixed) are optimized to achieve maximum possible bandwidth. As expected, the bandwidth is larger for a two-layer array than for a one-layer array.

The resonant frequency of an array varies significantly with the element spacing. Figure 7.5 shows the resonant frequency versus element spacing of a one-layer array. The resonant frequency is defined as the frequency that corresponds to the peak input resistance. The change in the resonant frequency is due to the change in mutual reactance between patch elements. Two resonances are observed when the element spacing exceeds about one wavelength. This is primarily due to the appearance of the grating lobes.

7.2.4 Active Element Patterns

The active element patterns are defined as the radiation patterns of an element in an infinite array environment when other elements of the array are match terminated. Expressions for the θ- and ϕ-components of the active element patterns are deduced in Chapter 3, and are given by

$$E_\theta^a(\theta, \phi) = j\frac{\sqrt{ab}}{\lambda_0} S_{21}^{\text{TM}}(0, 0) \qquad E_\phi^a(\theta, \phi) = -j\frac{\sqrt{ab}}{\lambda_0} \cos\theta\, S_{21}^{\text{TE}}(0, 0) \qquad (7.22)$$

where $a \times b$ is the cell size. The quantities $S_{21}^{\text{TE}}(0,0)$ and $S_{21}^{\text{TM}}(0,0)$ are the elements associated with the TE_{00} and TM_{00} modes in the $[S_{21}]$ sub matrix (or the transmission

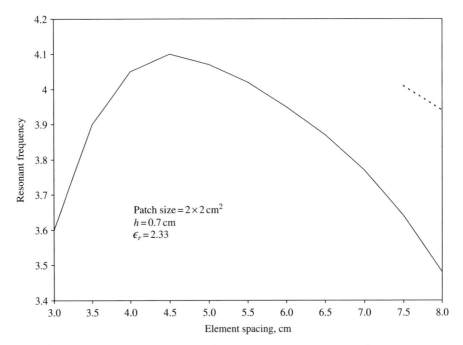

FIGURE 7.5 Resonant frequency (corresponds to peak input resistance) versus element spacing. Square grid is assumed. Dimensions are the same as in Figure 7.2.

matrix) of the overall GSM. The GSM corresponds to the (θ, ϕ) scan direction. The radiated fields are normalized with respect to the incident mode amplitude at the input port. The space factor $\exp(-jk_0 r)/r$ is suppressed in the far-field expression.

In order to employ (7.22) it is necessary to obtain the overall GSM of the array. To that end, we should first obtain the GSM of the probe layer. We convert the $[Z]$ matrix (GIM) of the probe layer to its GSM, denoted by $[S]$. The $[Z]$ matrix of the probe layer is given in (7.3). The $[S]$ matrix of the probe layer is given by

$$[S] = [Z_0'][Z + Z_0']^{-1}[Z - Z_0'][Z_0']^{-1} \tag{7.23}$$

where $[Z_0']$ is the characteristic impedance matrix given by

$$[Z_0'] = \begin{bmatrix} z_0^{\text{inp}} & [0] \\ [0] & [Z_0] \end{bmatrix} \tag{7.24}$$

where $[Z_0]$ represents the characteristic impedance matrix of the Floquet modes (diagonal). The first element z_0^{inp} is the characteristic impedance of the input transmission line with which the probe is attached. Typically its value is $50\,\Omega$. The GSMs of the probe layer and the patch layers are then combined to obtain the overall GSM of the probe-fed array.

As mentioned before, the radiated field is normalized with the incident voltage. To obtain the gain of the element, the far field should be normalized with the average incident power per unit solid angle. The average incident power in this case is

$$P_{\text{inc}} = \frac{1}{4\pi z_0^{\text{inp}}} \tag{7.25}$$

The normalized pattern thus becomes

$$\bar{E}_\theta^a(\theta, \phi) = j \frac{\sqrt{ab}}{\lambda_0} S_{21}^{\text{TM}}(0, 0) \sqrt{\frac{4\pi z_0^{\text{inp}}}{\eta}}$$

$$\bar{E}_\phi^a(\theta, \phi) = -j \frac{\sqrt{ab}}{\lambda_0} \cos\theta\, S_{21}^{\text{TE}}(0, 0) \sqrt{\frac{4\pi z_0^{\text{inp}}}{\eta}} \tag{7.26}$$

with $\eta = 377\Omega$. The above expressions are the gain of an element with respect to an isotropic source.

Equations (7.26) are employed to compute the copolar and cross-polar active element patterns. The computed results are plotted in Figure 7.6 for a one-layer patch array. The bore-sight gain corresponds to about 100% aperture efficiency, because no grating lobe exists. The cross-polar performance is very good along the

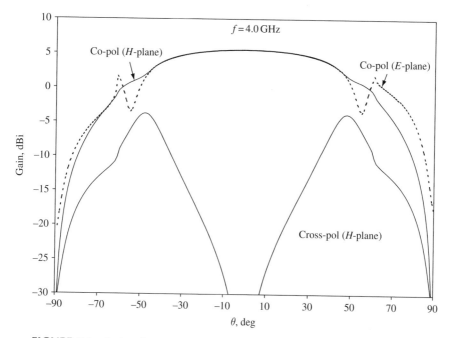

FIGURE 7.6 Active element pattern. Dimensions are the same as in Figure 7.2.

E-plane due to the symmetrical configuration of the probe feed with respect to the E-plane. The symmetry is lost with respect to the H-plane (due to the "probe offset" from the patch center); hence the cross-polar performances deteriorate, particularly at the far-off bore-sight locations. The observed blind spot is consistent with the reflection coefficient plot in Figure 7.4.

7.3 EMC PATCH ARRAY

The EMC patch array can be considered as a modified version of a two-layer patch array where the bottom layer patch is in the form of a thin strip. The upper layer patch is off centered from the lower layer strip to achieve an optimum coupling. Figure 7.7 depicts a unit cell of an EMC patch array. The analysis is very similar with that of a regular two-layer array, except that the GSM of the upper layer patch array needs a minor adjustment because the upper layer patch center is not aligned with the lower layer patch center. Suppose (x_c, y_c) is the coordinate of the upper layer patch center with respect to the lower layer patch center. The lower layer patch is assumed to be symmetrically located within a unit cell. Then the modified GSM of the upper layer patch should be

$$[S'_{mn}] = [P]^{-1}[S_{mn}][P]$$

where the prime denotes the submatrices ($m = 1, 2$ and $n = 1, 2$) of the modified GSM, $[P]$ is a diagonal matrix, and the nth diagonal element of $[P]$ is given by

$$P(n, n) = \exp(-jk_{xn}x_c - jk_{yn}y_c)$$

with k_{xn} and k_{yn} the wave numbers for the nth Floquet mode.

Figure 7.8 shows the input impedance of an EMC patch array element under Floquet excitation with respect to the bore-sight radiation. The impedance behavior is very similar to that of a two-layer array. Also plotted is the return loss with respect

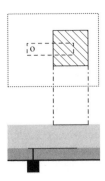

FIGURE 7.7   Unit cell of EMC patch array.

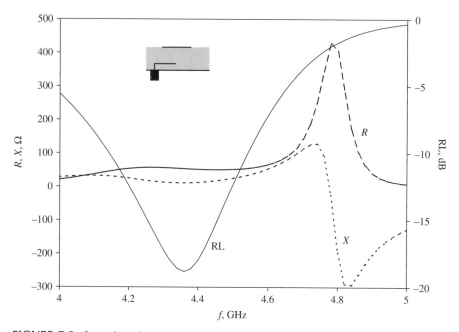

FIGURE 7.8 Input impedance and return loss of an EMC patch element in array environment. The array is under Floquet excitation for bore-sight beam. Dimensions: cell size $4 \times 4\,\mathrm{cm}^2$, strip size $2 \times 0.5\,\mathrm{cm}^2$, total substrate thickness 0.4 cm, strip height 0.1 cm, dielectric constant of substrates (both layers) 2.33, $x_c = 0.6$ cm, $y_c = 0$, probe diameter 0.1 cm, probe height 0.1 cm, probe located near the end of the strip.

to a 50-Ω feedline. Unlike a typical two-layer patch array, the second resonance is very sharp compared to the first resonance. However, the second resonance is unusable for the impedance mismatch problem.

7.4 SLOT-FED PATCH ARRAY

The configuration of a slot-fed patch element is shown in Figure 7.9a. A ground-plane slot excites the bottom-layer patch. A microstrip feedline on the opposite side of the patch layers excites the slot. Typically, the feedline substrate has a higher permittivity than that of the patch layer substrates. This choice essentially reduces the slot radiation in the feed side, which enhances the radiation efficiency of an element.

The modular representation of the slot-fed patch array of Figure 7.9a is shown in Figure 7.9b. The two blocks UPL and LPL represent the upper and lower patch layers, respectively. The SM/FM blocks represent the coupling between the slot modes and the Floquet modes at the two sides of the ground plane. The MF/SM block represents the coupling between the microstrip feedline and the slot modes. The GSM for an SM/FM block can be constructed using the procedure developed

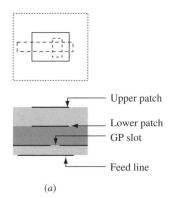

(*a*)

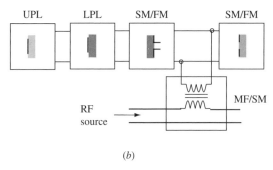

(*b*)

FIGURE 7.9 (*a*) Unit cell of a two-layer slot-coupled patch array. (*b*) Modular representation of the array.

for open-ended waveguide arrays in Chapter 3 (Section 3.6), because the slot modes essentially are rectangular waveguide modes of the same aperture size. For computing the input impedance experienced by the microstrip feedline, it is beneficial to derive an equivalent circuit representing the microstrip–slot transition. The equivalent circuit will also help understand the coupling mechanism. The following section will be devoted to derive the above equivalent circuit.

7.4.1 Microstripline–Slot Transition

Figure 7.10*a* shows a microstrip feedline with a slot on its ground plane. The length and width of the slot are *l* and *t*, respectively. The slot is placed at an angle 90° with the microstripline. The relation between slot voltage and the microstripline voltage will allow us to construct an equivalent circuit of the slot–feedline transition. We invoke the procedure detailed in [5]. Toward that end we first express the electric field on the slot aperture as

$$\vec{E}_{\text{slot}} = \sum_{n=1}^{N} A_n \vec{e}_n(x, y) \tag{7.27}$$

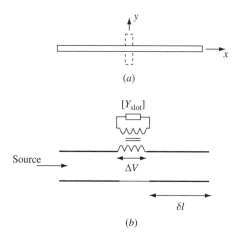

(a)

(b)

FIGURE 7.10 (a) Microstrip feedline with a ground-plane slot. (b) Equivalent circuit.

where $\vec{e}_n(x, y)$ is the nth modal function for the slot electric field. The slot modes are identical with the modes of a rectangular waveguide of dimension $l \times t$. The slot width is assumed to be very small compared to the wavelength of operation. Therefore, only the TE_{0n} modes are considered, ensuring that the x-component of the electric field is uniform along x.

To obtain the equivalent circuit of the slot, we first determine the reflected voltage in the microstripline in terms of the incident voltage. Toward that end we first express the electromagnetic fields of the microstripline before and after the slot discontinuity. We write the electromagnetic fields as

$$\vec{E}_1 = V_0\{\vec{e}_+(y, z)\exp(-j\beta x) + \Gamma \vec{e}_-(y, z)\exp(j\beta x)\} \tag{7.28a}$$

$$\vec{H}_1 = \frac{V_0}{Z_0}\{\vec{h}_+(y, z)\exp(-j\beta x) - \Gamma \vec{h}_-(y, z)\exp(j\beta x)\} \tag{7.28b}$$

$$\vec{E}_2 = V_0 T \vec{e}_+(y, z)\exp(-j\beta x) \tag{7.28c}$$

$$\vec{H}_2 = \frac{V_0}{Z_0} T \vec{h}_+(y, z)\exp(-j\beta x) \tag{7.28d}$$

where $\vec{E}_1$, $\vec{H}_1$ are the electromagnetic fields in the region $x < -t/2$; $\vec{E}_2$, $\vec{H}_2$ are the electromagnetic fields in the region $x > t/2$; V_0 is the incident voltage; Γ and T are the reflection and transmission coefficients due to the slot discontinuity; β is the propagation constant; and Z_0 is the characteristic impedance of the microstripline. The normalized electric and magnetic modal functions for the microstripline are $\vec{e}_+$ and $\vec{h}_+$, respectively, for propagation along the $+x$-direction and $\vec{e}_-$, $\vec{h}_-$ are the same quantities for propagation along the $-x$-direction. The y- and z-components of $\vec{e}_+$ and $\vec{e}_-$ are identical, but the x-components differ by a negative sign. This statement is true for $\vec{h}_+$ and $\vec{h}_-$ also. We will see shortly that only the y-component

of $\vec{h}_+$ is of importance. Using a spectral domain formulation [6] we derive the
y-component of $\vec{h}_+$ as

$$h_y(y, 0) = \frac{1}{2\pi} \int_{-\infty}^{\infty} \frac{\sin(k_y W/2)}{(\beta^2 + k_y^2)(k_y W/2)} \left[\frac{\beta^2 k_{0z}}{Y_1} + \frac{k_y^2 k_{1z}}{Y_2} \right] \exp(-jk_y y) \, dk_y \quad (7.29a)$$

with

$$Y_1 = k_{0z} \cos(k_{1z} h_f) + \frac{jk_{1z} \sin(k_{1z} h_f)}{\varepsilon_f} \qquad Y_2 = k_{1z} \cos(k_{1z} h_f) + jk_{0z} \sin(k_{1z} h_f) \tag{7.29b}$$

$$k_{0z} = \sqrt{k_0^2 - \beta^2 - k_y^2} \qquad k_{1z} = \sqrt{k_0^2 \varepsilon_f - \beta^2 - k_y^2}$$

In (7.29) W, h_f, and ε_f respectively are the line width, substrate height, and dielectric constant of the microstrip feedline. To determine the unknown reflection coefficient Γ, we invoke the reciprocity theorem. Toward that end, consider an auxiliary source located at $x = -\infty$ that produces the following electromagnetic fields:

$$\vec{E}_a = \vec{e}_+(y, z) \exp(-j\beta x) \tag{7.30a}$$

$$\vec{H}_a = \frac{1}{Z_0} \vec{h}_+(y, z) \exp(-j\beta x) \tag{7.30b}$$

Using reciprocity we write

$$\oiint_S \vec{E}_a \times \vec{H} \cdot \hat{n} \, dS = \oiint_S \vec{E} \times \vec{H}_a \cdot \hat{n} \, dS \tag{7.31}$$

where $\vec{E}$, $\vec{H}$ are the electromagnetic fields given in (7.28a)–(7.28d) and S is the closed surface formed by the planes $x = -t/2$, $x = t/2$, $y = -\infty$, $y = \infty$, $z = 0$, and $z = \infty$. Using (7.28) and (7.30) in (7.31) and then with a lengthy but straightforward algebraic manipulation, we arrive at the relation

$$\Gamma \int_0^\infty \int_{-\infty}^\infty \{\vec{e}_+ \times \vec{h}_+ + \vec{e}_- \times \vec{h}_-\} \cdot \hat{x} \, dy \, dz = -\frac{1}{V_0} \iint_{\text{Slot}} \vec{E}_{\text{slot}} \times \vec{h}_+ \cdot \hat{z} \, \exp(-j\beta x) \, dx \, dy \tag{7.32a}$$

The left-hand side of (7.32a) becomes 2Γ because the modal vectors are normalized. Thus we have

$$\Gamma = -\frac{1}{2V_0} \iint_{\text{Slot}} \vec{E}_{\text{slot}} \times \vec{h}_+ \cdot \hat{z} \, \exp(-j\beta x) \, dx \, dy \tag{7.32b}$$

For obtaining T, we use reciprocity once again, assuming the auxiliary source to be located at $x = \infty$. The auxiliary fields would thus be a propagating wave along the negative x-direction. Application of the reciprocity theorem once again yields

$$1 - T = -\frac{1}{2V_0} \iint\limits_{\text{Slot}} \vec{E}_{\text{slot}} \times \vec{h}_- \cdot \hat{z} \, \exp(j\beta x) \, dx \, dy \qquad (7.33)$$

Now consider the double integrals in (7.32b) and (7.33). Recall that the modal vectors $\vec{h}_+$ and $\vec{h}_-$ are functions of y and z, and they differ by their x-components only. The x-component of $\vec{h}_+$ and $\vec{h}_-$ do not contribute to those integrals because $\vec{E}_{\text{slot}}$ has only the x-component for a narrow slot. Furthermore, the slot field does not vary along x because of the narrow-slot assumption (equivalent to TE_{0n} types of waveguide modes). Therefore, the integrals are practically identical. Accordingly, we obtain

$$\Gamma = 1 - T \qquad (7.34)$$

From (7.28b), the modal current at $x = -t/2$ is (for $t << \lambda_0$)

$$I_1 = \frac{V_0}{Z_0} \left\{ \exp\left(\frac{j\beta t}{2}\right) - \Gamma \exp\left(-\frac{j\beta t}{2}\right) \right\} \approx \frac{V_0}{Z_0}(1 - \Gamma) \qquad (7.35)$$

Similarly, from (7.28d) the modal current at $x = t/2$ is

$$I_2 = \frac{V_0}{Z_0} T \exp\left(-\frac{j\beta t}{2}\right) \approx \frac{V_0}{Z_0} T \qquad (7.36)$$

Using (7.34) in (7.36) we find $I_1 = I_2$. This means that the current on the microstripline before and after the slot discontinuity remains unchanged, which is possible if the slot is equivalent to a series impedance. The impedance can be determined from the voltage drop across the discontinuity. The voltage drop is given by

$$\Delta V = (1 + \Gamma)V_0 - TV_0 = 2\Gamma V_0 \qquad (7.37)$$

We now express Γ in terms of the slot modal amplitudes. We use (7.32b) and (7.27) to obtain

$$\Gamma = \frac{\sum_{n=1}^N A_n R_n}{2V_0} \qquad (7.38)$$

where

$$R_n = -\iint\limits_{\text{Slot}} \vec{e}_n \times \vec{h}_+ \cdot \hat{z} \exp(-j\beta x) \, dx \, dy \qquad (7.39)$$

From (7.37) and (7.38) we write

$$\Delta V = \sum_{n=1}^{N} A_n R_n \tag{7.40}$$

The voltage drop ΔV is a linear combination of the modal voltages on the slot aperture. In matrix notation we can write (7.40) as

$$[\Delta V] = [R][A] \tag{7.41}$$

where $[R]$ is a row matrix and $[A]$ is a column matrix, each with N elements. Suppose Y_{eq} is the equivalent series admittance. The total reaction would be equal to $(\Delta V)^2 Y_{eq}$. Suppose $[Y_{slot}]$ is the admittance matrix of the slot. Then the self-reaction of the slot becomes $[A]^T[Y_{slot}][A]$. The reaction must be conserved; accordingly

$$(\Delta V)^2 Y_{eq} = [A]^T[Y_{slot}][A] \tag{7.42}$$

Using (7.41) in (7.42) we obtain

$$[A]^T\{[R]^T[Y_{eq}][R] - [Y_{slot}]\}[A] = [0] \tag{7.43}$$

Now (7.43) should be true for any arbitrary slot voltage amplitude vector $[A]$. Therefore the following relation must hold:

$$[R]^T[Y_{eq}][R] = [Y_{slot}] \tag{7.44}$$

Equation (7.44) yields a relation between the slot admittance matrix and the series admittance seen by the microstripline. From this relation and from (7.41) we can construct an equivalent circuit representing the coupling between the microstripline and the slot modes. The equivalent circuit consists of a generalized transformer, as shown in Figure 7.10b. The matrix $[R]$ represents the turns ratio of the transformer.

7.4.2 Input Impedance

The input impedance can be determined from the equivalent circuit of Figure 7.10 and the block diagram of Figure 7.9. The slot admittance matrix $[Y_{slot}]$ must be determined first. The slot admittance matrix essentially is the summation of two admittance matrices as below because the slot radiates on both sides (patch side and the feed side[3]) of the ground plane:

$$[Y_{slot}] = [Y_{slot}^{patch}] + [Y_{slot}^{feed}] \tag{7.45}$$

---

[3] To be rigorous, the feed-side radiation must include the effect of feedline obstruction, which is ignored in this analysis. The rationale for this is the small radiated energy in the feed side compared with that of the patch side.

The patch-side admittance matrix $[Y_{\text{slot}}^{\text{patch}}]$ is determined from the combined GSMs of the patch and dielectric layers and the slot mode–Floquet mode transition block. To include the ground-plane thickness, the combined GSM is modified by introducing appropriate phase terms. The phase terms are primarily the propagation delays of the slot modes through the "slot waveguide" across the ground plane [2]. The feed-side admittance matrix $[Y_{\text{slot}}^{\text{feed}}]$ is obtained from the combined GSMs of the slot mode–Floquet mode transition block in the feed side, the feed dielectric layer, and the dielectric–air interface. Equation (7.44) is then used to determine the equivalent series admittance.

In a slot-fed patch antenna, the feedline is extended beyond the slot to cancel out the offset reactance. Suppose the extended stub length beyond the slot center is δl (including the end effect). Then the input impedance seen by the microstrip feedline at the slot location becomes

$$Z_{\text{in}} = \frac{1}{Y_{\text{eq}}} - jZ_0 \cot(\beta \delta l) \tag{7.46}$$

The input impedances of one- and two-layer slot-fed patch arrays are shown in Figures 7.11 and 7.12, respectively. Similar to the probe-fed cases, the two-layer

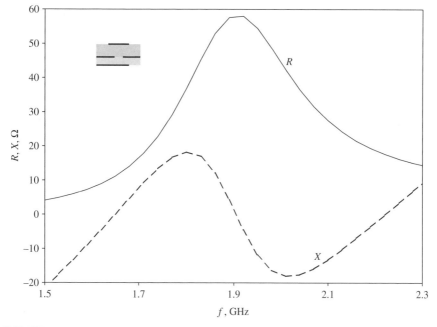

FIGURE 7.11 Input impedance of a one-layer slot-fed patch element in array environment: cell size 7×7 cm^2, slot length 3.2 cm, width 0.4 cm, patch size 5.8×5.8 cm^2, patch side dielectric constant 1.1, thickness 0.5 cm, feed-side dielectric constant 3.27, thickness 0.318 cm, microstripline width 0.8 cm, stub length 1.0 cm.

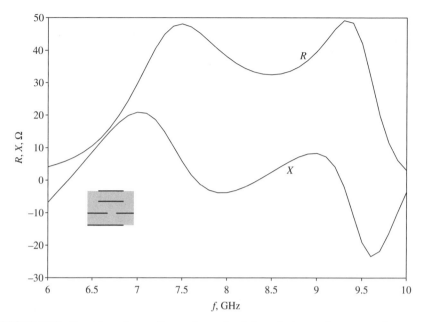

FIGURE 7.12 Input impedance of a two-layer slot-fed patch element in array environment: cell size $2.01 \times 2.01 \text{ cm}^2$; slot length 0.8 cm; width 0.12 cm; patch dimensions $1 \times 1 \text{ cm}^2$ (upper), $0.9 \times 0.9 \text{ cm}^2$ (bottom); dielectric constant 2.2 for all layers; layer thickness for all layers 0.159 cm; microstripline width 0.45 cm; stub length 0.4 cm.

array has wider bandwidth than the one-layer array. Figure 7.13 shows input resistance at resonance versus ground-plane thickness. The ground-plane thickness has a significant effect on the input impedance. The input resistance decays with ground-plane thickness. For a shorter slot the decay rate is higher than that of a longer slot. The bandwidth characteristic of a slot-fed patch array is very similar to that of the probe-fed patch array and will not be shown here.

7.4.3 Active Element Patterns

In order to determine the active element pattern it is convenient to obtain the GSM of the feedline–slot transition first. We consider port 1 as the microstripline input port and port 2 as the slot waveguide port. The feed-side radiation admittance of the slot, $[Y_{\text{slot}}^{\text{feed}}]$, must be connected in parallel at port 2 (following Figure 7.9b $[Y_{\text{slot}}^{\text{feed}}]$ represents the input impedance matrix of the SM/FM block at the extreme right). In order to simplify our analysis this parallel admittance is included as a part of the MF/SM block. For obtaining $[S_{11}]$ of our new MF/SM block we match terminate port 2. With port 2 matched, the input impedance in port 1 should be

$$Z_{\text{in}} = jX_{\text{stub}} + \frac{[RR^T]^2}{[R][Y_{\text{slot}}^{\text{feed}} + Y_{\text{slot}}^{0}][R]^T} \tag{7.47}$$

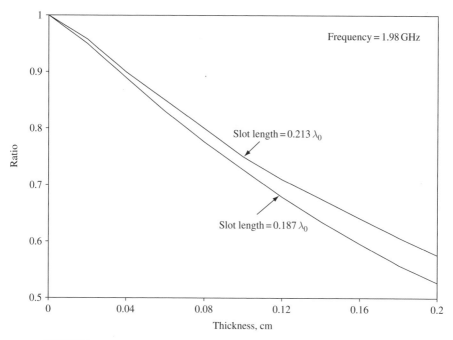

FIGURE 7.13 Input resistance at resonance versus ground-plane thickness.

where $[Y_{slot}^0]$ is the characteristic admittance matrix of the slot waveguide and X_{stub} is the reactance of the microstripline open stub. From (7.47), $[S_{11}]$ is obtained as

$$S_{11} = \frac{Z_{in} - Z_0}{Z_{in} + Z_0} \tag{7.48}$$

The $[S_{21}]$ matrix is deduced as

$$[S_{21}] = \frac{1 + S_{11}}{Z_{in}} [Y_{slot}^{feed} + Y_{slot}^0]^{-1} [R]^T \tag{7.49}$$

For determining $[S_{22}]$ and $[S_{12}]$ we need to match terminate port 1 and excite from port 2. Under this condition, the input admittance matrix seen by port 2 is

$$[Y_{in}] = \frac{1}{Z_0 + jX_{stub}} [R]^T [R] + [Y_{slot}^{feed}] \tag{7.50}$$

From the input admittance matrix $[Y_{in}]$ in (7.50), the $[S_{22}]$ and $[S_{12}]$ submatrices are obtained as

$$[S_{22}] = [Y_{slot}^0 + Y_{in}]^{-1} [Y_{slot}^0 - Y_{in}] \tag{7.51}$$

$$[S_{12}] = \frac{Z_0}{Z_0 + jX_{stub}} [R][I + S_{22}] \tag{7.52}$$

The constituent submatrices of the GSM are deduced. The GSM is then combined with the GSMs of the patch layers and the slot waveguide module to obtain the overall GSM of the structure. From the overall GSM we can obtain the active element patterns using (7.26). It is important to mention that the active element patterns thus obtained are for the upper half space (patch side) only. For the feed-side patterns, the same process should be repeated after replacing $[Y_{slot}^{feed}]$ by $[Y_{slot}^{patch}]$ and vice-versa.

The active element patterns of a one-layer patch array are shown in Figure 7.14. Unlike a probe-fed patch array, the cross-polarization fields do not occur for both the E- and H-plane scans, because the slot is placed symmetrically. In the 45° plane, the cross-polarization level is 15 dB below the copolarization level within 60° off-bore-sight scan. Therefore, for a single linear polarization application a center-fed slot-coupled array could be a better choice than a probe-fed patch array with respect to the cross-polar performance. For circular polarization or for dual-polarization applications, dual-offset slots are commonly used and the cross-polar performance should be similar to that of a probe-fed patch array. Also, observe that no blind spot occurs in this particular case because of the combined effects of small element spacing and low-permittivity substrate material. The bore-sight gain

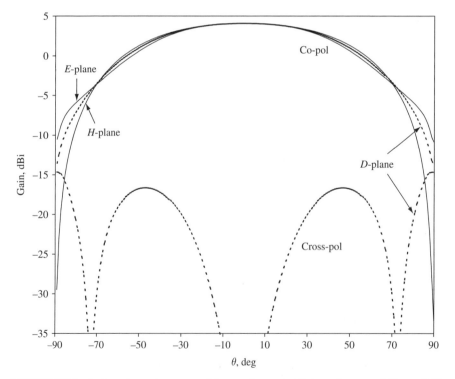

FIGURE 7.14 Active element pattern of the one-layer slot-fed patch array of Figure 7.11 at $f = 1.98$ Ghz.

is 4.06 dBi, which is about 0.22 dB lower than the gain corresponding to 100% aperture efficiency. This 0.22-dB reduction in gain is due to the slot radiation in the feed side. This power loss in the back lobe can be estimated from the ratio of the equivalent conductances experienced by the dominant slot mode at the two sides of the ground plane. The result indicates that a microstripline-fed slot-coupled array is less efficient than a probe-fed array. Electrically smaller slots improve the radiation efficiency. Also, a high-dielectric-constant substrate in the feed side reduces the back-lobe radiation. However, a high-dielectric-constant substrate could cause scan blindness for some scan angles.

7.5 STRIPLINE-FED SLOT-COUPLED ARRAY

In order to eliminate the undesired back-lobe radiation another ground plane is typically added behind the feed substrate, so that the feedline essentially becomes an asymmetric stripline, as shown in Figure 7.15. This additional ground plane not only eliminates the back-lobe radiation but also isolates the other layers consisting of active components such as power amplifiers and phase shifters. The analysis of this structure is very similar to that of a microstripline-fed slot-coupled array discussed in Section 7.4. The radiation characteristics in terms of radiation patterns and bandwidth are very similar with that of a microstripline-fed patch.

An important difference between the microstripline feed and the stripline feed should be noted. Because of the two conducting planes the slots will excite parallel-plate modes, which could cause scan blindness. The number of parallel-plate modes increases with the separating distance between the planes; hence the number of blind spots increases with the separating distance. In order to alleviate the parallel-plate-mode problems, ground vias are used around the slot to suppress the parallel-plate modes, as shown in Figure 7.15. A detailed analysis of the structure with mode

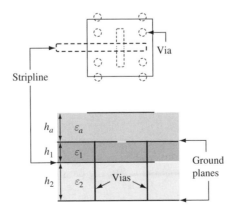

FIGURE 7.15 Configuration of a stripline-fed, slot-coupled patch element with mode suppressing vias.

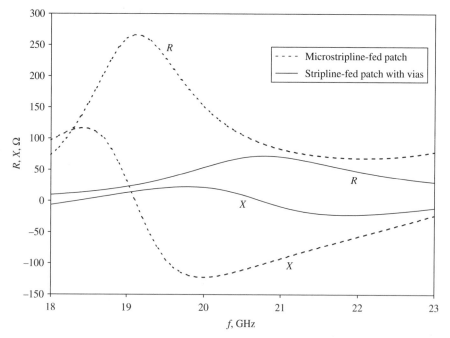

FIGURE 7.16 Floquet impedances of a microstripline-fed slot-coupled array and stripline-fed slot-coupled array with mode suppressing vias: cell size 0.876×0.876 cm, patch dimension $= 0.35 \times 0.68$, slot size 0.29×0.025, patch side substrate thickness 0.084 cm, dielectric constant 2.53, $h_1 = 0.038$, $\varepsilon_1 = 9.8$, $h_2 = 0.0635$ cm, $\varepsilon_2 = 3.27$, 50-Ω feedlines for both cases. Stub lengths to compensate offset reactance near resonance.

suppressing vias is presented in [7]. The mode suppressing vias have profound impacts on the input impedance and resonant frequency of the patch elements. The effect of vias on the input impedance is depicted in Figure 7.16. It can be noted that vias significantly modify the input impedance and the resonant frequency of an element in array environment.

7.6 FINITE PATCH ARRAY

It is instructive to examine the impedance match of patch elements in a "real finite array." We compute the active reflection coefficients of the individual patch elements using the procedure developed in Section 4.7. To compute the active reflection coefficients we first obtain the scattering parameters of the "finite array"[4] (see Section 4.3). We assume that the scattering parameters of a real finite array are

---

[4] We use the same definition of a "finite array" as in Chapter 4, that is, an infinite number of physical elements with a finite number of excited elements.

the same as that of a finite array. Suppose $[S]$ is the scattering matrix of the real finite array. This scattering matrix essentially relates the reflected voltage vector with the incident voltage vector of the elements. The relation is given by

$$[V^-] = [S][V^+] \tag{7.53}$$

where $[V^-]$ is the reflected voltage vector and $[V^+]$ is the incident voltage vector of order N, where N is the total number of elements in the finite array. The (m, n) element of the $[S]$ matrix represents mutual coupling between the mth and nth elements of the array that can be expressed in terms of the Floquet reflection coefficients of an element (see Section 4.3).

Figures 7.17 and 7.18 show the active return loss of the elements in an (11×11)-element patch array with uniform and tapered distributions, respectively. Microstripline-fed slot-coupled patch elements were considered. For the tapered array, Gaussian amplitude distributions with a 10-dB taper for both planes were considered. For the plots, elements were numbered according to the numbering scheme depicted in Figure 4.7a. For each amplitude distribution, three scan angles were considered, including bore sight, 30° off bore sight along the E-plane, and 30° off bore sight along the H-plane, respectively. The elements were designed to have

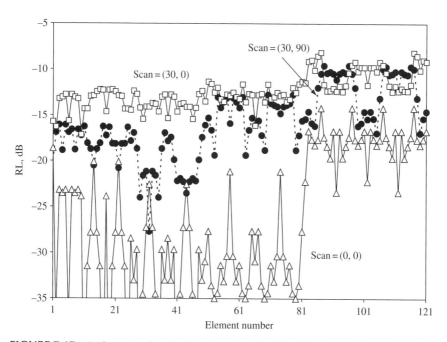

FIGURE 7.17 Active return loss for the elements of an 11×11-element slot-coupled patch array. The element-numbering scheme is the same as in Figure 4.7. The cell size and element dimensions are given in Figure 7.11. Uniform amplitude distribution for the array was considered.

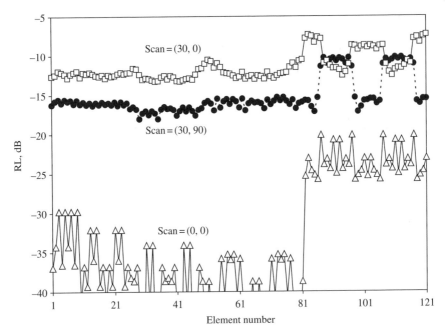

FIGURE 7.18 Active return loss for the array of Figure 7.17 with Gaussian amplitude distributions in both planes. The edge taper was 10 dB for both planes.

bore-sight match under Floquet excitation. It is found that the active return loss varies from element to element. In particular, the edge elements have noticeably different return losses than the rest of the array elements. For the bore-sight scan, the return loss for most of the elements lies below $-20\,dB$. The return loss deteriorates for the off-bore-sight scans. The effect is more prominent for the E-plane scan than for the H-plane scan because the mutual coupling is tighter between the E-plane elements than between the H-plane elements.

REFERENCES

[1] A. K. Bhattacharyya, "A Modular Approach for Probe-Fed and Capacitively Coupled Multilayered Patch Array," *IEEE Trans.*, Vol. AP-45, No. 2, pp. 193–202, Feb. 1997.

[2] A. K. Bhattacharyya, "A Numerical Model for Multilayered Microstrip Phased-Array Antennas," *IEEE Trans.*, Vol. AP-44, No. 10, pp. 1386–1393, Oct. 1996.

[3] R. F. Harrington, *Time-Harmonic Electromagnetic Fields*, McGraw-Hill, New York, 1961.

[4] R. E. Collin, *Field Theory of Guided Waves*, IEEE, New York, 1991.

[5] J. E. Eaton, L. J. Eyges, and G. G. Macfarlane, "Linear Array Antennas and Feeds," in S. Silver (Ed.), *Microwave Antenna Theory and Design*, McGraw-Hill, New York, 1949, Chapter 9.

[6] A. K. Bhattacharyya, in *Electromagnetic Fields in Multilayered Structures*, Artech House, Norwood, MA, 1994, "Analysis of Planar Transmission Lines," Chapter 6.

[7] A. K. Bhattacharyya, "Analysis of Stripline-Fed Slot-Coupled Patch Antennas with Vias for Parallel-Plate Mode Suppression," *IEEE Trans.*, Vol. AP-46, No. 4, pp. 538–545, April 1998.

BIBLIOGRAPHY

Aberle, J. T., and D. M. Pozar, "Analysis of Infinite Arrays of One- and Two-Probe-Fed Circular Patches," *IEEE Trans.*, Vol. AP-38, No. 4, pp. 421–432, Apr. 1990.

Gay-Balmaz, P., J. A. Encinar, and J. R. Mosig, "Analysis of Multilayer Printed Arrays by a Modular Approach Based on the Generalized Scattering Matrix," *IEEE Trans.*, Vol. AP-48, No. 1 , pp. 26–34, Jan. 2000.

Gianola, P., D. Finotto, and A. P. Ansbro, "S-parameter Analysis of Multislot Microstrip Patch Arrays: Spectral Domain Approach," *IEE Proc.*, Vol. MAP-143, No. 5, pp. 430–436, Oct. 1996.

Gonzalez de Aza, M. A., J. Zapata, and J. A. Encinar, "Broad-Band Cavity-Backed and Capacitively Probe-Fed Microstrip Patch Arrays," *IEEE Trans.*, Vol. AP-48, No. 5, pp. 784–789, May 2000.

Haddad, P. R., and D. M. Pozar, "Analysis of an Aperture Coupled Microstrip Patch Antenna with a Thick Ground Plane," *IEEE APS Symp. Dig.*, Vol. 2, pp. 932–935, 1994.

Herd, J. S.,"Modeling of Wideband Proximity Coupled Microstrip Array Elements," *Electron. Lett.*, Vol. 26, pp. 1282–1284, Aug. 1990.

Jassim, A. M., and H. D. Hristov, "Cavity Feed Technique for Slot-Coupled Microstrip Patch Array Antenna," *IEE Proc.*, Vol. MAP-142, No. 6, pp. 452–456, Dec. 1995.

Jackson, D. R., and P. Manghani, "Analysis and Design of a Linear Array of Electromagnetically Coupled Microstrip Patches," *IEEE Trans.*, Vol. AP-38, No. 5, pp. 754–759, May 1990.

Liu, C. C, A. Hessel, and J. Shmoys, "Performance of Probe-Fed Microstrip-Patch Element Phased Arrays," *IEEE Trans.*, Vol. AP-36, No. 11, pp. 1501–1509, Nov. 1988.

Mailloux, R. J., J. F McIlvenna, and N. P. Kernweis, "Microstrip Array Technology," *IEEE Trans.*, Vol. AP-29, No. 1, pp. 25–37, Jan. 1981.

Pozar, D. M. " Analysis of an Infinite Phased Array of Aperture Coupled Microstrip Patches," *IEEE Trans.*, Vol. AP-37, No. 4, pp. 418–425, Apr. 1989.

Pozar, D. M., and D. H. Schaubert, "Analysis of an Infinite Array of Rectangular Microstrip Patches with Idealized Probe Feed," *IEEE Trans.*, Vol. AP-32, No. 10, pp. 1101–1107, Oct. 1984.

Skrivervik, A. K., and J. R. Mosig, "Analysis of Printed Array Antennas," *IEEE Trans.*, Vol. AP-45, No. 9, pp. 1411–1418, Sept. 1997.

Warehouse, R. B., "Improving the Scan Performance of Probe-Fed Microstrip Patch Arrays on High Dielectric Constant Substrate," *IEEE Trans.*, Vol. AP-43, No. 7, pp. 705–712, July 1995.

Warerhouse, R. B., "The Use of Shorting Posts to Improve the Scanning Range of Probe-Fed Microstrip Patch Phased Arrays," *IEEE Trans.*, Vol. AP-44, No. 3, pp. 302–309, Mar. 1996.

Yakovlev, A. B., S. Ortiz, M. Ozkar, A. Mortazawi, and M. B. Steer, "A Waveguide-Based Aperture-Coupled Patch Amplifier Array-Full-Wave System Analysis and Experimental Validation," *IEEE Trans.*, Vol. MTT-48, No. 12, pp. 2692–2699, Dec. 2000.

Yang, H.-Y., N. G. Alexopoulos, P. M. Lepeltier, and G. J. Stern, "Design of Transversely Fed EMC Microstrip Dipole Arrays Including Mutual Coupling," *IEEE Trans.*, Vol. AP-38, No. 2, pp. 145–151, Feb. 1990.

PROBLEMS

7.1 Deduce an expression for Z_{11} of an infinite array of probes assuming current distribution of

$$I_z = I_1 \cos\left(\frac{\pi z}{2h}\right)$$

where I_1 is the input current at $z = 0$ and h is the probe height. Assume that the probes are immersed in a dielectric layer of dielectric constant ε_r. Port 2 is defined on the $z = h$ plane. The cell size of the array is $a \times b$ in a rectangular lattice.

7.2 Obtain the generalized impedance matrix of the probe of problem 7.1. The reference plane for port 1 is at $z = 0$ and for port 2 is at $z = h$. Assume N Floquet modes for the impedance matrix.

7.3 Consider an infinite array of dual-probe-fed patch elements. The unit cell dimensions are $a \times b$ and the array is on rectangular grid. Assume that the probes are located at (x_1, y_1), and (x_2, y_2), respectively, in a cell with respect to the cell center. Assuming uniform current distributions on the probes, obtain the Z_{11} matrix of the probe feed.

7.4 Find Z_{11} of an infinite array of probes of diameter d, assuming that:

(a) The source current is a surface current.
(b) The source current is a volume current uniformly distributed over the cross section of the probe.

Show that the two solutions are equivalent for small d.

7.5 Consider two infinite arrays of uniform line current sources one above another. The source currents are given by

$$\vec{J}_1 = I_1\hat{y}\sum_{n=-\infty}^{\infty}\delta(x-na)\delta(z)\exp(-jn\psi)$$

$$\vec{J}_2 = I_2\hat{y}\sum_{n=-\infty}^{\infty}\delta\left(x-na-\frac{a}{2}\right)\delta(z-h)\exp(-jn\psi)$$

Show that the mutual reactions (as defined by the integrals below) between two elements are not equal for these two sources, that is

$$\iiint\limits_{Source\ 2} \vec{E}_1 \cdot \vec{J}_2 dV \neq \iiint\limits_{Source\ 1} \vec{E}_2 \cdot \vec{J}_1 dV$$

where the volume integrations are performed inside a unit cell.[5]

7.6 Consider two infinite arrays of uniform line current sources one above another. The source currents are given by

$$\vec{J}_1 = I_1 \hat{y} \sum_{n=-\infty}^{\infty} \delta(x - na)\delta(z) \exp(-jn\psi)$$

$$\vec{J}_2 = I_2 \hat{y} \sum_{n=-\infty}^{\infty} \delta\left(x - na - \frac{a}{2}\right)\delta(z - h) \exp(jn\psi)$$

(a) Obtain the electric fields $\vec{E}_1$ and $\vec{E}_2$ produced by $\vec{J}_1$ and $\vec{J}_2$, respectively.

(b) Show that

$$\iiint\limits_{Source\ 2} \vec{E}_1 \cdot \vec{J}_2 dV = \iiint\limits_{Source\ 1} \vec{E}_2 \cdot \vec{J}_1 dV$$

(c) Use the following definitions of mutual impedance:

$$Z_{12} = -\frac{1}{I_1 I_2} \iiint \vec{E}_2 \cdot \vec{J}_1 \, dx \, dy \, dz$$

$$Z_{21} = -\frac{1}{I_1 I_2} \iiint \vec{E}_1 \cdot \vec{J}_2 \, dx \, dy \, dz$$

where the integration is performed in a unit cell of cross-sectional area $a \times b$. Show that $Z_{12} = Z_{21}$, that is the $[Z]$ matrix becomes symmetrical if the phase progression of source 2 is exactly opposite to that of source 1. Notice that in the region $0 < z < h$, the two fields $\vec{E}_1$ and $\vec{E}_2$ travel in the opposite directions of one another.

7.7 The reflection coefficient matrix of a patch layer is $[S_{11}]$ with respect to a set of Floquet modes as defined in Chapter 3. The reflection coefficient matrix can also be obtained with respect to another set of Floquet modes with shifted reference axes (parallel shift only). Suppose the origin of the xy-coordinate

---

[5] One should not consider this relation as a violation of the reciprocity theorem, because the integrations are carried over only two elements of the array, not on the entire array sources.

system moves at (x_c, y_c) with reference to the original coordinate system. Then show that the modified reflection matrix is

$$[S'_{11}] = [P][S_{11}][P]^{-1}$$

where $[P]$ is a diagonal matrix and the diagonal elements are given by

$$P(n, n) = \exp(-jk_{xn}x_c - jk_{yn}y_c)$$

with k_{xn} and k_{yn} as the wave numbers for the nth Floquet mode. Also show that a similar relation is valid for $[S_{12}]$, $[S_{21}]$, and $[S_{22}]$.

7.8   Deduce(7.33).

7.9   In analyzing a slot-coupled patch array, only one slot mode (TE_{01}) is assumed for the slot field. The computed patch-side and feed-side admittances seen by the slot are $0.0143 + j0.0175$ mho and $0.00018 - j0.00561$ mho, respectively. The turn ratio of the equivalent transformer is 0.803.

   (a) Find the equivalent series impedance seen by the microstripline.
   (b) If the characteristic impedance of the microstripline is 50Ω, find the stub length (in microstripline guide wavelength) to compensate the reactive part of the impedance. Ignore the end effect of the stub.
   (c) Compute the percentage of the total radiated power that goes to the back-lobe radiation. Compute the gain loss due to back-lobe radiation.

7.10 A simplified model to incorporate the effect of vias of Figure 7.15 is to assume a lossless cavity with conducting walls through the vias. With this assumption, deduce the feed-side admittance seen by the slot, assuming that the slot has the TE_{01} mode only.

7.11 In problem 7.10, assume a slot length of $0.25\lambda_0$ and width of the slot is $0.01\lambda_0$. The feed-side dielectric constant is 2.6. Feed-side dielectric thickness is $0.2\lambda_0$ (which is equal to the distance between the two ground planes of the stripline feed). Obtain the feed-side reactance for the following two cases:

   (a) The dimension of the simplified cavity is $0.4\lambda_0 \times 0.2\lambda_0$.
   (b) The dimension of the simplified cavity is $0.6\lambda_0 \times 0.4\lambda_0$.

CHAPTER EIGHT

Array of Waveguide Horns

8.1 INTRODUCTION

Waveguide horn arrays are commonly used for producing scan beams and contour beams for a variety of high-frequency applications, such as radar, tracking, and satellite communications. They are also used as primary feeds for reflector antennas in communication satellites to produce multiple spot beams. The waveguide horn elements have the following useful features:

(a) Horns are fairly easy to fabricate.
(b) Horns have reasonable bandwidth performances, typically 5–25% return loss bandwidth.
(c) The size of a horn aperture is independent of the operating frequency; thus the element gain can be adjusted as required.
(d) A horn element has very low RF loss; thus it can be used at very high frequencies such as millimeter-wave bands.
(e) Horns have large power handling capacities.

Compared to a microstrip patch array, a horn array has higher profile but may be of comparable weight. In this chapter we primarily focus on the radiation characteristics of waveguide horn arrays. The chapter begins with linearly flared square horn arrays. We briefly discuss the method of analysis followed by the return loss and scan characteristics of such arrays. Numerical results reveal that a grazing lobe may cause scan blindness for an array of small-aperture horns. For an array of large-aperture horns, a grazing mode typically causes a minor dent or dip in the active element pattern, but serious scan blindness does not generally happen. A simplified

Phased Array Antennas. By Arun K. Bhattacharyya
© 2006 John Wiley & Sons, Inc.

model based on two-mode theory is presented to explain these differential effects. It is found that under Floquet excitation the aperture admittance matrix becomes singular if a TM_z grazing mode (mode associated with a grazing lobe or end-fire lobe) exists. A procedure is presented to alleviate the singularity problem. It is shown that under certain conditions horn array structures support surface wave and leaky wave modes. The dips and nulls in active element patterns are explained from a leaky wave perspective. The "super gain" phenomenon of an array element is demonstrated and explained in light of leaky wave coupling. For blindness removal, the wide-angle impedance matching (WAIM) concept is discussed. We then present the characteristics of multimodal step horns of rectangular and circular apertures. Of special mention are the high-efficiency horn elements, which appear to be very promising for array applications.

8.2 LINEARLY FLARED HORN ARRAY

The geometry of a linearly flared horn with rectangular cross section is shown in Figure 8.1a. The input waveguide dimension is $a_1 \times b_1$ and the aperture dimension is $a_2 \times b_2$. The GSM approach is typically invoked to analyze such a horn structure. The analysis begins by dividing the entire horn into several sections of uniform waveguides, as shown in Figure 8.1b. For a reasonable accuracy, the number of sections per wavelength is kept between 10 and 15. This number may be increased for very large flared angles.

The modified horn structure in Figure 8.1b can be treated as a cascade connection of several small uniform waveguide sections and waveguide junctions. The GSM of each individual waveguide section can be obtained using the procedure developed in Section 6.5. For analyzing a junction of two waveguides, one typically employs

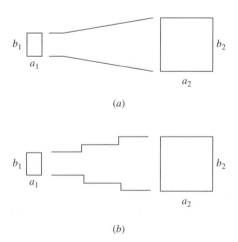

FIGURE 8.1 (a) Linearly flared horn. (b) Equivalent horn structure with discretized sections.

either a mode matching procedure [1] or a MoM analysis [2]. The latter is very similar with that detailed in Section 3.6 for analyzing waveguide mode–Floquet mode transitions. For the problem at hand the Floquet modes should be replaced by the larger waveguide modes. The individual GSMs are then combined to obtain the overall GSM of the flared horn. The overall GSM of the horn essentially relates the input waveguide modes to the aperture waveguide modes.

In order to analyze an infinite array of such flared horns under Floquet excitation, the modal transformation between the waveguide modes and Floquet modes must be taken into account. The GSM of an aperture essentially represents such a modal transformation, which is deduced in Section 3.6. The overall GSM of a horn element in an infinite array environment is the combined GSM of the horn and the GSM of the aperture.

8.2.1 Return Loss Characteristics

The return loss characteristics of an infinite array of horns are studied and the results are presented in Figures 8.2a, b. The array elements are excited with uniform amplitudes and linearly progressed phases (Floquet excitation). The input return loss of an element with respect to the TE_{01} incident mode versus the scan angle is plotted for three different scan planes. The return loss varies widely with the scan angle. For the $0.65\lambda_0$ (λ_0 is the wavelength in free space) square horn array in a square grid and of cell size $0.7\lambda_0 \times 0.7\lambda_0$, the return loss for the bore-sight beam is about -14 dB. The major portion of this reflection takes place at the aperture end because of the mismatch between the waveguide modes and Floquet modes. The return loss generally improves for larger aperture horn arrays because the fundamental Floquet modes and the dominant waveguide mode (the TE_{01} mode) match well for a larger aperture. As shown in Figure 8.2b, the return loss for the bore-sight beam is better than -19 dB for the $1.49\lambda_0$ square horn array.

The return loss curves have several dips and peaks, and some of them appear as sharp spikes. These dips and peaks are primarily due to the grazing lobes, also known as end-fire lobes. In Figure 8.2a the peak near the $23°$ scan angle on the E-plane is due to the grating lobe at the $-90°$ location. This grating lobe corresponds to the $TM(-1, 0)$ Floquet mode (to avoid a negative subscript, we place the mode indexes inside parentheses). The grazing lobe causes strong mutual coupling[1] between the elements, which in this case leads to scan blindness because the return loss becomes near 0 dB. The scan angle associated with a grazing lobe can be easily identified from the circle diagram of the array shown in Figure 8.3. A grazing lobe occurs when the scan location on the $k_x k_y$-plane lies on a Floquet mode circle. For rectangular grids with cell size $a \times b$, the scan angles at $E, H,$

---

[1] The coupling due to a grazing lobe may, for some cases, be viewed as leaky wave coupling, which will be discussed in a later section.

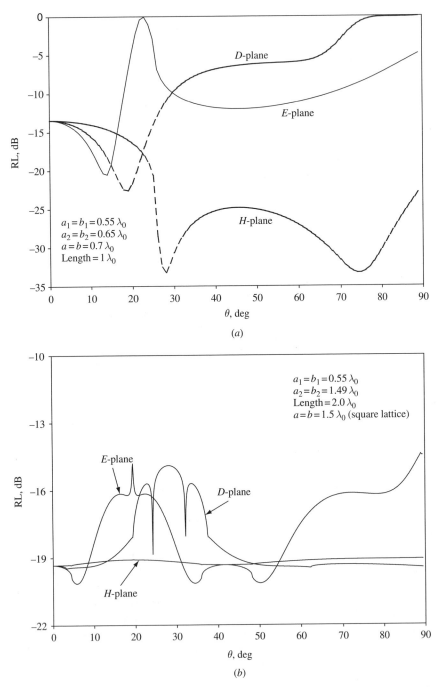

FIGURE 8.2 Floquet return loss of flared horns in square lattice: (*a*) horn aperture size $0.65\lambda_0 \times 0.65\lambda_0$, length $1\lambda_0$, cell size $0.7\lambda_0 \times 0.7\lambda_0$; (*b*) horn aperture size $1.49\lambda_0 \times 1.49\lambda_0$, length $2\lambda_0$, cell size $1.5\lambda_0 \times 1.5\lambda_0$.

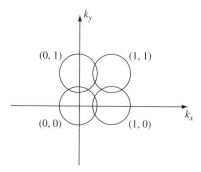

FIGURE 8.3 Circle diagram for array of cell size $0.7\lambda_0 \times 0.7\lambda_0$. Sections of the higher order mode circles inside (0,0) mode circle are associated with grazing grating lobes. Four circles correspond to (0, 0), (1, 0), (0, 1), and (1, 1) Floquet modes.

and D (45°) planes corresponding to the TE_{mn} or TM_{mn} grazing lobes can be computed from the following equations:

$$\theta_E = -\sin^{-1}\left[\frac{m\lambda_0}{a} \pm \sqrt{1 - \left(\frac{n\lambda_0}{b}\right)^2}\right] \tag{8.1}$$

$$\theta_H = -\sin^{-1}\left[\frac{n\lambda_0}{b} \pm \sqrt{1 - \left(\frac{m\lambda_0}{a}\right)^2}\right] \tag{8.2}$$

$$\theta_D = -\sin^{-1}\left[\lambda_0\frac{m/a + n/b}{\sqrt{2}} \pm \frac{\sqrt{2 - \lambda_0^2(m/a - n/b)^2}}{\sqrt{2}}\right] \tag{8.3}$$

It is seen that a return loss peak does not necessarily occur exactly at the scan angle corresponding to a grazing lobe. For the flared horn array considered in Figure 8.2a the scan angle corresponding to the return loss peak is at 23°, but the scan angle corresponding to a grazing lobe is at 25.4°. The reason for such a difference will be explained latter.

It is interesting to observe that the E-plane grazing lobe causes high reflection loss that sometimes leads to scan blindness. However, no major blindness occurs for the H-plane scan, particularly for a small-aperture horn array, where the horn's aperture electric field has only one component (say x-component). It will be seen later that only the TM_z grazing Floquet modes are responsible for high reflection loss. The TM_z Floquet modes are not excited for the H-plane scan; as a result no large reflection occurs at this plane.

The return loss curves in Figure 8.2b correspond to a horn array of $1.5\lambda_0$ element spacing. The aperture size of a horn element is $1.49\lambda_0 \times 1.49\lambda_0$. The grazing lobes for the E-plane scan occur at 4.5°, 19.5°, 19.5°, and 36°, associated with the $\text{TM}(-1, -1)$, $\text{TM}(-2, 0)$, $\text{TM}(-1, 0)$, and $\text{TM}(-2, 1)$ Floquet modes, respectively. The Floquet impedance being a periodic function of $\psi_x(= k_0 a \sin\theta)$ with a periodicity 2π, the grazing lobes reoccur at a regular interval in the $\sin\theta$-space.

It is observed from Figures 8.2*a*, *b* that although several grazing lobes occur for the *E*-plane scan cases, many of them could not cause serious blindness problems because the magnitudes of the return loss are much below the 0-dB level. This implies that occurrence of a grazing lobe does not always mean scan blindness. In fact, a severe blindness (near 0 dB return loss) only occurs in the case of small apertures. For larger apertures, the reflection loss curve shows peaks (or spikes) at multiple scan angles but the magnitudes are not high enough to cause blindness. In Section 8.3.3 we will present an explanation for these differential effects.

8.2.2 Active Element Pattern

The active element pattern essentially is the embedded element pattern in an infinite array environment. It is defined as the radiation pattern of an element when all other elements of the infinite array are match terminated at the input ports. The array pattern, including mutual coupling effects, is equal to the active element pattern multiplied by the array factor that involves the complex amplitude distribution. The array gain at a given scan angle is directly proportional to the active element gain at that angle. The active element pattern essentially is the "true signature" of an array in regard to far-field radiation.

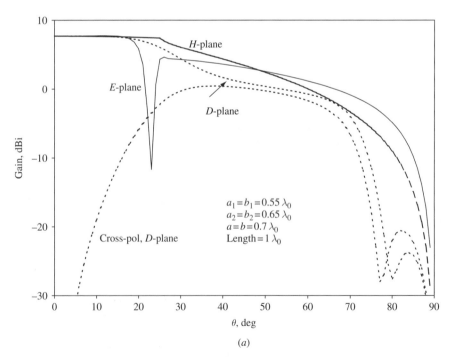

(a)

FIGURE 8.4 Active element patterns of flared horns: (*a*) horn aperture size $0.65\lambda_0 \times 0.65\lambda_0$, length $1\lambda_0$; (*b*) horn aperture size $1.49\lambda_0 \times 1.49\lambda_0$, length $2\lambda_0$.

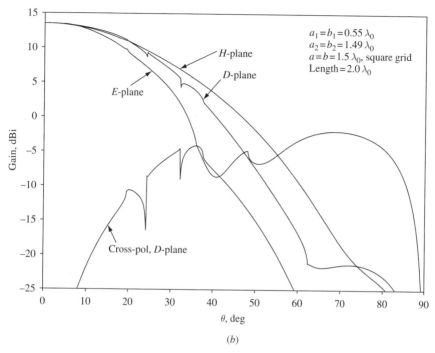

FIGURE 8.4 (Continued).

The active element pattern can be computed from the GSM of the infinite array as deduced in Section 3.5. We normalize the radiated field with respect to the incident power such that the far-field intensity represents the active element gain.

Figures 8.4a, b depict the active element patterns for the two arrays considered in Figures 8.2a, b, respectively. In Figure 8.4a we observe a deep null near the 23° scan angle on the E-plane. Comparing with Figure 8.2a, we notice that the location of the null matches the location of the peak return loss. The active element gain almost remains constant (ignoring the cos θ variation) until the first grating lobe appears in the visible region. This characteristic is typical for small horns. The array with the larger aperture elements does not have any null as such, but small dips and dents are found on the active element patterns (see Figure 8.4b). The dips and dents are due to the grazing lobes. An abrupt change of the impedance behavior for the Floquet mode associated with the grazing lobe results in such a dip or dent.

8.3 GRAZING LOBES AND PATTERN NULLS

We observed that a grazing lobe could cause a deep null or a small dip in the active element pattern, which is primarily due to strong mutual coupling between the elements of the array. The depth of a null or the gain loss depends on the aperture size. The smaller the aperture size, the larger is the gain loss. In this section we investigate the root cause of the gain loss and its relation to the aperture size.

The null depth is directly related to the input reflection coefficient, which is determined by the aperture mismatch with the free space. In the following section we formulate the coupling between waveguide modes and Floquet modes to determine the aperture admittance.

8.3.1 Aperture Admittance Formulation

Let us consider an infinite array of rectangular horns with $a_1 \times b_1$ as the element aperture size. The unit cell size is $a \times b$ and the unit cells are arranged in a grid defined by the grid parameters $[a, b, \gamma]$. Suppose that under a Floquet excitation the transverse electric field on the horn aperture at $z = 0$ is given by

$$\vec{E} = \sum_{n=1}^{N} V_{gn} \vec{e}_{gn}(x, y) \tag{8.4}$$

where V_{gn} is the modal voltage for the nth waveguide mode and $\vec{e}_{gn}$ is the normalized modal voltage vector. Notice, we use a *single index* to represent a waveguide mode for simplicity. The expressions for the modal vectors are deduced in Chapter 3. We now expand the aperture field in terms of Floquet modes supported by the infinite array. Thus we write

$$\sum_{m=1}^{\infty} V_{fm} \vec{e}_{fm}(x, y) = \sum_{n=1}^{N} V_{gn} \vec{e}_{gn}(x, y) \tag{8.5}$$

where V_{fm} is the modal voltage for the mth Floquet mode[2] and $\vec{e}_{fm}$ is the corresponding modal vector. Using the orthogonality relation we obtain

$$V_{fm} = \sum_{n=1}^{N} R(m, n) V_{gn} \tag{8.6}$$

with

$$R(m, n) = \iint_{\text{Unit cell}} \vec{e}_{gn} \cdot \vec{e}_{fm}^{*} \, dx \, dy \tag{8.7}$$

The transverse magnetic field at $z = 0$ can be expressed in terms of Floquet modal voltages as

$$\vec{H} = \sum_{m=1}^{\infty} V_{fm} Y_{0m}^{f} \vec{h}_{fm}(x, y) \tag{8.8}$$

---

[2] Note that for simplicity of presentation we here use a single index instead of two indexes to represent a Floquet mode (and a waveguide mode as well).

In the above equation Y^f_{0m} is the characteristic admittance of the mth Floquet mode and $\vec{h}_{fm}$ is the modal current vector. The above magnetic field can be expressed in terms of the modal currents of the "waveguide aperture" as (essentially the magnetic field continuity condition)

$$\sum_{i=1}^{N} I_{gi}\vec{h}_{gi}(x, y) = \sum_{m=1}^{\infty} V_{fm}Y^f_{0m}\vec{h}_{fm}(x, y) \qquad (8.9)$$

where I_{gi} is the unnormalized[3] modal current on the horn aperture with respect to the ith waveguide mode and $\vec{h}_{gi}$ is the corresponding modal current vector. To determine I_{gi} we use the orthogonality relation for the waveguide modes once again and obtain

$$I_{gi} = \sum_{m=1}^{\infty} V_{fm}Y^f_{0m}R^*(m, i) = \sum_{m=1}^{\infty} I_{fm}R^*(m, i) \qquad (8.10)$$

In the above I_{fm} is the unnormalized modal current for the mth Floquet mode. Substituting the expression for V_{fm} from (8.6) into (8.10) we obtain

$$I_{gi} = \sum_{n=1}^{N}\sum_{m=1}^{\infty} R(m, n)Y^f_{0m}R^*(m, i)V_{gn} = \sum_{n=1}^{N} Y(i, n)V_{gn} \qquad (8.11)$$

with

$$Y(i, n) = \sum_{m=1}^{\infty} Y^f_{0m}R(m, n)R^*(m, i) \qquad (8.12)$$

Equation (8.11) yields the relation between aperture voltage and aperture current; therefore $Y(i, n)$ represents the mutual admittance between the ith and the nth waveguide modes with respect to a Floquet excitation of the infinite array.

8.3.2 Equivalent Circuit

Equations (8.6) and (8.10) lead to the equivalent circuit depicted in Figure 8.5. The equivalent circuit essentially represents coupling between waveguide modes and Floquet modes. The circuit consists of a generalized transformer where the turns ratio is a matrix. The primary side of the transformer corresponds to the horn aperture that supports several waveguide modes and the secondary side corresponds to an infinite number of Floquet modes. The quantity $R(m, n)$ represents the voltage ratio of the transformer, which relates the modal voltages between the mth Floquet mode and the nth waveguide mode (can also be viewed as the coupling coefficient

---

[3] The normalized modal current is the unnormalized modal current multiplied by the characteristic admittance of the mode.

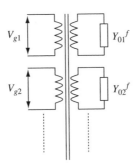

FIGURE 8.5 Equivalent circuit representing coupling between waveguide modes and Floquet modes.

between these two modes). The matrix $[R]$ constituted by the elements $R(m, n)$ is the turns-ratio matrix of the transformer. Notice from (8.10) that the current vectors at the two ports are related through the complex conjugate of $[R]^T$, which should be remembered for impedance transformation from one side to the other side of the transformer.

In order to include the excitation source we use Thevenin's equivalent circuit concept [3] with reference to the aperture plane. Thevenin's equivalent circuit consists of a voltage source and a series impedance. The magnitude of the voltage source is equal to the open-circuit voltage at the aperture plane and the series impedance is equal to the input impedance seen from the aperture toward the original source with the original voltage source replaced by a short. For a horn with N modes, Thevenin's voltage is equivalent to a voltage vector of order N and Thevenin's impedance is represented by an impedance matrix of order $N \times N$. For simplicity we consider a single-stepped horn as shown in Figure 8.6. Also, we assume that, except for the dominant mode, no higher order mode exists at the input waveguide. These assumptions decouple the waveguide modes in the aperture section; thus Thevenin's impedance matrix becomes diagonal. This simplifies Thevenin's equivalent circuit as shown in Figure 8.6. In the equivalent circuit V_0 is the source voltage of the incident mode before the step junction and $n_1, n_2, \dots$ are the voltage ratios for the smaller to larger waveguide modes. The larger waveguide voltages should be computed at the horn aperture under open-circuit condition. The input impedance seen by the input voltage source, V_0, can be calculated from this equivalent circuit.

8.3.3 Reflection Loss at Grazing Lobe Condition

We will present a quantitative analysis of the reflection coefficient at the horn's input in the case of a grazing lobe. With reference to the equivalent circuit of Figure 8.6, we assume that the Floquet mode 2 corresponds to a grazing lobe and it is of TM_z type. Thus Y_{02}^f becomes infinitely large because $k_z = 0$ for a grazing Floquet mode. This makes $Y(i, n)$ in (8.12) infinitely large, making the $[Y]$ matrix

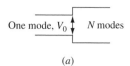

(a)

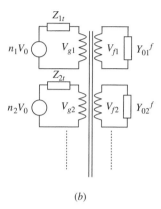

(b)

FIGURE 8.6 (a) Step horn. (b) Thevenin's equivalent circuit with respect to the aperture of the step horn array.

singular.[4] To circumvent the singularity problem in the aperture admittance matrix, we use the following approach. We separate out the singular term that causes the elements of the $[Y]$ matrix infinitely large. We know that if Y_{02}^f becomes infinitely large (short-circuit condition), then V_{f2} becomes zero. Now V_{f2} is linearly related to the waveguide voltages V_{gi}'s; therefore, the magnitudes of the waveguide voltages will adjust among themselves to satisfy $V_{f2} = 0$. Therefore, for the grazing lobe, we would have the following relation:

$$V_{f2} = 0 = \sum_{n=1}^{N} R(2, n) V_{gn} \tag{8.13}$$

For Y_{02}^f infinitely large, the current through the admittance Y_{02}^f becomes indeterminate. To alleviate this situation, we assume that as Y_{02}^f approaches infinity, V_{f2} approaches zero, keeping the current through Y_{02}^f finite. Equation (8.10) thus can be written as

$$I_{gi} = \sum_{m=1, m \neq 2}^{\infty} V_{fm} Y_{0m}^f R^*(m, i) + I_{f2} R^*(2, i) \tag{8.14}$$

---

[4] Note that a TE_z grazing mode does not cause a singularity problem in $[Y]$.

From the equivalent circuit in Figure 8.6 and using (8.6) in (8.14), we obtain

$$\frac{n_i V_0 - V_{gi}}{Z_{it}} = \sum_{k=1}^{N} \sum_{m=1,\neq 2}^{\infty} V_{gk} Y_{0m}^f R(m,k) R^*(m,i) + I_{f2} R^*(2,i) \qquad i = 1, 2, \ldots, N$$

(8.15)

Equations (8.13) and (8.15) represent a system of $N+1$ linear equations with $N+1$ unknowns. The unknown quantities are $V_{gi}, i = 1, 2, \ldots, N$ and I_{2f}. The unknown voltages can be solved using matrix inversion. Equation (8.15) can be written as

$$\frac{n_i V_0}{Z_{it}} = \sum_{k=1}^{N} V_{gk} Y'(i,k) + I_{f2} R^*(2,i)$$

(8.16)

with

$$Y'(i,k) = \sum_{m=1,m\neq 2}^{\infty} Y_{0m}^f R(m,k) R^*(m,i) + \frac{\delta_{ik}}{Z_{it}}$$

(8.17)

The solutions for I_{f2} and $V_{gk}, k = 1, 2, \ldots, N$ are thus expressed as

$$\begin{bmatrix} V_{g1} \\ V_{g2} \\ \vdots \\ V_{gN} \\ I_{f2} \end{bmatrix} = V_0 \begin{bmatrix} Y'(1,1) & Y'(1,2) & \cdots & Y'(1,N) & R^*(2,1) \\ Y'(2,1) & Y'(2,2) & \cdots & Y'(2,N) & R^*(2,2) \\ \vdots & \vdots & \ddots & \vdots & \vdots \\ Y'(N,1) & Y'(N,2) & \cdots & Y'(N,N) & R^*(2,N) \\ R(2,1) & R(2,2) & \cdots & R(2,N) & 0 \end{bmatrix}^{-1} \begin{bmatrix} n_1/Z_{1t} \\ n_2/Z_{2t} \\ \vdots \\ n_N/Z_{Nt} \\ 0 \end{bmatrix}$$

(8.18)

The square matrix on the right-hand side can be inverted because the elements have no singularity. Thus the waveguide modal voltages on the aperture can be computed for the grazing lobe condition.

It is interesting to observe that, for $N = 1$, $V_{g1} = 0$ (because the elements of the first row and first columns of the inverted matrix become zero), implying $0\,dB$ return loss. This is a hypothetical situation, because however small the aperture size might be, the aperture field must contain more than one mode to satisfy the continuity of the fields. A more realistic situation, perhaps, would be considering $N = 2$. For $N = 2$, (8.18) becomes

$$\begin{bmatrix} V_{g1} \\ V_{g2} \\ I_{f2} \end{bmatrix} = V_0 \begin{bmatrix} Y'(1,1) & Y'(1,2) & R^*(2,1) \\ Y'(2,1) & Y'(2,2) & R^*(2,2) \\ R(2,1) & R(2,2) & 0 \end{bmatrix}^{-1} \begin{bmatrix} n_1/Z_{1t} \\ n_2/Z_{2t} \\ 0 \end{bmatrix}$$

(8.19)

The solution for V_{g1} is given by

$$V_{g1} = V_0 \frac{n_1 |R(2,2)|^2/Z_{1t} - n_2 R(2,2) R^*(2,1)/Z_{2t}}{Y'(1,1)|R(2,2)|^2 + Y'(2,2)|R(2,1)|^2 - R(2,1)R^*(2,2)Y'(1,2) - R^*(2,1)R(2,2)Y'(2,1)}$$

(8.20)

The input admittance experienced by the dominant mode at the aperture plane is

$$Y_{\text{inp}}^{(1)} = \frac{n_1 V_0 - V_{g1}}{Z_{1t} V_{g1}} \qquad (8.21)$$

Substituting the expression for V_{g1} from (8.20) and then using the expressions for Y' from (8.17) we derive

$$Y_{\text{inp}}^{(1)} = \frac{n_1 Y_{01}^f |\Delta|^2 + n_1 Y_{2t} |R(2,1)|^2 + n_2 Y_{2t} R^*(2,1) R(2,2)}{Z_{1t} R(2,2)[n_1 Y_{1t} R^*(2,2) - n_2 Y_{2t} R^*(2,1)]} \qquad (8.22a)$$

with

$$\Delta = R(1,1)R(2,2) - R(1,2)R(2,1)$$
$$Y_{it} = \frac{1}{Z_{it}} \qquad i = 1,2 \qquad (8.22b)$$

From (8.22a) it is apparent that $R(m,n)$'s play important roles in determining the aperture admittance in the case of a grazing mode. We now assume that only one mode is incident on the aperture. As mentioned before, we still need to consider the other higher order mode to satisfy continuity of the aperture fields. Since the higher order mode is produced at the aperture, we have $n_2 = 0$ in Figure 8.6. The input admittance seen by the incident mode then becomes

$$Y_{\text{inp}}^{(1)} = \frac{Y_{01}^f |\Delta|^2 + Y_{2t} |R(2,1)|^2}{|R(2,2)|^2} = Y_{01}^f \left| R(1,1) - \frac{R(1,2)R(2,1)}{R(2,2)} \right|^2 + Y_{2t} \left| \frac{R(2,1)}{R(2,2)} \right|^2 \qquad (8.23)$$

Now consider a small aperture that supports only one propagating mode while the other mode is evanescent. A component of the higher order modal field changes the polarity (at least once) within the small-aperture region (in a rectangular waveguide the aperture field distribution of a higher order mode, the TE_{11} mode for instance, passes through a null). This makes the coupling coefficients $R(1,2)$ and $R(2,2)$ very small compared to $R(1,1)$ and $R(2,1)$ due to phase cancellation. To illustrate, denote TE_{01} and TE_{11} modes as mode 1 and mode 2, respectively, on an open-ended waveguide of aperture size $0.25\lambda_0 \times 0.65\lambda_0$. Denote $TM(0,0)$ and $TM(-1,0)$ Floquet modes as radiating modes 1 and 2, respectively. The cell size is $0.65\lambda_0 \times 0.65\lambda_0$ with a 90° grid angle. For the E-plane scan at about 25° off bore sight, we compute $|R(1,1)| = 0.55$, $|R(1,2)| = 0.04$, $|R(2,1)| = 0.49$, and $|R(2,2)| = 0.09$. This difference in magnitudes between $R(2,1)$ and $R(2,2)$ makes the imaginary part of the input admittance [note that the last term of (8.23) is imaginary because Y_{2t} is imaginary for a lossless waveguide] very large, resulting in a high reflection loss. As the aperture size increases, $|R(2,2)|$ becomes comparable to $|R(2,1)|$, and the imaginary part of the input admittance approaches Y_{2t}, which is not too large to cause high reflection loss. Therefore, the depth of the null in the radiation patterns generally improves with an increase in the aperture size, as observed in Figure 8.4*b*. It should be mentioned that the "two-mode" theory is a crude approximation for a large aperture; however, the rudimentary analysis gives an idea about the trend of the reflection loss versus aperture dimensions at the grazing lobe situation.

The relation between the return loss and aperture size can also be comprehended qualitatively. The grazing modal voltage is linearly related to the waveguide modal voltages through coupling coefficients [see (8.6)]. For a large aperture, the higher order waveguide modes moderately couple with the grazing mode. The number of waveguide modes being large, a large number of adjustable parameters (in this case modal amplitudes) exists to satisfy the zero-voltage situation associated with the TM_z grazing Floquet mode. The dominant waveguide modal amplitude need not be small in this situation. Contrary to this, in the case of a small aperture the higher order waveguide modes weakly couple, but the dominant mode strongly couples with the grazing mode. Therefore, to satisfy the zero voltage for the grazing mode, the higher order waveguide modal amplitudes must be very large while the dominant modal amplitude must be very small. This results in a large reflection coefficient with respect to the dominant mode.

8.4   SURFACE AND LEAKY WAVES IN AN ARRAY

In principle an array of horns cannot support a surface wave when it radiates in the visible region. This can be comprehended in the following manner. In the $z > 0$ region, a surface wave field must decay with z; therefore k_{zs} (the wave number of the surface wave along z) must be imaginary. Because k_{zs} is imaginary $k_{xs}^2 + k_{ys}^2$ must be greater than k_0^2. This implies that the (k_{xs}, k_{ys}) coordinate point must lie outside the visible region represented by the (0,0) mode circle of the $k_x k_y$-plane.

A horn array structure may support surface wave modes if the array elements are phased to radiate in the "invisible" region, however. In other words, if the phase progression is set such that the corresponding (k_{xs}, k_{ys}) point lies outside the mode circles, then occurrence of a surface wave mode is possible. However, an additional condition needs to be satisfied. The Floquet impedance must have a null at the (k_{xs}, k_{ys}) point. The real part of the Floquet impedance is always zero for a scan angle in the invisible region. Thus, the imaginary part needs to be zero. Typically, for a two-dimensional array, the surface wave poles have multiple loci because surface waves can propagate in various directions. Table 8.1 shows few points on a surface wave pole locus of an open-ended waveguide array of element aperture size $0.12\lambda_0 \times 0.81\lambda_0$ and cell size $0.16\lambda_0 \times 0.85\lambda_0$ (rectangular grid). The length of each element is $0.2\lambda_0$. The surface wave pole loci exist every $2\pi/a$ apart along the k_x-axis and $2\pi/b$ along the k_y-axis, as sketched in Figure 8.7. Furthermore, from symmetry consideration an additional set of loci exists, which is an exact mirror image with respect to the k_y-axis of the aforementioned set.

TABLE 8.1   Surface Wave Poles ($Z^F = 0$) of Array of Open-Ended Waveguide

k_{xs}	7.308	7.454	7.762	8.055	8.283	8.428	8.476
k_{ys}	3.696	3.086	2.458	1.848	1.238	0.610	0.00

Note: Aperture size $0.12\lambda_0 \times 0.81\lambda_0$, cell size $0.16\lambda_0 \times 0.85\lambda_0$, waveguide length $0.2\lambda_0$, $\lambda_0 = 1$.

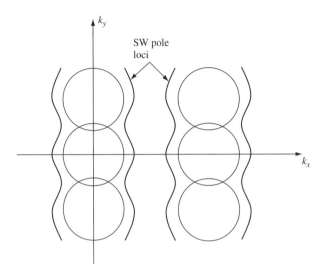

FIGURE 8.7   Surface wave (SW) pole loci ($Z^F = 0$) for an open-ended waveguide array. The surface wave poles exist in the invisible region of the $k_x k_y$-plane. The circles represent Floquet modes.

We have seen that a surface wave mode may exist if an invisible region exists in the $k_x k_y$-plane. If the array element spacing is such that the Floquet mode circles occupy the entire $k_x k_y$-plane, then the surface wave propagation condition cannot be satisfied. For instance, if the element spacing is more than one-half wavelength, then the entire k_x-axis is completely occupied by the mode circles. Such an array structure cannot have surface wave propagation along x. In a rectangular grid array a surface wave mode will not exist if the following condition is satisfied:

$$\frac{1}{a^2} + \frac{1}{b^2} \leq \frac{4}{\lambda_0^2} \tag{8.24}$$

The above condition is deduced based on the fact that the Floquet mode circles cover the entire (k_x, k_y)-plane if at least three nearby circles meet at a common point.

It should be mentioned that although existance of a zero of Floquet impedance in the invisible region satisfies the propagation condition of a surface wave mode, in reality a surface wave mode does not propagate under a Floquet excitation. Since the source internal impedance is finite, zero Floquet impedance results in a total reflection, leaving no power for the surface wave mode. In order to have the surface wave propagation, only a finite number of horns should be excited while the remaining horns should be short circuited at the input locations. In the following section we consider a special case of surface wave propagation.

8.4.1 Surface Wave

In this section we determine the surface wave fields produced by a single horn element when all other horns in an infinite array are short circuited at the input end. Under such a short-circuit condition, the current induced by the excited horn to a neighboring horn's input is proportional to the mutual admittance between the two horns. Thus the mutual admittance between the horns gives an idea of wave propagation along the surface of the horn array.

In Chapter 4 we have seen that the mutual admittance between the zeroth and mth elements of an infinite array with rectangular grid is related to the Floquet impedances through the following integral:

$$Y(m, 0) = \frac{1}{4\pi^2} \int\limits_{-\pi}^{\pi} \int\limits_{-\pi}^{\pi} \frac{1}{Z^F(\psi_x, \psi_y)} \exp\left(-\frac{jx\psi_x}{a} - \frac{jy\psi_y}{b}\right) d\psi_x\, d\psi_y \qquad (8.25)$$

In the above equation, Z^F is the Floquet impedance, ψ_x and ψ_y are the phase differences between the adjacent elements along x- and y-directions, respectively, (x, y) are the relative coordinates of the mth element with respect to the zeroth element, and $a \times b$ is the cell size. We can modify the above integral with the following substitutions:

$$\psi_x = ak_x \qquad \psi_y = bk_y \qquad (8.26)$$

Upon substitution we obtain

$$Y(m, 0) = \frac{ab}{4\pi^2} \int\limits_{-\pi/a}^{\pi/a} \int\limits_{-\pi/b}^{\pi/b} \frac{1}{Z^F(k_x, k_y)} \exp\{-jk_x x - jk_y y\}\, dk_x\, dk_y \qquad (8.27)$$

We have seen that the Floquet impedance Z^F may have zeros on the $k_x k_y$-plane if an invisible region exists on the (k_x, k_y)-plane, that is, if the array cell dimensions violate the condition in (8.24). A singularity of the integral signifies existence of a surface wave mode. Notice that Z^F is a function of two variables, k_x and k_y, that are considered to be complex in nature for handling the singularity of the integral. Because the singular points lie on a curve, we can consider that the singularities of Z^F occur with respect to only one variable, say k_x, and that Z^F is continuous with respect to the other variable, k_y. Then to extract the surface wave contribution to the mutual admittance, one can apply the singularity extraction method with respect to the complex variable k_x. Of course, the singularity variable can be selected arbitrarily without altering the end result.

For the surface wave mutual admittance between far-off elements, one can derive a closed-form expression as follows. We apply the following coordinate transformations and then use k_ρ as the variable for singularity.

$$k_x = k_\rho \cos \alpha \qquad k_y = k_\rho \sin \alpha$$
$$x = \rho \cos \phi \qquad y = \rho \sin \phi$$
(8.28)

With this transformation (8.27) becomes

$$Y(m,0) = \frac{ab}{4\pi^2} \int\limits_{k_\rho} \int\limits_0^{2\pi} \frac{T(k_\rho, \alpha)}{Z^F(k_\rho, \alpha)} \exp\{-jk_\rho\rho \cos(\alpha - \phi)\} k_\rho \, dk_\rho \, d\alpha \qquad (8.29)$$

In (8.29) we introduced the function $T(k_\rho, \alpha)$, which is unity inside the shaded region and zero outside the shaded region of Figure 8.8. This would allow us to uniquely specify the limit of k_ρ from zero to R, where R represents the radius of the circle around the shaded rectangle.

In (8.29) we now consider the integration with respect to α. For a large value of ρ we can apply the saddle point method to obtain the asymptotic form. The result is

$$Y(m,0) \approx \frac{ab}{2\pi} \int\limits_{k_\rho} \sqrt{\frac{2k_\rho}{\pi\rho}} \frac{\cos(k_\rho\rho - \pi/4)}{Z^F(k_\rho, \phi)} T(k_\rho, \phi) \, dk_\rho \qquad (8.30)$$

Now we apply the singularity extraction method to find the contribution of the surface wave mode. Suppose Z^F has a zero at $k_\rho = k_{\rho s}$. For a surface wave propagating in the outward direction we have to assume that $k_{\rho s}$ has an infinitesimal negative imaginary part, so that the zero lies below the real axis of k_ρ. The surface

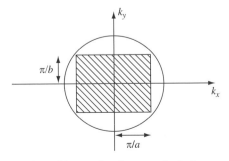

FIGURE 8.8 Region of integration for mutual admittance computation.

wave contribution to the mutual admittance is obtained by integrating on a small segment near the singularity of the integrand. We thus have

$$Y_s(m,0) \approx \frac{ab}{2\pi} T(k_{ps}, \phi) \int_{k_{ps}-\Delta}^{k_{ps}+\Delta} \sqrt{\frac{2k_\rho}{\pi\rho}} \frac{\cos(k_\rho\rho - \pi/4)}{Z^F(k_\rho, \phi)} dk_\rho \qquad (8.31)$$

We will now derive a closed-form expression for the surface wave admittance for a large ρ. Notice that for a large ρ the square-root term can be regarded as a constant, as compared with the highly oscillatory cosine function. Therefore, for a small Δ, we can have the following approximation for $Y_s(m,0)$ in (8.31):

$$Y_s(m,0) \approx \frac{ab}{2\pi} T(k_{ps}, \phi) \sqrt{\frac{2k_{ps}}{\pi\rho}} \int_{k_{ps}-\Delta}^{k_{ps}+\Delta} \frac{\cos(k_\rho\rho - \pi/4)}{Z^F(k_\rho, \phi)} dk_\rho \qquad (8.32)$$

We apply the singularity extraction method for the above integral [4] and write

$$\int_{k_{ps}-\Delta}^{k_{ps}+\Delta} \frac{\cos(k_\rho\rho - \pi/4)}{Z^F(k_\rho, \phi)} dk_\rho = \int_{k_{ps}-\Delta}^{k_{ps}+\Delta} \frac{\cos(k_{ps}\rho - \pi/4)}{Z^F(k_\rho, \phi)} dk_\rho$$
$$+ \int_{k_{ps}-\Delta}^{k_{ps}+\Delta} \frac{\cos(k_\rho\rho - \pi/4) - \cos(k_{ps}\rho - \pi/4)}{Z^F(k_\rho, \phi)} dk_\rho \qquad (8.33)$$

In the close vicinity of k_{ps}, Z^F can be approximated as $Z^F(k_\rho, \phi) \approx (k_\rho - k_{ps})Z^{F'}(k_{ps}, \phi)$, where the prime denotes the partial derivative with respect to k_ρ. The first integral on the right-hand side of (8.33) becomes

$$I_1 = \int_{k_{ps}-\Delta}^{k_{ps}+\Delta} \frac{\cos(k_{ps}\rho - \pi/4)}{Z^F(k_\rho, \phi)} dk_\rho = -\pi j \frac{\cos(k_{ps}\rho - \pi/4)}{Z^{F'}(k_{ps}, \phi)} \qquad (8.34)$$

The last integral of (8.33) can be expressed as

$$I_2 = -2 \int_{k_{ps}-\Delta}^{k_{ps}+\Delta} \frac{\sin[(k_{ps}+k_\rho)\rho/2 - \pi/4]\sin[(k_\rho - k_{ps})\rho/2]}{Z^F(k_\rho, \phi)} dk_\rho \qquad (8.35)$$

We substitute $k_\rho - k_{ps} = \mu$ and express the above integral as

$$I_2 = -2 \int_{-\Delta}^{\Delta} \frac{\sin(\mu\rho/2 + k_{ps}\rho - \pi/4)\sin(\mu\rho/2)}{Z^F(\mu + k_{ps}, \phi)} d\mu \qquad (8.36)$$

Now, for small μ, the denominator of the integrand is equal to $\mu Z^{F'}(k_{\rho s}, \phi)$. Expanding the first sine function of the numerator we obtain

$$I_2 = -2 \int_{-\Delta}^{\Delta} \frac{\cos(k_{\rho s}\rho - \pi/4)\sin^2(\mu\rho/2)}{\mu\, Z^{F'}(k_{\rho s}, \phi)}\, d\mu - \int_{-\Delta}^{\Delta} \frac{\sin(k_{\rho s}\rho - \pi/4)\sin(\mu\rho)}{\mu\, Z^{F'}(k_{\rho s}, \phi)}\, d\mu$$

(8.37)

The first integral vanishes because the integrand is an odd function of μ. For a large ρ, I_2 thus becomes

$$I_2 = -\frac{\sin(k_{\rho s}\rho - \pi/4)}{Z^{F'}(k_{\rho s}, \phi)} \int_{-\infty}^{\infty} \frac{\sin(\mu\rho)}{\mu\rho}\, d(\mu\rho) = -\pi\frac{\sin(k_{\rho s}\rho - \pi/4)}{Z^{F'}(k_{\rho s}, \phi)}$$

(8.38)

Using these results in Y_s we finally obtain

$$Y_s(m, 0) = -j\frac{ab}{2}T(k_{\rho s}, \phi)\frac{\exp[-j(k_{\rho s}\rho - \pi/4)]}{Z^{F'}(k_{\rho s}, \phi)}\sqrt{\frac{2k_{\rho s}}{\pi\rho}}$$

(8.39)

Expectedly, the surface wave admittance varies inversely with the square root of the element spacing because when only one element is excited, the surface wave becomes a cylindrical wave at the far-off region.

It should be noted that the surface wave coupling may not occur between all the elements in an array, because $Z^F(k_\rho, \phi)$ may not have zeros for all values of ϕ. Further, apart from the aperture size and element spacing, the existence of surface wave poles depends on the input location with respect to the aperture (the axial length of the horn). This is consistent with the surface wave propagation characteristics on a corrugated surface [5] because short-circuited horn elements create a corrugated surface – like structure. The propagation of a surface wave mode depends on the depth of the corrugation, which in this case becomes the axial length of the horn elements.

If the array elements are match terminated (instead of short-circuit terminations), then the terminating loads perturb the surface wave propagation. The surface wave turns into a leaky wave due to the power dissipation at the loads. This can also be understood from the analytical expression of the scattering parameter $S(m, 0)$ in Section 4.5. The scattering parameter will never have a "propagating wave" – like expression because the Floquet reflection coefficient can never have a pole.

As explained before, a true surface wave does not exist when an infinite array is phased to radiate in the visible region, although the array structure may be capable of supporting a surface wave. Nevertheless, the surface wave pole plays an important role in determining the mutual admittance between the array elements. Figures 8.9a, b show the mutual conductance and susceptance, respectively, between the E-plane array elements. For obtaining the mutual admittance, the expression in (8.25) was computed numerically. A singularity extraction method

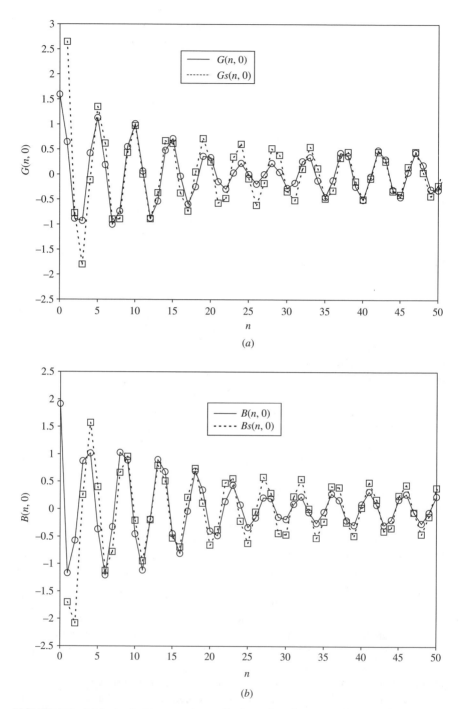

FIGURE 8.9 Mutual admittance and contribution of surface wave between E-plane elements with $a = 0.16\lambda_0$, $b = 0.85\lambda_0$, $a_1 = 0.12\lambda_0$, $b_1 = 0.81\lambda_0$, length $0.2\lambda_0$: (a) mutual conductance; (b) mutual susceptance.

was applied with respect to ψ_x. The symmetry property of $Z^F(\psi_x, \psi_y)$ was exploited to reduce the computation time. About 200×200 grid points in the $\psi_x \psi_y$-plane was used. Also plotted is the contribution of the surface wave deduced in (8.39). Expectedly, the mutual admittance between two far-off elements is dominated by the surface wave coupling.

8.4.2 Leaky Wave

We have seen that if the Floquet impedance has zeros in the $k_x k_y$-plane, then the array structure is capable of supporting surface wave modes. A surface wave propagates only when a finite number of elements are excited while the remaining elements are short circuited at the input ends. Now consider a different scenario. Suppose the imaginary part of the Floquet impedance is zero while the real part has a small positive value, making the return loss near 0 dB. This happens when a grazing lobe occurs in an array of electrically small aperture dimensions, as shown in Section 8.3.3. In such a case the Floquet impedance can be considered to have a complex zero with a small imaginary part. The complex zero of the Floquet impedance is associated with a leaky wave. Unlike a surface wave, a leaky wave radiates as it propagates. The leaky wave radiation helps in explaining the dips and nulls in the element pattern [6]. The leaky wave propagation can be well understood from its contribution to the mutual admittance. In order to estimate the mutual admittance contributed by the leaky wave, we follow the same procedure as used in the case of surface wave admittance.

The most crucial part of computing leaky wave coupling is determination of the complex leaky wave pole (a complex zero of Z^F). To search for a complex zero one needs computations of Z^F with respect to complex values of k_ρ. Furthermore, one requires a two-dimensional search, which is computationally extensive. To circumvent this, we utilize the principle of analytic continuation of a complex function. We fit a polynomial function for Z^F utilizing a set of data points on the real axis of k_ρ and then determine the complex zeros of the polynomial. The polynomial should be constructed utilizing the data points that are in the close vicinity to the "leaky wave pole." A leaky wave pole generally exists at a nearby point of the real axis of k_ρ where the real part of Z^F has a small value while the imaginary part passes though a zero. For fitting a second-degree polynomial the Floquet impedance in that neighborhood can be expressed as

$$Z^F(k_\rho, \alpha) \approx A + Bk_\rho + Ck_\rho^2 \qquad (8.40)$$

The parameters A, B, and C can be determined from three selective values of Z^F on the real axis of k_ρ. The leaky wave poles are the complex zeros of Z^F in (8.40). If both zeros of Z^F estimated using (8.40) are far off from the range of k_ρ data points that were utilized in estimating A, B, and C, then one needs to reiterate the process (by selecting a different region of the k_ρ-axis) until a zero lies in close vicinity. An alternative approach could be fitting a higher degree polynomial to cover up a wider range.

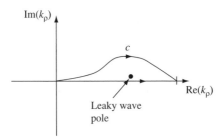

FIGURE 8.10 Leaky wave pole of an array and the deformed contour for integration in complex k_ρ-plane.

In order to compute the leaky wave admittance, one needs to deform the path of the k_ρ-integral of (8.30) to capture the complex pole. The path c in Figure 8.10 represents the deformed path assuming that the imaginary part of the leaky wave pole is positive. The asymptotic form of the leaky wave admittance can be expressed as

$$Y_l(m,0) = \mp j\frac{ab}{2}T(k_{\rho l}, \phi)\frac{\exp[\mp j(k_{\rho l}\rho - \pi/4)]}{Z^{F'}(k_{\rho l}, \phi)}\sqrt{\frac{2k_{\rho l}}{\pi\rho}} \qquad (8.41a)$$

where $k_{\rho l}$ represents a leaky wave pole. The negative sign corresponds to a pole lying below the real axis, and the positive sign corresponds to a pole above the real axis. The derivative of the Floquet impedance can be estimated from (8.40) as

$$Z^{F'}(k_{\rho l}, \alpha) \approx B + 2Ck_{\rho l} \qquad (8.41b)$$

The leaky wave poles were computed for an array of an open-ended square waveguide of $0.65\lambda_0$ aperture length and $0.7\lambda_0$ element spacing in a square lattice. The axial length of an element was $0.25\lambda_0$. The leaky wave poles were computed in the vicinity of $k_\rho = 2.1$ for $\lambda_0 = 1$. The poles were obtained as $1.647 - 0.685j$ and $2.153 + 0.074j$. The second pole is dominating because the associated leaky wave has a lower decay rate than that of the first pole. Moreover, the first pole is ignored because it is somewhat far from the search region.

It is instructive to explore the propagation characteristics of the leaky wave under consideration. The imaginary part of the complex pole being positive, the positive sign should be used in the argument of the exponent. As a result, the phase of the mutual admittance increases with the radial distance ρ. The induced current being proportional to the mutual admittance, the leaky wave radiates in the $-\rho$ direction (see Figure 8.11). However, for a pole with negative imaginary part, the radiation may occur along the $+\rho$ direction.

Figures 8.12a, b show the mutual conductance and susceptance, respectively, versus element number along the E-plane of an infinite array. The leaky wave contributions are also computed and plotted. The two sets of results match for the elements that are far apart because the coupling between such far-apart elements is dominated by the leaky wave propagation.

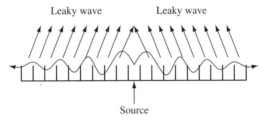

Source

FIGURE 8.11   Radiation direction of a leaky wave that has a pole with positive imaginary part.

8.4.3 Supergain Phenomenon

We now estimate the contribution of the leaky wave to the radiation pattern of an excited element in an infinite array while other elements are terminated by short circuits. The radiation pattern is equal to the active element pattern divided by $\{1 + \Gamma^F(\theta, \phi)\}$, where $\Gamma^F(\theta, \phi)$ represents the Floquet reflection coefficient with respect to the scan angle (θ, ϕ). Figure 8.13 shows the computed radiation pattern.

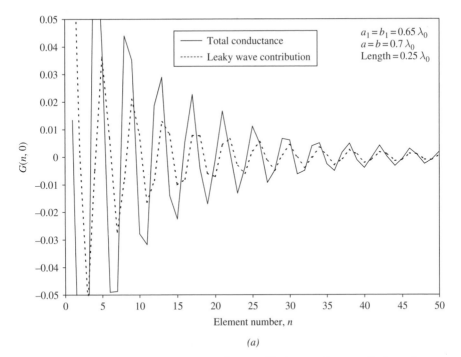

(a)

FIGURE 8.12   Mutual admittance and contribution of leaky wave between E-plane elements with $a = b = 0.7\lambda_0, a_1 = b_1 = 0.65\lambda_0$ length $0.25\lambda_0$: (*a*) mutual conductance; (*b*) mutual susceptance.

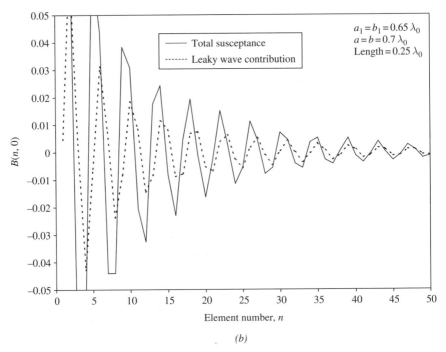

(b)

FIGURE 8.12   (Continued).

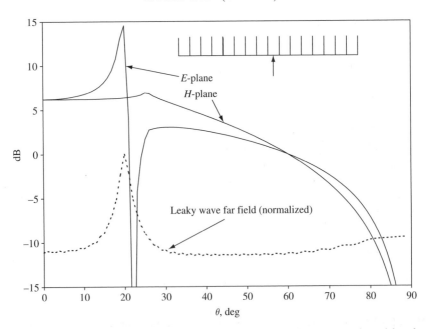

FIGURE 8.13   Supergain pattern of an element with other elements terminated by short circuits. The parameters are the same as in Figure 8.12. The leaky wave pattern is also shown.

The "supergain" characteristic near $20°$ on the E-plane is observed, which is due to the leaky wave radiation. To verify, we estimate the radiation pattern of the induced currents caused by leaky wave coupling only. The leaky wave pattern along the E-plane is approximated using the following expression:

$$E_l \approx \sum_{m=-\infty}^{-1} Y_l(m,0) \exp(jmak_0 \sin\theta) + \sum_{m=1}^{\infty} Y_l(m,0) \exp(jmak_0 \sin\theta) \quad (8.42)$$

Notice, the $m = 0$ term was excluded to eliminate the direct radiation from the excited element. Also, radiations from off-E-plane elements are ignored. The leaky wave pattern shows peak field intensities near $\pm20°$ as shown in Figure 8.13. The leaky wave radiation direction θ_l is consistent with the phase of the leaky wave mutual admittance because $k_0 \sin\theta_1 \approx -\text{Real}(k_{pl})$. Also, a sharp phase swing of the leaky wave radiated field occurs near the peak intensity, implying that a low-intensity point (sometimes a null) also exists in the total field pattern near the peak-intensity point, as seen in Figure 8.13.

8.5 WIDE-ANGLE IMPEDANCE MATCHING

We have seen that for an array of small aperture horns the Floquet return loss is generally poor primarily due to the mismatch problem between the aperture and the free space. The horn array of aperture size $0.65\lambda_0$ considered in Figure 8.2a shows a return loss of about $-14\,\text{dB}$ for the bore-sight scan. The return loss improves significantly if the array aperture is loaded with a dielectric layer. For the above array a dielectric sheet of thickness $0.36\lambda_0$ and dielectric constant 2.0 improves the return loss to $-24\,\text{dB}$ for the bore-sight scan. However, for an off-bore-sight scan the return loss still remains poor. Figure 8.14 shows the return loss performance of the array. The blind spot caused by the grazing lobe (or leaky wave) still exists but moves closer to the bore sight (from $23°$ to $20°$ for the E-plane scan) due to the dielectric layer loading.

From numerical results it is found that by adjusting the thickness of the dielectric sheet it is possible to move the blind spot away from its present position. For the above array, a dielectric sheet of thickness $0.06\lambda_0$ with dielectric constant 2.0 removes the blind spot completely from the visible region of the E-plane scan. This phenomenon is depicted in Figure 8.15. The return loss for the bore-sight scan improves slightly, but not as much as that in Figure 8.14. Therefore, a dielectric loading can possibly be designed to move the blind spot far away from the bore sight. Such a dielectric sheet is commonly known as a WAIM sheet [7]. The principle for WAIM can be explained from the input admittance characteristics of an array near the grazing lobe scan.

8.5.1 WAIM: Input Admittance Perspective

The principle of WAIM can be understood from the input admittance behavior of an array. We have seen before that the impedance matching is not required for a

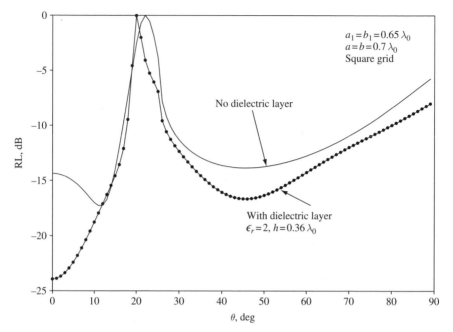

FIGURE 8.14 Floquet return loss with and without dielectric layer loading.

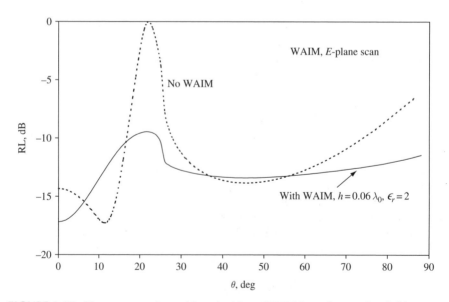

FIGURE 8.15 Floquet return loss with and without WAIM layer for $a = b = 0.7\lambda_0$, $a_1 = b_1 = 0.65\lambda_0$.

larger aperture because the match is good usually. Therefore we will restrict our discussion to small apertures only, where the "two-mode" approximation is valid, at least for the purpose of understanding the basic principle of WAIM. We assume two modes (TE_{01} and TE_{11}) on the horn aperture and two Floquet modes in the free-space region. We consider the E-plane scan case. For the x-directed aperture field, the TE_z Floquet modes do not exist for the E-plane scan; therefore, we consider two TM_z Floquet modes. The aperture admittance matrix in this case becomes a 2×2 matrix. The input admittance seen by the TE_{01} mode (mode 1) can be derived using (8.11) and Figure 8.6 as

$$Y_{inp}^{(1)} = Y(1, 1) - \frac{Y(1, 2)Y(2, 1)}{Y(2, 2) + Y_{2t}} \tag{8.43}$$

where Y_{2t} is Thevenin's admittance for the TE_{11} mode (mode 2), which, in this case, is equal to the characteristic admittance of that mode, because we consider only the TE_{01} mode as the mode of incidence on the aperture. The TE_{11} mode is produced at the aperture to satisfy the continuity condition. The expression for $Y(m, n)$ is derived in (8.12). Substituting the expressions for $Y(1, 1)$, $Y(1, 2)$, $Y(2, 1)$, and $Y(2, 2)$ from (8.12) and then with some algebraic manipulation, we derive

$$Y_{inp}^{(1)} = \frac{Y_{01}^f |R(1, 1)|^2 \{Y_{2t} + Y_{02}^f |R(2, 2) - R(1, 2)R(2, 1)/R(1, 1)|^2\} + |R(2, 1)|^2 Y_{02}^f Y_{2t}}{\{Y_{2t} + Y_{02}^f |R(2, 2)|^2\} + Y_{01}^f |R(1, 2)|^2} \tag{8.44}$$

Now, Y_{2t} is negative imaginary because the TE_{11} mode is below cutoff. For no WAIM layer, Y_{01}^f and Y_{02}^f become equal to the characteristic admittances of the Floquet modes 1 and 2 (the $TM(0, 0)$ and $TM(-1, 0)$ modes), respectively. We assume that Floquet mode 2 is responsible for the grazing lobe at $\theta = -90°$. Suppose the grazing grating lobe appears when the array scans at an angle θ_g. Therefore, at a scan angle below θ_g, Y_{02}^f is positive imaginary, and at a scan angle above θ_g, Y_{02}^f is real. Furthermore, for the E-plane scan $R(1, 1)$ and $R(2, 1)$ are real numbers with the same sign; $R(1, 2)$ and $R(2, 2)$ are imaginary numbers[5] with opposite signs. Thus (8.44) can be expressed as

$$Y_{inp}^{(1)} = \frac{Y_{01}^f |R(1, 1)|^2 \{Y_{2t} + Y_{02}^f (|R(2, 2)| + \delta)^2\} + |R(2, 1)|^2 Y_{02}^f Y_{2t}}{\{Y_{2t} + Y_{02}^f |R(2, 2)|^2\} + Y_{01}^f |R(1, 2)|^2} \tag{8.45}$$

where δ is real and positive given by

$$\delta = \left| \frac{R(1, 2)R(2, 1)}{R(1, 1)} \right| \tag{8.46}$$

---

[5] Consequential from the Fourier transforms of even and odd functions.

Notice that for $\theta < \theta_g$ the quantities inside the second brackets in both the numerator and denominator are purely imaginary and change monotonically with the scan angle θ, because Y_{02}^f increases monotonically with θ. If at the bore sight $|Y_{02}^f||R(2,2)|^2 < |Y_{2t}|$, then the above two imaginary quantities will have zeros at two different scan angles, say at θ_1 and θ_2, respectively. The imaginary part of the numerator becomes zero at a smaller scan angle than that of the denominator ($\theta_1 < \theta_2$), because of the additional term δ. For a small aperture $|R(1,2)|$ being small, the denominator is dominated by the imaginary part; thus the real part of $Y_{\text{inp}}^{(1)}$ becomes close to zero at $\theta = \theta_1$. At that scan angle, the input admittance is almost purely reactive; thus a high reflection occurs (see Figure 8.16). Therefore, the maximum reflection occurs even before appearance of the grazing lobe. At $\theta = \theta_2$, both the real and imaginary parts of $Y_{\text{inp}}^{(1)}$ are large, because the denominator is small. Also, the imaginary part has a peak near θ_2. The real part reaches its peak when the imaginary parts of both numerator and denominator become positive (phase angles are close to each other), which happens at a scan angle between θ_2 and θ_g.

Now if a WAIM sheet consisting of a dielectric layer (or more than one layer) is loaded on the aperture, then Y_{01}^f and Y_{02}^f become the input admittances (that include the WAIM layer) of the Floquet modes with reference to the aperture plane. For a lossless dielectric layer, Y_{02}^f remains imaginary for $\theta < \theta_g$; however Y_{01}^f becomes complex. The angles θ_1, θ_2, and θ_g move closer to the bore sight. For an optimum dielectric layer, it may be possible that at the bore sight $|Y_{02}^f||R(2,2)|^2 > |Y_{2t}|$, so that

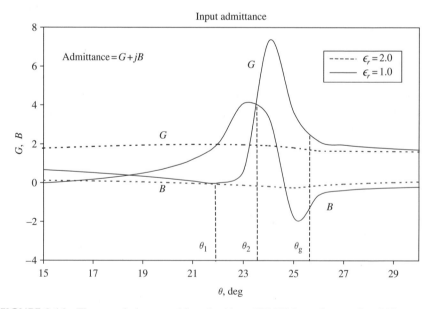

FIGURE 8.16 Floquet admittance with and without WAIM layer for $a = b = 0.7\lambda_0$, $a_1 = b_1 = 0.65\lambda_0$.

θ_1 does not exist. Under such a condition, the real part of $Y_{inp}^{(1)}$ remains reasonable and the blind spot does not exist. For very small aperture dimensions, however, it may not be possible to eliminate the zero of the numerator. This is primarily due to $|Y_{2t}|$, which becomes so large that the condition $|Y_{02}^f||R(2,2)|^2 > |Y_{2t}|$ will never besatisfied at the bore sight.

The condition that $|Y_{02}^f||R(2,2)|^2 < |Y_{2t}|$ can be applied to estimate the dielectric thickness of a WAIM layer. Recall that the above relation is deduced using only two Floquet modes. The other higher order Floquet modes together yield a capacitive reactance which increases with the dielectric constant. To estimate the actual thickness of the dielectric layer, the effects of the higher order Floquet modes must be included. It is found that in most cases the actual thickness (determined using several Floquet modes) is lower than the estimated thickness using the two-mode theory, because the capacitive reactance of the higher order modes is ignored in the latter case.

Table 8.2 shows the numerical data for the optimum thickness of the WAIM layer to be used in order to eliminate blindness. Nominally, blind spot occurs at about the 23° scan angle (which varies with the aperture dimension, as discussed before) from the bore sight along the E-plane. A WAIM layer removes the blind spot, making the array usable for scanning to $\pm 50°$ off bore sight. The cell size of the array was $0.7\lambda_0 \times 0.7\lambda_0$ in a square lattice. Notice that below a certain aperture dimension the WAIM layer does not serve the purpose of blindness removal, which is consistent with the two-mode theory. For some aperture dimensions, two optimum values of the dielectric thickness exist. However, in most cases, the input return loss does not vary appreciably if an intermediate thickness is chosen. Figure 8.17 shows the variation of the worst case input return loss with dielectric thickness for two different aperture dimensions.

TABLE 8.2   E-Plane Dimension of Aperture Versus WAIM Layer Thickness

E-Plane Dimension of Aperture (a)	Dielectric Thickness (h)	Worst Case Return Loss (dB) in 50° Scan Angle in E-Plane
0.35	Not possible	
0.40	0.32	−5.7
0.45	0.31	−10.5
0.50	0.15 or 0.32	−7.6 or −11.0
0.55	0.11 or 0.29	−8.7 or −8.9
0.60	0.08 or 0.26	−9.3 or −7.5
0.65	0.07 or 0.23	−9.6 or −6.7

Note: The dielectric constant of the WAIM layer is 2.0. The H-plane dimension of the aperture is $0.65\lambda_0$; the cell size is $0.7\lambda_0 \times 0.7\lambda_0$ in a square lattice. All dimensions are normalized with respect to the wavelength.

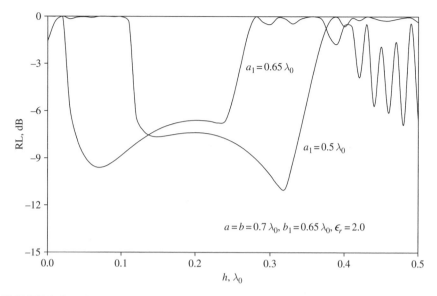

FIGURE 8.17 Worst case Floquet return loss within 50° scan angle versus WAIM layer thickness for two different element aperture sizes.

8.6 MULTIMODAL RECTANGULAR/SQUARE HORN ELEMENTS

For low scan applications, it is desirable that the horn apertures be as large as possible to minimize the number of elements in a phased array. For example, the maximum scan angle is less than 9° to scan the entire earth for a phased array on a geostationary satellite. From the grating lobe constraint, the maximum possible element spacing in a square grid array is given by

$$a = \frac{\lambda_0}{2 \sin 9°} = 3.1 \lambda_0 \qquad (8.47)$$

The above element spacing ensures that no grating lobe enters into the scan region. The above large-aperture horn can be made using linearly flared horns. Such horns have low aperture efficiency, partly due to the cosine aperture distribution along the H-plane and partly due to the large phase error caused by the flare angle. Also, the E-plane pattern is narrower than the H-plane pattern due to the uniform aperture field along the E-plane. For some applications, symmetrical patterns are desirable, particularly for circularly polarized radiation. Multimodal step horns can be designed to achieve a symmetrical element pattern and/or high aperture efficiency. In the following sections we consider two different types of multimodal square horns.

8.6.1 Potter Horn

For a circularly polarized radiated field with low cross-polarization level a Potter horn [8] with square aperture may be used. A square Potter horn has cosine aperture field distributions in both planes, resulting in a circularly symmetric radiation pattern in the bore-sight region. In a square Potter horn, the aperture field is dominated by the TE_{01}, TE_{21}, and TM_{21} waveguide modes. The percentages of the total power carried by the above modes are 82, 3.6, and 14.4%, respectively. In order to generate the above modes a step discontinuity should be placed at a location of the horn where the TE_{21} and TM_{21} modes can propagate. The relative amplitudes of the higher order modes determine the step sizes while the relative phases determine the length of the horn.

Figure 8.18 shows the active element patterns of a square Potter horn of $3.1\lambda_0$ aperture. The horn is designed using a single step, as shown in the inset of Figure 8.18. The side-lobe levels are below 22 dB of the beam peak. The aperture efficiency is about 69%. The E, H, and D (diagonal) patterns in the main-lobe region are very close to each other, which is a desirable feature for a dual linear polarization and for circular polarization applications. The dips in the element patterns are due to the grazing lobes.

8.6.2 High-Efficiency Horn

Square Potter horns yield symmetrical radiation patterns with low cross-polarization levels. However, Potter horns have low aperture efficiency. It is desirable that the aperture efficiency is as high as possible to minimize the number of elements for a desired array gain. For achieving high aperture efficiency the aperture field distribution should be as uniform as possible. This can be accomplished if the horn aperture consists of the TE_{01}, TE_{03}, TE_{05}, ... modes, respectively [9]. If the

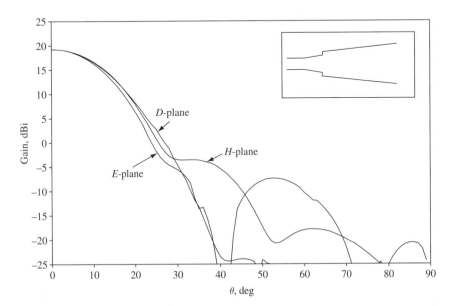

FIGURE 8.18 Active element pattern cuts of square Potter horn of aperture size $3.1\lambda_0 \times 3.1\lambda_0$.

horn aperture supports the above three modes, then they should carry approximately 86.9, 9.6, and 3.5% of the total power. No TM mode should exist on the aperture. In order to generate the desired modes, step discontinuities are placed in appropriate locations. The size of the steps and the distances between the steps are adjusted to achieve the above modal fields on the aperture in right amplitudes and phases. For single-polarization (x-polarization, for example) applications, discontinuities should be placed only on the E-plane walls. For dual linear polarization or circular polarization applications, discontinuities should be placed in both E- and H-plane walls. However, this type of discontinuity will also excite the undesired TE_{21} and TM_{21} modes. An optimum design will minimize the undesired modes through phase cancellations.

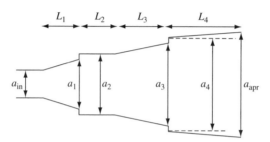

FIGURE 8.19 Geometry of a square high-efficiency horn. Approximate design dimensions are $a_{in} = 0.636$, $a_1 = 1.460$, $a_2 = 1.771$, $a_3 = 2.440$, $a_4 = 2.656$, $L_1 = 1.686$, $L_2 = 1.223$, $L_3 = 1.177$, and $L_4 = 3.356 + 2.119(a_{apr} - 3)$ for $3 \leq a_{apr} \leq 4.5$. All dimensions are in λ_0.

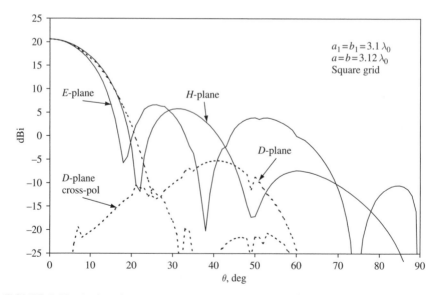

FIGURE 8.20 Active element pattern cuts of square high-efficiency horn of aperture size $3.1\lambda_0 \times 3.1\lambda_0$.

Figure 8.19 shows a structure of a high-efficiency horn with two steps. These horns can be designed to have aperture efficiency over 85% within 5% bandwidth if the aperture size is less than $4.5\lambda_0$. Approximate design dimensions are given in the figure caption. The aperture efficiency can be improved to some extent by fine tuning these dimensions using an optimizer routine.

Figure 8.20 shows the radiation patterns of a high-efficiency square horn of aperture size $3.1\lambda_0 \times 3.1\lambda_0$. The aperture efficiency of this horn is about 94%. Expectedly, the peak side lobe is about $-14\,\mathrm{dB}$ below the main lobe. The cross-polarization components for such horns are generally 30 dB below the copolarization components within a half-power beam width.

8.7 MULTIMODAL CIRCULAR HORN ELEMENTS

Multimodal circular horn elements are particularly suitable for triangular lattice structures. Like square horn elements, generally two types of designs are possible. For low cross-polarization patterns, a Potter-type design [8] may be used. In such a horn, the aperture field consists of a balanced hybrid mode, which is a combination of the TE_{11} and TM_{11} modes at about 84:16 power ratio. This modal combination with proper relative phase yields almost equal-amplitude tapers along the E- and H-planes, resulting in a circularly symmetric and low side-lobe pattern. For a circular geometry, identical E- and H-plane patterns ensure a good cross-polarization performance.

To generate the TM_{11} mode, a single step is used near the cross section of the horn where the TM_{11} mode propagates. Figure 8.21 shows the active element pattern of a Potter horn in triangular lattices. The cross-polar level is about 40 dB below the copolar level within 9° scan angle. The aperture efficiency is 73%.

A multimode circular horn can be designed that exhibits high aperture efficiency and circularly symmetric radiation pattern in the main-lobe region. In this design, a uniform aperture field distribution is realized by exciting appropriate circular waveguide modes. It is found that in order to obtain a uniform aperture field only TE modes should exist on the aperture with appropriate amplitude and phase distributions [9]. Multiple steps on the horn wall produce the desired modes. The structure of a circular high-efficiency horn is very similar to that of a square horn, as depicted in Figure 8.19. The design dimensions of a circular version of the high-efficiency horn are [9] $r_{\mathrm{in}} = 0.365$, $r_1 = 0.79$, $r_2 = 0.94$, $r_3 = 1.29$, $r_4 = 1.46$, $L_1 = 1.72$, $L_2 = 0.86$, $L_3 = 1.46$, $L_4 = 2.895 + 1.497(2r_{\mathrm{apr}} - 3)$ for $3 \le 2r_{\mathrm{apr}} \le 5$, where r stands for radius. All dimensions are in λ_0.

Figure 8.22 shows the active element pattern of a high-efficiency horn of aperture diameter $3\lambda_0$. The aperture efficiency is 93%, which is much higher than that of the Potter horn. The cross-polar level is about 30 dB below the copolar level within 9° scan angle. For a low scan array, high-efficiency horn elements seem to be preferable than Potter horn elements, because the former employs less elements than the latter.

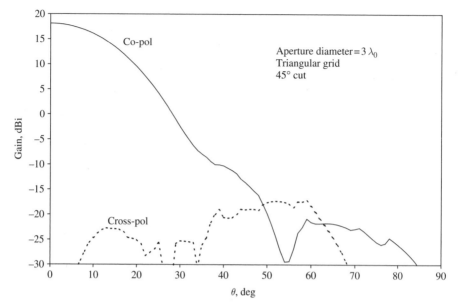

FIGURE 8.21   Active element pattern cut of circular Potter horn of aperture diameter $3\lambda_0$.

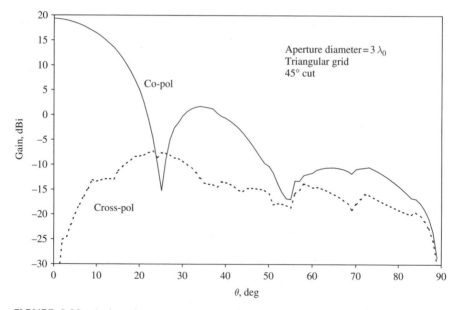

FIGURE 8.22   Active element pattern cut of circular high-efficiency horn of aperture diameter $3\lambda_0$.

REFERENCES

[1] G. L. James, "Analysis and Design of TE_{11} to HE_{11} Corrugated Cylindrical Waveguide Mode Converters," *IEEE Trans. Microwave Theory Tech.*, Vol. MTT-29, pp. 1059–1066, Oct. 1981.

[2] A. K. Bhattacharyya, "Multimode Moment Method Formulation for Waveguide Discontinuities," *IEEE Trans. Microwave Theory Tech.*, Vol. MTT-42, pp. 1567–1571, Aug. 1994.

[3] E. C. Jordan and K. G. Balman, *Electromagnetic Waves and Radiating Systems*, 2nd ed., Prentice-Hall, Englewood Cliffs, NJ, 1968.

[4] A. K. Bhattacharyya, *Electromagnetic Fields in Multilayered Structures*—Theory and Applications, Norwood, MA, Artech House, 1994.

[5] A. K. Bhattacharyya, "High-Q Resonances Due to Surface Waves and Their Effects on the Performances of Corrugated Horns," *IEEE Trans. Antennas Propagat.*, Vol. AP-49, pp. 555–566, Apr. 2001.

[6] G. H. Knittel, A. Hessel, and A. A. Oliner, "Element Pattern Nulls in Phased Arrays and Their Relation to Guided Waves," *Proc. IEEE*, Vol. 56, pp. 1822–1836, Nov. 1968.

[7] E. G. Magill and H. A. Wheeler, "Wide-Angle Impedance Matching of a Planar Array Antenna by a Dielectric Sheet," *IEEE Trans. Antennas Propagat.*, Vol. AP-14, pp. 49–53, Jan. 1966.

[8] P. D. Potter, "A New Horn Antenna with Suppressed Sidelobes and Equal Beamwidths," *Microwave.*, Vol. VI, pp. 71–78, June 1963.

[9] A. K. Bhattacharyya and G. Goyette, "A Novel Horn Radiator with High Aperture Efficiency and Low Cross-Polarization and Applications in Arrays and Multibeam Reflector Antennas," *IEEE Trans. Antennas Propagat.*, Vol. AP-52, pp. 2850–2859, Nov. 2004.

BIBLIOGRAPHY

Amitay, N., and V. Galindo, "Characteristics of Dielectric Loaded and Covered Circular Waveguide Phased Arrays," *IEEE Trans. Antennas Propagat.*, Vol. AP-17, pp. 722–729, Nov. 1969.

Amitay, N., and M. J. Gans, "Design of Rectangular Horn Arrays with Oversized Aperture Elements," *IEEE Trans. Antennas Propagat.*, Vol. AP-29, pp. 871–884, Nov. 1981.

Baccarelli, P., P. Burghignoli, F. Frezza, A. Galli, and P. Lampariello, "Novel Modal Properties and Relevant Scanning Behavior of Phased Arrays of Microstrip Leaky Wave Antennas," *IEEE Trans. Antennas Propagat.*, Vol. AP-51, pp. 3228–3238, Dec. 2003.

Byron, E. V., and J. Frank, "Lost Beams from a Dielectric Covered Phased Array Aperture," *IEEE Trans. Antennas Propagat.*, Vol. AP-16, pp. 496–499, July 1968.

Chen, C-C., "Broad-Band Impedance Matching of Rectangular Waveguide Phased Arrays," *IEEE Trans. Antennas Propagat.*, Vol. AP-21, pp. 298–302, May 1973.

Guglielmi, M., and D. R., Jackson, "Broadside Radiation from Periodic Leaky-Wave Antennas," *IEEE Trans. Antennas Propagat.*, Vol. AP-41, pp. 31–37, Jan. 1993.

Jackson, D. R., and A. A Oliner, "A Leaky-Wave Analysis of the High-Gain Printed Antenna Configuration," *IEEE Trans. Antennas Propagat.*, Vol. AP-36, pp. 905–910, July 1988.

Lechtreck, L. W., "Effects of Coupling Accumulation in Antenna Arrays," *IEEE Trans. Antennas Propagat.*, Vol. AP-16, pp. 31–37, Jan. 1968.

Lee, S. W., "Aperture Matching for an Infinite Circular Polarized Array of Rectangular Waveguides," *IEEE Trans. Antennas Propagat.*, Vol. AP-19, pp. 332–342, May 1971.

Mailloux, R. J., "Surface Waves and Anomalous wave Radiation Nulls on Phased Arrays of TEM Waveguides with Fences," *IEEE Trans. Antennas Propagat.*, Vol. AP-20, pp. 160–166, Mar. 1972.

Skobelev, S. P., and P-S. Kildal, "Blindness Removal in Arrays of Rectangular Waveguides Using Dielectric Loaded Hard Walls," *IEEE Trans. Antennas Propagat.*, Vol. AP-46, pp. 546–550, Apr. 1998.

PROBLEMS

8.1 Determine the scan angles associated with a grazing grating lobe for an array of square horns of aperture dimension $1.2\lambda_0$ (λ_0 = wavelength in free space) in a square lattice. The element spacing is $1.4\lambda_0$ in both planes. The scan planes are given by (a) $\phi = 0$ and (b) $\phi = 30°$. Assume that the principal polarization is x-directed, that is, $\phi = 0$ corresponds to the E-plane.

8.2 Repeat problem 8.1 for a triangular lattice with the following lattice parameters: $a = b = 1.4\lambda_0$, $\gamma = \tan^{-1}(2) = 63.4°$.

8.3 It is known that the blind spot nearest the bore sight on the $\phi = 45°$ plane occurs at $\theta = 15°$. Assuming a square lattice and using the periodic nature of the Floquet impedance, find the other possible blind spots on the $\phi = 45°$ scan plane. Element spacing $= 1.3\lambda_0$.

8.4 Using the equivalent circuit of Figure 8.6, deduce an expression for the input impedance seen by the voltage source V_0. Assume only two waveguide modes on the aperture and two Floquet modes for the radiated fields. Assume $R(m, n)$ as the coupling between the nth waveguide mode to the mth Floquet mode. Set $n_1 = 1$ and $n_2 = 0$ to the final result. Verify the expression in (8.23) assuming Y_{02}^f approaches infinity.

8.5 Using a circle diagram, show that for a rectangular grid an invisible region in the (k_x, k_y)-plane will not exist if the condition in (8.24) holds. Obtain the equivalent condition for an equilateral triangular grid structure.

8.6 Using the saddle point integration method for large ρ, establish the relation in (8.30) starting from (8.29). (*Note*: The integrand has two saddle points within the range of α.)

8.7 An array of rectangular waveguides has the following normalized Floquet impedances:

$\psi_x = ak_x$ (rad)	$\psi_y = bk_y$ (rad)	$Z^F(\psi_x, \psi_y)$
2.749	2.749	$0.0022j$
2.880	2.749	$-0.0007j$
2.749	2.880	$-0.0085j$
2.880	2.880	$-0.0111j$

The grid size is $0.305\lambda_0 \times 0.7\lambda_0$. Determine the surface wave mutual conductance between the two elements that are about $10\lambda_0$ apart at a direction for which the Floquet impedance has a zero. [*Hints:* Use (8.39).]

8.8 An infinite array of square horns and aperture dimension $0.65\lambda_0 \times 0.65\lambda_0$ in a square grid of size $0.7\lambda_0 \times 0.7\lambda_0$, has the following Floquet impedances for the E-plane scan:

ψ_x (rad)	$Z^F(\psi_x, \psi_y)$ for $\psi_y = 0$
1.445	$0.25 - 0.044j$
1.476	$0.19 + 0.023j$
1.508	$0.14 + 0.101j$

Following the procedure described in Section 8.4.2, determine the leaky wave pole. Obtain an expression for the mutual admittance associated with the leaky wave coupling.

8.9 The aperture electric field for a rectangular aperture is given by

$$\vec{E}^{\text{aper}} = A\,\hat{x}\sin\left(\frac{\pi x}{a}\right)\sin\left(\frac{\pi y}{b}\right)$$

where a and b are the aperture dimensions and the origin of the xy-coordinate system is located at one corner of the horn. Determine the modal power ratio and relative phases for the first three fundamental modes. Assume that a and b are sufficiently large so that the characteristic impedances for the first three modes are approximately $377\,\Omega$.

8.10 Repeat problem 8.9 if the aperture field is assumed to be uniform for gain maximization given by $\vec{E}^{\text{aper}} = A\,\hat{x}$.

8.11 The orthogonal modal vectors of a circular waveguide of radius b are given by

$$\vec{e}_m^{\text{TE}}(\rho, \phi) = -\hat{\rho}\,\frac{J_1(k'_m\rho/b)}{\rho}\cos\phi + \hat{\phi}\left(\frac{k'_m}{b}\right)J'_1\left(\frac{k'_m\rho}{b}\right)\sin\phi$$

$$\vec{e}_m^{\text{TM}}(\rho, \phi) = -\hat{\phi}\,\frac{J_1(k_m\rho/b)}{\rho}\sin\phi + \hat{\rho}\left(\frac{k_m}{b}\right)J'_1\left(\frac{k_m\rho}{b}\right)\cos\phi$$

where k_m and k'_m are the mth zeros of $J_1(x)$ and $J'_1(x)$, respectively. Here, $J_1(x)$ is the Bessel function of first order and its prime denotes a derivative with respect to the argument. By expanding the uniform aperture field $\vec{E}^{\text{aper}} = A\hat{x}$ in a Fourier–Bessel series, show that the amplitudes for the TM modes are identically zero. Determine the amplitude ratio for the first three TE modes.

CHAPTER NINE

Frequency-Selective Surface, Polarizer, and Reflect-Array Analysis

9.1 INTRODUCTION

This chapter presents the principles of operations of three passive printed array structures. They include the frequency-selective surface (FSS), screen polarizer, and printed reflect array. The FSS structures are commonly used as antenna radomes to eliminate undesired frequency components. They are also used as subreflectors in dual-band Cassegrain reflector systems. Screen polarizers essentially are the polarization converters that are commonly used in phased arrays and other antenna systems. Printed reflect arrays represent the microstrip equivalence of parabolic or shaped reflectors, producing pencil or shaped beams.

In the first part of the chapter, we explore important electrical characteristics of one- and two-layer FSS structures. This is followed by an analysis and performance study of a horn antenna loaded with an FSS. In the second part, we perform the analysis of a meander line polarizer and present the return loss, axial ratio, and scan characteristics of a two-layer polarizer. We present numerical results of a multilayer polarizer that shows enhanced bandwidth performance. In the last part of the chapter, we perform the analysis of a printed reflect-array antenna followed by gain and bandwidth studies for linearly polarized radiation. Different configurations for circularly polarized reflect arrays are presented next. Bandwidth enhancement methods of reflect-array antennas are discussed. The chapter ends with a brief discussion on contour-beam reflect arrays.

Phased Array Antennas. By Arun K. Bhattacharyya
© 2006 John Wiley & Sons, Inc.

9.2 FREQUENCY-SELECTIVE SURFACE

An FSS essentially is a filter in the shape of a screen that transmits signals at one frequency band and reflects signals at the other frequency band. The former band is known as the transmit band and the latter band is known as the reflect band. A typical FSS may have multiple transmit bands and reflect bands. In filter terminology, these bands are called pass band and stop band, respectively. A signal with frequency lying between the transmit band and the reflect band experiences partial transmission and partial reflection. A typical FSS structure consists of an array of patches printed on a thin dielectric layer. The patch size and unit cell size are dependent upon the transmit- and reflect-band frequencies [1]. A complementary structure with apertures on the conductor is also used as an FSS. Depending on applications, single- or multiple-layer FSS structures can be designed.

The principle of operation of an FSS is very similar to that of a waveguide filter. An infinite array of conducting patches is equivalent to a shunt susceptance in regard to a plane wave incidence. The magnitude of the susceptance is a function of the operating frequency. Operation of a single-layer FSS relies on the magnitude of the susceptance. A transmit band exists if the susceptance is low because that would allow transmission of a signal in that band. The susceptance is high in a reflect band. For a multilayer FSS the separating distances between the layers also play important roles in deciding the transmit- and reflect-band frequencies.

Typically, Galerkin's MoM is employed to obtain the GSMs of the individual layers and then the GSMs are cascaded to characterize the entire multilayer FSS [1]. The GSM of a patch array has been deduced in Chapter 6. For identical periodicities of the layers, the cascading formula is also deduced in Chapter 6. For nonidentical periodicities, the cascading process is more involved and will be treated in Chapter 10. For most practical applications, however, identical periodicities are commonly used.

9.2.1 Reflection and Transmission Characteristics

The reflection and transmission characteristics of a single-layer FSS can be estimated from the equivalent susceptance of a patch element. For a very low frequency, a patch array is equivalent to a shunt capacitance with respect to a plane wave incidence. This shunt capacitance is primarily due to opposite line charges at the nearest edges of two consecutive patch elements. The magnitude of the capacitance can be determined from the GSM of the patch layer. The magnitude may vary with the incident angle as well as the reference polarization (TE or TM)[1] of the incident plane wave.

Figure 9.1 shows equivalent capacitance (normalized with respect to the free-space admittance) of square patch elements versus frequency for normal incidence.

---

[1] The TE and TM modes are transverse to the FSS normal.

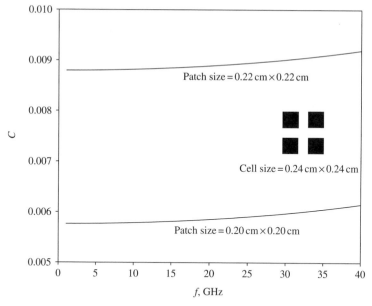

FIGURE 9.1 Equivalent shunt capacitance of an array of square patches in square grid for normal incidence: cell size $0.24 \times 0.24 \, \text{cm}^2$, capacitance normalized with the free-space admittance and computed as $C = B/(2\pi f)$, $B =$ normalized susceptance, $f =$ frequency in gigahertz.

Square-grid arrays are assumed. The capacitance is fairly flat in the low-frequency region and then gradually increases. For a given frequency, the capacitance is larger for bigger patch elements. This is expected, because as the element size increases, the interelement spacing decreases, and hence the capacitance increases.

The performance of an FSS can be evaluated from its return loss characteristics over the band of interest. For a transmit band, the return loss should be far below 0 dB. For a reflect band, the return loss should be very close to 0 dB. Figure 9.2 shows return loss characteristics of one-layer FSS screens of square-patch elements. For normal incidence, the return loss is below $-20 \, \text{dB}$ at low frequencies. Thus a transmit band exists in the low-frequency region. The return loss increases with frequency. At high frequency, the FSS structure can be used in the reflection mode. However, at high frequency grating lobes may appear, particularly for an oblique incidence. It is worth mentioning that contrary to a one-layer patch FSS, a one-layer aperture FSS shows a reflect band at the low-frequency region.

A two-layer FSS can be designed to have both transmit and reflect bands below the grating lobe cutoff frequency, as depicted in Figure 9.3. A transmit band is observed near 14 GHz. The FSS also shows a reflect band near 30 GHz (below the grating lobe cutoff). The reflection loss at the reflect band is about $-0.7 \, \text{dB}$. By increasing the number of layers and using appropriate element shape, the slope of the band edges can be controlled and the performance at the reflect band can

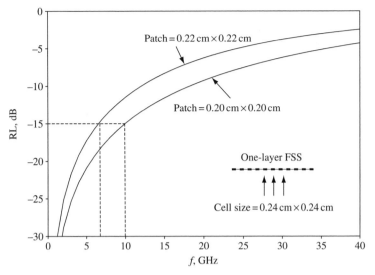

FIGURE 9.2   Return loss characteristics of one-layer FSS screens with square patch elements in a square grid. Cell size $0.24 \times 0.24 \, \text{cm}^2$.

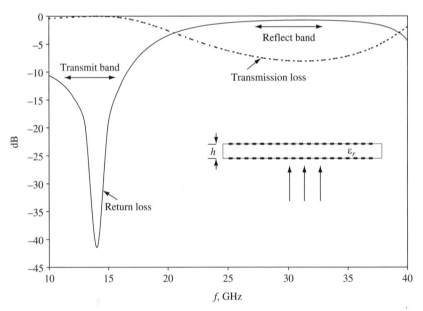

FIGURE 9.3   Return loss and transmission loss of a two-layer FSS of square patches: cell size $0.24 \times 0.24 \, \text{cm}^2$, patch size $0.22 \times 0.22 \, \text{cm}^2$, $h = 0.4 \, \text{cm}$, $\varepsilon_r = 1.2$, square grid.

be improved. It is, however, preferable to keep the minimum number of layers to avoid excitations of guided wave modes (for example surface wave modes or parallel-plate modes).

We compare the performances of a dual-band FSS for normal and oblique incidences. Figure 9.4 shows the return loss and transmission loss versus frequency for two incident angles of a two-layer FSS with ring elements. The TE incident mode was considered for this exercise. The FSS was designed for a transmit band between 6 and 7.5 GHz and a reflect band between 17.5 and 20 GHz. Compared with the normal incidence it is found that the performance at the transmit band does not change significantly for the 15° incident angle. However, at the reflect band, the return loss deteriorates near the upper edge of the band.

To examine the scan performance further, the transmission loss at 7.5 GHz and the return loss at 20 GHz are plotted in Figure 9.5 against scan angle for the 22.5° scan plane. At 7.5 GHz the transmission loss is better than 0.05 dB within the 40° scan angle for the TE incident mode. For the TM incident mode, the transmission loss is good within the 67° scan angle. At 20 GHz frequency, the return loss is better than 0.05 dB within the 14° scan angle for both TE and TM modes. The return loss significantly deteriorates near the 30° scan angle due to the appearance of a grazing lobe and a surface wave excitation (for a substrate with low dielectric constant, grazing lobe and surface wave resonances occur almost at the same scan angle).

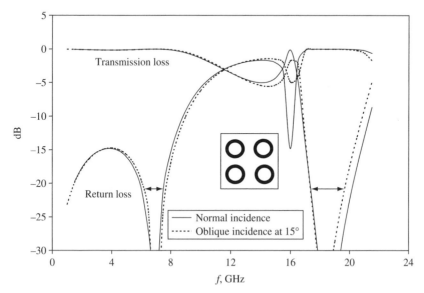

FIGURE 9.4 Return loss and transmission loss versus frequency of a two-layer, dual-band FSS with ring elements with respect to the TE incident mode: cell size $1.02 \times 1.02 \, \text{cm}^2$, inner radius of ring 0.25 cm, outer radius 0.36 cm, substrate thickness $(h) = 1.02$ cm, $\varepsilon_r = 1.2$, square grid.

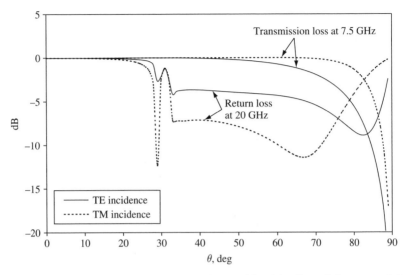

FIGURE 9.5   Scan performance of the ring FSS considered in Figure 9.4 at transmit-band and reflect-band frequencies for 22.5° scan plane.

9.2.2   Cross-Polarization Performance

The cross-polarization performance of an FSS operating in a transmit band can be evaluated from its transmission matrix (T-matrix). The T-matrix relates the incident and transmitted modal voltage vectors as

$$
\begin{bmatrix} V_{tr}^{TE} \\ V_{tr}^{TM} \end{bmatrix} = \begin{bmatrix} T(TE, TE) & T(TE, TM) \\ T(TM, TE) & T(TM, TM) \end{bmatrix} \begin{bmatrix} V_{inc}^{TE} \\ V_{inc}^{TM} \end{bmatrix} \tag{9.1}
$$

where V_{tr}^{TE} and V_{tr}^{TM} represent the transmitted modal voltages for the TE and TM modes, respectively. The quantities in column vector on the right-hand side are the incident modal voltages. The 2×2 matrix is the T-matrix of the FSS. The elements of the T-matrix are the transmission coefficients of the fundamental TE and TM modes. The T-matrix becomes diagonal if an FSS does not introduce any cross-polarization to the transmitted field.

For the TE plane wave incidence, the transmitted fields can be determined from (9.1). The transmitted electric field vector is given by

$$
\vec{E}_{tr} = V_{tr}^{TE} \vec{e}^{TE} + V_{tr}^{TM} [\vec{e}^{TM} + \hat{z} e_z] \tag{9.2}
$$

where $\vec{e}^{TE}$ and $\vec{e}^{TM}$ are the transverse components of the modal voltage vectors for the TE and TM modes, respectively, and e_z is the z-component of the TM modal vector. The expressions for the modal vectors are deduced in Chapter 3 [see (3.91)], and in spherical coordinates these quantities are given by

$$\vec{e}^{\text{TE}} = -\frac{\hat{\phi}}{\sqrt{ab}} \qquad \vec{e}^{\text{TM}} + \hat{z}e_z = \frac{\hat{\theta}}{\sqrt{ab}\cos\theta} \tag{9.3}$$

where $a \times b$ represents the unit cell size. The transmitted electric field thus can be written as

$$\vec{E}_{\text{tr}} = -\frac{\hat{\phi}\,V_{\text{tr}}^{\text{TE}}}{\sqrt{ab}} + \frac{\hat{\theta}\,V_{\text{tr}}^{\text{TM}}}{\cos\theta\sqrt{ab}} \tag{9.4}$$

The copolarization component is the ϕ-component of the transmitted field, because the TE incident mode is assumed. The θ-component of the transmitted field produced by the FSS can be considered as the cross-polarization component. The cross-polar isolation (CPI) is defined as the ratio of the cross-polarization to the copolarization field intensities. In decibels, the isolation is given by

$$\text{CPI} = 20\log\left(\left|\frac{V_{\text{tr}}^{\text{TM}}}{V_{\text{tr}}^{\text{TE}}\cos\theta}\right|\right) \tag{9.5}$$

Using (9.1) in (9.5) we obtain

$$\text{CPI} = 20\log\left(\left|\frac{T(\text{TM, TE})}{T(\text{TE, TE})\cos\theta}\right|\right) \tag{9.6}$$

For the TM incident mode,

$$\text{CPI} = 20\log\left(\left|\frac{T(\text{TE, TM})\cos\theta}{T(\text{TM, TM})}\right|\right) \tag{9.7}$$

It is intuitive that the CPI should be very good along a plane of symmetry, because the cross-polarized field must be antisymmetric with respect to a plane of symmetry. For square elements in a square grid there are four such planes: $\phi = 0$, 45°, 90°, 135° planes. Expectedly, the cross-polar isolation would be worst between two planes of symmetry. In Figure 9.6 we have plotted the isolation at the $\phi = 22.5°$ plane. At 7.5 GHz, the isolation is below -40 dB for all scan angles. Near the 90° scan angle, the isolation for the TM incident mode is better than that of the TE incident mode, because the cross-polarization counterpart for the TM incident mode (which is the TE mode) cannot exist at the 90° scan angle. This is primarily due to the metallic boundary of the conducting rings. At 20 GHz, the isolation is better than -25 dB below the 27° scan angle. The isolation significantly deteriorates in the 30° scan region due to grazing lobe and surface wave mode excitation. This result is consistent with the return loss behavior in Figure 9.5. Notice, no grazing lobe or surface wave resonance occurs at 7.5 GHz, because the element spacing is only about a quarter wavelength.

In regard to the cross-polarization isolation, a hexagonal grid is preferable to a square grid, because the former has six planes of symmetry as opposed to four

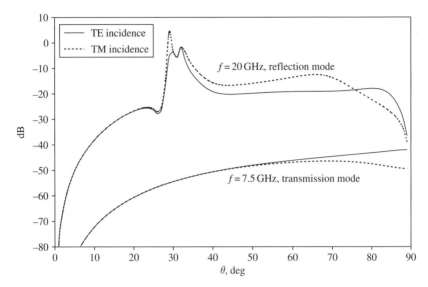

FIGURE 9.6 Cross-polar isolation at 22.5° scan plane for the FSS considered in Figure 9.4.

planes. The cross-polarization performance of an FSS due to a circularly polarized incident field would be very similar to that of a linearly polarized incident field as long as the magnitudes of the CPI are of same order for the TE and TM incident modes.

9.2.3 FSS-Loaded Antenna

The secondary patterns of an antenna loaded with an FSS can be determined from the primary patterns of the antenna and the transmission (or reflection) characteristics of the FSS. Figure 9.7 is a schematic of a horn loaded with an FSS. To obtain the secondary pattern we assume that (a) the FSS is extended to infinity in the transverse directions and (b) the reflected field from the FSS does not perturb the

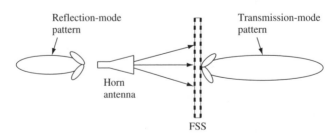

FIGURE 9.7 Schematic of a horn loaded with an FSS.

primary radiated fields of the antenna. Suppose the electric field emanating from the antenna is given by

$$\vec{E}(x, y, z) = \int\limits_{-\infty}^{\infty} \int\limits_{-\infty}^{\infty} V_{\text{TE}}(k_x, k_y)[\hat{x}k_y - \hat{y}k_x] \exp(-j\vec{k} \cdot \vec{r}) \, dk_x \, dk_y$$

$$+ \int\limits_{-\infty}^{\infty} \int\limits_{-\infty}^{\infty} V_{\text{TM}}(k_x, k_y) \left[\hat{x}k_x + \hat{y}k_y - \frac{\hat{z}(k_x^2 + k_y^2)}{k_z} \right]$$

$$\times \exp(-j\vec{k} \cdot \vec{r}) \, dk_x \, dk_y \qquad (9.8)$$

For brevity we use the notation $\vec{k} \cdot \vec{r}$ to represent $xk_x + yk_y + zk_z$ in (9.8). The first and second integrals on the right-hand side of (9.8) represent the TE$_z$ and TM$_z$ fields, respectively. The transmitted wave amplitudes are modified by the transmission coefficients of the FSS. From (9.1) and (9.8) we obtain the transmitted field through the FSS as

$$\vec{E}'(x, y, z) = \int\limits_{-\infty}^{\infty} \int\limits_{-\infty}^{\infty} [V_{\text{TE}} T(\text{TE}, \text{TE}) + V_{\text{TM}} T(\text{TE}, \text{TM})][\hat{x}k_y - \hat{y}k_x] \exp(-j\vec{k} \cdot \vec{r}) \, dk_x \, dk_y$$

$$+ \int\limits_{-\infty}^{\infty} \int\limits_{-\infty}^{\infty} [V_{\text{TE}} T(\text{TM}, \text{TE}) + V_{\text{TM}} T(\text{TM}, \text{TM})] \left[\hat{x}k_x + \hat{y}k_y - \frac{\hat{z}(k_x^2 + k_y^2)}{k_z} \right]$$

$$\times \exp(-j\vec{k} \cdot \vec{r}) \, dk_x \, dk_y \qquad (9.9)$$

In order to obtain the far field we use the asymptotic form of the following exponential (see Section 3.5.2):

$$\exp(-j\vec{k} \cdot \vec{r}) \approx j2\pi k_0 \delta(u - k_x)\delta(v - k_y) \frac{\exp(-jk_0 r)}{r} \cos \theta \qquad (9.10)$$

with

$$u = k_0 \sin \theta \cos \phi \qquad v = k_0 \sin \theta \sin \phi \qquad (9.11)$$

Using (9.10) in (9.9) we obtain

$$\vec{E}'(r, \theta, \phi) \approx C \frac{\exp(-jk_0 r)}{r} \{-[V_{\text{TE}} T(\text{TE}, \text{TE}) + V_{\text{TM}} T(\text{TE}, \text{TM})] \cos \theta \, \hat{\phi}$$

$$+ [V_{\text{TE}} T(\text{TM}, \text{TE}) + V_{\text{TM}} T(\text{TM}, \text{TM})]\hat{\theta}\} \qquad (9.12)$$

where C represents a constant factor. Excluding the unimportant constant and r-dependent terms, we can write

$$E'_\phi = -V_{\text{TE}} T(\text{TE}, \text{TE}) \cos \theta - V_{\text{TM}} T(\text{TE}, \text{TM}) \cos \theta \qquad (9.13)$$

From Chapter 3, Section 3.5 we know that $V_{TE} \cos \theta$ is proportional to E_ϕ and V_{TM} is proportional to E_θ. Ignoring the constant factor again, we obtain

$$E'_\phi = T(TE, TE)E_\phi - T(TE, TM) \cos \theta \, E_\theta \qquad (9.14)$$

Similarly the θ-component can be expressed as

$$E'_\theta = -\frac{T(TM, TE)E_\phi}{\cos \theta} + T(TM, TM)E_\theta \qquad (9.15)$$

In matrix format we write

$$\begin{bmatrix} E'_\phi \\ E'_\theta \end{bmatrix} = \begin{bmatrix} T(TE, TE) & -T(TE, TM) \cos \theta \\ -\dfrac{T(TM, TE)}{\cos \theta} & T(TM, TM) \end{bmatrix} \begin{bmatrix} E_\phi \\ E_\theta \end{bmatrix} \qquad (9.16)$$

Suppose the principal polarization is along the x-direction. Then according to Ludwig's third definition of copolarization and cross-polarization [2], we have

$$E_{cr} = E_\theta \sin \phi + E_\phi \cos \phi \qquad E_{co} = E_\theta \cos \phi - E_\phi \sin \phi \qquad (9.17)$$

If we use the above definition of copolarization and cross-polarization and use (9.16), we can write

$$\begin{bmatrix} E'_{cr} \\ E'_{co} \end{bmatrix} = \begin{bmatrix} \sin \phi & \cos \phi \\ \cos \phi & -\sin \phi \end{bmatrix} \begin{bmatrix} T(TE, TE) & -T(TE, TM) \cos \theta \\ -\dfrac{T(TM, TE)}{\cos \theta} & T(TM, TM) \end{bmatrix}$$
$$\times \begin{bmatrix} \sin \phi & \cos \phi \\ \cos \phi & -\sin \phi \end{bmatrix} \begin{bmatrix} E_{cr} \\ E_{co} \end{bmatrix} \qquad (9.18)$$

Figure 9.8 shows the copolar and cross-polar patterns of two Potter horns loaded by the FSS described in Figure 9.4. The horns are designed to operate at transmit and reflect bands of the FSS. The aperture diameter of a horn is about $3\lambda_0$ with respect to its operating frequency. The horn apertures are assumed parallel to the FSS surface. The feed blockage is ignored for the reflected fields. At the transmit-band frequency the secondary patterns do not differ from the primary patterns. At the reflect-band frequency the secondary patterns significantly differ from the primary patterns near the 30° scan angle. As explained before, this deterioration of the copolar and cross-polar performances is due to the grazing lobe and the surface wave mode.

The above analysis assumes the FSS surface to be of infinite extent along the transverse plane. For a finite FSS, a physical optics approximation [3] can be employed to obtain the first-order effect. For the transmit band the differential electric field (electric field with FSS loading minus electric field of the horn without FSS) at the finite surface of the FSS is determined assuming the FSS is of infinite extent. A superposition of the primary horn patterns and the radiation patterns of

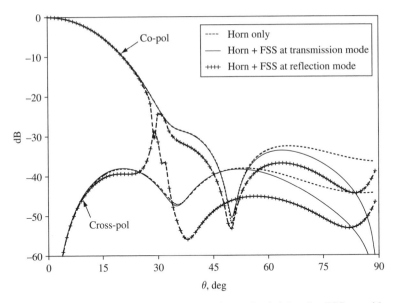

FIGURE 9.8 Radiation patterns of two Potter horns loaded by the FSS considered in Figure 9.4. Diameter of each horn is $3\lambda_0$ at the operating frequency.

the differential equivalent magnetic current yields the secondary patterns with finite size FSS.

The present analysis may not be applied for an FSS-loaded array antenna because the reflected fields of the FSS will affect the primary radiation emanating from the array. We will consider this problem in Chapter 10.

9.3 SCREEN POLARIZER

A screen polarizer converts a linearly polarized wave to a circularly polarized wave and vice-versa. It consists of two or more printed line screens (typically meander line or stripline screens) separated by dielectric layers called spacers. Figure 9.9 shows a polarizer with two meander line screens. The principle of operation of a

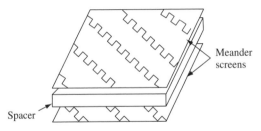

FIGURE 9.9 Exploded view of a two-layer meander line screen polarizer.

screen polarizer is as follows. Consider, for instance, the meander line polarizer in Figure 9.9. A linearly polarized incident wave with electric field vector at a 45° angle with the meander line axis will have two orthogonal components, one parallel to the meander line axis and the other perpendicular to the meander line axis. A meander line surface behaves differently with these two field components. To the parallel component, the meander line surface is equivalent to a shunt inductance and to the perpendicular component it is equivalent to a shunt capacitance. Thus the two transmitted field components will experience two different phase shifts. If two screens are placed in such a way that the total transmission coefficients for the two orthogonal components have equal magnitudes with 90° phase difference, then the emerging wave will be circularly polarized. The meander line parameters and the spacer thickness are optimized to have a circularly polarized transmitted field.

9.3.1 Analysis

The analysis of a meander line polarizer can be performed using the GSM approach, as in the case of a patch array. There are two important differences between the analysis of a patch array and a meander line array. In the case of a meander line polarizer, the meander lines are etched on a thin film of substrate material (typically 2–3 mils thick). The dielectric constant of the spacer is generally lower than that of the film. As a result one requires a very large number of modes to incorporate the effect of the film, because many evanescent modes do not decay within the thin layer. In order to circumvent such a situation, the higher order modes that are not considered in the GSM can be included appropriately through equivalent input admittances that accounts for the thin layer and the spacer layer [4]. The other difference is that, the meander lines being connected from cell to cell, the continuity of the induced current must be maintained. The basis functions should be selected accordingly. One such set of entire domain basis functions is

$$\vec{i}_n(l, t) = \hat{l} \frac{\exp(j 2\pi n l / L)}{\sqrt{1 - (2t/w)^2}} \qquad n = 0, \pm 1, \pm 2, \pm 3, \ldots, \pm N \qquad (9.19)$$

for $0 < l < L$ and $-w/2 < t < w/2$. In the above equation, L is the length of the meander line conductor within a unit cell and w is the width of the conductor, l represents the distance of a point along the length of the conductor, and t is the distance along the width of the conductor. The unit vector $\hat{l}$ is either $\hat{x}$ or $\hat{y}$, depending on the location of l. The vector $\hat{l}$ is assumed to change its direction at a diagonal location of each bend of a meander line.

9.3.2 Meander Susceptance

A meander line is equivalent to a shunt susceptance with respect to a plane wave incidence. For horizontal polarization the susceptance is capacitive, and for vertical polarization it is inductive. The equivalent susceptance, B, essentially accommodates the effects of higher order evanescent Floquet modes. In order to design a polarizer, one should have an idea about the magnitudes of the susceptances. This susceptance is determined from the GSM of a meander line layer. Suppose Γ is the reflection coefficient of a plane wave incidence, which is equal to the diagonal element corresponding to the dominant Floquet mode in $[S_{11}]$. The normalized input admittance experienced by a plane wave incident on a meander line plane is given by

$$G + jB = \frac{1-\Gamma}{1+\Gamma} \qquad (9.20)$$

The imaginary part on the right-hand side is thus equal to the equivalent shunt susceptance.

Figure 9.10 shows the equivalent susceptance of a meander line screen versus frequency with the meander depth as a parameter. We consider normal incidences for the plot. As expected, the susceptance is capacitive for the x-polarized plane wave incidence and inductive for the y-polarized plane wave incidence. The depth of the meander controls the magnitude of the susceptance. The shunt capacitance is more sensitive to the meander depth than the shunt inductance. For a wide-band design, the magnitudes of the capacitance and the inductance should be of same order to balance the amplitudes of x- and y-components of the transmitted field.

9.3.3 Return Loss and Axial Ratio

The return loss and axial ratio characteristics of a meander line polarizer can be determined from the overall GSM of the structure. In a polarizer, the number of

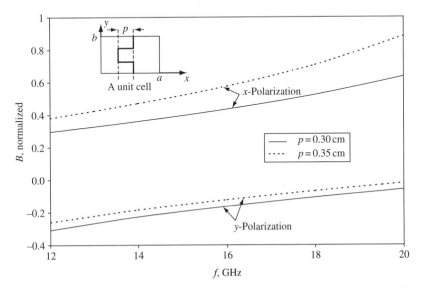

FIGURE 9.10 Equivalent susceptance of a meander line screen versus frequency with the meander depth, p, as a parameter: $a = 1.016$ cm, $b = 0.432$ cm, strip width 0.025 cm, film thickness 0.0254 cm, $\varepsilon_r = 2.2$.

meander line screens should be at least 2 in order to have good return loss and axial ratio performances. A single meander line screen may be designed to have good axial ratio, but the return loss would be poor. In order to achieve 90° phase difference between x- and y-components by a single screen, the equivalent susceptance (for both polarizations) of the meander line should be large, which results in a high reflection loss. The second layer essentially minimizes the reflected field by phase cancellation, like a quarter-wave transformer in a waveguide circuit.

In order to obtain the GSM of a two-layer polarizer, one needs to compute the individual GSMs of the two screens and the spacer between the screens and cascade them in proper sequence. The return loss and axial ratio can be determined from the overall GSM of the polarizer.

Suppose $V_{\text{inc}}^{\text{TE}}$ and $V_{\text{inc}}^{\text{TM}}$ are the incident modal voltages for the TE and TM modes, respectively. Then the reflected and transmitted voltages are related through the matrix relation in (9.1). The transmitted field is expressed in (9.4) in terms of transmitted modal voltages. We rewrite the above equation ignoring the factor $1/\sqrt{ab}$ on the right hand side, which is unimportant for the axial ratio computation:

$$\vec{E}_{\text{tr}} = -\hat{\phi}\, V_{\text{tr}}^{\text{TE}} + \frac{\hat{\theta}\, V_{\text{tr}}^{\text{TM}}}{\cos\theta} \tag{9.21}$$

The above transmitted field can be decomposed into lcp and rcp components:

$$\vec{E}_{\text{tr}} = V_{\text{tr}}^{\text{lcp}} \frac{\hat{\theta} + j\hat{\phi}}{\sqrt{2}} + V_{\text{tr}}^{\text{rcp}} \frac{\hat{\theta} - j\hat{\phi}}{\sqrt{2}} \tag{9.22}$$

Comparing (9.21) and (9.22) we obtain

$$V_{\text{tr}}^{\text{lcp}} = \frac{V_{\text{tr}}^{\text{TM}}/\cos\theta + jV_{\text{tr}}^{\text{TE}}}{\sqrt{2}} \qquad V_{\text{tr}}^{\text{rcp}} = \frac{V_{\text{tr}}^{\text{TM}}/\cos\theta - jV_{\text{tr}}^{\text{TE}}}{\sqrt{2}} \tag{9.23}$$

Suppose the intended polarization is left-hand circular polarization. Then the axial ratio, AR, is defined as

$$\text{AR} = 20\log \frac{|V_{\text{tr}}^{\text{lcp}}| + |V_{\text{tr}}^{\text{rcp}}|}{|V_{\text{tr}}^{\text{lcp}}| - |V_{\text{tr}}^{\text{rcp}}|} \tag{9.24}$$

For the return loss computation one needs to find the reflected modal voltages. The reflected voltages are related to the incident voltages through $[S_{11}]$. The reflected power can be determined from the reflected voltage. The return loss of the polarizer is the ratio between the reflected power and the incident power.

The axial ratio and return loss characteristics of two-layer polarizers are shown in Figure 9.11 for normal incidence. The polarization of the incident field is linear with the electric field vector inclined at a 45° angle with the polarizer axis. Identical line widths were considered for both polarizers. The stripline polarizer (that is, for $p = 0$) shows a smaller bandwidth than that of the meander line polarizer. The 2-dB

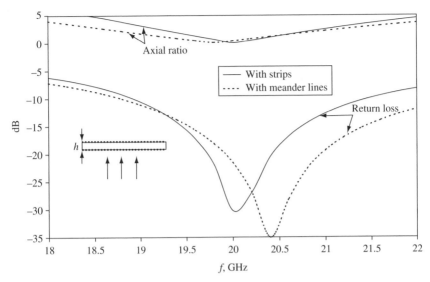

FIGURE 9.11 Axial ratio and return loss of two-layer polarizers for normal incidence: $a = 0.35$ cm, $b = 0.7$ cm, strip width 0.025 cm, spacer $\varepsilon_r = 1.1$, $p = 0.1$ cm (meander lines), $h = 0.492$ cm (meander polarizer), $h = 0.528$ cm (strip polarizer).

axial ratio bandwidths are 7% and 10%, respectively. The meander line polarizer also exhibits wider return loss bandwidth than that of the stripline polarizer.

The bandwidth of a polarizer can be further enhanced using multiple meander line screens. Figure 9.12 shows the axial ratio and return loss of a four-layer screen polarizer. The total width of the polarizer was about $1\lambda_0$ with respect to the center frequency of the band. The line widths and the spacer thickness were adjusted to perform in a wider band. The 2-dB axial ratio bandwidth is over 25% for normal incidence.

9.3.4 Scan Characteristics

The scan characteristics of the polarizers considered in Figures 9.11 and 9.12 are presented in Figure 9.13. We consider the $90°$ scan plane (along the meander line axis). The return loss and axial ratio characteristics of the two-layer stripline polarizer gradually deteriorate with the scan angle. However, there is no abrupt change in the return loss. The two-layer meander line polarizer shows sharp deterioration of the axial ratio near the $80°$ scan angle. The meander line polarizer supports slow wave modes. For about an $80°$ scan angle, the phase slope of the incident wave matches the propagation constant of a slow wave, causing a resonance. This phenomenon is very similar with the scan blindness in a patch array. The blind angle can be estimated from the meander line structure. Notice that two nearby meander lines (one above the other) at two different layers form a transmission line supporting guided modes. The propagation constant of the fundamental quasi-TEM mode is approximately $\beta = k_0\sqrt{\varepsilon_r}$, where ε_r is the dielectric constant of the spacer.

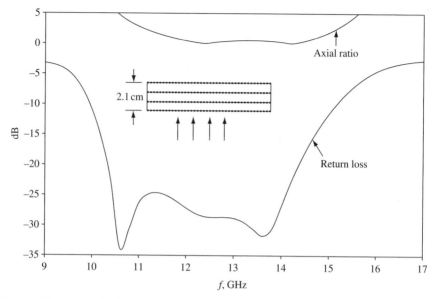

FIGURE 9.12 Axial ratio and return loss versus frequency of a four-layer polarizer: $h = 2.1$ cm, spacer $\varepsilon_r = 1.1$, $a = b = 1.27$ cm.

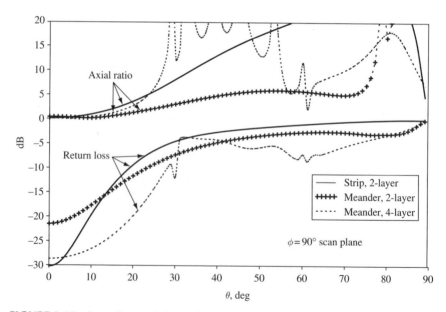

FIGURE 9.13 Scan characteristics of the polarizers considered in Figures 9.11 and 9.12.

The phase difference between two corresponding points on a line that are situated on two adjacent cells is given by

$$\Delta\varphi = k_0\sqrt{\varepsilon_r}(b+2p) \qquad (9.25)$$

where $b + 2p$ is the meander length, L, in a unit cell. The effective propagation constant of the fundamental mode along y is thus

$$\beta_y = \frac{\Delta\varphi}{b} = \frac{k_0\sqrt{\varepsilon_r}(b+2p)}{b} \qquad (9.26)$$

According to the Floquet theorem, the propagation constants of other guided modes are

$$\beta_{yn} = \frac{k_0\sqrt{\varepsilon_r}(b+2p)}{b} + \frac{2n\pi}{b} \qquad n = \pm1, \pm2, \pm3, \ldots \qquad (9.27)$$

At resonance the following condition must be satisfied:

$$\beta_{yn} = k_0 \sin\theta \qquad (9.28)$$

Setting $n = -1$ we obtain

$$|\sin\theta| = \frac{|\lambda_0 - (b+2p)\sqrt{\varepsilon_r}|}{b} \qquad (9.29)$$

Substituting $b = 0.7$cm, $p = 0.1$cm, $\lambda_0 = 1.5$cm, and $\varepsilon_r = 1.1$, we obtain $|\theta| = 52.6°$. However, the resonance is observed near the 80° scan angle. This difference is primarily due to approximation in the transmission line length along the meander conductor. The actual line length should be shorter than $b+2p$ (should be between b and $b+2p$) because of four 90° bends along the line. However (9.29) yields an empirical relation between the meander line parameters and the guided wave resonance angle.

The dashed curves in Figure 9.13 correspond to a four-layer meander line polarizer. The first resonance occurs near the 30° scan angle for 12.5 GHz frequency. Several guided modes are supported by the four-conductor transmission line structure, made by four meander line screens. The meander line parameters differ from layer to layer. To estimate the scan angle for resonance we substitute the average value of p in (9.29). The estimated scan angle is about 18°, which is much lower than the observed scan angle (about 30°). As before, this discrepancy can be attributed to the approximation in the transmission line length. It is found numerically that by increasing the values of p in four layers, the resonant angle move closer to the bore sight, which is consistent with (9.29). Further, the dielectric width is very insensitive to the resonant angle. These observations confirm that the guided mode essentially is the quasi-TEM mode supported by the meander conductors. For the stripline polarizer, no such resonance occurs for the 90° scan plane, because for a plane wave incidence, no higher order guided mode exists besides the fundamental mode [that is, n cannot take any value except zero in (9.27)]. From (9.27) and (9.28) we see that $n = 0$ yields the imaginary scan angle for resonance.

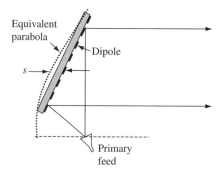

FIGURE 9.14 Schematic of a reflect array.

9.4 PRINTED REFLECT ARRAY

A printed reflect array can be considered as the "microstrip equivalent" of a conventional parabolic reflector antenna. It consists of an array of dipoles or patches of variable sizes printed on a thin substrate and backed by a ground plane, as shown in Figure 9.14. The array is space fed by a primary feed (typically a horn element). The array operates in the reflection mode, similar to a parabolic reflector. The feed is placed either at a symmetrical location, as in a center-fed reflector, or at an offset location, as in an offset-fed reflector. However, the offset arrangement shown in Figure 9.14 is often preferable to minimize the feed blockage.

The array elements act as passive phase shifters, which essentially cancel out the quadratic phase error of the incident fields emanating from the primary feed. This phase adjustment results in a planar phase distribution for the reflected fields; thus a high-gain secondary pattern results. In a parabolic reflector, the parabolic shape of the surface essentially does this necessary phase adjustment.

The required phase delay created by a patch element can be estimated from the distance between the patch and the equivalent parabola that yields a similar secondary pattern. The focus of the equivalent parabola is located at the feed's phase center and its axis is parallel to the secondary-beam direction. A patch element should create a phase delay that corresponds to twice the distance ($2s$) between the patch and the corresponding point on the parabolic surface. The above phase delay is realized by selecting appropriate dimensions of a printed element. Therefore, in order to design a reflect array it is crucial to understand the relationship between the reflected phase and physical dimensions of an element. This will be considered in the following section.

9.4.1 Phase Characteristics

Consider an infinite array of printed dipoles (or patches with narrow widths) of identical sizes backed by a ground plane, as shown in Figure 9.15. Suppose a plane wave is incident upon the infinite array. For a lossless situation, a 100% reflection of

FIGURE 9.15   Infinite array of identical dipoles printed on a dielectric-coated ground plane.

the RF field occurs from the surface, irrespective of the size of the dipole elements. The reflected fields may be directed in several directions, depending on the number of propagating Floquet modes. We assume a small cell size such that the structure does not support higher order Floquet modes besides the fundamental modes. The reflected Floquet modes propagate along the specular reflection direction. The phase of the reflected field with respect to the incident field can be computed from the overall GSM of the array structure. The overall GSM is determined from the GSM of the dipole and the reflection matrix of the dielectric-coated ground plane.

Figure 9.16 shows the phase variation of the reflected wave with respect to the dipole length. The polarization direction of the incident wave is assumed to be along the dipole length. The phase versus length of the dipole is nonlinear and the phase saturates after a certain length of the dipole. The nonlinearity can be improved by increasing the substrate thickness, but a thicker substrate has other limitation as we shall see shortly. The phase swing within the feasible limits of the dipole length

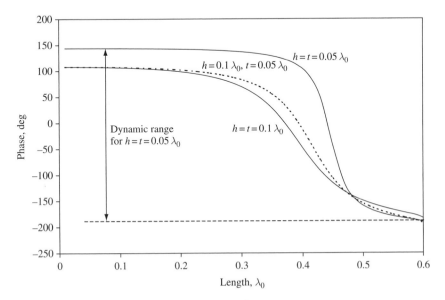

FIGURE 9.16   Phase of reflected wave versus dipole length for normal incidence: $a = 0.6\lambda_0$, $b = 0.3\lambda_0$, $\varepsilon_r = 1.10$.

is the available dynamic range of the phase. For a perfect operation the dynamic range should be at least 360° in order to implement the phase corrections necessary for producing a pencil or shaped beam. However, as can be seen, for a very thin substrate the dynamic range is near 360°, but for a thicker substrate it is significantly lower than 360°. Therefore, a reflect array with thin dielectric substrate exhibits good gain performance because of good dynamic range. However, the bandwidth is poor because of nonlinearity of the phase[2]. A thicker substrate yields somewhat lower gain but wider bandwidth.

A simple circuit model allows understanding the relation between the substrate thickness and the dynamic range. A dipole in an array environment is equivalent to a shunt susceptance in the transmission line equivalent circuit, as shown in Figure 9.17. The susceptance is positive and its magnitude increases with the dipole length until a resonance occurs[3]. After the resonance, the susceptance abruptly becomes negative, as shown in Figure 9.18. Suppose B is the equivalent normalized

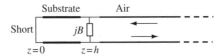

FIGURE 9.17   Equivalent transmission line model of a dipole array.

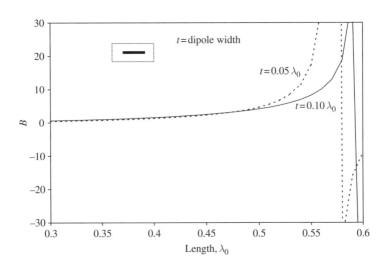

FIGURE 9.18   Dipole susceptance (Normalized) versus dipole length: cell size $0.6\lambda_0 \times 0.3\lambda_0$, rectangular grid.

---

[2] Another factor that limits the bandwidth is the phase variation of the incident field with frequency.
[3] A resonance may not occur for a thick dipole array or a patch array.

susceptance of the dipole. Then the equivalent admittance seen by the incident Floquet mode is

$$Y = jB - jY_{0s} \cot(\beta h) \qquad (9.30)$$

where β is the propagation constant of the incident Floquet mode inside the substrate, h is the substrate thickness, and Y_{0s} is the characteristic admittance (normalized with respect to the characteristic admittance of the incident Floquet mode in free space) of the Floquet mode in the dielectric substrate. The reflection coefficient is given by

$$\Gamma = \frac{1 - Y}{1 + Y} \qquad (9.31)$$

Substituting Y from (9.30) into (9.31) we obtain the phase of Γ as

$$\psi = 2 \tan^{-1}(Y_{0s} \cot \beta h - B) \qquad (9.32)$$

For $l = 0$ (l = dipole length), $B = 0$. Thus the phase angle becomes

$$\psi(l = 0) = 2 \tan^{-1}(Y_{0s} \cot \beta h) = 2 \tan^{-1}(\cot \beta' h) = \pi - 2\beta' h \qquad (9.33)$$

where the effective propagation constant β' takes care of the factor Y_{0s}. Notice, the π term is due to the 180° phase change of the reflected wave at the ground plane and $-2\beta' h$ term represents the phase lag due to the dielectric layer plus the reflection at the air–dielectric interfce.

As the dipole length increases, B increases in the positive direction, because the dipole becomes more capacitive. At an intermediate length, B becomes equal to $Y_{0s} \cot \beta h$, making $\psi = 0$ or 2π. We discard 2π and retain the zero value for ψ to maintain the continuity of ψ with l. Thus the phase lags as l increases. As we increase l further, B continues to increase, becoming very large just before the resonant length (see Figure 9.18). At this point, the value of ψ should be close to $-\pi$ as opposed to π to be compliant with the phase-lag trend. Immediately after passing the resonant length, the polarity of B changes (becomes inductive) while keeping the magnitude large. The value of ψ remains close to $-\pi$. For l close to the unit cell dimension, a, B remains inductive but with a smaller magnitude. The corresponding phase angle becomes

$$\psi(l = a-) = 2 \tan^{-1}(|B| + Y_{0s} \cot \beta h) = -\pi - 2\beta'' h \qquad (9.34)$$

where

$$\cot \beta'' h = |B| + Y_{0s} \cot \beta h \qquad (9.35)$$

The dynamic range is given by

$$|\Delta \psi| = |\psi(l = a-) - \psi(l = 0)| = 2\pi - 2(\beta' - \beta'')h \qquad (9.36)$$

Comparing (9.33) and (9.35) we see that $\beta'' < \beta'$ because of the additional term $|B|$ in (9.35). Thus the dynamic range is always less than 2π, and it decreases with the thickness of the substrate. The dynamic range also decreases with the dipole width t, because for $l = a-$, the magnitude of B in (9.35) increases with t. If the dipole is not too thin compared with b, then $|B|$ is large and β'' becomes negligibly small. For instance, with $a = 0.6\lambda_0$, $b = 0.3\lambda_0$, $h = 0.05\lambda_0$, and $t = 0.1\lambda_0$, B becomes -63 for $l \approx 0.6\lambda_0$, making $\beta'' \approx 0.05\beta'$. The magnitude of β'' reduces further for thicker dipoles and patch elements. For most practical cases, the dynamic range can be considered as

$$|\Delta\psi| \approx 2\pi - 2\beta' h \qquad (9.37)$$

The above result is consistent with the data plotted in Figure 9.16.

9.4.2 Design and Performance

In order to design a reflect-array antenna the following points need to be considered. The typical bandwidth of a reflect array is only a few percent. For a wider band design, the phase linearity with the dipole length must be improved. We observed in Figure 9.16 that the linearity can be improved by increasing the dielectric thickness. However, the phase dynamic range decreases with the dielectric thickness. A small dynamic range causes low directive gain of the secondary beam, because some of the intended phase delay cannot be realized. Therefore, for operating in a wider band, the directive gain should be compromised. Conversely, a high-gain reflect array may be possible to design using a thin dielectric substrate, but bandwidth will be narrow.

Another important point should be considered in the design process. For obtaining a pencil beam, the relative orientation of the reflect-array surface should be such that the desired beam direction aligns with the specular reflection direction. This will minimize the beam scanning with the frequency because in a plane reflector the linear part of the phase front of the reflected fields remains stationary only in the specular reflection direction. In another direction, this phase front changes its slope with frequency, which results in beam scanning and beam broadening. Also, from a cross-polarization point of view, it is preferable to use the specular reflection direction as the beam direction because the cross-polarization contributed by the feed will be minimum in this direction.

It should be mentioned that a practical reflect array is not infinite and the elements are not of identical size. Nevertheless, we use the phase characteristics associated with infinite arrays of identical elements. Because the dimensions of the reflect-array elements change gradually, the "locally periodic" assumption is reasonable.

A reflect array with printed dipoles is designed and the radiation pattern of the reflect array is computed. The dipoles are printed on a substrate of dielectric constant 1.1 and thickness $0.05\lambda_0$. The width of the printed dipoles is $0.05\lambda_0$. The unit cell size is $0.5\lambda_0 \times 0.6\lambda_0$ and the size of the reflect array is $12.5\lambda_0 \times 14.4\lambda_0$. Figure 9.19

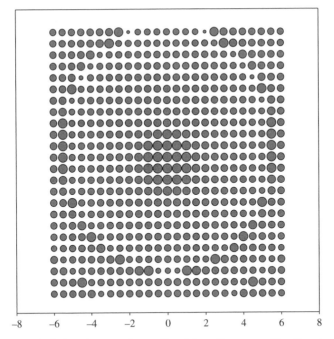

FIGURE 9.19 Dipole lengths at various cells of the reflect array (the diameter of a circle represents the dipole length at that location). The dimensions are in λ_0.

shows the length of the dipoles at various points of the reflect array (in the figure the diameter of a circle represents the dipole length at that location). Figure 9.20 shows a pattern cut through the beam peak along the vertical plane. A Gaussian feed with about -10 dB edge taper was considered. The computed gain is 31.3 dBi (neglecting the spill over loss) that occurs at $33.5°$ off the reflector axis. The 3-dB beamwidth is about $5°$. The aperture efficiency is about 58%. This somewhat low efficiency is due to (a) nonuniform illumination, (b) off-axis beam direction, and (c) limited phase dynamic range. The side-lobe level is below -20 dB, which is due to the amplitude taper of the feed.

Figure 9.21 shows the directive gain versus normalized frequency of a spot beam with substrate thickness as a parameter. As expected, the peak gain decreases with an increase in the substrate thickness as the phase dynamic range shrinks. The decrease in the peak gain with substrate thickness can be estimated from the available phase dynamic range given by (9.37). The shortage of dynamic range from full $360°$ is approximately equal to $2\beta' h$. This shortage will cause a maximum phase error of $\beta' h$ for some elements of the reflect array for which the desired phase lies outside the dynamic range. This phase error is not random in a strict sense and does not happen for all the elements. If we empirically assume that $\beta' h/2$ is the rms phase error, then the average gain of the reflect array can be estimated from

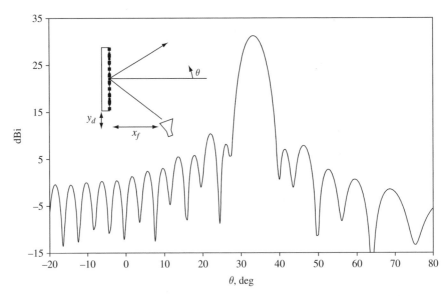

FIGURE 9.20 Pattern cut through the beam peak along the vertical plane of a reflect array: cell size $0.6\lambda_0 \times 0.5\lambda_0$, $h = t = 0.05\lambda_0$, $\varepsilon_r = 1.10$, 600 cells, 10-dB feed taper, $x_f = 12\lambda_0$, $y_d = 1\lambda_0$.

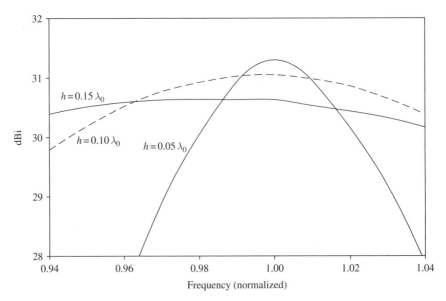

FIGURE 9.21 Directive gain of a reflect array versus normalized frequency of a spot beam with substrate thickness, h, as a parameter: cell size $0.6\lambda_0 \times 0.5\lambda_0$, dipole width $0.05\lambda_0$, $\varepsilon_r = 1.1$.

Ruze's equation [5] as

$$G = G_0 \left[1 - \left(\frac{\beta' h}{2} \right)^2 \right] \qquad (9.38)$$

where G_0 is the gain of the reflect array if no phase error exists. Using the above formula, we obtain the gain loss as 0.12, 0.47, and 1.06 dB for $h = 0.05\lambda_0$, $0.1\lambda_0$, and $0.15\lambda_0$, respectively. The computed gain losses were 0.1, 0.34, and 0.80 dB, respectively. This difference is primarily due to the crude approximation in the rms phase error. However, for a small h, the formula yields an estimation of gain loss from an ideal situation.

In Figure 9.21 we observe that the bandwith improves with the dielectric thickness because the linearity of the phase for the printed elements improves, as depicted in Figure 9.16. For $h = 0.05\lambda_0$, the 1-dB-gain bandwidth is about 4%. The above 1-dB bandwidth improves to about 8% for $h = 0.1\lambda_0$.

9.4.3 Circular Polarization

For circularly polarized radiation, two different types of configurations are possible. In a type 1 configuration the phase can be realized by varying the size of the printed elements. Because a circular polarization consists of both vertical and horizontal polarizations, the length and the width of the elements should be varied simultaneously to realize the desired phase distribution for both polarizations. Thus, thin dipoles cannot be used for this purpose, but patch elements are suitable. In the type 1 configuration (see Figure 9.22a), both the vertical and horizontal polarization components undergo equal amount of phase shifts. Therefore, the secondary pattern will have opposite sense of polarization rotation compared to the feed polarization because of reflection. This means that a type 1 configuration fed with an lcp feed will produce an rcp secondary pattern and vice versa as that happens for a typical parabolic reflector antenna.

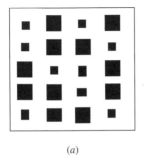 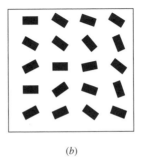

(a)   (b)

FIGURE 9.22 Schematic of circular polarization reflect arrays: (a) type 1, Phase shift by size variation of square patch elements; (b) type 2, phase shift by angular rotation of rectangular patch elements.

In a type 2 configuration, the desired phase distribution is realized by rotating the printed patches [6] instead of changing their dimensions. In this case the patch elements have almost identical dimensions (may vary slightly to accommodate incident angle variations) with different angular orientations, as shown in Figure 9.22b. This configuration requires special patch elements that could provide 180° phase difference between the reflection coefficients of the TE and TM modes. It will be shown that such an element yields 2α additional phase shift to the reflected signal for a physical rotation of α (thus a 180° physical roation yields 360° phase shift, which is intuitively correct, because a 180° physical rotation means no rotation in regard to the reflected phase angle). Furthermore, the sense of polarization of the secondary pattern becomes identical with that of the feed pattern. That is, a type 2 configuration radiates an lcp secondary pattern if the feed radiation is lcp, for instance.

In order to prove the above phase-shift concept by rotation, consider a rectangular patch element that has 180° phase difference between the TE and TM mode reflection coefficients. We assume that the TE mode is x-polarized and the TM mode is y-polarized. Such a patch element can be designed by adjusting its length and width. Suppose an lcp signal is incident upon the patch element. The incident field can be expressed as

$$\vec{E}^{\text{inc}} = A[\hat{x} + j\hat{y}] \tag{9.39}$$

Suppose the patch is oriented at an angle α with the x-axis as shown in Figure 9.23. Then the field components along the x'- and y'-axes of the patch can be written as

$$E_{x'}^{\text{inc}} = A(\cos\alpha + j\sin\alpha) = A\exp(j\alpha)$$
$$E_{y'}^{\text{inc}} = A(-\sin\alpha + j\cos\alpha) = jA\exp(j\alpha) \tag{9.40}$$

Assuming $\Gamma_{x'y'} = \Gamma_{y'x'} = 0$, we obtain reflected field components as

$$E_{x'}^{\text{ref}} = \Gamma_{x'x'}E_{x'}^{\text{inc}} = \Gamma_{x'x'}A\exp(j\alpha)$$
$$E_{y'}^{\text{ref}} = \Gamma_{y'y'}E_{y'}^{\text{inc}} = j\Gamma_{y'y'}A\exp(j\alpha) \tag{9.41}$$

In the above Γ's represent reflection coefficients. The reflected electric field vector with respect to the xy reference system thus becomes

$$\vec{E}^{\text{ref}} = \hat{x}[E_{x'}^{\text{ref}}\cos\alpha - E_{y'}^{\text{ref}}\sin\alpha] + \hat{y}[E_{x'}^{\text{ref}}\sin\alpha + E_{y'}^{\text{ref}}\cos\alpha] \tag{9.42}$$

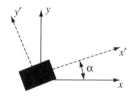

FIGURE 9.23 Patch orientation with the reference axis.

Substituting the expressions for $E_{x'}^{\text{ref}}$ and $E_{y'}^{\text{ref}}$ from (9.41) into (9.42) we obtain

$$\vec{E}^{\text{ref}} = \Gamma_{x'x'} A (\hat{x} - j\hat{y}) \exp(2j\alpha) \qquad (9.43)$$

In the above derivation we substitute $\Gamma_{y'y'} = -\Gamma_{x'x'}$, because 180° phase difference between two reflection coefficients is assumed.

From (9.43) we see that the reflected field changes the sign of the y-component. Because the reflected field propagtes along the $-z$-direction, the reflected signal remains lcp. Furthermore, the additional factor (besides the $\Gamma_{x'x'}$ term) $\exp(2j\alpha)$ is responsible for the physical rotation of the patch. This proves that the phase of the reflected field is advanced by 2α if the patch is rotated by α. If an rcp signal is considered, then for a physical rotation of α the corresponding phase shift becomes -2α. Because two different phase shifts occur for two polarizations (lcp and rcp), a dual circular polarization reflect array cannot be designed using type 2 configuration (which is possible with type 1 configuration, however). For a square patch $\Gamma_{y'y'} = \Gamma_{x'x'}$; thus no such phase shift can be achieved by a physical rotation.

A rectangular patch element or circular patch element with a notch can be designed to have 180° phase difference between the TE and TM mode reflection coefficients. Figure 9.24 shows the phase versus frequency of a rectangular patch that can be used for a type 2 configuration. Notice that unlike the type 1 configuration, the type 2 configuration is *not limited* by the phase dynamic range, because the full 360° phase can be realized by physical rotation of the elements. Therefore, there is no loss of the peak gain due to the phase dynamic range. However, the

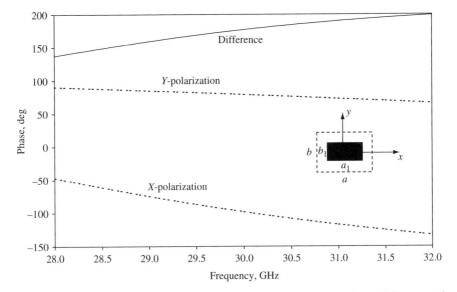

FIGURE 9.24 Phase versus frequency of a rectangular patch that can be used for a type 2 configuration: $a = b = 0.6\lambda_0$, $a_1 = 0.41\lambda_0$, $b_1 = 0.24\lambda_0$, $h = 0.1\lambda_0$, $\varepsilon_r = 1.1$, $\lambda_0 = 1$ cm.

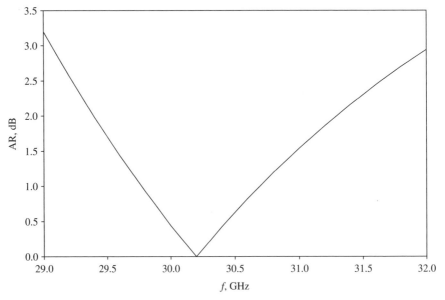

FIGURE 9.25   Axial ratio versus frequency. Dimensions are the same as in Figure 9.24.

180° phase difference cannot be maintained over a large bandwidth; thus the axial ratio deteriorates due to the phase error. Figure 9.25 shows the axial ratio versus frequency introduced by a patch element. An ideal feed with no cross-polarization is assumed. The axial ratio is calculated using the formula

$$AR(dB) = 20\log\left\{\tan\left(\frac{\pi}{4}+|\frac{\varphi}{2}|\right)\right\}$$ (9.44)

where the phase error is

$$\varphi = |\angle\Gamma_{xx} - \angle\Gamma_{yy}| - 180$$ (9.45)

From Figure 9.25 we notice that the axial ratio of the secondary pattern decides the bandwidth of the type 2 configuration. On the other hand, the phase linearity and the phase dynamic range limit the bandwidth of the type 1 configuration.

9.4.4   Bandwidth Enhancement

We have seen that the gain and bandwidth of a reflect array are limited by two major factors: (a) smaller than 360° dynamic range and (b) a nonlinear relationship between the phase and the size of the printed elements. It is proven in Section 9.4.1 that for a one-layer structure the dynamic range is always less than 360°. For a two-layer printed structure, however, the dynamic range can be increased beyond 360°. It has been shown that in a two-layer square patch element the phase linearity can be improved significantly if the patch lengths are in a given ratio [7]. Figure 9.26

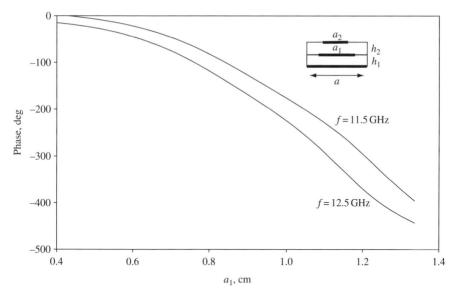

FIGURE 9.26 Phase versus patch length of a two-layer patch array: $a = b = 1.4$ cm, $b_1 = a_1$, $b_2 = a_2$, $a_2 = 0.7a_1$, $h_1 = h_2 = 0.3$ cm, $\varepsilon_{r1} = \varepsilon_{r2} = 1.05$.

shows the phase versus length of a two-layer patch array. The upper layer patch length is 70% shorter than that of the lower layer patch in this case. The dynamic range is over 400° and the phase is fairly linear with the length of the patch. Over 16% bandwidth was achieved for this two-layer reflect array with respect to the 1.5-dB-gain variation.

Another method of improving the phase linearity is using a slot-coupled patch reflect array [8]. In this configuration the patch dimensions remain fixed, but the coupled stripline (or microstripline) dimensions are varied. Figure 9.27 shows the phase variation with the stripline length. The patch and slot dimensions should be designed to have a good match with the stripline that is hypothetically match loaded at both ends. For the reflect-array mode, the strip ends are left open. As a result the reflected phase varies almost linearly with the length of the strip. One drawback of this structure (for the microstripline version) is that the slots will radiate in the back-lobe region, causing a gain loss for the secondary patterns. This can be alleviated by placing another ground plane as shown in the inset of Figure 9.27. In that case the feedline will operate in the asymmetric stripline mode.

9.4.5 Contour-Beam Reflect Array

Reflect-array structures can be designed to produce contour or shaped beams. The design procedure involves "phase-only" optimization for the desired beam shape (we will discuss important optimization algorithms in Chapter 11). The gain and bandwidth performances of a shaped beam depend on the dynamic range of the

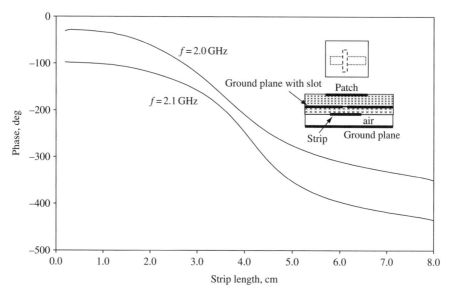

FIGURE 9.27   Phase variation with the stripline length of a slot-coupled patch: $a = b = 8$ cm, $h_a = 0.5$ cm, $h_f = 0.318$ cm, $a_1 = b_1 = 5.8$ cm, strip width 0.82 cm, $\varepsilon_{ra} = 1.1$, $\varepsilon_{rf} = 2.56$, slot 2.9 cm × 0.5 cm.

phase and the phase linearity, as in the case of a pencil beam. The performance of a shaped-beam reflect array with three-layer patch elements with about 10% bandwidth is reported [9].

REFERENCES

[1] R. Mittra, C. H. Chan and T. Cwik, "Techniques for Analyzing Frequency Selective Surfaces—A Review," *Proc. IEEE*, Vol. 76, No. 12, pp. 1593–1615, Dec. 1988.

[2] A. C. Ludwig, "The Definition of Cross Polarization," *IEEE Trans. Antennas Propagat.*, Vol. AP-21, pp. 116–119, Jan. 1973.

[3] Y. Rahmat-Samii and A. N. Tulintseff, "Diffraction Analysis of Frequency Selective Reflector Antennas," *IEEE Trans. Antennas Propagat.*, Vol. AP-41, No. 4, pp. 476–487, Apr. 1993.

[4] A. K. Bhattacharyya and T. J. Chwalek, "Analysis of Multilayered Meander Line Polarizer," *Int. J. Microwave Millimeter-Wave CAE*, Vol. 7, No. 6, pp. 442–454, July 1997.

[5] J. Ruze, "Antenna Tolerance Theory—A Review," *Proc. IEEE*, Vol. 54, pp. 633–640, Apr. 1966.

[6] J. Huang and R. J. Pogorzelski, "A Ka-Band Microstrip Reflectarray with Elements Having Variable Rotation Angles," *IEEE Trans. Antennas Propagat.*, Vol. AP-46, No. 5, pp. 650–656, May 1998.

[7] J. A. Encinar, "Design of Two Layer Printed Reflectarrays Using Patches of Variable Size," *IEEE. Trans. Antennas Propagat.*, Vol. AP-49, No. 10, pp. 1403–1410, Oct. 2001.

[8] A. K. Bhattacharyya, "Multi-Layered Patch Antenna," U.S. Patents Number 5,990,836, Nov. 23, 1999.

[9] J. A. Encinar and J. A. Zornoza," Three-Layer Printed Reflectarrays for Contoured Beam Space Applications," *IEEE Trans. Antennas Propagat.*, Vol. AP-52, No. 5, pp. 1138–1148, May 2004.

BIBLIOGRAPHY

Bialkowski, M. E., A.W. Robinson, and H. J. Song, "Design, Development, and Testing of X-Band Amplifying Reflectarrays," *IEEE Trans. Antennas Propagat.*, Vol. AP-50, No. 8, pp. 1065–1076, Aug. 2002.

Chu, R. S., and K. M. Lee, "Analytical Model of a Multilayered Meander Line Polarizer Plate with Normal and Oblique Plane-Wave Incidence," *IEEE Trans. Antennas Propagat.*, Vol. AP-35, pp. 652–661, June 1987.

Han, C., C. Rodenbeck, J. Huang, and K. Chang, "A C/Ka Dual Frequency Dual Layer Circularly Polarized Reflectarray Antenna with Microstrip Ring Elements," *IEEE Trans. Antennas Propagat.*, Vol. AP-52, No. 11, pp. 2871–2876, Nov. 2004.

Javor, R. D., X-D., Wu and K. Chang, "Design and Performance of a Microstrip Reflectarray Antenna," *IEEE Trans. Antennas Propagat.*, Vol. AP-43, No. 9, pp. 932–939, Sept. 1995.

Ko, W. L., and R. Mittra, "Scattering by a Truncated Periodic Array," *IEEE Trans. Antennas Propagat.*, Vol. AP-36, No. 4, pp. 496–503, Apr. 1988.

Martynyuk, A. E., J. I. M. Lopez, and N. A. Martynyuk, "Spiraphase-Type Reflectarrays Based on Loaded Ring Slot Resonators," *IEEE Trans. Antennas Propagat.*, Vol. AP-52, No. 1, pp. 142–153, Jan. 2004.

Munk, B. A., R. G. Kouyoumjian, and L. Peters, Jr., "Reflection Properties of Periodic Surfaces of Loaded Dipoles," *IEEE Trans. Antennas Propagat.*, Vol. AP-19, pp. 612–617, Sept. 1971.

Pozar, D. M., S. D. Targonski, and R. Pokuls, "A Shaped-Beam Microstrip Patch Reflectarray," *IEEE Trans. Antennas Propagat.*, Vol. AP-47, No. 7, pp. 1167–1173, July 1999.

Terret, C., J. R. Levrel, and K. Mahdjoubi, "Susceptance Computation of a Meander Line Polarizer Layer," *IEEE Trans. Antennas Propagat.*, Vol. AP-32, pp. 1007–1011, Sept. 1984.

Wu, T. K. (Ed.), *Frequency Selective Surface and Grid Array*, Wiley, New York, 1995.

Wu, T-K., and S-W. Lee, "Multiband Frequency Selective Surface with Multiring Patch Elements," *IEEE Trans. Antennas Propagat.*, Vol. AP-42, No. 11, pp. 1484–1490, Nov. 1994.

Young, L., L. A. Robinson, and C. A. Hacking, "Meander-Line Polarizer," *IEEE Trans. Antennas Propagat.*, Vol. AP-21, pp. 376–379, May 1973.

PROBLEMS

9.1 The transmission coefficients of a two-layer ring FSS for a scan angle $\theta = 60°$, $\phi = 45°$ are as follows: $T(\text{TE, TE}) = -0.28 - 0.76j$, $T(\text{TM, TM}) = -0.99j$, $T(\text{TE, TM}) = T(\text{TM, TE}) = 0$. Determine the copolar and cross-polar transmitted fields if a lcp plane wave represented by $\vec{E}_{\text{inc}} = \hat{\theta} + j\hat{\phi}$ is incident upon the FSS at $\theta = 60°$ and $\phi = 45°$. Determine the axial ratio of the transmitted field and the transmission loss of the FSS at this scan angle.

9.2 The far-field radiation pattern of a horn antenna is given by $\vec{E} = [\hat{\theta} f(\theta, \phi) + \hat{\phi} g(\theta, \phi)] \exp(-jk_0 r)/r$. The horn is loaded with an FSS at a far-field distance. The horn axis is perpendicular to the FSS surface as in Figure 9.7. The transmission coefficients of the FSS are given by $T(\mathrm{TE}, \mathrm{TE}) = T_1(\theta, \phi)$, $T(\mathrm{TM}, \mathrm{TM}) = T_2(\theta, \phi)$, $T(\mathrm{TE}, \mathrm{TM}) = T(\mathrm{TM}, \mathrm{TE}) = 0$. Show that the transmitted field by the infinite FSS is given by

$$\vec{E}_{\mathrm{fss}} = [\hat{\theta} f(\theta, \phi) T_2(\theta, \phi) + \hat{\phi} g(\theta, \phi) T_1(\theta, \phi)] \exp(-jk_0 r)/r.$$

Then obtain the differential electric field on a finite rectangular FSS of dimensions $A \times B$ and formulate the far field of the horn–FSS combination as the superposition of the horn radiation and radiation by the equivalent magnetic current obtained from the differential field on the finite FSS surface.

9.3 The normalized susceptances of a meander line screen are 0.094 and −1.67 for x- and y-polarizations, respectively. A polarizer is made with two such screens separated by a spacer of width $0.34\lambda_0$ and the spacer material has the dielectric constant of unity. Using the loaded transmission line model shown in Figure P9.3, obtain the axial ratio of the transmitted field if the incident field is linearly polarized with polarization angle 45° with the x–axis.

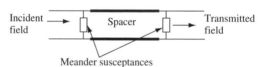

FIGURE P9.3 Transmission line model of a two-layer polarizer.

9.4 A two-layer polarizer is designed with meander line screens having triangular wave patterns (zigzag lines as opposed to square wave lines) as shown in Figure P9.4. Estimate the blind angle location if the scan plane aligns with the meander line axis. The dielectric constant of the spacer is 1.2. The periodicity along the meander axis is $0.4\lambda_0$. The height of the triangle is $0.2\lambda_0$.

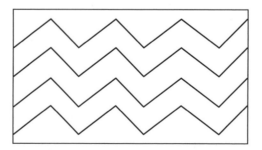

FIGURE P9.4 Polarizer screen with zigzag lines.

9.5 For a reflect array of type 2 configuration, deduce an expression for the cross-polarization amplitude if $\angle\Gamma_{y'y'} - \angle\Gamma_{x'x'} = 180 + \varphi$. Assume that the incident field is lcp. Also show that if $\varphi = 180°$, then for the circular polarization operation the concept "phase shift by rotation" becomes invalid.

9.6 Show that for an rcp incident field the phase shift is -2α if the physical rotation is α. Assume $\angle\Gamma_{y'y'} - \angle\Gamma_{x'x'} = 180$.

CHAPTER TEN

Multilayer Array Analysis with Different Periodicities and Cell Orientations

10.1 INTRODUCTION

Thus far we have analyzed multilayer structures having equal cell size and identical cell orientations for the layers; hence the layers have identical Floquet modal functions. However, this is not the case if the layers have different cell sizes and/or different cell orientations.[1] A typical example of this kind of structure is a patch array antenna loaded with an FSS. The cell size of the radiating patch layer is generally determined by the maximum scan angle. On the other hand, the cell size of the FSS layer is determined independently by the passband and stop-band specifications. This may result in two different cell sizes for the two layers. A more complex multilayer array structure is a patch array loaded with a polarizing screen. For this structure, the patch layer and the polarizer layer have different cell sizes and the cell orientations are nonparallel between the layers (typically, patch rows and the polarizer grids are at a 45° angle).

In this chapter, we present a methodology for analyzing general multilayer array structures [1]. The array layers may have different lattice structures with different periodicities and axis orientations. In our analysis, we first obtain the global unit cell. The global unit cell structure is dependent upon the cell structures and the relative array axes orientations of the individual layers. We then find the GSMs

---

[1] Actually, a Floquet modal function changes its identity (modal indices) from one layer to another layer if the grid structures are not identical.

Phased Array Antennas.   By Arun K. Bhattacharyya
© 2006 John Wiley & Sons, Inc.

of the individual layers with respect to the global unit cell. Construction of the "global GSM" for each individual layer is necessary in order to validate the GSM cascading rule for analyzing multiple layers. It is found that a set of local GSMs (GSMs with local unit cell and local coordinate system of an individual layer) can be used to construct the global GSM for that layer. This is possible because a Floquet mode associated with a given lattice structure can represent a Floquet mode associated with a different lattice structure. The mode numbers (modal indices for two-dimensional periodic structures), however, will be different for the two different lattice structures. A mapping relation between the modes associated with a local cell and the global cell allows constructing the global GSM for the layer using the elements of local GSMs. We establish such mapping relations for different cases with increasing complexity. Once we determine the global GSMs for all the individual layers, we can obtain the overall GSM of the structure applying the GSM cascading formulas derived in Chapter 6.

To illustrate the methodology, we consider two different examples for numerical analysis. The first example is a two-layer patch array with different periodicities of the layers. The lower layer patches are excited by probe feeds. The input impedances seen by the driven patch elements are determined. The second example is a patch array loaded with a two-layer polarizing screen. We present detailed electrical characteristics including input match, axial ratio bandwidth, and scan performance.

10.2 LAYERS WITH DIFFERENT PERIODICITIES: RECTANGULAR LATTICE

Consider a multilayer array of printed elements as shown in Figure 10.1, where the layers have different periodicities. The row elements and column elements for the layers are parallel to the x- and y-directions, respectively. Suppose $a \times b$ is the overall cell size (we will refer to it as the global cell) of the structure, where a and b simply are the lowest common multiple (lcm) of the different periods along x- and y-directions, respectively. Thus, a global cell will accommodate a finite number of local cells of any layer of the multilayer structure. The number of cells will differ, in general, from layer to layer. The Floquet mode supported by the global cell is given by the following vector modal function:

$$\vec{\psi}_{pq}(x, y, z) = \vec{A}_{pq} \exp(-jk_{xp}x - jk_{yq}y)\{\exp(-jk_{zpq}z) + \Gamma \exp(jk_{zpq}z)\} \quad (10.1)$$

where $\vec{A}_{pq}$ is a constant vector. Expressions for $\vec{A}_{pq}$ with respect to the TE_z and TM_z fields are given in Chapter 3. For a rectangular lattice, the wave numbers are given by

$$k_{xp} = \frac{2p\pi + \phi_x}{a} \qquad k_{yq} = \frac{2q\pi + \phi_y}{b} \qquad k_{zpq} = \sqrt{k_1^2 - k_{xp}^2 - k_{yq}^2} \quad (10.2)$$

where k_1 is the propagation constant in a given medium and ϕ_x and ϕ_y are the phase differences per unit cell dimension along x- and y-directions, respectively. We will use the symbols $\{\phi_x, \phi_y\}$ to represent the above *differential phase pair*.

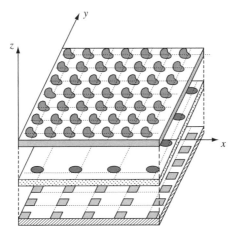

FIGURE 10.1 Three-layer printed array structure with different periodicities and parallel array axis orientations.

Suppose the mode in (10.1) is incident upon the ith patch layer having the local cell size $a_i \times b_i$. The incident phase differences between two adjacent local elements of the ith layer along x- and y-directions would be respectively

$$\phi_{xi} = \frac{(2p\pi + \phi_x)a_i}{a} \qquad \phi_{yi} = \frac{(2q\pi + \phi_y)b_i}{b} \qquad (10.3)$$

The reflected and transmitted modes supported by the ith layer due to the incident field given in (10.1) can be represented by the following local modal function:

$$\vec{\psi}_{mn}^{(i)}(x, y) = \vec{A}_{mn}^{(i)} \exp(-jk_{xm}^{(i)}x - jk_{yn}^{(i)}y) \qquad (10.4)$$

The z-dependent factor is suppressed, which is unimportant with respect to the present context. The wave numbers in (10.4) are given by

$$k_{xm}^{(i)} = \frac{2m\pi + \phi_{xi}}{a_i} \qquad k_{yn}^{(i)} = \frac{2n\pi + \phi_{yi}}{b_i} \qquad (10.5)$$

where m and n are the local mode index pair. Henceforth, a Floquet mode with m and n as mode index pair will be denoted as the (m, n) mode. Substituting ϕ_{xi} and ϕ_{yi} from (10.3) into (10.5) we have the following expressions for the wave numbers associated with the reflected and transmitted fields from the ith layer:

$$k_{xm}^{(i)} = \frac{2\pi(m\alpha + p) + \phi_x}{a} \qquad k_{yn}^{(i)} = \frac{2\pi(n\beta + q) + \phi_y}{b} \qquad (10.6)$$

with

$$a = \alpha a_i \qquad b = \beta b_i \qquad (10.7)$$

where α and β are integers.

From (10.6) we see that if the global (p, q) mode is incident upon the ith layer, then the elements of the ith layer will produce the $(p+m\alpha, q+n\beta)$ global modes, where m and n both run from $-\infty$ to ∞. Interestingly, for a single global mode incidence, the reflected and transmitted global mode indices appear only at discrete intervals. These intervals are α and β for x and y indices, respectively. Moreover, the $(p+i\alpha, q+j\beta)$ global incident mode (i, j integers) will create the same set of reflected and transmitted modes. The amplitudes of the modes could be different, however. Furthermore, the intermediate reflected and transmitted modes will be created if the global incident modes have modal indices like $(p+i, q+j)$, where $i < \alpha, j < \beta$. The next step is to establish a one-to-one mapping between the global modal indices and the local modal indices of the ith layer. This mapping relation allows representing the local GSM of the ith layer with respect to the global modal index system. This will facilitate applying the multimode cascading rules to analyze a multilayer structure.

From (10.5) and (10.6) we observe that the $(p+m\alpha, q+n\beta)$ mode with respect to the global cell is equivalent to the (m, n) mode with respect to the ith local cell. The differential phase pair for the above cells are $\{\phi_x, \phi_y\}$ and $\{(2p\pi + \phi_x)/\alpha, (2q\pi + \phi_y)/\beta\}$, respectively. Therefore, setting different values of p from zero to $\alpha - 1$ and q from zero to $\beta - 1$, one can cover the entire "global modal space" using local modes. In order to fill in the entire global modal space one needs to compute the local GSMs of the ith layer with the following set of differential phase pairs:

$$\{\phi_{xi}, \phi_{yi}\} = \left\{\frac{\phi_x}{\alpha}, \frac{\phi_y}{\beta}\right\}, \left\{\frac{2\pi + \phi_x}{\alpha}, \frac{\phi_y}{\beta}\right\}, \left\{\frac{\phi_x}{\alpha}, \frac{2\pi + \phi_y}{\beta}\right\}, \dots,$$
$$\left\{\frac{2(\alpha - 1)\pi + \phi_x}{\alpha}, \frac{2(\beta - 1)\pi + \phi_y}{\beta}\right\}$$

The corresponding global mode index pairs are

$$(j, k) = (m\alpha, n\beta), (m\alpha + 1, n\beta), (m\alpha, n\beta + 1), \dots, ((m+1)\alpha - 1, (n+1)\beta - 1)$$

where m and n both run from $-\infty$ to ∞. Note that to construct the global GSM for the ith layer, the local GSMs should be computed $\alpha\beta$ number of times. The following simple steps should be followed in order to analyze a multilayer structure:

Step 1: Find the global cell size $a \times b$ of the structure, which basically is the lcm of individual layers' cell dimensions.

Step 2: Find the global differential phase pair $\{\phi_x, \phi_y\}$ from the input excitations or scan direction, for example.

Step 3: Consider the ith layer and find the local cell size $a_i \times b_i$. Define $\alpha = a/a_i$ and $\beta = b/b_i$. Select two integers p and q where $0 \le p \le \alpha - 1$ and

$0 \leq q \leq \beta - 1$. Set the differential phase pair $\{(2p\pi + \phi_x)/\alpha, (2q\pi + \phi_y)/\beta\}$. Compute the corresponding local GSM. Rename the local GSM element indices (m, n) globally as $(p + m\alpha, q + n\beta)$. Repeat this for other sets of p and q. This involves $\alpha\beta$ times local GSM computations associated with $\alpha\beta$ differential phase pairs.

Step 4: Repeat step 3 for other layers in the structure. Then combine the individual layer GSMs to find the overall GSM of the structure.

It should be mentioned that in order to increase the computational efficiency and numerical accuracy, the total number of global Floquet modes must be decided first. This number depends on the element spacings in wavelength and the separation between the adjacent layers. To comply with the total-number-of-global-mode requirement, the number of local modes in step 3 may be considered approximately as $1/\alpha\beta$ times the total number of global modes. Therefore, although the GSMs of an individual layer are computed $\alpha\beta$ times, the dimensions of the local GSMs can be reduced by the same factor.

10.2.1   Patch-Fed Patch Subarray

A patch-fed patch subarray [2] is considered as an example of a multilayer (two layers in this case) array structure with different periodicities. Figure 10.2*a* shows the geometry of a unit cell of the structure. The array consists of two patch array layers backed by a ground plane. Two dielectric layers separate the patch layers and the ground plane. Probes excite the lower layer patches. Each lower layer patch

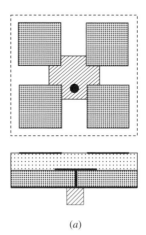

(*a*)

FIGURE 10.2 (*a*) Two-layer patch-fed patch subarray. (*b*) Input impedance–frequency plot of the subarray. Subarray *E*-plane dimension 7.2 cm, *H*-plane 6.4 cm, upper layer patches 2.0 cm × 1.6 cm, lower layer patch 2.13 cm × 2.0 cm. Substrate thickness 0.16 cm for each layer, dielectric constant for both layers 2.55. Probe located at 0.8 cm from the center along *E*-plane.

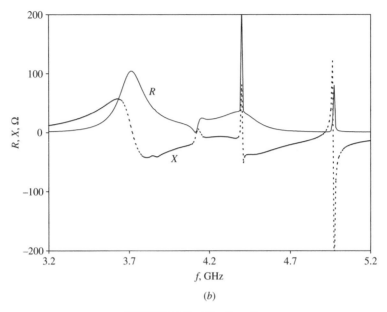

FIGURE 10.2 (Continued).

electromagnetically couples four upper layer patches. The driven patch in the lower layer and four coupled patches in the upper layer make a subarray of an infinite array. The lower layer patches have the cell size $a \times b$ while the upper layer patches have the cell size $a/2 \times b/2$. Both layers have rectangular lattice structures. In a global unit cell, the driven patch is symmetrically located below four upper layer patches. This arrangement reduces the number of feed points by a factor of 4; as a result the complexity of the beam forming circuitry is reduced significantly.

To analyze the infinite array, we first compute the local GSMs of the upper layer and then construct the global GSM of the layer following the steps prescribed before. For the upper layer, we compute four local GSMs with differential phase pairs $\{\phi_x/2, \phi_y/2\}$, $\{\phi_x/2 + \pi, \phi_y/2\}$, $\{\phi_x/2, \phi_y/2 + \pi\}$, $\{\phi_x/2 + \pi, \phi_y/2 + \pi\}$, respectively, where $\phi_x = k_0 a \sin\theta \cos\phi$, $\phi_y = k_0 b \sin\theta \sin\phi$, (θ, ϕ) being the scan direction in the spherical coordinate system. We consider $a/2 \times b/2$ as the local cell size for the upper layer. The local mode indices are then converted to global mode indices. Because the cell centers between the two layers are misaligned, an appropriate phase factor is introduced for each element of the global GSM. To compute the GSM of the lower layer, we use $\{\phi_x, \phi_y\}$ as the differential phase pair and $a \times b$ as the cell size. The two global GSMs are combined to obtain the overall GSM of the patch layers. To determine the input impedance seen by the probe feed, the overall GSM is converted to the $[Z]$ matrix. The $[Z]$ matrix of the probe feed is obtained separately using the reaction theorem and Floquet modal analysis as detailed in Chapter 7. Combining the $[Z]$ matrices, the input impedance seen by the probe is obtained.

Figure 10.2*b* shows the variation of input impedance with frequency. About 120 Floquet modes were used to obtain a reasonable convergence. High-Q resonances are observed for the infinite array. The high-Q resonances near 4.4 and 5 GHz are due to the parallel-plate modes that couple with higher order Floquet modes. The uncharacteristic behavior of the array impedance near 4.16 GHz is due to the appearance of the first grating lobe. The general shapes of the impedance curves are similar to that of an isolated subarray published by Legay and Shafai [2] (see also Figure 4 of [1]).

10.3 NONPARALLEL CELL ORIENTATIONS: RECTANGULAR LATTICE

The foregoing section assumed rectangular lattice structures for all layers and parallel cell orientations between the layers. This section considers a more general case; the layers have rectangular lattice structures but the cell orientations are nonparallel, as shown in Figure 10.3*a*.

As before, we assume that the global cell size of the structure is $a \times b$ and the local cell size for the *i*th layer is $a_i \times b_i$. We assume that the global cell is also rectangular. The x'-axis of the *i*th layer is inclined at an angle ψ with the global x-axis of the structure. The local modal function of the *i*th layer can be expressed as

$$\zeta(x, y) = B_{mn} \exp(-jk'_{xm}x' - jk'_{yn}y') \tag{10.8}$$

with

$$k'_{xm} = \frac{2m\pi + \phi'_{xi}}{a_i} \qquad k'_{yn} = \frac{2n\pi + \phi'_{yi}}{b_i} \tag{10.9}$$

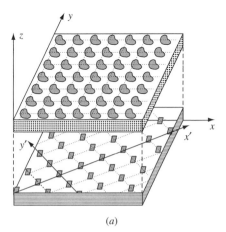

(a)

FIGURE 10.3 (*a*) Two-layer printed array structure (rectangular lattices) with different periodicities and nonparallel array axis orientations. (*b*) Floquet local mode index *mn*-plane. (*c*) Floquet global mode index *pq*-plane.

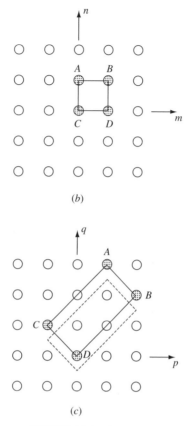

FIGURE 10.3 (Continued).

We employ the following coordinate transformation for axis rotation:

$$x' = x \cos \psi + y \sin \psi \qquad y' = -x \sin \psi + y \cos \psi \qquad (10.10)$$

Substituting (10.10) into (10.8), we obtain

$$\zeta(x, y) = B_{mn} \exp(-jk_{xm}x - jk_{yn}y) \qquad (10.11)$$

with

$$k_{xm} = \frac{2m\pi \cos \psi}{a_i} - \frac{2n\pi \sin \psi}{b_i} + \frac{\phi'_{xi} \cos \psi}{a_i} - \frac{\phi'_{yi} \sin \psi}{b_i} \qquad (10.12)$$

$$k_{yn} = \frac{2m\pi \sin \psi}{a_i} + \frac{2n\pi \cos \psi}{b_i} + \frac{\phi'_{xi} \sin \psi}{a_i} + \frac{\phi'_{yi} \cos \psi}{b_i} \qquad (10.13)$$

Recall that a_i and b_i are the periodicities of the ith layer along x'- and y'-directions, respectively. Now consider the periodicity of the ith layer along x- and y-directions of the global coordinate system. The periodicity along x would be the lcm of $a_i/\cos\psi$ and $b_i/\sin\psi$.[2] Similarly, the periodicity along y would be the lcm of $a_i/\sin\psi$ and $b_i/\cos\psi$. Since $a \times b$ is the global cell size, the following relations must hold:

$$a = \frac{m_1 a_i}{\cos\psi} = \frac{n_1 b_i}{\sin\psi} \qquad b = \frac{m_2 a_i}{\sin\psi} = \frac{n_2 b_i}{\cos\psi} \qquad (10.14)$$

where m_1, m_2, n_1, n_2 are all integers. Using (10.14) in (10.12) and (10.13) we obtain

$$k_{xm} = \frac{2\pi(mm_1 - nn_1) + (m_1 \phi'_{xi} - n_1 \phi'_{yi})}{a}$$

$$k_{yn} = \frac{2\pi(mm_2 + nn_2) + (m_2 \phi'_{xi} + n_2 \phi'_{yi})}{b} \qquad (10.15)$$

Suppose the above wave number corresponds to the (p, q) global mode with a differential phase pair $\{\phi_x, \phi_y\}$. Then, we can write

$$2\pi p + \phi_x = 2\pi(mm_1 - nn_1) + (m_1 \phi'_{xi} - n_1 \phi'_{yi})$$

$$2\pi q + \phi_y = 2\pi(mm_2 + nn_2) + (m_2 \phi'_{xi} + n_2 \phi'_{yi}) \qquad (10.16)$$

Equation (10.16) essentially relates the local mode indices (m, n) to the global modal indices (p, q). For given ϕ_x and ϕ_y, which are typically determined by the scan angle, each global mode index pair has a corresponding local mode index pair and vice-versa. In other words, every local mode, represented by a coordinate point on the mn-plane, has an image point on the pq-plane, as shown in Figures 10.3b, c. Since m and n take all possible integer values, a unit cell in the mn-plane in general maps into a parallelogram in the pq-plane, as shown in Figure 10.3c. The area of the parallelogram can be found from the Jacobian relation of (10.16), which is given by

$$\Delta p \Delta q = \frac{\partial(p, q)}{\partial(m, n)} \Delta m \Delta n \qquad (10.17a)$$

Since $\Delta m = \Delta n = 1$, from (10.14), (10.16), and (10.17a) we obtain

$$\Delta p \Delta q = m_1 n_2 + m_2 n_1 = ab/(a_i b_i). \qquad (10.17b)$$

The above result is not surprising because the number of discrete coordinate points inside the parallelogram should be equal to the ratio between the global and local cell areas. If we excite the ith layer with a differential phase pair $\{\phi'_{xi}, \phi'_{yi}\}$ that satisfies (10.16), then the global modes will occur at discrete intervals. Each differential

---

[2] For all practical purposes $a_i/\cos\psi$ and $b_i/\sin\psi$ can be considered as rational numbers; hence an lcm is possible to obtain.

phase pair associates with only one global mode point within each parallelogram. The remaining mode points correspond to other differential phase pairs. To generate the other global modes we should use appropriate sets of $\{\phi'_{xi}, \phi'_{yi}\}$. To fill in the entire global modal space one can follow a systematic procedure as below:

Step 1: Estimate the global cell size $a \times b$ and the global coordinate axes for the structure. Consider the ith layer. Use (10.14) to find m_1, n_1, m_2, and n_2, which depends on a_i, b_i, a, b, and ψ. Find ϕ_x and ϕ_y from the excitation condition of the array or other conditions relevant to the function of the structure.

Step 2: Construct a parallelogram in the pq-plane. The four corner points of the parallelogram are $(0,0)$, (m_1, m_2), $(-n_1, n_2)$ and $(m_1 - n_1, m_2 + n_2)$, respectively. Locate all the discrete points (with integer coordinates) inside the parallelogram.

Step 3:

(a) Set $m = n = p = q = 0$. Then use (10.16) to obtain ϕ'_{xi} and ϕ'_{yi}. Obtain the local GSM of the ith layer for a differential phase pair $\{\phi'_{xi}, \phi'_{yi}\}$. Rename all the (m, n) modal indices of the local GSM elements by (p, q), where p and q are computed using (10.16). Note that (p, q) occurs at discrete intervals only.

(b) Select another discrete point of coordinate (p, q) inside the parallelogram constructed in step 2. Then set the equations

$$p = mm_1 - nn_1 \qquad q = mm_2 + nn_2 \qquad (10.18)$$

and find solutions for m and n. At least one of them (m or n) must be non-integer. Retain only the integer parts as the solutions for m and n. Use these integers for m and n in (10.16) to obtain a new set of ϕ'_{xi} and ϕ'_{yi}. With this new $\{\phi'_{xi}, \phi'_{yi}\}$ repeat step 3(a). If some points lie on the boundary of the parallelogram, shift the parallelogram in appropriate directions (no rotation allowed) so that no point lies on the boundary. Then consider all the points inside the shifted parallelogram (parallelogram with dashed lines in Figure 10.3c).

Step 4: Repeat step 3(b) for other discrete points inside the parallelogram. The procedure involves a total of $m_1 n_2 + m_2 n_1$ local GSM computations for constructing the global GSM of the ith layer.

Step 5: Follow steps 1–4 for the other layers and then combine the global GSMs to obtain the overall GSM of the structure.

10.3.1 Patch Array Loaded with Screen Polarizer

A slot-fed patch array antenna loaded with a two-layer polarizing screen is an example of a multilayer structure with different periodicities and nonparallel axis orientations. The exploded view of the structure is shown in Figure 10.4. The patch array is designed to radiate a linearly polarized wave with its polarization direction along the x-axis. The screen polarizer converts the fields from linear polarization to circular polarization. The screen polarizer consists of two printed meander line

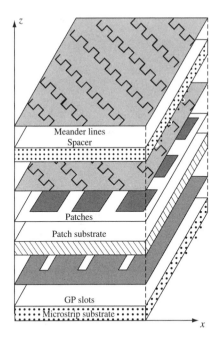

FIGURE 10.4 Exploded view of a slot-coupled patch array loaded by a meander line polarizing screen for circular polarization. GP = Ground Plane.

screens with a spacer between the screens. The polarizer is placed above the patch surface at a finite distance from the patch metallization. The distance is adjusted to optimize the electrical performance of the antenna. In Chapter 9 we presented the analysis of isolated polarizers. Here we will present an analysis of the patch–polarizer combination.

In our present structure, the periodicity of the patch elements is 0.99 cm in both directions. For the screen polarizer, the spacing between two grid lines is 0.35 cm and the grids are oriented at an angle of 135° with the E-plane of the patch array. The meander periodicity along the length is 0.7 cm. The x'-axis of the meander line makes a 45° angle with the x-axis of the patch. Using (10.14) we find $m_1 = m_2 = 2$ and $n_1 = n_2 = 1$. Following the steps described above, we obtain the GSM of the structure with respect to the global modes. From the GSM, we determine the input impedance and axial ratio of the array.

Figure 10.5a shows the Floquet reflection coefficients with and without the screen polarizer. Two different polarizers (meander line and stripline polarizers) are considered for this study. The reflection coefficients are computed for boresight beams. As can be observed, polarizers reduce the bandwidth of the array. The stripline polarizer introduces high-Q resonances, which are due to coupling between stripline modes and higher order Floquet modes. These resonance effects are subdued by raising the polarizer height from the patch surface. The meander line polarizer does not show any such resonance in this particular case. We will explain

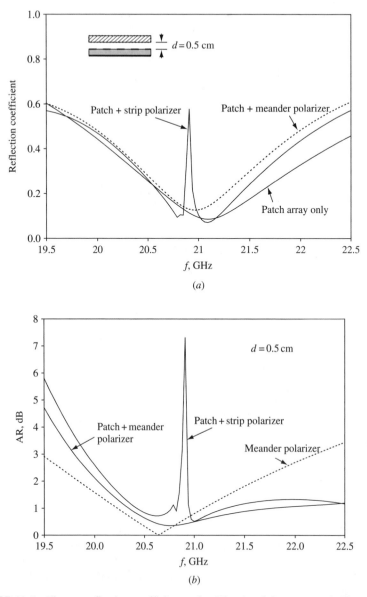

FIGURE 10.5 Floquet reflection coefficient and axial ratio of the structure in Figure 10.4. Patch cell size 0.99 cm × 0.99 cm. Patch size 0.35 cm × 0.6 cm. Slot size 0.27 cm × 0.025 cm. Patch substrate thickness 0.081 cm, feed substrate thickness 0.038 cm. Dielectric constant for patch substrate (also feed substrate) 2.55. Fifty-ohms microstrip feedlines excite slots. Meander line periodicity 0.35 cm × 0.7 cm. Line width 0.035 cm, meander depth 0.1 cm. Meander line axis 135° with x-axis of patch array. Polarizer spacer thickness 0.502 cm, dielectric constant of spacer 1.1. Strip width for the stripline polarizer 0.025 cm, spacer thickness 0.5 cm. Polarizer is at 0.5 cm above the patch surface. (*a*) Reflection coefficient versus frequency. (*b*) Axial ratio versus frequency.

these phenomena later using a circle diagram. Figure 10.5*b* shows the axial ratio–frequency plot. The axial ratio degrades significantly at the resonance locations.

To examine the scan performance of the complex array structure, we compute the reflection coefficient seen by the microstrip feedline at various scan planes. Figures 10.6*a, b* show the scan performance of the patch array without and with meander line polarizer loading. The scan planes are 0, 45°, 90°, and 135°, respectively. Several blind spots are found. These blind spots are due to the trapped guided modes that exist between the polarizer screen and the patch ground plane. There are three different types of guided modes responsible for scan blindness: (a) surface wave modes, (b) meander line guided modes, and (c) planar waveguide modes. The occurrence of a blind spot can be explained through the Floquet mode circle diagram of Figure 10.7. Each circle (with radius equal to k_0) is associated with a Floquet mode supported by the array structure. The family of curves (almost straight lines with small periodic perturbations) in the diagram represents the wave number loci of meander line guided modes.[3] The meander mode loci were obtained

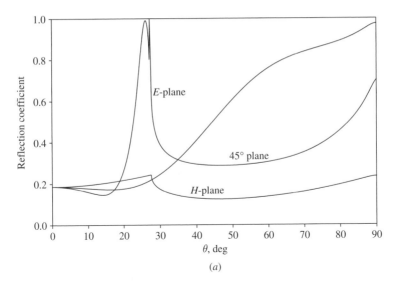

(a)

FIGURE 10.6 (*a*) Reflection coefficient versus scan angle of the slot-coupled patch array (no polarizer) described in Figure 10.5. (*b*) Reflection coefficient of the patch array with meander line polarizer loading. Separation between patch surface and inner surface of polarizer is 0.5 cm. Frequency $f = 20.7$ GHz. The resonances are explained by the circle diagram in Figure 10.7.

---

[3] If β is the propagation constant of a guided mode, then the guided mode will be coupled to a Floquet mode that has a wave number of $(2n\pi/b \pm \beta)$, where b represents the periodicity of the meander line along the propagation direction and n is an integer. Therefore to examine the coupling between all possible Floquet modes and the meander line mode, we should plot two sets of loci, one being the mirror image of the other, as shown in Figure 10.7.

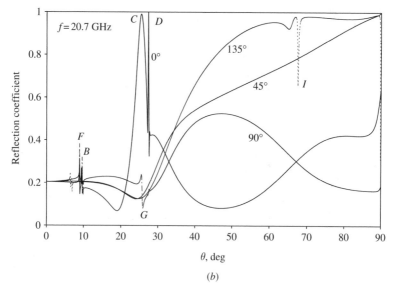

(b)

FIGURE 10.6 (Continued).

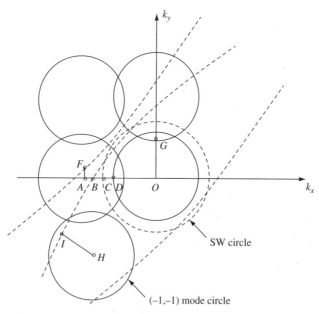

FIGURE 10.7 Floquet mode circle diagram. Each "solid-line circle" corresponds to a Floquet mode. The circle with dashed lines corresponds to the TM_0 surface wave (SW) mode. The family of curves represents the propagation constant loci of the meander line guided modes. The points indicated by capital letters correspond to the resonances in Figure 10.6b.

from an independent analysis of meander line structure. The loci are oriented at
45° with the k_x-axis, because the meander line axis is at 135° with the x-axis of the
patch array.

In Figure 10.7, the circle with slightly larger radius concentric with the (0,0)
mode represents the surface wave mode. A guided mode locus crossing a Floquet
mode circle is an indication of coupling between the modes, causing a resonance
and subsequent scan blindness. Each point of the guided mode locus lying inside a
Floquet mode circle corresponds to a blind spot. For example, the point B represents
a blind spot due to coupling between a meander line mode and the $(-1, 0)$ Floquet
mode. The direction of the blind spot can be determined from the vector $\overrightarrow{AB}$. In
this case the blind spot is located at $\phi = 0$ (E-plane) and $\theta = \sin^{-1}(|AB|) \approx 9°$
(for a stripline polarizer, this spot moves toward the bore sight [1]). Likewise,
the vector $\overrightarrow{AC}$ represents another blind spot on the E-plane due to the surface
wave resonance. Similarly, the vector $\overrightarrow{HI}$ represents a blind spot at $\phi = 135°$ and
$\theta \approx 63°$. In Figure 10.7, we marked the points corresponding to the blind spots of
Figure 10.6. Notice that some of the blind spots resulted from the coupling between
the patch array modes and polarizer guided wave mode. For instance, the blind
spot near 9° scan angle *will not* occur in an isolated meander line polarizer for a
plane wave incidence. Also, the intensity of the above blindness can be reduced or
eliminated by reducing the coupling between the meander mode and the Floquet
mode. This can be accomplished by elevating the meander line polarizer to a larger
height [1].

10.4 LAYERS WITH ARBITRARY LATTICE STRUCTURES

The procedure in Section 10.3 can be generalized for different lattice structures at
different layers. A general lattice structure of the ith layer is shown in Figure 10.8.
The unit cell area is $a_i \times b_i$, where a_i is the element spacing along the row (x'-axis),
b_i is the perpendicular distance between two consecutive rows, and γ_i is the lattice
angle. The three parameters a_i, b_i, and γ_i can describe any arbitrary unit cell. The
global cell is determined from the local cells. Suppose the global unit cell has
a lattice angle γ and the cell area is $a \times b$. The row axis (x-axis) of the global
cell makes an angle ψ with the x'-axis of the local cells. The local wave numbers
associated with the (m, n) local mode of the ith layer can be expressed as

$$k'_{xm} = \frac{2m\pi + \phi'_{xi}}{a_i} \qquad k'_{yn} = \frac{2n\pi + \phi'_{yi}}{b_i} - \frac{2m\pi}{a_i \tan \gamma_i} \qquad (10.19)$$

It should be pointed out that, although the wave number along y is a function of both
m and n, for simplicity we keep the same notation as in the case of a rectangular
lattice. The global wave numbers corresponding to the (p, q) global mode are

$$k_{xp} = \frac{2p\pi + \phi_x}{a} \qquad k_{yq} = \frac{2q\pi + \phi_y}{b} - \frac{2p\pi}{a \tan \gamma} \qquad (10.20)$$

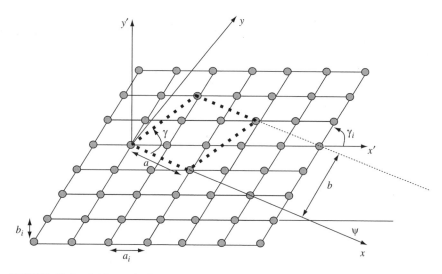

FIGURE 10.8 Arbitrary lattice structure of a layer. Also shown are the global unit cell (parallelogram of thick dashed line) and the relative axis orientation. Area of the global unit cell is $a \times b$ and $a_i \times b_i$ is the area of a local unit cell.

Following the procedure detailed in Section 10.3, we can find the relation between the local and the global mode indices. The final relation becomes

$$2\pi p + \phi_x = 2\pi(mm_1 - nn_1) + (\alpha_1 \phi'_{xi} + \beta_1 \phi'_{yi})$$
$$2\pi q + \phi_y + \frac{b\phi_x}{a \tan \gamma} = 2\pi(mm_2 + nn_2) + (\alpha_2 \phi'_{xi} + \beta_2 \phi'_{yi}) \qquad (10.21)$$

$$\alpha_1 = \frac{a \cos \psi}{a_i} \qquad \beta_1 = \frac{-a \sin \psi}{b_i}$$
$$\alpha_2 = \frac{b \cos(\gamma - \psi)}{a_i \sin \gamma} \qquad \beta_2 = \frac{b \sin(\gamma - \psi)}{b_i \sin \gamma} \qquad (10.22)$$

where m_1, n_1, m_2, n_2 are four integers and their expressions are

$$m_1 = \frac{a \sin(\psi + \gamma_i)}{a_i \sin \gamma_i} \qquad (10.23a)$$

$$n_1 = \frac{a \sin \psi}{b_i} \qquad (10.23b)$$

$$m_2 = \frac{b \sin(\psi + \gamma_i - \gamma)}{a_i \sin \gamma \sin \gamma_i} \qquad (10.23c)$$

$$n_2 = \frac{b \sin(\gamma - \psi)}{b_i \sin \gamma} \qquad (10.23d)$$

To obtain the global GSM from local GSMs of the ith layer one should use (10.21) instead of (10.16) and then follow steps 1–5 of Section 10.3. Notice that, setting $\gamma_i = \gamma = 90°$, (10.21) modifies to (10.16).

10.5 SUMMARY

The primary objective of this chapter was to present a systematic analysis of a multilayer array with different periodicities and cell orientations of the layers. The analysis relies on modal mapping from local cells to a global cell. This modal mapping allows representing an individual layer's Floquet modes in terms of global Floquet modes. This representation facilitates a direct application of the GSM method that is otherwise prohibited for a multilayer structure, where the cell size and orientation vary from layer to layer. The theory is demonstrated through numerical examples of practical structures. More numerical results featuring multilayer scan arrays can be found in [1].

REFERENCES

[1] A. K. Bhattacharyya, "Analysis of Multilayer Infinite Periodic Array Structures with Different Periodicities and Axes Orientations," *IEEE Trans. Antennas Propagat.*, Vol. 48, No. 3, pp. 357–369, Mar. 2000.

[2] H. Legay and L. Shafai, "New Stacked Microstrip Antenna with Large Bandwidth and High Gain," *Proc. IEE*, Vol. MAP-141, No. 3, pp. 199–204, June 1994.

BIBLIOGRAPHY

Bhattacharyya, A. K., "Scan Characteristics of Polarizer Loaded Horn Array," *IEEE APS Symp. Dig.*, Vol. 1, pp. 102–105, July 2000.

Chu, R. S., and K. M. Lee, "Analytical Model of a Multilayered Meander Line Polarizer Plate with Normal and Oblique Plane-Wave Incidence," *IEEE Trans. Antennas Propagat.*, Vol. AP-35, pp. 652–661, June 1987.

Das, N. K., and A. Mohanty, "Infinite Array of Printed Dipoles Integrated with a Printed Strip Grids for Suppression of Cross-Polar Radiation—Part I: Rigorous Analysis," *IEEE Trans.*, Vol. AP-45, pp. 960–972, June 1997.

Vacchione, J. D., and R. Mittra, "A Generalized Scattering Matrix Analysis for Cascading FSS of Different Periodicities," *IEEE APS Int. Symp. Dig.*, Vol. 1, pp. 92–95, 1990.

PROBLEMS

10.1 A plane wave is incident upon a linear array along x with element spacing a. The incident electric field is given by

$$\vec{E}_{inc}(x, z) = \hat{x} A \exp(-jx - j\beta z)$$

Show that the x-component of the transmitted (and reflected) Floquet modal field can be expressed as

$$E_{\text{trns}}^x(x, z) = \sum_n B_n \exp\left[\frac{-j(2n\pi + a)x}{a}\right] \exp(-jk_{zn}z)$$

where $k_{zn} = \sqrt{k_0^2 - (2n\pi/a + 1)^2}$ with k_0 the wave number in free space.

10.2 A Floquet mode, represented by the modal function $\exp(-j2\pi x/a)$, is incident on a linear periodic array of periodicity b along the x-direction. Obtain expressions for the reflected modes produced by the array for the following three cases: (a) $b = a$, (b) $b = a/2$, and (b) $b = 2a$.

10.3 A printed dipole array is loaded with an FSS. The unit cell size of the printed array is $3 \times 3\,\text{cm}^2$ and the unit cell size of the FSS is $1 \times 1\,\text{cm}^2$. Both arrays (dipole and FSS) have rectangular lattice structures with parallel axis orientations. Suppose the array is excited to radiate in the bore-sight direction. If the TE_{10} global modal fields are incident upon the FSS surface, then express the modal functions for the reflected fields from the FSS.

10.4 In problem 10.3, suppose the $(1,0)$ local mode is represented by the modal function $\exp[-j(2\pi + 2\pi/3)x]$ that is equivalently represented by the (m, n) global mode. Find m and n.

10.5 Determine the differential phase pairs for the local array of $2 \times 1\,\text{cm}^2$ cell size to represent its characteristics with respect to a global cell size of $4 \times 3\,\text{cm}^2$. Assume normal incidence and parallel axis orientations.

10.6 Determine the differential phase pairs to be considered in the local array to determine the necessary scattering parameters with respect to the global cell, as shown in Figure P10.6. The small circles represent the local elements and the lines represent the global grids. Notice, a global cell captures two full local elements.

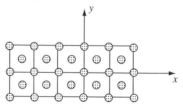

FIGURE P10.6 Local and global grids of an array structure.

10.7 Deduce (10.22) and (10.23a)–(10.23d).

CHAPTER ELEVEN

Shaped-Beam Array Design: Optimization Algorithms

11.1 INTRODUCTION

This chapter is concerned with shaped-beam array designs and various optimization algorithms that are commonly employed for accomplishing such designs. Shaped-beam arrays have potential applications in satellite communications. An on-board satellite antenna is desired to radiate maximum amount of power in a well-defined illumination region while maintaining very good RF isolations with other regions. This requirement is satisfied by a shaped- or contour-beam pattern that matches with the edge contour of the desired illumination region. Such a shaped beam can be designed by optimizing the amplitude and phase of the array elements. For obtaining a shaped beam, *phase-only* optimization is more practical than *phase-and-amplitude* optimizations, because the former allows a predetermined power distribution, which is a desirable feature in an active array antenna.

At the very beginning of the shaped-beam array design process, one must estimate the size of the array aperture and the size of an array element. The size of the array predominantly decides how closely one can accomplish a desired beam shape. From the implementation cost point of view smaller arrays are preferable, but from the performance point of view, bigger arrays are preferable. The allowable tolerances in terms of gain ripple and RF isolations from nonilluminated regions essentially decided the array size. Furthermore, for a given array size, a smaller element size is preferable because that would increase the number of elements, thereby increasing the degrees of freedom to the optimization process. Again, to minimize the cost the element size should be large because that would reduce the element count, reducing the complexity of the feed circuitry and the number of active components.

Phased Array Antennas. By Arun K. Bhattacharyya
© 2006 John Wiley & Sons, Inc.

However, larger elements have other undesirable effects, which will be discussed in appropriate places. Like the array size determination, the element size determination is a compromise between cost and performance.

In the first part of the chapter we discuss the array size and the element size and their effects on a shaped beam in terms of ripple heights inside the coverage region. Guidelines for selecting these two important parameters are presented. Woodward's superposition method for pattern synthesis is presented next. In the second part, mathematical foundations of three optimization algorithms that are commonly used for beam shaping are presented. They include the gradient search algorithm, conjugate match algorithm, and successive projection algorithm. Design examples of one- and two-dimensional shaped beam are shown to demonstrate the effectiveness of the algorithms. Shaped-beam synthesis using phase-only optimizations is also presented. Effects of amplitude tapers are discussed. The genetic algorithm and its operating principle are briefly mentioned. A general guideline for the array size determination is presented at the end.

11.2   ARRAY SIZE: LINEAR ARRAY

Determination of the array size is the primary task for a shaped-beam designer. As mentioned in the introduction, larger arrays are preferable from the performance viewpoint, but smaller array are preferable, particularly for satellite and other airborne applications, due to low weight and low implementation cost. There is no analytical formula for deciding the array size as such; however the following one-dimensional example of a simple shaped beam provides a guideline for determining the array size.

Consider a one-dimensional continuous source distribution function $f(x)$. The far-field pattern of the source is given by the Fourier integral of the source distribution:

$$F(u) = \int_x f(x) \exp(jxu) dx \tag{11.1}$$

with

$$u = k_0 \sin \theta \tag{11.2}$$

where θ is the observation angle from the bore sight and k_0 is the wave number in free space. Now consider that the desired beam shape is given by

$$F(u) = \begin{cases} 1 & |u| \le u_0 \\ 0 & |u| > u_0 \end{cases} \tag{11.3}$$

The source distribution function can be determined by the following inverse Fourier transform:

$$f(x) = \frac{1}{2\pi} \int_u F(u) \exp(-jxu) \, du$$

$$= \frac{1}{2\pi} \int_{-u_0}^{u_0} \exp(-jxu) \, du$$

$$= \frac{u_0}{\pi} \operatorname{sinc}(xu_0) \tag{11.4}$$

where $\operatorname{sinc}(y) = \sin(y)/y$. Notice that $f(x)$ exists for $-\infty < x < \infty$; therefore, to have the exact desired pattern, the source should be infinitely long. However, an infinitely long source does not exist in reality. Let us now examine what happens if the source length is truncated. We define the truncated source as

$$f_T(x) = \begin{cases} \dfrac{u_0}{\pi} \operatorname{sinc}(xu_0) & |x| \le a/2 \\ 0 & |x| > a/2 \end{cases} \tag{11.5}$$

The far-field pattern for the truncated source would be given by

$$F_T(u) = \frac{u_0}{\pi} \int_{-a/2}^{a/2} \operatorname{sinc}(xu_0) \exp(jxu) \, dx \tag{11.6}$$

Substituting $y = xu_0$, the above integral can be expressed as

$$F_T(u) = \frac{2}{\pi} \int_0^{au_0/2} \operatorname{sinc}(y) \cos\left(\frac{uy}{u_0}\right) dy \tag{11.7}$$

In Figure 11.1, $|F_T(u)|$ is plotted against u for various values of the source length a. As can be noted, the truncated sources produce ripples in the far-field patterns. The ripple height decreases with a. We will estimate the ripple heights within the coverage region defined by $-u_0 < u < u_0$. We must ignore the regions near $u = \pm u_0$ because $F_T(\pm u_0)$ deviates significantly from unity even for a large value of a. Furthermore, this region is dominated by the Gibbs phenomenon [1, p. 622] due to the sharp edge of the desired beam. A reasonable choice, perhaps, is considering 90% of the coverage region defined by $-0.9u_0 < u < 0.9u_0$. Figure 11.2 shows the variation of the ripple height (inside 90% coverage region) with the normalized length of the source. Expectedly, the ripple height decreases as a increases. This plot gives an empirical relation between the source length, coverage region, and lowest far-field intensity within the coverage region.

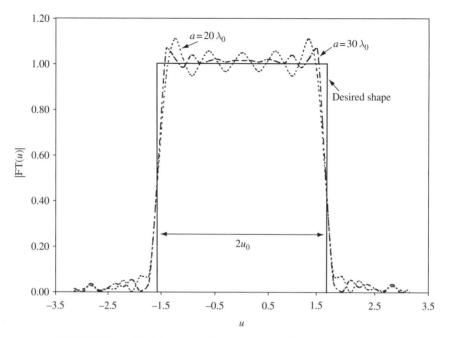

FIGURE 11.1 $F_T(u)$ versus u for various a with $u_0 = \pi/2$, $\lambda_0 = 1$.

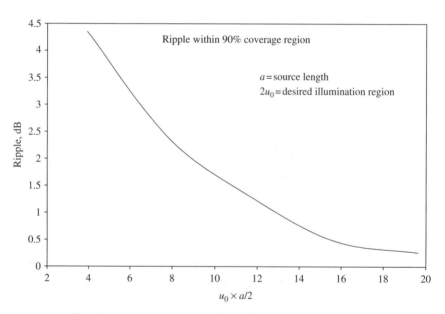

FIGURE 11.2   Variation of ripple height with source length.

It should be mentioned that we use the source function in (11.5) to generate the far-field patterns. For a given source length it may be possible to have an optimized distribution which yields a better match with the desired patterns with smaller ripple heights. However, Figure 11.2 may be used as a preliminary guideline for the array length estimation with respect to a minimum acceptable gain within a coverage region.

11.3 ELEMENT SIZE

In the previous section we have seen that the source length affects the performance in the coverage region. We considered a continuous source distribution to study this effect. An array source, however, can be viewed as a sum of multiple discrete sources, as shown in Figure 11.3. The normalized aperture distributions of the individual source elements are identical (we ignore the aperture field perturbation due to mutual coupling). The element size essentially is equal to the length of a discrete source.

The maximum distance between two consecutive elements can be determined using Shannon's sampling theorem [2] because the far field is the Fourier spectrum of the source distribution. Accordingly, a set of samples with sampling interval $\Delta x = \pi/u_0$ will reconstruct the desired beam shape defined in (11.3), provided that

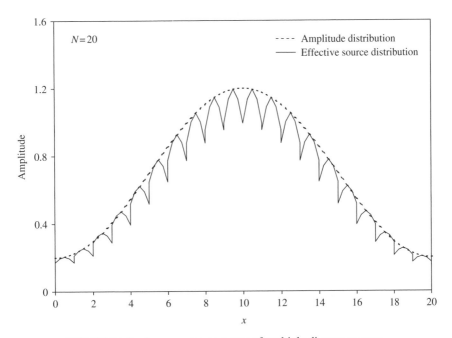

FIGURE 11.3 Array source as a sum of multiple discrete sources.

the beam is confined within the range $|u| < u_0$. However, such a "band-limited" type of beam should have an infinitely long source. Thus, the above sampling interval is valid only for an infinitely long source. In practice, a source has a finite length. Theoretically, a finite-length source has an infinite spectrum width. It follows that to reconstruct the error-free pattern of a finite source by samples, the sampling interval should be zero, which is equivalent to the continuous source itself. A nonzero sampling interval causes error in the pattern. It is intuitive that the larger the sampling interval, the larger is the error. An allowable error in the pattern determines the sampling interval.

In order to quantify the effect of the sampling interval and element size in a shaped-beam array, we consider an array of finite length. We consider a continuous source of length $a = 10\lambda_0$ ($\lambda_0 =$ wavelength in free space) with a source distribution given in (11.5) with $u_0 = \pi/(2\lambda_0)$ and assume its radiation pattern as the desired pattern. Now we consider a uniformly spaced N-element array of length $10\lambda_0$. The amplitude of an element is set equal to the amplitude of the continuous distribution at the center point of the element. The array factor for such an array would be given by

$$A_f(u) = \sum_n f_T(x_n) \exp(jux_n) \tag{11.8}$$

where x_n is the center's coordinate of the nth element. The element pattern is given by

$$E_p(u) = \int_{-\Delta x/2}^{\Delta x/2} e_p(x) \exp(jux) \, dx \tag{11.9}$$

where $e_p(x)$ is the aperture distribution of an element and $\Delta x(= 10\lambda_0/N)$ is the element size. For simplicity we assume uniform aperture distribution of an element, so the element pattern becomes

$$E_p(u) = \int_{-\Delta x/2}^{\Delta x/2} \exp(jux) dx = \Delta x \, \mathrm{sinc}(\tfrac{1}{2}u\Delta x) \tag{11.10}$$

The radiation pattern of the array thus becomes

$$F_T^{\mathrm{arr}}(u) = A_f(u)E_p(u) = \Delta x \frac{u_0}{\pi} \mathrm{sinc}\left(u\frac{\Delta x}{2}\right) \sum_n \mathrm{sinc}(x_n u_0) \exp(jux_n) \tag{11.11}$$

Figure 11.4 compares the pattern of a continuous source with that of an equivalent array of six elements ($N = 6$). Also plotted in the figure is the array factor. The array factor essentially is the radiation pattern of Dirac delta samples. The major portion of the desired beam exists within $|u| < \pi/(2\lambda_0)$; therefore the sampling interval

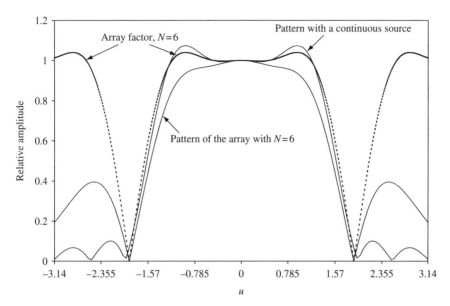

FIGURE 11.4   Array factor, radiation pattern of the array, and the radiation pattern of a continuous source of identical distributions ($a = 10\lambda_0$, $\lambda_0 = 1$).

should be smaller than $2\lambda_0$. We use a sampling interval of $1.67\lambda_0$. As can be seen, with this sampling interval the array factor matches the continuous source pattern reasonably well in the coverage region. A grating lobe exists outside the coverage region. However, the radiation pattern of the array deviates significantly from the continuous source pattern, particularly near the edge of the coverage region. The difference near the edge of the coverage region is primarily due to the element pattern roll-off.

In order to alleviate this effect, one may reduce the element size while keeping the sampling interval unchanged. However, the aperture efficiency of the array will be reduced because of unused spaces between elements. A smaller sampling interval is a better proposition because that would reduce the edge taper of the element pattern as well as move the grating lobe further away. As a result, the array pattern closely follows the continuous source pattern within and outside the coverage region. Figure 11.5 shows the patterns for a sampling interval of $1\lambda_0$ ($N = 10$). The continuous source pattern and the array factor are practically indistinguishable, especially in the coverage region. The array pattern has a small deviation near the edge because of the element pattern roll-off.

We thus see that the element size plays an important role in the array pattern. Therefore, in order to optimize an array distribution for a shaped beam, the element size should be taken into account, unless the element size is too small or the coverage region is too small. As a thumb rule, the element pattern will have negligible effect if $|\Delta x u_0| < \pi/2$.

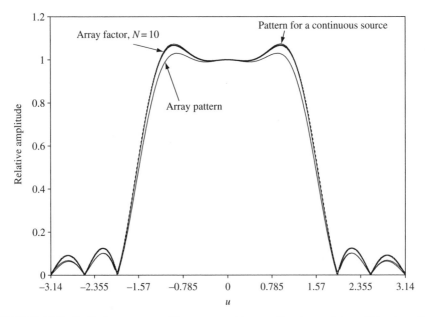

FIGURE 11.5   Radiation patterns with respect to $1\lambda_0$ sampling interval $(a = 10\lambda_0, \lambda_0 = 1)$.

11.4   PATTERN SYNTHESIS USING SUPERPOSITION (WOODWARD'S METHOD)

A straightforward method of realizing a shaped beam is by superposing multiple pencil beams inside the illuminated region. This is pictorially shown in Figure 11.6. A uniform amplitude distribution yields a pencil beam. By adding a linear phase taper, the beam can be positioned at an arbitrary location. Toward that end we set M sample points that are uniformly spaced in the u-domain within the coverage region. Then we produce M beams with beam peaks at M sample locations. The superposition of M such pencil beams will produce a composite beam that matches with the desired shape in the coverage region. The corresponding amplitude distribution is obtained from a linear superposition of the source distributions associated with each pencil beam. This method of synthesis is known as Woodward's method [3].

To demonstrate the methodology we consider the example in the previous section. Suppose the desired beam shape is given by $F(u)$ defined in (11.3). We consider an array of N elements. The element spacing is assumed as Δx. For a uniform excitation, the radiation pattern of the array will be given by

$$F_{\text{uni}}(u, u_1) = 1 + \exp[j(u - u_1)\,\Delta x] + \exp[2j(u - u_1)\,\Delta x] + \cdots$$
$$+ \exp[j(N-1)(u - u_1)\,\Delta x] \tag{11.12}$$

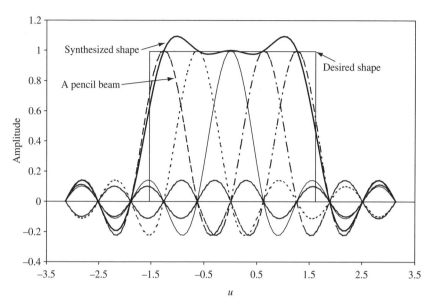

FIGURE 11.6 Woodward's synthesis using multiple pencil beams ($N = 10$, $a = 10\lambda_0$, $u_0 = \pi/2$, $\lambda_0 = 1$).

where $u_1 \Delta x$ is the phase difference between the adjacent elements. The right-hand side can be expressed as

$$F_{\text{uni}}(u, u_1) = \exp\left[\frac{j(N-1)\,\Delta x(u-u_1)}{2}\right] \frac{\sin[N\,\Delta x(u-u_1)/2]}{\sin[\Delta x(u-u_1)/2]} \qquad (11.13)$$

Notice that the peak value of $F_{\text{uni}}(u, u_1)$ is N and that occurs at $u = u_1$. Furthermore, $F_{\text{uni}}(u, u_1) = 0$ at $u = u_1 \pm m\,\Delta u$, where $\Delta x \Delta u = 2\pi/N$ and m is an integer smaller than N. This property of $F_{\text{uni}}(u, u_1)$ can be utilized to construct an arbitrary shaped beam in the following way. An arbitrary pattern function $F(u)$ defined in the coverage region can be approximated in terms of $F_{\text{uni}}(u, u_m)$ as follows:

$$F(u) \approx \frac{1}{N} \sum_m F(u_m) F_{\text{uni}}(u, u_m) \qquad (11.14)$$

The above relation is exact at $u = u_m$ subject to the condition that $\Delta x \Delta u = 2\pi/N$, where $\Delta u = u_{m+1} - u_m$. Using superposition, we can determine the excitation coefficient of the nth element as follows:

$$A_n = \frac{1}{N} \sum_m F(u_m) \exp(-jn\,\Delta x\,u_m) \qquad (11.15)$$

The above amplitude distribution will ensure that the far-field pattern will match the desired pattern $F(u)$ at the sample points given by $u = u_m$. It should be mentioned

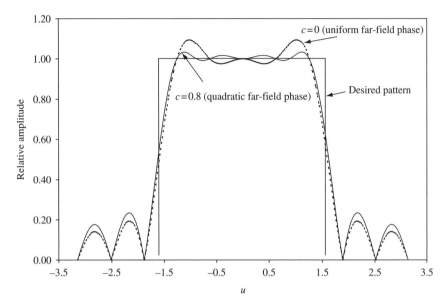

FIGURE 11.7   Synthesized patterns with respect to assumed far-field phase.

that for a desired pattern magnitude $|F(u)|$ several amplitude distributions can be obtained by introducing an arbitrary phase in $F(u)$. For a quadratic phase distribution, $F(u)$ can be expressed as

$$F(u) = |F(u)| \exp(jcu^2) \qquad (11.16)$$

where c is a constant. Figure 11.7 shows two such patterns, one with $c = 0$ and the other with $c = 0.8$. Note that with $c = 0.8$ (this value of c was found by trial and error) the synthesized pattern becomes very close to the desired pattern (with reduced ripples).

It should be emphasized that in order to apply this method the following two conditions need to be considered:

(i) $\Delta x \Delta u = 2\pi/N$
(ii) $\Delta x < \pi/u_0$

The first condition ensures that the synthesized pattern will coincide with the desired pattern at N discrete points. The second condition ensures that no overlap between the desired beam and the grating beam occurs. The first condition yields a relation between array length and sampling interval in the u-domain ($\Delta u = 2\pi/a$), while the second condition sets the maximum limit for the element spacing. One may use a smaller value of Δu than $2\pi/a$, but the synthesized pattern will not change appreciably. A reasonable choice for Δx is $\pi/(2u_0)$, where $2u_0$ is the width of the coverage region in the u-domain.

It is very straightforward to extend this approach for a two-dimensional array for a shaped beam. The element size in that case should be decided by the maximum span of the shaped beam along the u- and v-directions.

Woodward's method essentially yields amplitudes and phases of array elements. For many practical applications element amplitudes cannot be changed because they are fixed by the amplifiers' output power. In such cases phase-only optimization for beam shaping is the only viable option. The following two sections consider two algorithms that may be applied for phase-only optimizations as special cases.

11.5 GRADIENT SEARCH ALGORITHM[1]

The gradient search algorithm (GSA) is widely used to optimize a function with multiple independent variables [1, 4]. For a shaped-beam array problem, the independent variables, in general, are the amplitudes and phases of the elements. The function that needs to be optimized is commonly known as the cost function or objective function. For a shaped-beam antenna problem, the cost function could be considered as the sum of the square of the difference between the desired gain and the achieved gain at a few selective far-field points [5]. Minimizing the magnitude of a cost function ensures that the achieved pattern shape closely resembles the desired pattern shape.

In this section we first discuss the mathematical foundation of the GSA. We then demonstrate the application of this algorithm for a few array synthesis problems.

11.5.1 Mathematical Foundation

Suppose $f(x_1, x_2, x_3 \ldots x_n)$ represents a cost function of n independent variables. The objective is to find a set of numbers $(y_1, y_2, y_3, \ldots, y_n)$ such that $f(y_1, y_2, y_3, \ldots, y_n)$ has a local minimum in the neighborhood of the coordinate point $(y_1, y_2, y_3, \ldots, y_n)$ of an n-dimensional space. In order to search for the desired set, we choose a starting solution as $(x_1^{(0)}, x_2^{(0)}, x_3^{(0)}, \ldots, x_n^{(0)})$. We know that the maximum rate of change of a function f occurs if the coordinate point moves parallel to the gradient vector of f. If f is positive definite, then the magnitude of the function decreases most rapidly along $-\nabla f$. Thus, f becomes smaller at a point $(x_1^{(1)}, x_2^{(1)}, x_3^{(1)}, \ldots, x_n^{(1)})$ if

$$x_i^{(1)} = x_i^{(0)} - \rho \frac{\partial f}{\partial x_i} \qquad i = 1, 2, 3, \ldots n \qquad (11.17)$$

In the above $\partial f / \partial x_i$ is the partial derivative of the function f with respect to x_i at the point $(x_1^{(0)}, x_2^{(0)}, x_3^{(0)}, \ldots, x_n^{(0)})$ and ρ is a positive number associated with

---

[1] Also known as the steepest descent algorithm.

the distance between two successive points along the gradient path. If the desired distance is considered as δl, then ρ should be given by

$$\rho^2 \sum_{i=1}^{n} \left| \frac{\partial f}{\partial x_i} \right|^2 = (\delta l)^2 \tag{11.18}$$

The value of δl must be decided before the optimization process begins.

Once the new set for x_i's is obtained, the same procedure is repeated considering the new point as the initial point. This process is repeated several times until the values of f reach a saturation, that is, f does not decrease appreciably any more. In fact, near a saturation point the magnitude of f oscillates between multiple values. If this situation occurs, then δl should be reduced in magnitude and the process repeated with the previous solution as the starting solution. The final value of δl should correspond to the desired accuracy of the solution.

11.5.2 Application of GSA for Array Synthesis

Consider an array of N elements uniformly spaced with element spacing Δx. The array pattern for this linear array is given by

$$F(u) = E_p(u) \sum_{n=1}^{N} A_n \exp(jn \, \Delta x \, u - j\psi_n) \tag{11.19}$$

where A_n represents the amplitude (a positive real number), ψ_n is the phase lag for the nth element, and $E_p(u)$ is the element pattern. We assume a unit pulse function of length Δx as a source element, which yields $E_p(u) = \Delta x \mathrm{sinc}(\Delta x \, u/2)$. The power radiated by the "unit pulse source" is Δx. For the gain optimization one must consider the normalized pattern of the array with respect to the input power. The total power is proportional to the sum of the square of the amplitudes. Thus the normalized pattern is given by

$$\hat{F}(u) = \frac{E_p(u)}{\sqrt{\Delta x \sum_{n=1}^{N} A_n^2}} \sum_{n=1}^{N} A_n \exp(jn \, \Delta x \, u - j\psi_n) \tag{11.20}$$

Notice, the term inside the square-root sign represents the total radiated power calculated in the x-domain. Suppose the desired shape is represented by $F_d(u)$. The achievable upper limit for $F_d(u)$ can be estimated from the well-known Parseval relation [1, p. 223], which can be expressed as

$$\int_{-u_0}^{u_0} |F_d(u)|^2 du \le 2\pi \tag{11.21}$$

For a desired uniform far-field in the coverage region, the upper limit for $|F_d(u)|$ should be equal to $\sqrt{\pi/u_0}$. For other shapes, (11.21) can be used for the upper limit at each point of the shaped beam (see problem 11.5).

It may not be possible to find a set of amplitudes and phases to fit the desired shape at every point of the beam. Therefore we will select a set of M discrete points on the far-field pattern and try to match these points. In other words we will minimize the error between the array factor and the desired pattern shape at the M discrete points. The distance Δu between two consecutive far-field points should be kept about $2\pi/a$, where a is the array length.[2] We thus set a cost function as

$$f(A_1, A_2, \ldots, A_N, \psi_1, \psi_2, \ldots, \psi_N) = \sum_{i=1}^{M} \left| |F_d(u_i)| - |\hat{F}(u_i)| \right|^2 \qquad (11.22)$$

Notice, that the right-hand side is a function of $2N$ independent variables (N amplitudes and N phases) that needs to be minimized. One may use the real and imaginary parts of the excitation coefficients as the independent variables instead of amplitudes and phases. In that case all the independent variables will have the same unit (volt or ampere).

It should be mentioned that the cost function could be set as the sum of the pth power of the errors, instead of the second power as in (11.22). For a large value of p, the emphasis goes more on the point that has the largest deviation, because the cost function will be dominated by the largest error.

The above optimization method was applied to an array of 10 elements spaced $1\lambda_0$ apart. The desired beam shape is considered as that defined in (11.3). To obtain the optimum solution for amplitude and phase distributions, we considered a starting solution as uniform excitation and uniform phase distribution. For optimization, we considered real and imaginary parts as independent variables. The partial derivatives of the cost function were obtained numerically. The initial step size for excitations was 0.01 and that was refined again to obtain the final solution. Figure 11.8 shows the synthesized array pattern. The element pattern taper was ignored for the computation. Figures 11.9a, b show the optimum amplitude and phase distributions. It is found that the starting solution plays an important role for a fast convergence. The optimum solution may be different for different starting solutions. For some starting solutions, convergence may not be even reached. The excitation set in (11.15) is found to be a very effective starting solution.

11.5.3 Phase-Only Optimization

For large arrays, it is often preferred to use a predetermined amplitude distribution. This is primarily due to the difficulty in realizing arbitrary voltage distribution in a passive power divider network. In the case of an active power divider network used in an active array, it may not be practical to set arbitrary power levels to

---

[2] Generally, a smaller Δu than $2\pi/a$ does not help in improving the pattern shape. This is a consequence of the sampling theorem.

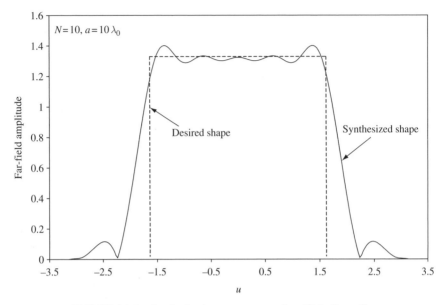

FIGURE 11.8   Synthesized array pattern using GSA ($\lambda_0 = 1$).

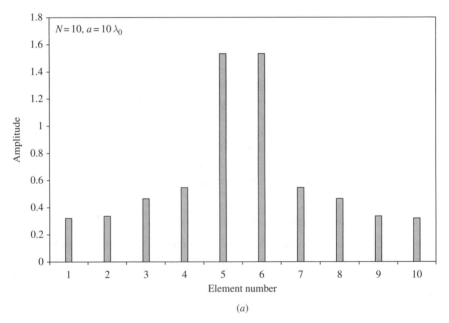

(*a*)

FIGURE 11.9   Amplitude and phase distributions for the synthesized pattern in Figure 11.8: (*a*) amplitude; (*b*) phase.

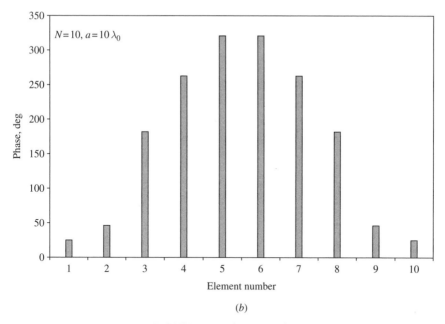

(b)

FIGURE 11.9   (Continued).

the amplifiers. Therefore, the phase-only optimization is commonly employed in a shaped-beam array [6]. Typically, the amplitude distribution is selected based on the side-lobe levels and ripples in the main beam and from the effective isotropic radiated Power (EIRP) of the array.

In order to employ the phase-only optimization, the cost function is set as in (11.22). In this case the cost function has only N independent variables[3] (N phases only). The rest of the procedure is the same as before.

Figure 11.10 shows the optimized shaped-beam patterns for three amplitude tapers of the excitation coefficients. A Gaussian amplitude distribution was considered. The edge tapers for the array excitation were 0 dB (uniform amplitudes), 10 dB, and 21 dB, respectively. The desired beam shape was a uniform far-field distribution in the coverage region. As can be seen, the 21-dB-amplitude taper yields the best synthesized pattern that is very close to the desired shape. The other amplitudes have ripples in the beam shapes. The uniform amplitude distribution had the largest ripples (about 2.9 dB peak to peak). Table 11.1 shows the amplitude taper versus peak-to-peak ripple in the coverage region.

The cause of the ripples in the coverage region can be explained analytically. Suppose a continuous distribution that exists from $-\infty$ to ∞ in the x-domain has a smooth radiation pattern with no ripple in it. It can be shown that if the above continuous source is truncated at a point, the radiation pattern of the truncated

---

[3] Strictly speaking, $N - 1$ independent variables, because it is the relative phase that matters.

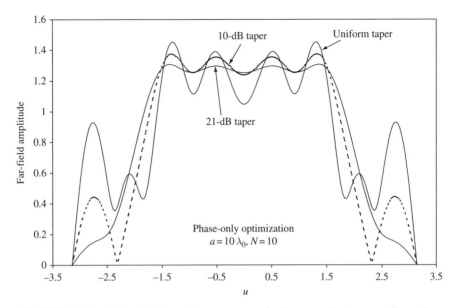

FIGURE 11.10 Optimized shaped-beam patterns for three amplitude tapers ($\lambda_0 = 1$).

TABLE 11.1 Amplitude Edge Taper Versus Ripple in Far-Field Pattern of Shaped Beam

Edge Taper (dB)	Peak-to-Peak Ripple (dB)
0	2.88
10	0.86
14	0.53
21	0.34

distribution will have ripples in the far field. To demonstrate this, consider a symmetric complex source function $f(x)$ that has a smooth far-field pattern $F(u)$. Suppose the source is truncated at $\pm x_1$. Then the far field of the truncated source would be given by

$$F_t(u) = \int_{-x_1}^{x_1} f(x)\exp(jux)\ dx$$

The above integral can be expressed as

$$F_t(u) = \int_{-\infty}^{\infty} f(x)\exp(jux)\ dx - \int_{x_1}^{\infty} f(x)\exp(jux)\ dx - \int_{-\infty}^{-x_1} f(x)\exp(jux)\ dx$$

$$= F(u) - T_1(u) - T_2(u)$$

where $T_1(u)$ and $T_2(u)$ represent the second and third integrals. As postulated, the first integral $F(u)$ is a smooth function of u with no ripple. With a change of variables, the sum of second and third integrals can be expressed as

$$T_1(u) + T_2(u) = \int_{x_1}^{\infty} f(x)\exp(jux)\,dx + \int_{x_1}^{\infty} f(-x)\exp(-jux)\,dx$$

Because $f(x)$ is a symmetric function of x, after a few simple steps, we obtain

$$T_1(u) + T_2(u) = 2\cos(ux_1)\int_0^{\infty} f(y+x_1)\cos(uy)\,dy - 2\sin(ux_1)$$

$$\times \int_0^{\infty} f(y+x_1)\sin(uy)\,dy$$

A realizable source distribution must have a finite amount of power; therefore, for a large value of x, the source function either decays monotonically or may be a decaying oscillatory type. Therefore, for a large value of x_1, the contribution comes near the $x = x_1$ region. Furthermore, in the bore-sight region u is small; therefore, the first term contributes more than the second term. Thus near the bore-sight region we write

$$T_1(u) + T_2(u) \approx 2\cos(ux_1)\int_0^{\infty} f(y+x_1)\,dy$$

Now, $T_1(u) + T_2(u)$ is an oscillatory function causing ripples. The ripple amplitude depends on the portion of the area of a continuous source outside the truncated region. This area is directly related to the edge taper at the truncation, because $f(x_1)$ is directly related to the edge taper. The ripple height near the bore sight can be estimated as

$$R = -20\log\left[1 - \frac{T_1(0) + T_2(0)}{F(0)}\right]$$

Figure 11.11 shows the ripple height versus edge taper for a Gaussian distribution. The results generally agree with that presented in Table 11.1. This plot can be used to estimate the ripple height in a shaped beam for a given source taper.

11.5.4 Contour Beams Using Phase-Only Optimization

For a contour beam, phase-only optimization is commonly used, keeping the amplitude distribution unchanged. The basic procedure for beam shaping is very similar to that of the one-dimensional case described in the previous section. In the case of a two-dimensional shaped beam, the desired gain of the array is set at two-dimensional

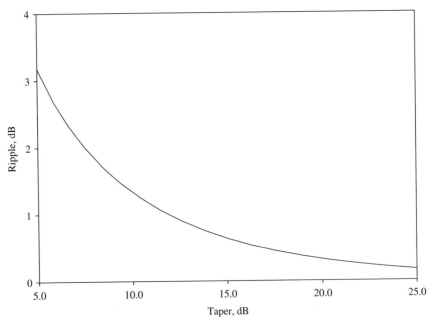

FIGURE 11.11   Ripple height versus edge taper for a Gaussian distribution.

grid points. For the far-field pattern, the uv-coordinate system is commonly used. The variables are defined as

$$u = k_0 \sin\theta \cos\phi \qquad v = k_0 \sin\theta \sin\phi \tag{11.23}$$

Notice that the far field is related to the two-dimensional Fourier transform of the source distribution. For the two-dimensional array, the far field of N elements can be expressed as

$$F(u, v) = E_p(u, v) \sum_{n=1}^{N} A_n \exp(jx_n u + jy_n v - j\psi_n) \tag{11.24}$$

where $E_p(u, v)$ is the element pattern, (x_n, y_n) is the coordinate point of the nth element, and A_n and ψ_n are the amplitude and phase lag, respectively, of the element. If the amplitudes are normalized with respect to an input power of 4π, then (11.24) yields the directivity of the array. Thus the normalization relation is given by

$$4\pi = \left[\iint_{xy} |e_p(x, y)|^2 dx\, dy \right] \sum_{n=1}^{N} A_n^2 \tag{11.25}$$

where $e_p(x, y)$ is the aperture field distribution of an element.

Equation (11.22) is used as the cost function in regard to the GSA. The rule discussed in Section 11.5.2 applies for selecting sampling intervals Δu and Δv for the far-field points. Notice that there are N unknown parameters ($\psi_i, i = 1, 2, \ldots, N$) to be determined for the desired beam shape. For a large number of array elements this could be a very time-consuming process because each iteration demands computation of N partial derivatives. To alleviate this problem we reduce the number of variables to fewer unknowns. Various expansion schemes for the phase distribution may be applied. One such scheme is using a polynomial approximation as

$$\psi_i = \psi_i(x_i, y_i) = \sum_{m=0}^{K-1} \sum_{n=0}^{L-1} P_{mn} x_i^m y_i^n \tag{11.26}$$

where (x_i, y_i) is the coordinate of the ith element of the array. In the above the P_{mn}'s are unknown parameters to be determined. From the above equation we see that by setting the values of K and L, we can reduce the number of unknowns from N to $K \times L$. For instance, if $K = 4$ and $L = 4$, then the number of P_{mn}'s becomes only 16. For a very complex gain contour, these numbers can be set larger.

In order to set the desired gain of a shaped beam, the following guidelines should be followed. It is understood that the peak gain in the coverage region should be less than the maximum possible gain of the array, which happens in the case of a pencil beam. The expectable gain inside a contour can be estimated from a parameter called the gain area product (GAP) of the beam. The GAP of a beam is defined as the product of the minimum gain in the coverage region and the solid angle covered by the desired beam. Mathematically,

$$\text{GAP} = |F_{\min}|^2 \times \Delta\Omega \tag{11.27}$$

where $|F_{\min}|$ represents the minimum gain inside the coverage region and $\Delta\Omega$ is the solid angle covered by the contour beam. The maximum possible value of the GAP is 4π square radians ($\approx 41,253$ square degrees) and that occurs for a beam that has a constant gain inside the coverage region and zero gain outside the coverage region (see problem 11.6). To generate such a beam one needs an infinitely large array. For a practical array, the typical GAP is of the order of 20,000 square degrees or less. The following numerical examples yield an idea of the GAP that can be expected from an array.

Figure 11.12 shows an array of 377 elements in a square grid. The desired beam shape in the uv-coordinate system is shown in Figure 11.13. The coverage area of this beam in a spherical coordinate system can be computed as

$$\Delta\Omega = \iint_{\theta,\phi} \sin\theta\, d\theta\, d\phi = \frac{1}{k_0^2} \iint_{u,v} \frac{du\, dv}{\sqrt{1 - u^2 - v^2}} \tag{11.28}$$

If the origin of the uv-coordinate is located inside the coverage region and the coverage area is small, then inside the coverage region u and v would be small. Then the integral can be approximated as

$$\Delta\Omega = \frac{1}{k_0^2} \iint_{u,v} du\, dv \tag{11.29}$$

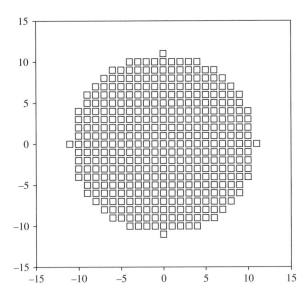

FIGURE 11.12 Array of 377 elements in a square grid.

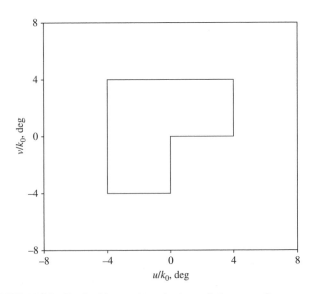

FIGURE 11.13 Desired beam shape in the scaled uv-coordinate system.

Therefore for small beams, the beam area in spherical coordinates is proportional to the beam area in the uv-coordinate system. The lowest gain can be estimated using the GAP number and $\Delta\Omega$.

Figures 11.14a–c show the optimized shaped beams with square elements of lengths $1\lambda_0$, $1.5\lambda_0$ and $2\lambda_0$, respectively. A Gaussian amplitude distribution is

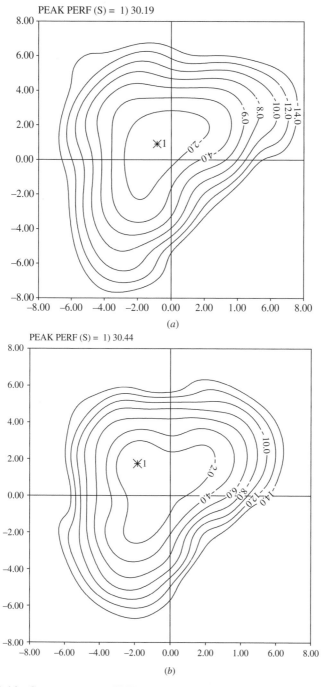

FIGURE 11.14 Contour patterns of 377 square elements with three different element lengths (EL); both axes in degrees: (*a*) EL = $1\lambda_0$, (*b*) EL = $1.5\lambda_0$, (*c*) EL = $2\lambda_0$. The beam shapes are optimized employing GSA (phase-only optimization). Axes labels are same as that of Figure 11.13.

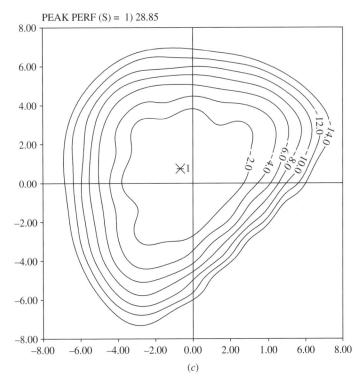

FIGURE 11.14 (Continued).

assumed with an edge taper of 15 dB. For the optimization process we used a polynomial function for the phase distribution given in (11.26) with $K = L = 4$. From the contour plot we see that with respect to the 26-dBi-gain contour, the beam size is the biggest for the $2\lambda_0$ size elements array. Therefore, the GAP increases with the element size of the array. The GAP with respect to the 26-dBi-gain contour is computed for the three designs and tabulated in Table 11.2. This gives an idea about the expected GAP for a given size of an array with respect to a complex beam shape.

TABLE 11.2   Array Size Versus GAP of Array of 377 Square Elements

Element Size (wavelengths)	Array Size (square wavelengths)	GAP (square degrees) for 26-dBi-Gain Contour
1.0	377	14,992
1.5	848	16,066
2.0	1508	16,587

Note: The array geometry is shown in Figure 11.12.

11.6 CONJUGATE MATCH ALGORITHM

The conjugate match algorithm is another iterative algorithm commonly used for beam shaping in phased array antennas. The algorithm relies on the maximum array gain theorem (see Section 1.2.4), which states that the maximum far-field intensity occurs at a point if the excitation coefficients are proportional to the complex conjugates of the elements' field intensities (normalized with respect to the incident power) at that point [7]. This algorithm can also be viewed as a modified version of the GSA. The conjugate match algorithm essentially optimizes the field strength instead of the gain. Thus the cost function becomes linear with respect to the unknown excitation coefficients. As a result, the partial derivatives of the cost function have simple closed-form expressions. We will present the mathematical background of the conjugate match algorithm from a gradient search point of view.

The radiated field of an array for given amplitude and phase distribution is given in (11.24), which is rewritten here for convenience:

$$F(u, v) = E_p(u, v) \sum_{n=1}^{N} A_n \exp(jx_n u + jy_n v - j\psi_n) \qquad (11.30)$$

We now modify this equation to accommodate the phase term within the excitation as

$$F(u, v) = E_p(u, v) \sum_{n=1}^{N} B_n \exp(jx_n u + jy_n v) \qquad (11.31)$$

with

$$B_n = A_n \exp(-j\psi_n) \qquad (11.32)$$

Suppose the far-field amplitudes computed using (11.31) differ from the desired far-field amplitudes at a few selective points inside the coverage region. Our task is to modify the excitation coefficients in order to improve the far-field match. Since $F(u, v)$ is a linear equation of B_n's we write

$$\Delta F(u, v) = E_p(u, v) \sum_{n=1}^{N} \exp(jx_n u + jy_n v) \Delta B_n \qquad (11.33)$$

We now assume that each element is contributing equally to increase or decrease the field strength at the (u, v) point. Then we write

$$\frac{\Delta F(u, v)}{N} = E_p(u, v) \exp(jx_n u + jy_n v) \Delta B_n \qquad (11.34)$$

which yields

$$\Delta B_n = \frac{\Delta F(u, v)}{N[E_p(u, v) \exp(jx_n u + jy_n v)]} \qquad (11.35)$$

Notice that the incremental excitation of the nth element is proportional to the complex conjugate of its radiated field (hence the name conjugate match). The above equation gives an improved excitation coefficient if $\Delta F(u, v)$ is known. In practice $\Delta|F|$, the magnitude difference between the desired and achieved field intensities, is the only known quantity. The phase of ΔF is unknown. Typically, one resolves this problem by selecting the phase of ΔF as that of $F(u, v)$ obtained in the previous iteration. Mathematically one obtains

$$\Delta B_n = \frac{\Delta|F(u, v)|\exp(j\xi)}{N[E_p(u, v)\exp(jx_n u + jy_n v)]} \tag{11.36}$$

where ξ represents the phase of $F(u, v)$ achieved in the previous iteration. In the case of simultaneous amplitude and phase optimization, the amplitudes must be renormalized at the beginning of each iteration process by adjusting the total power to 4π as per (11.25). The updated excitation would be given by

$$B_n^{\text{new}} = \frac{B_n + \Delta B_n}{\sqrt{\sum_m |B_m + \Delta B_m|^2}} \tag{11.37}$$

This normalization is valid if $e_p(x, y)$ is normalized with respect to 4π radiated power, that is, if $E_p(u, v)$ in (11.24) represents the element gain pattern.

For phase-only optimization we only update the phase part, keeping the magnitude unchanged. Therefore, the updated excitation for the phase-only optimization becomes

$$B_n^{\text{new}} = A_n \frac{B_n + \Delta B_n}{|B_n + \Delta B_n|} \tag{11.38}$$

It is worth mentioning that for monotonic convergence we use only a fraction (typically $\frac{1}{2}$ or $\frac{1}{4}$) of ΔB_n in (11.36).

The above expressions for complex amplitudes can be used at a point in the uv-space where the achieved gain differs from the desired gain. In practice, however, the point with the largest difference is considered for a given iteration. With a new set of excitation coefficients we again compute the gain and search for the new point where the gain difference is the largest. We keep repeating this process until a satisfactory gain distribution is reached.

We applied the phase-only optimization process to generate a shaped beam with 377 elements. A reasonable convergence was achieved with about 2000 iterations (each new phase computation is considered to be an iteration). Figure 11.15 shows the contour pattern with $1.5\lambda_0$ square elements. The GAP with respect to the 26-dBi-gain contour was 17,206 square degrees, which is about 6% more than that obtained using the GSA. This difference is primarily due to the limited number of parameters used for the gradient search optimization process [recall, we used $K = L = 4$ in (11.26), which is equivalent to 16 parameters only]. With larger values of K and L, the GAP in the gradient search process is expected to improve because

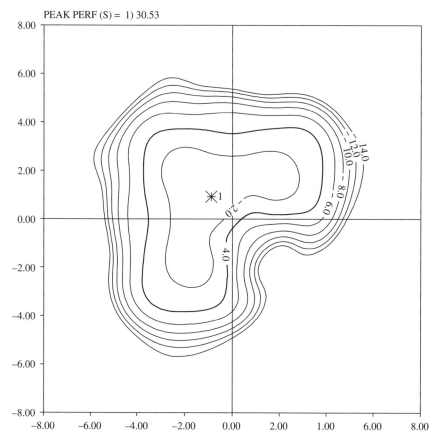

FIGURE 11.15 Contour pattern of 377 square elements of element length $1.5\lambda_0$ (both axes in degrees). The beam shape is optimized using the conjugate match algorithm. Axes labels are same as that of Figure 11.13.

the phase relation between the elements will not be constrained by the fixed order of the polynomial.

The conjugate match algorithm is computationally much faster than a GSA, because there is no partial derivative computation involved in the process. However, the algorithm is limited to the phased array beam shaping only. On the other hand, the GSA is slow but very versatile and is applicable for any optimization process.

11.7 SUCCESSIVE PROJECTION ALGORITHM

The GSA and the conjugate match algorithm are widely used for array pattern synthesis; however, many other algorithms are available in the open literature for beam shaping. For example, the successive projection algorithm (SPA) has been used for such a purpose [8, 9]. The SPA originated from the projection matrix method used in linear algebra to obtain the best-fit solution for a system of linear equations [10]. In the SPA a projection vector concept is invoked to minimize the difference between the desired and achieved patterns. We will explain the mathematical foundation of this algorithm in light of array beam synthesis.

Consider a one-dimensional array problem. Suppose the desired far field of the array is expressed by the complex function $F_d(u)$. For an N-element array the far-field pattern is given by (ignoring the element pattern factor)

$$F(u) = \sum_{n=1}^{N} A_n \exp(jx_n u) \tag{11.39}$$

The objective is to determine A_n's such that $F(u)$ becomes very close to the desired pattern $F_d(u)$. To estimate A_n's we select M sample points $(M > N)$ and write

$$F_d(u_i) = \sum_{n=1}^{N} A_i \exp(jx_n u_i) \qquad i = 1, 2, \dots, M \tag{11.40}$$

We write the above equation in vector form as

$$
\begin{bmatrix} F_d(u_1) \\ F_d(u_2) \\ \vdots \\ F_d(u_M) \end{bmatrix}
= A_1 \begin{bmatrix} \exp(jx_1 u_1) \\ \exp(jx_1 u_2) \\ \vdots \\ \exp(jx_1 u_M) \end{bmatrix}
+ A_2 \begin{bmatrix} \exp(jx_2 u_1) \\ \exp(jx_2 u_2) \\ \vdots \\ \exp(jx_2 u_M) \end{bmatrix}
+ \cdots + A_N \begin{bmatrix} \exp(jx_N u_1) \\ \exp(jx_N u_2) \\ \vdots \\ \exp(jx_N u_M) \end{bmatrix}
$$

$$\tag{11.41}$$

From (11.41) we notice that the vector $[F_d]$ on the left-hand side is represented as a linear combination of N independent vectors, which, for the brevity of presentation, we name $[E_1], [E_2], \dots, [E_n]$, respectively. For solutions of unknown terms A_1, $A_2, \dots, A_N$, a necessary condition is that $[F_d]$ must lie on the space spanned by the vectors $[E_1], [E_2], \dots, [E_n]$. Otherwise an exact solution is not possible. For $M > N$, it is quite possible that $[F_d]$ lies outside the space of $[E_n]$'s. This is pictorially shown in Figure 11.16 for $N = 2$ and $M = 3$. The best possible solution in the least squares

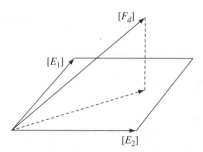

FIGURE 11.16 Vector space description of the projection method.

error sense is obtained by replacing the $[F_d]$ vector by its projection vector onto the space spanned by the vectors $[E_1]$ and $[E_2]$. The dashed line denotes this projection vector. The projection vector can be determined easily using simple vector algebra (see problem 11.7).

The problem at hand, however, is slightly different than that considered above. In our problem, the $[F_d]$ vector is not completely known because only the magnitudes are specified by the desired power pattern. The phase is still unknown. The phase of $[F_d]$ is introduced using the successive approximation method. Toward that end, we assume a trial solution for the unknown scalar A_n as $A_n^{(0)}$ $(n = 1, 2, \ldots, N)$. The trial values must satisfy the power normalization condition in (11.25). Using these trial values of A_n's we compute the right-hand side of (11.41) and use the phase term as the phase of $[F_d]$. We call this modified vector $[F_d^{(0)}]$. The error vector is given by

$$[\varepsilon^{(0)}] = [F_d^{(0)}] - A_1^{(0)}[E_1] - A_2^{(0)}[E_2] - \cdots - A_N^{(0)}[E_N] \qquad (11.42)$$

If this error vector becomes orthogonal to the space spanned by the vector $[E_n]$ $(n = 1, 2, \ldots, N)$, then we have the best possible solution. Therefore our objective is to make the error vector orthogonal to the space of $[E]$ vectors. To accomplish that, we determine the projection of the error vector and refine the amplitudes A_n. The projection of the error vector onto the $[E]$ vector space allows finding a new set of A_n. Accordingly, we write

$$\Delta A_1^{(0)}[E_1] + \Delta A_2^{(0)}[E_2] + \cdots + \Delta A_N^{(0)}[E_N] = [\varepsilon^{(0)}] - [\perp] \qquad (11.43)$$

where $[\perp]$ represents the component of the $[\varepsilon^{(0)}]$ vector that is perpendicular to the $[E]$ vector space. Taking the scalar product with $[E_1^*]$ on both sides of (11.43) we have

$$\Delta A_1^{(0)}[E_1] \cdot [E_1^*] + \Delta A_2^{(0)}[E_2] \cdot [E_1^*] + \cdots + \Delta A_N^{(0)}[E_N] \cdot [E_1^*] = [\varepsilon^{(0)}] \cdot [E_1^*] \quad (11.44)$$

Notice that the $[E]$ vectors have equal magnitudes (lengths); therefore. the $[E_m] \cdot [E_n^*]$ will be maximum if $m = n$. Furthermore, for a large number of sampling points, $[E_m]$ and $[E_n]$ will be nearly orthogonal for $m \neq n$. With these assumptions we can write

$$\Delta A_1^{(0)} \approx \frac{[\varepsilon^{(0)}] \cdot [E_1^*]}{[E_1] \cdot [E_1^*]} \qquad (11.45)$$

In general we obtain

$$\Delta A_n^{(0)} \approx \frac{[\varepsilon^{(0)}] \cdot [E_n^*]}{[E_n] \cdot [E_n^*]} \qquad n = 1, 2, \ldots, N \qquad (11.46)$$

Therefore the new excitation coefficients after first iteration are given by

$$A_n^{(1)} = A_n^{(0)} + \Delta A_n^{(0)} \qquad n = 1, 2, \ldots N \qquad (11.47)$$

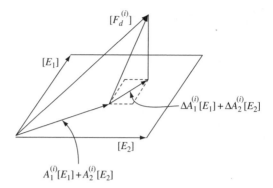

FIGURE 11.17   Vector space diagram of the successive projection method.

We now renormalize the new values of A_n. With these new values of A_n we use (11.41) to update the phase for $[F_d]$. Then we return to (11.42) and repeat the entire process until we meet our error criteria. The concept of this successive projection process is shown pictorially in Figure 11.17 for two unknowns.

11.7.1   Successive Projection and Conjugate Match

The conjugate match method becomes a special case of successive projection if the following approximation is used. In this approximation, we consider only the dominant component in the error vector in (11.46). This means we fix the largest error point, that is, the far-field point where the achieved gain differs most from the desired gain. With this consideration, (11.46) becomes

$$\Delta A_n^{(0)} \approx \frac{\varepsilon^{(0)}(m)\exp(-jx_n u_m)}{[E_n]\cdot[E_n^*]} \qquad n = 1, 2, \ldots, N \qquad (11.48)$$

In the above equation $\varepsilon^{(0)}(m)$ represents the mth component of $[\varepsilon^{(0)}]$ which has the largest magnitude. We notice from (11.41) that $[E_n]\cdot[E_n^*] = M$. Therefore (11.48) becomes

$$\Delta A_n^{(0)} \approx \frac{\varepsilon^{(0)}(m)\exp(-jx_n u_m)}{M} \qquad n = 1, 2, \ldots, N \qquad (11.49)$$

The right-hand side of (11.49) is similar to that of (11.35) for a linear array, which is based on the conjugate match method. There are two apparent differences found in (11.49). First, the element pattern factor does not appear in (11.49) because it is ignored for simplicity of presentation. More importantly, (11.49) has the term M in the denominator instead of N. In Fact, the factor N in (11.35) was chosen with an assumption that each element contribute equally to improve the gain at a point. This consideration is somewhat arbitrary and one could as well choose a different number than N. This selection may lead to a different convergence rate, but the basic computation scheme remains unchanged.

11.8 OTHER OPTIMIZATION ALGORITHMS

For array synthesis, several other optimization algorithms are being used; however, most of them are modified versions of the gradient search or successive projection algorithm (in fact we have shown the conjugate match algorithm is a special case of the GSA and SPA). These algorithms are proven to be very efficient computationally. One drawback is that these algorithms may yield a solution associated with a local minimum if a good starting solution is not used. The genetic algorithm has drawn considerable attention from antenna designers [11, 12]. This algorithm is based on a "natural selection process" that always evolves to an improvement of the solution. This algorithm is computationally very slow but generally yields a global minimum solution.

11.9 DESIGN GUIDELINES OF A SHAPED BEAM ARRAY

We have presented various optimization algorithms and have demonstrated their effectiveness in the design of one- and two-dimensional shaped beam arrays. In this section we provide a general guideline for the array size determination with respect to a desired GAP of a shaped beam. We consider circular shaped flat-top beams; however, the results may be utilized for nearly circular beams also.

It can be shown that for a circular array and for a circularly shaped beam, the GAP is an explicit function of the product of the array radius and beam radius in the uv-domain. This is true under the assumption that a particular type of amplitude distribution is used, for instance, a Gaussian distribution with fixed edge taper. Figure 11.18 shows the GAP versus au_0, where a is the array radius and u_0 is the beam radius in the uv-plane. The plot is generated using an array of 317 square elements and optimizing the GAP within the coverage region. A Gaussian amplitude distribution with 10 dB edge taper was used. It is found that the GAP lies within $\pm5\%$ of the presented data for different starting solutions. Furthermore, the GAP does not vary appreciably if the edge taper increases to 15 dB. The plot can be utilized to estimate the array size for accomplishing a minimum gain within the coverage region. For example, if one wants a beam radius of 3° with a minimum gain of 28 dBi, then the GAP becomes approximately 17,840. From the plot we notice that the above GAP corresponds to $au_0 \approx 4$. Since $u_0 = 2\pi/\lambda_0 \sin(3°)$, the array radius a becomes approximately $12\lambda_0$. Thus an array of radius $12\lambda_0$ can produce a circular shaped beam of minimum gain of 28 dBi within a 3° radius. This plot is generated for circular shaped beams; however, the data can be utilized for nearly circular shaped beams. For a complex beam shape the array size would be somewhat larger with respect to a given GAP. The element size, assuming closely spaced elements, can be determined using the criterion $\Delta x\, u_0 \leq \pi$. The smaller

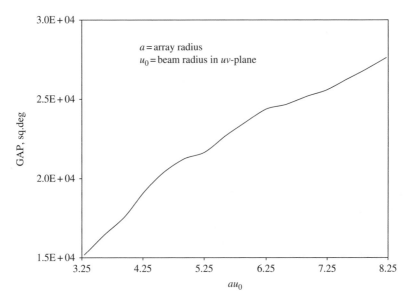

FIGURE 11.18 GAP versus au_0 for circular flat-top beams. Array amplitude taper 10 dB with Gaussian distribution.

the element size, the better is the performance, because that would minimize the power loss in the grating lobe region. Once the array size and the element size are estimated, the phase distribution can be determined using an optimizer.

REFERENCES

[1] E. Kreyszig, *Advanced Engineering Mathematics*, 6th ed., Wiley, New York, 1988.

[2] J. C. Hancock, *An Introduction to the Principles of Communication Theory*, McGraw-Hill, New York, 1961.

[3] C. A. Balanis, *Antenna Theory Analysis and Design*, Wiley, New York, 1982, p. 673.

[4] W. H. Press, B. P. Flannery, S. A. Teukolsky, and W. T. Vetterling, *Numerical Recipes in Pascal*, Cambridge University Press, Cambridge, 1989.

[5] A. R. Cherrette, S. W. Lee, and R. J. Acosta, "A Method for Producing Shaped Contour Radiation Pattern Using a Single Shaped Reflector and a Single Feed," *IEEE Trans.*, Vol. AP-37, No. 6, pp. 698–706, June 1989.

[6] L. I. Vaskelainen, "Phase Synthesis of Conformal Array Antennas," *IEEE Trans.*, Vol. AP-48, No. 6, pp. 987–991, June 2000.

[7] P. T. Lam, S-W. Lee, D. C. D. Chang, and K. C. Lang, "Directivity Optimization of a Reflector Antenna with Cluster Feeds: A Closed-Form Solution," *IEEE Trans.*, Vol. AP-33, No. 11, pp. 1163–1174, Nov. 1985.

[8] H. Elmikati and A. A. Elsohly, "Extension of Projection Method to Nonuniformly Linear Antenna Arrays," *IEEE Trans.*, Vol. AP-32, No. 5, pp. 507–512, May 1984.

[9] G. T. Poulton, "Power Pattern Synthesis Using the Method of Successive Projections," *IEEE APS Symp.* Dig., Vol. 24, pp. 667–670, 1986.

[10] G. Strang, *Linear Algebra and Its Applications*, Saunders, Philadelphia, 1988.

[11] D. W. Boeringer and D. H. Werner, "Particle Swarm Optimization Versus Genetic Algorithms for Phased Array Synthesis," *IEEE Trans.*, Vol. AP-52, No. 3, pp. 771–779, Mar. 2004.

[12] J. M. Johnson and Y. Rahmat-Samii, "Genetic Algorithm in Engineering Electromagnetics," *IEEE APS Maga.*, Vol. 39, No. 4, pp. 7–25, Aug. 1997.

BIBLIOGRAPHY

Botha, E., and D. A. McNamara, "A Contoured Beam Synthesis Technique for Planar Antenna Arrays with Quadrantal and Centro-Symmetry," *IEEE Trans.*, Vol. AP-41, No. 9, pp. 1222–1230, Sept. 1993.

Bucci, O. M., G. D'Elia, G. Mazzaarella, and G. Panariello, "Antenna Pattern Synthesis: A New General Approach," *Proc. IEEE*, Vol. 82, No. 3, pp. 358–371, Mar. 1994.

Bucci, O. M., G. Mazzarella, and G. Panariello, "Reconfigurable Arrays by Phase-Only Control," *IEEE Trans.*, Vol. AP-39, No. 7, pp. 919–925, July 1991.

Chakraborty, A., B. N. Das, and G. S. Sanyal, "Beam Shaping Using Nonlinear Phase Distribution in a Uniformly Spaced Array," *IEEE Trans.*, Vol. AP-30, No. 5, pp. 1031–1034, Sept. 1982.

Elliott, R. S., and G. J. Stern, "Footprint Patterns Obtained by Planar Arrays," *IEE Proc.*, Pt. H, Vol. 137, pp. 108–112, Apr. 1990.

Holland, J. H., "Genetic Algorithm," *Sci. Am.*, pp. 66–72, July 1992.

Klein, C. A., "Design of Shaped-Beam Antennas through Minimax Gain Optimization," *IEEE Trans.*, Vol. AP-32, No. 9, pp. 963–968, Sept. 1984.

PROBLEMS

11.1 Show that if au_0 approaches infinity, then $F_T(u)$ approaches unity if $u < u_0$, $F_T(u) = \frac{1}{2}$ if $u = u_0$ and $F_T(u) = 0$ if $u > u_0$, where $F_T(u)$ is given in (11.7). *Hint*: Use $\int_0^\infty (\sin\theta/\theta)\, d\theta = \pi/2$.

11.2 Estimate the element size of a two-dimensional circular flat-top beam. The coverage region is defined by the circle $u^2 + v^2 = \frac{1}{4}$ on the uv-plane. Using Figure 11.2, estimate the array size and then the number of elements of the array for 1.0 dB ripple within the coverage region. Employing Woodward's method, obtain the amplitude and phase for the array elements.

11.3 Prove that at a point (x, y, z) a function $f(x, y, z)$ changes its value at its maximum rate if x, y, and z move along a direction parallel to the vector ∇f at (x, y, z).

11.4 Prove that $\int\limits_{-\infty}^{\infty} |F_d(u)|^2 \, du = 2\pi$, where $F_d(u)$ is the Fourier transform of $f_d(x)$ with $\int\limits_{x} |f_d(x)|^2 dx = 1$.

11.5 The desired pattern is given by the function

$$F_d(u) = \begin{cases} A\sqrt{1+u^2} & |u| \leq 1 \\ 0 & |u| > 0 \end{cases}$$

Using (11.21) find A and then obtain the expected normalized fields at $u = 0$ and at $u = 1$.

11.6 For a lossless antenna if $F(\theta, \phi)$ represents the gain function, then prove that the following relation is true:

$$\int\limits_{0}^{\pi} \int\limits_{0}^{2\pi} |F(\theta, \phi)|^2 \sin\theta \, d\theta \, d\phi = 4\pi$$

Hence prove that if F_{min} is the minimum intensity inside a coverage region $\Delta\Omega$, then

$$|F_{min}|^2 \times \Delta\Omega \leq 4\pi$$

Also argue that equality occurs if $F(\theta, \phi)$ becomes constant within the coverage region $\Delta\Omega$ and becomes zero outside the coverage region.

11.7 In Figure 11.17, assume that the projection vector of $[F_d]$ onto the plane spanned by $[E_1]$ and $[E_2]$ is given by $A_1[E_1]+A_2[E_2]$, where A_1 and A_2 are two scalar constants. Determine A_1 and A_2 using the condition that the error vector must be perpendicular to the plane of $[E_1]$ and $[E_2]$, that is, $[F_d - A_1 E_1 - A_2 E_2] \cdot [E_1^*] = [F_d - A_1 E_1 - A_2 E_2] \cdot [E_2^*] = 0$.

CHAPTER TWELVE

Beam Forming Networks in Multiple-Beam Arrays

12.1 INTRODUCTION

One of the most important tasks in a phased array design is designing the beam forming network (BFN). A BFN essentially provides the necessary amplitudes and phases to the radiating elements to produce the desired beams. In particular, for a multiple-beam array, an appropriate BFN is vital in order to distribute the signals to the radiating ports properly.

In this chapter we present operating principles and designs of most common types of BFNs. We start with a simple BFN using passive power divider circuits. Next we present the Butler matrix BFN, which is designed based on the underlying principle of fast Fourier transform (FFT). Toward that pursuit we first establish a correlation between Fourier transformation (FT) and the input–output relation of a beam former. An FFT algorithm allows a minimum number of computations in performing the FT of a function. It is shown that the number of components (power dividers, combiners, and phase shifters) in a BFN design can be minimized applying the principle of the FFT. A modified version of the Butler BFN using hybrids, called hybrid matrix BFN, is also shown. We then present the operation of a Blass matrix BFN, which is followed by a Rotman lens BFN. Design equations for the Rotman lens profile are obtained using the principle of geometrical optics. Spectral domain analysis is conducted for obtaining the scattering parameters of a planar multiport Rotman lens. The chapter ends with a brief discussion of digital beam formers and optical beam formers. The principles of operations, advantages, and limitations are discussed.

Phased Array Antennas. By Arun K. Bhattacharyya
© 2006 John Wiley & Sons, Inc.

12.2   BFN USING POWER DIVIDERS

The simplest type of beam former utilizes power dividers, power combiners, and phase shifters. Figure 12.1 shows such a beam former for four simultaneous beams and four array elements. The number of beams and the number of array elements need not be equal, however. This type of BFN is capable of producing beams at predetermined locations in a multibeam array.

Observe that this particular configuration requires 4 power dividers (D), 4 power combiners (C), and 12 phase shifters (represented by small circles). Considering only 4 beams, the number of components is large. To minimize the RF loss and cost and size, every effort should be made to reduce the number of components. In the following section we present an elegant design process that minimizes the number of components.

12.3   BUTLER MATRIX BEAM FORMER

The Butler matrix beam former is a very common type of beam former used in practice. This type of beam former is used when the array needs to transmit (or receive) multiple beams simultaneously in prefixed directions. Each beam is associated with an input signal at the input port, commonly known as the *beam port*, of the BFN as depicted in Figure 12.2. The number of array elements at the *array port* is usually equal to 2^m, where m is a positive integer. The beams are mutually orthogonal, that is, each beam power can be controlled independently even if the beams have the same carrier frequency. In the following sections we will deduce the condition for beam orthogonality. Then we consider BFN realization for orthogonal beams.

12.3.1   Orthogonal Beams

A Butler matrix BFN is employed to produce multiple orthogonal beams (sometimes called component beams), that is, the beams do not couple. This orthogonality is ensured through mutually orthogonal excitation vectors at the array port. In Figure 12.2, the BFN produces four simultaneous beams A, B, C, and D. We will

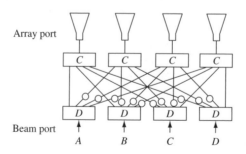

FIGURE 12.1   Simple BFN.

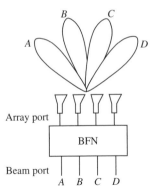

FIGURE 12.2   Schematic of a Butler BFN.

develop the theory based on a four-beam BFN; however, the analysis can be extended easily for a larger number of beams. Suppose the excitation vectors at the array port are $[a_1, a_2, a_3, a_4]$, $[b_1, b_2, b_3, b_4]$, $[c_1, c_2, c_3, c_4]$, and $[d_1, d_2, d_3, d_4]$ associated with beams A, B, C and D, respectively. Then the beams are orthogonal if

$$\sum_{i=1}^{4} a_i b_i^* = \sum_{i=1}^{4} a_i c_i^* = \sum_{i=1}^{4} c_i b_i^* \ldots = 0 \tag{12.1}$$

A Butler matrix BFN produces uniform amplitude excitation for each beam, so that the complex excitation vectors can be expressed as

$$
\begin{aligned}
[a_1, a_2, a_3, a_4] &= [1, \ \exp(-j\psi_a), \ \exp(-j2\psi_a), \ \exp(-j3\psi_a)] \\
[b_1, b_2, b_3, b_4] &= [1, \ \exp(-j\psi_b), \ \exp(-j2\psi_b), \ \exp(-j3\psi_b)] \\
[c_1, c_2, c_3, c_4] &= [1, \ \exp(-j\psi_c), \ \exp(-j2\psi_c), \ \exp(-j3\psi_c)] \\
[d_1, d_2, d_3, d_4] &= [1, \ \exp(-j\psi_d), \ \exp(-j2\psi_d), \ \exp(-j3\psi_d)]
\end{aligned}
\tag{12.2}
$$

where ψ_a is the phase difference between the adjacent antenna elements for beam A. Similarly, ψ_b, ψ_c, and ψ_d are the above phase differences for the other three beams. If (12.1) needs to be satisfied, then we must have

$$\sum_{n=0}^{3} \exp[jn(\psi_b - \psi_a)] = 0$$

$$\sum_{n=0}^{3} \exp[jn(\psi_c - \psi_a)] = 0 \tag{12.3}$$

$$\sum_{n=0}^{3} \exp[jn(\psi_d - \psi_a)] = 0$$

The above summations can be expresses in closed forms. The first one can be written as

$$\frac{\exp[4j(\psi_b - \psi_a)] - 1}{\exp[j(\psi_b - \psi_a)] - 1} = 0 \tag{12.4}$$

implying

$$4(\psi_b - \psi_a) = 2m\pi \tag{12.5}$$

where m should not be any multiple of 4 in order to make the denominator of (12.4) different from zero. We consider $m = 1$ for now, so that we have

$$\psi_b = \psi_a + \frac{1}{2}\pi \tag{12.6}$$

Similarly, from the second equation of (12.3) we will have

$$\psi_c = \psi_a + \frac{1}{2}k\pi \tag{12.7}$$

where k is an integer. We *should not* use $k = 1$ in this case, in order to have ψ_c different from ψ_b. We consider $k = 2$ and get

$$\psi_c = \psi_a + \pi \tag{12.8}$$

In a similar fashion we have

$$\psi_d = \psi_a + \frac{3}{2}\pi \tag{12.9}$$

Notice that ψ_a, ψ_b, ψ_c, ψ_d are in arithmetic progression and the common difference is $\pi/2$ for the four-beam case. It is straightforward to show that for N orthogonal beams this phase difference should be $2\pi/N$.

An interesting observation can be made from the above results of orthogonal beams. The phase angles being in arithmetic progression, the beams are equally spaced in the $\sin\theta$-space. The beam locations for the beams are given by

$$k_0 d \sin\theta_a = \psi_a \qquad k_0 d \sin\theta_b = \psi_b \qquad k_0 d \sin\theta_c = \psi_c \qquad k_0 d \sin\theta_d = \psi_d$$
$$\tag{12.10}$$

where d is the array element spacing and k_0 is the wave number in free space. For N orthogonal beams the distance between the adjacent beam peaks in $\sin\theta$-space is

$$\Delta \sin\theta_{\text{peak}} = \frac{2\pi}{Nk_0 d} \tag{12.11}$$

Figure 12.3 shows the radiation patterns of four orthogonal beams in $\sin\theta$-space. The crossover point between two adjacent beams occurs at about 3.9 dB below the beam peak. This level varies slightly with N.

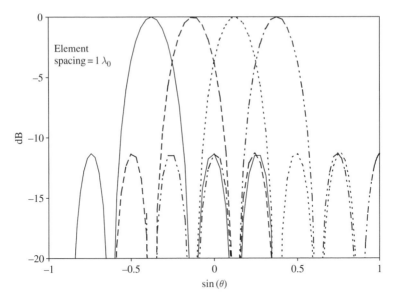

FIGURE 12.3   Radiation patterns of four orthogonal beams.

12.3.2   Fourier Transform and Excitation Coefficients

For orthogonal beams the array port voltages and beam port voltages are related through Fourier transformation. To establish this fact, let us determine the excitation voltages at the array ports. The total voltage vector at the array port must be equal to the summation of the individual array port vectors given in (12.2). Therefore, for the nth array element ($n = 0, 1, 2, 3$) the excitation voltage should be given by

$$E_n = V_a \exp(-jn\psi_a) + V_b \exp(-jn\psi_b) + \cdots \qquad (12.12)$$

We introduce the coefficients V_a, V_b, V_c, and V_d to allow independent power levels for the beams. For simplicity we assume $\psi_a = 0$. This assumption will cause beam A at bore-sight. With this assumption and using (12.6)–(12.9) we obtain the excitation coefficient for the nth array element as

$$E_n = V_a + V_b \exp\left(-jn\frac{\pi}{2}\right) + V_c \exp\left(-j2n\frac{\pi}{2}\right) + V_d \exp\left(-j3n\frac{\pi}{2}\right) \qquad (12.13)$$

Denoting $\pi/2$ as ω, we write[1]

$$E_n = V_a + V_b \exp(-jn\omega) + V_c \exp(-2jn\omega) + V_d \exp(-3jn\omega) \qquad (12.14)$$

---

[1] Not to be confused with the frequency.

The right-hand side of (12.14) can be represented by the FT of a series of Dirac delta functions given by

$$f(x) = V_a\delta(x) + V_b\delta(x-1) + V_c\delta(x-2) + V_d\delta(x-3) \tag{12.15}$$

The following definition of the FT is used:

$$F(k) = \int_x f(x)\exp(-jk\,x)\,dx \tag{12.16}$$

Notice, the desired excitation coefficients E_0, E_1, E_2, and E_3 are equally spaced samples of the Fourier spectrum $F(k)$. Mathematically we can write

$$E_n = F(n\omega) \tag{12.17}$$

In (12.15), $f(x)$ essentially corresponds to the input voltages at the beam port. We thus establish that for orthogonal beams the excitation voltages at the array port and at the beam port are related via the FT. This is pictorially shown in Figure 12.4. Notice, $F(k)$ is a periodic and continuous function of k, but we need $F(k)$ only at four discrete points of interval $\pi/2$ on the k-axis. The primary objective of a BFN

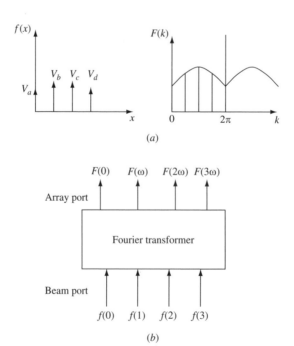

FIGURE 12.4 (a) Relation between beam port voltages and array port voltages. (b) Block diagram representing the beam port and array port.

is thus to simulate the FT through an RF circuit. In the following section we will consider an algorithm for performing the FT in an efficient manner. An efficient algorithm will help reduce the number of RF components, as we shall see shortly.

12.3.3 FFT Algorithm

We have seen that an "antenna excitation voltage" in (12.14) is a summation of four terms. Each term (except the first term) is a product of two factors: a complex number and a complex exponent. If one has to compute all four excitation voltages, then one must perform $3 \times 4 = 12$ multiplication operations. For N beams, the required number of multiplication operations becomes $N(N - 1)$. Now consider that an RF circuit, as an analog computer, performs the above computation. Multiplication of a complex exponent signifies a phase shift, which can be realized by a phase shifter. Thus, the RF circuit should have as many phase shifters as the number of multiplications. It will be seen that computations of antenna excitation voltages employing the FFT would reduce the number of multiplication operations. It implies that a BFN design with reduced number of phase shifters may be accomplished following the guideline of an FFT algorithm.

In order to understand the operation of an FFT algorithm we consider the input voltage of the nth antenna element given in (12.14). Recall that $\omega = \pi/2$; therefore, E_n repeats every four intervals with respect to the subscript n, that is,

$$E_n(\omega) = E_{n+4}(\omega) \tag{12.18}$$

This can also be understood from Figure 12.4a. Furthermore, the sum of first and the third terms of E_n in (12.14), denoted by the symbol P_n, repeats every two intervals, that is,

$$P_{n+2} = P_n \qquad P_n = V_a + V_c \exp(-2jn\omega) \tag{12.19}$$

Similarly, if we take out the common factor between the second and fourth terms of E_n, then the remaining part, Q_n, also repeats every two intervals. This periodic property of the partial sums of the input voltage is exploited to minimize the number of operations. Let us now utilize this property to manipulate E_n. We can write E_n alternately as

$$E_n = V_a + V_c \exp(-2jn\omega) + \exp(-jn\omega)[V_b + V_d \exp(-2jn\omega)]$$
$$= P_n + \exp(-jn\omega)Q_n \tag{12.20}$$

As mentioned before, $P_{n+2} = P_n$ and $Q_{n+2} = Q_n$. We thus obtain

$$E_0 = P_0 + Q_0 \qquad E_1 = P_1 + WQ_1$$
$$E_2 = P_0 + W^2 Q_0 \qquad E_3 = P_1 + W^3 Q_1 \tag{12.21}$$

with $W = \exp(-j\omega)$.

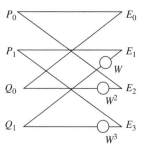

FIGURE 12.5   Signal flow graph from PQ-domain to E-domain.

We can now draw a signal flow graph from the PQ-domain to the E-domain, as shown in Figure 12.5. The circles represent multipliers and the multiplication factors are shown next to the circles. The four quantities $P_n, Q_n (n = 0, 1)$ are expressed in terms of the input voltage in (12.20). Using that we express

$$P_0 = V_a + V_c \qquad P_1 = V_a + V_c W^2$$
$$Q_0 = V_b + V_d \qquad Q_1 = V_b + V_d W^2 \tag{12.22}$$

From (12.22) we can construct the signal flow graph from the V-domain to the PQ-domain, as in Figure 12.6. Combining Figures 12.5 and 12.6 we obtain the final signal flow graph of the FFT algorithm as depicted in Figure 12.7.

In Figure 12.7 it is interesting to notice that the FFT algorithm requires only 5 multiplication operations as opposed to 12. This reduction of multiplication

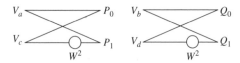

FIGURE 12.6   Signal flow graph from V-domain to PQ-domain.

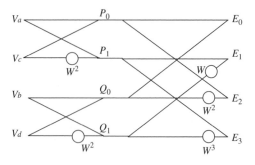

FIGURE 12.7   Signal flow graph of a four-point FFT.

operations is more dramatic for a larger number of beams. For instance, for a BFN of eight beams the corresponding eight-point FFT algorithm reduces the multiplication operations to only 17, as opposed to $56(= 8 \times 7)$ operations. Figure 12.8 shows the signal flow graph for the eight-beam case. For the eight beams $W = \exp(-j\pi/4)$. It can be shown that for 2^m beams, the FFT algorithm requires only $2^m(m-1)+1$ operations.

From the signal flow graph of the eight-point FFT we notice that the output terminals of two four-point FFT blocks are connected in a systematic fashion to obtain the desired eight output terminals. The signal flow diagram for the 2^m-point FFT essentially consists of two 2^{m-1}-point FFT blocks followed by a set of direct and cross-connected branches. This concept can be used iteratively to construct the signal flow graph for the 2^m-point FFT with an arbitrarily large m.

The FFT algorithm can be extended to the number of beams that are different from 2^m. To illustrate this, consider an example of 12 beams. Then $E_n(n = 0, 1, 2, \ldots, 11)$ can be expressed as

$$E_n = V_a + V_b W^n + \cdots + V_l W^{11n} \tag{12.23}$$

with $W = \exp(-j\pi/6)$. We rearrange the right-hand side of (12.23) by separating odd and even terms as below:

$$E_n = P_n + W^n Q_n$$
$$P_n = V_a + V_c W^{2n} + \cdots + V_k W^{10n} \tag{12.24}$$
$$Q_n = V_b + V_d W^{2n} + \cdots + V_l W^{10n}$$

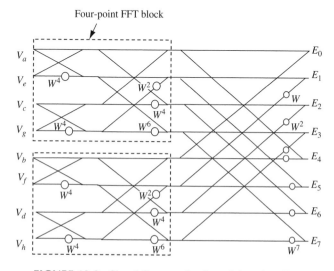

FIGURE 12.8 Signal flow graph of an eight-point FFT.

Notice, $P_{n+6} = P_n$ and $Q_{n+6} = Q_n$, that is, there are only six distinct P's and Q's. Now separating the even and odd terms in P_n we can write

$$P_n = R_n + W^{2n}S_n$$
$$R_n = V_a + V_e W^{4n} + V_i W^{8n} \qquad (12.25)$$
$$S_n = V_c + V_g W^{4n} + V_k W^{8n}$$

Now, $R_{n+3} = R_n$ and $S_{n+3} = S_n$. Similarly

$$Q_n = T_n + W^{2n}U_n$$
$$T_n = V_b + V_f W^{4n} + V_j W^{8n} \qquad (12.26)$$
$$U_n = V_d + V_h W^{4n} + V_l W^{8n}$$

For this case also we have $T_{n+3} = T_n$ and $U_{n+3} = U_n$. Equations (12.24), (12.25), and (12.26) yield the signal flow graph shown in Figure 12.9.

Thus the FFT algorithm can be applied for an array where the number of beams can be expressed as a product of two numbers. For more details the reader is referred to a standard reference of the FFT algorithm [1].

12.3.4 FFT and Butler Matrix

We have seen that an FFT algorithm can reduce the number of multiplication operations significantly for computing the FT of a signal represented by a series of

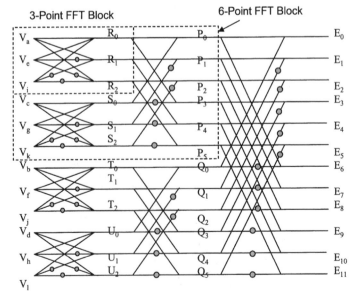

FIGURE 12.9 A 12-point FFT.

Delta functions. In a phased array, this series of Delta functions can be considered as RF signals producing multiple beams. The signal flow graph of an FFT algorithm can be followed to construct a BFN producing multiple beams. The BFN should consist of phase shifters and power divider and power combiner circuits. For instance, consider the signal flow graph in Figure 12.7. A multiplication operation with a complex exponent is equivalent to a phase shift. Therefore, the circles in Figure 12.7 can be considered as phase shifters. A junction point of three lines is equivalent to either a power divider or a power combiner, depending on the number of incoming and outgoing lines or ports. A simple equal-split power divider or a Wilkinson power combiner replaces such a junction. The equivalent circuit representation of the FFT signal flow graph is shown in Figure 12.10. The square blocks are either $1:2$ power dividers or $2:1$ power combiners that can be easily identified by the number of input and output ports. A BFN that is designed based on the FFT algorithm is known as the Butler matrix, after J. L. Butler [2].

12.3.5 Hybrid Matrix

The BFN in Figure 12.10 can be further simplified with less components using four-port hybrids. In addition, a BFN using hybrids is preferable, because a hybrid has better port-to-port isolation than a Wilkinsons' power divider. As a result the effective loss in a hybrid is lower than that of a typical power divider/combiner circuit. A hybrid has two input ports and two output ports. A single hybrid can replace two power dividers and two power combiners. The input–output relation of a hybrid is shown in Figure 12.11. The equivalent hybrid matrix BFN is shown in Figure 12.12. Notice that the number of components is reduced to 11, compared to 21 in Figure 12.10.

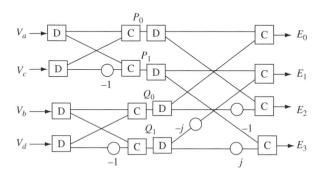

FIGURE 12.10 Butler matrix BFN.

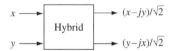

FIGURE 12.11 Input–output relation of a hybrid.

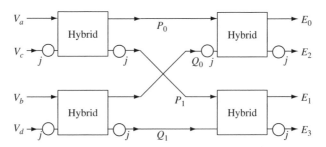

FIGURE 12.12   Butler BFN using hybrids.

It should be mentioned that if the number of beams is less than the number of array elements, then the number of beam ports should be reduced. This can be accomplished by match terminating the "unused" ports. For some cases, it is possible to eliminate some hybrids at the beam port section. For example, for producing only two beams (A and C), the left bottom hybrid may be eliminated. This elimination is not always possible, for example, in the case of beams B and C.

It is straightforward to implement a Butler BFN using regular hybrids if the number of array ports is 2^m, where m is an integer. For array ports different from 2^m one may need to use hybrids with multiple input ports. In most cases, 2^m array ports are used in order to avoid multiport hybrids, which are difficult to implement.

The Butler matrix is generally implemented using rectangular or coaxial waveguides for high-power applications [3]. For low-power applications, printed transmission lines are used [4].

12.3.6 Modified Butler BFN for Nonuniform Taper

We have seen that a Butler BFN is inherently limited to uniform taper illumination as given in (12.2). However, with additional power divider and combiner circuits a cosine taper is possible to achieve. To demonstrate, we consider the array of four elements producing four orthogonal beams. From (12.14), the array port voltages are expressed as

$$E_0 = V_a + V_b + V_c + V_d$$
$$E_1 = V_a + V_b \exp(-j\omega) + V_c \exp(-2j\omega) + V_d \exp(-3j\omega)$$
$$E_2 = V_a + V_b \exp(-2j\omega) + V_c \exp(-4j\omega) + V_d \exp(-6j\omega)$$
$$E_3 = V_a + V_b \exp(-3j\omega) + V_c \exp(-6j\omega) + V_d \exp(-9j\omega)$$

$$(12.27)$$

Recall, $\omega = \pi/2$ for four beams. Now suppose we excite two adjacent beam ports, say ports A and B, with equal magnitude and with relative phase 1.5ω. With reference to Figure 12.2, the excitation voltages at beam ports A and B are V_a and $V_a \exp(1.5j\omega)$, respectively. Temporarily assume that the remaining beam ports are not excited. Then substituting $V_b = V_a \exp(1.5j\omega)$ and $V_c = V_d = 0$ in (12.27) we obtain the excitation voltages at the antenna ports as

$$E_0 = V_a + V_a \exp(1.5j\omega) = 2V_a \cos(0.75\omega) \exp(0.75j\omega)$$
$$E_1 = V_a + V_a \exp(0.5j\omega) = 2V_a \cos(0.25\omega) \exp(0.25j\omega)$$
$$E_2 = V_a + V_a \exp(-.5j\omega) = 2V_a \cos(0.25\omega) \exp(-0.25j\omega)$$
$$E_3 = V_a + V_a \exp(-1.5j\omega) = 2V_a \cos(0.75\omega) \exp(-0.75j\omega)$$

$$(12.28)$$

Notice that the above special excitations at the beam ports result in a cosine taper at the array port. The phase difference between two adjacent antenna elements is 0.5ω. Thus the beam peak of the "composite" beam will be located between the two "component" beams of Figure 12.2. The gain of the composite beam is about 0.9 dB lower than that of a component beam because of the cosine taper. The peak side lobe for the composite beam is 23 dB below the beam peak, which is about 10 dB lower than that of a component beam. In a similar fashion we can produce other composite beams with cosine taper. For another such composite beam we excite the beam port B with V_b and the beam port C by $V_b \exp(1.5j\omega)$. The excitation circuit in the beam port region can be realized using 3-dB power dividers, power combiners, and phase shifters. For the four-beam case phase shifters should provide 135° phase shift. A block diagram of the input circuitry is shown in Figure 12.13.

It should be pointed out that the beams with cosine taper are not orthogonal, as the excitation vectors do not satisfy the orthogonality condition given in (12.1). This implies that if the signals have identical carrier frequencies, then the beams will couple. As a consequence, the individual beam power cannot be controlled independently. However, the selective beams, which are two beam widths apart, are decoupled because their excitation vectors are mutually orthogonal.

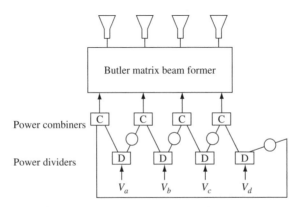

FIGURE 12.13 Modified Butler BFN for cosine taper.

12.3.7 Beam Port Isolation

The orthogonal BFN has an important characteristic that allows one to design a lossless BFN with perfect isolation between the beam ports. To illustrate this, consider a passive, lossless network with N input ports and N output ports as shown in Figure 12.14. The beams are assumed to have identical carrier frequencies. The GSM can be expressed as

$$\text{GSM} = \begin{bmatrix} S_{11} & S_{12} \\ S_{21} & S_{22} \end{bmatrix} \tag{12.29}$$

In the above S_{ij} are the $N \times N$ submatrices. For orthogonal beams, the excitation vectors at the array port are orthogonal. Suppose $V_1, V_2, \ldots, V_N$ are the incident voltages at the input (beam) ports. Then the excitation vectors at the array port corresponding to the mth beam port voltage would be given by

$$[A_m] = [S_{21}] \begin{bmatrix} 0 \\ 0 \\ \vdots \\ V_m \\ \vdots \\ 0 \end{bmatrix} \tag{12.30}$$

Since $[A_m]$ and $[A_n]$ are orthogonal vectors for $m \neq n$, we have

$$[A_m^*]^T [A_n] = P_m \delta_{mn} \tag{12.31}$$

where P_m represents the power radiated by the mth beam and δ_{mn} is the Kronecker delta. Using (12.30) in (12.31) we obtain

$$[0\,0 \quad \cdots \quad V_m^* \quad \cdots \quad 0\,0][S_{21}^*]^T[S_{21}][0\,0 \quad \cdots \quad V_n \quad \cdots \quad 0\,0]^T = P_m \delta_{mn} \tag{12.32}$$

If (12.32) is true for all m and n, then $[S_{21}^*]^T[S_{21}]$ must be a diagonal matrix. This is possible if the column vectors of $[S_{21}]$ are mutually orthogonal, that is, if $[S_{21}]$ is an orthogonal matrix.

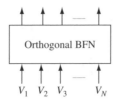

FIGURE 12.14 Orthogonal BFN.

The beam port isolation can be examined from the characteristics of the $[S_{11}]$ matrix. From power conservation of a lossless passive network we can deduce a relation between $[S_{21}]$ and $[S_{11}]$ matrices. The total radiated power is given by

$$P_{\text{rad}} = [V^*]^T [S_{21}^*]^T [S_{21}][V] \tag{12.33}$$

where $[V] = [V_1 \quad V_2 \quad V_3 \quad \cdots \quad V_N]^T$ represents the incident voltage vector (column vector). The total reflected power is given by

$$P_{\text{ref}} = [V^*]^T [S_{11}^*]^T [S_{11}][V] \tag{12.34}$$

The total incident power is

$$P_{\text{inc}} = [V^*]^T [V] \tag{12.35}$$

For a lossless network, $P_{\text{rad}} + P_{\text{ref}} - P_{\text{inc}} = 0$. Thus from (12.33), (12.34), and (12.35) we obtain

$$[V^*]^T \{[S_{11}^*]^T [S_{11}] + [S_{21}^*]^T [S_{21}] - [I]\}[V] = 0 \tag{12.36}$$

where $[I]$ represents an identity matrix. The above relation must be true for all V_i's. That happens when

$$[S_{11}^*]^T [S_{11}] + [S_{21}^*]^T [S_{21}] = [I] \tag{12.37}$$

We have seen before that $[S_{21}]$ is an orthogonal matrix. It follows that $[S_{11}^*]^T [S_{11}]$ must be diagonal in order to satisfy (12.37). For $[S_{11}^*]^T [S_{11}]$ to be a diagonal matrix, $[S_{11}]$ must be an orthogonal matrix, that is, the column vectors of $[S_{11}]$ must be mutually orthogonal. This means that the reflected voltage vectors corresponding to each beam port excitation are mutually orthogonal.

We prove that for a lossless orthogonal BFN $[S_{11}]$ matrix must be orthogonal. However, the orthogonality of the $[S_{11}]$ matrix does not ensure isolation between the beam ports because $[S_{11}]$ is not necessarily a diagonal matrix. It follows that in a lossless orthogonal BFN the beam ports are not necessarily decoupled. On the other hand, if $[S_{11}]$ is diagonal, then the BFN must have isolated beam ports and will produce orthogonal beams as well, because $[S_{21}]$ becomes an orthogonal matrix [see (12.37)]. The Butler matrix utilizing perfect hybrids belongs to such a BFN.

For a lossless nonorthogonal BFN, such as the BFN for cosine-taper contiguous beams, $[S_{21}]$ is not an orthogonal matrix; thus $[S_{11}]$ is not orthogonal. Thus $[S_{11}]$ is not a diagonal matrix because it is not even orthogonal. This result signifies that a nonorthogonal, lossless BFN cannot have isolated beam ports. On the other hand, allowing finite RF loss, the beam ports can be made isolated. To that end one can replace each power combiner of Figure 12.13 by a four-port hybrid with an output port terminated by a matched load. This arrangement will have perfect beam port isolation at the cost of 3 dB power loss at the terminated ports.

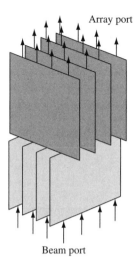

FIGURE 12.15   Conceptual three-dimensional BFN using stacked two-dimensional BFNs.

The above discussion for nonorthogonal beams is valid if the beams have identical carrier frequencies. If the beams have different carrier frequencies, then the beams do not mix. In that case, each power combiner in Figure 12.13 essentially combines two signals with different career frequencies. Such a power combiner acts like a diplexer, where the two input ports can be matched perfectly. Thus, a perfectly isolated beam port for cosine taper is possible in this case.

12.3.8 Three-Dimensional BFN

A three-dimensional BFN can be constructed by stacking several two-dimensional Butler BFNs as shown in Figure 12.15. The vertical and horizontal stacks scan the beams in two mutually orthogonal directions. The combined BFN can span the two-dimensional far-field space.

12.4 BLASS MATRIX BFN

The Blass matrix for a multiple-beam array consists of transmission line sections and multiple directional couplers. The phase shift is realized though the line length. Figure 12.16 shows a schematic of a Blass matrix BFN for three beams with four antenna elements. The circles with crosses are directional couplers.

The transmission line sections along the radial directions have equal lengths. The transmission line sections along the circumferencial directions have different lengths but make an equal angle at the center of the circles. The open ends of the transmission lines are match terminated as shown in the figure. A directional coupler has stronger coupling with the straight-through arm than with the branch arms.

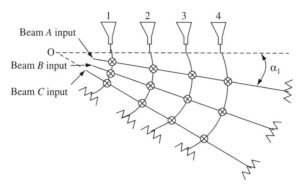

FIGURE 12.16   Schematic of a Blass matrix BFN for three beams with four antenna ports.

The operational principleal of this BFN is as follows. Suppose beam port A is excited. Suppose the radial line connected with beam port A port makes an angle α_1 with the dotted line. Then the phase of the signal at the input of antenna 1 is

$$\psi_{1A} = \psi_0 + r_1 \alpha_1 \xi \qquad (12.38)$$

where ξ represents the propagation constant of the transmission line, ψ_0 represents the combined phase due to the radial line section and the phase introduced by the coupler at the branch arm, and r_1 is the radius of the arc connecting antenna 1. The phase at the input of antenna 2 would be

$$\psi_{2A} = \psi_0 + d\xi + (r_1 + d)\alpha_1 \xi \qquad (12.39)$$

where d represents the transmission line length between two couplers along the radial direction. It is assumed that the coupler does not introduce additional phase shift to its straight arm (if it does, that can be compensated by transmission line length without affecting the analysis). Similarly, the phase at the inputs of antennas 3 and 4 are

$$\psi_{3A} = \psi_0 + 2\,d\xi + (r_1 + 2d)\alpha_1 \xi$$
$$\psi_{4A} = \psi_0 + 3\,d\xi + (r_1 + 3d)\alpha_1 \xi \qquad (12.40)$$

Notice, the phases at the antenna input are in arithmetic progression as desired. The phase difference between two successive antenna inputs is

$$\Delta\psi_A = (1 + \alpha_1)\,d\xi \qquad (12.41)$$

The location of the beam peak with respect to the array normal is given by

$$\theta_A = \sin^{-1}\left(\frac{\Delta\psi_A}{k_0 d}\right) = \sin^{-1}\left[\frac{(1 + \alpha_1)\xi}{k_0}\right] \qquad (12.42)$$

In a similar fashion, we can obtain the location of beams B and C.

It is important to point out that the beam direction depends on the ratio ξ/k_0, which is equal to the phase velocity of the RF signal in the transmission line. For TEM transmission lines (coaxial lines, for example) the phase velocity being independent of frequency (assuming a nondispersive dielectric-filled TEM line), the beam location does not change with frequency. In other words, a Blass matrix BFN with TEM lines is a true time delay matrix. On the other hand, if the phase velocity is a function of frequency, as in a waveguide transmission line, then the beam location changes with frequency. Several modified versions of the basic Blass matrix configuration can be found in the literature.

In a Blass matrix, the amplitude taper of the excitation coefficient can be controlled by suitably selecting coupling factors of the directional couplers. The circuit efficiency (=useful power/input power) for a beam is less than 100% because a finite amount of power is dumped in the matched load. In order to minimize the amplitude and phase error caused by the unintended signal paths, the directional couplers should have low coupling values with the side arms. This arrangement will reduce the circuit efficiency, but the radiation patterns of the beams will have minimum perturbation.

A systematic design procedure of a Blass matrix BFN (with modified configuration) is presented by Mosca et al. [5]. The procedure involves obtaining a matrix relation between the beam port and antenna port voltages, optimizing the desired antenna port excitation vectors with minimum power loss at the terminated loads. Over 80% circuit efficiency of the BFN for 9 beams in a 20 element linear array (element spacing $0.5\lambda_0$) is achieved. The beam width for each beam was 5°, covering 45° far-field space. The coupling values (for the side arms) for the directional couplers varied from −7 dB to about −29 dB.

12.5 ROTMAN LENS

The Rotman Lens beam former operates on the principle of geometrical optics. Typically, such a BFN is used for a wide-band operation. The schematic of a two-dimensional Rotman lens is shown in Figure 12.17. It consists of a cavity formed by two surfaces that may be called beam port and array port surfaces. The beam and array ports are situated along the periphery of the two surfaces as shown. Usually the beam port surface has a larger curvature than the array port surface. The beam port surface consists of several radiators (assuming the transmit mode operation). Each radiator corresponds to a beam. The array port surface consists of several receiving elements that are connected to the array elements through transmission lines.[2] The line lengths of the transmission lines are usually kept identical.

The principle of operation is as follows. Suppose the lens is operating in the transmit mode and only one radiator on the beam port surface is excited. Each

---

[2] For receive mode operation, the elements on the array port surface become transmitting elements and the elements on the beam port surface become receiving elements.

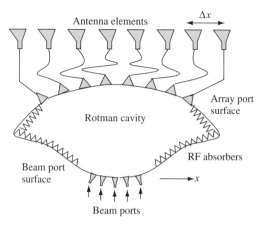

FIGURE 12.17   Schematic of a Rotman lens.

receiving element in the array port will receive a signal from the beam port radiator. The phase of the received signal is directly proportional to the patch length
between the beam port radiator and the receiving element. Now if the array port
surface is designed such that the path length varies linearly (that is, with a constant slope) with the locations of the receiving elements on the array port surface,
then the array elements will have a linear phase distribution resulting in a scanned
beam. The slope of the phase distribution changes with the location of the excited
element. Thus each beam port element is associated with a beam at a unique scan
angle.

Design of a Rotman lens includes essentially (a) design of the array port
and the beam port surfaces and (b) design of the beam port radiators and the
array port elements. A good design of the port surfaces will minimize the phase
error for all beams, that is, the lens will provide good phase linearity at the
array port. Theoretically it is impossible to obtain perfect phase linearity for all
beams. The beam port radiators and antenna port elements are designed to minimize the reflection loss and to maintain good isolation between the beam ports.
In the following section we will consider a simple technique to design the lens
surfaces.

12.5.1   Rotman Surface Design

Figure 12.18 shows the beam port surface, S_b, and the array port surface, S_a, of a
Rotman lens cavity. Typically a Rotman lens has three foci, that is, its behavior is
near perfect for three focus beams (beams corresponding to the ports at the foci).
The foci are denoted as F_1, F_2, and F_3. The beam port surface is usually a circular
arc that passes through all three foci. Suppose the coordinates of F_1, F_2, and F_3
are $(-a, b)$, $(0,0)$, and (a, b), respectively. Suppose the primary focal length is R,
which is the distance from the apex A of the array port surface S_a to the primary

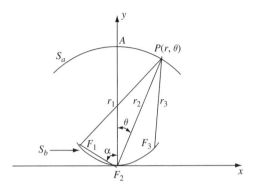

FIGURE 12.18 Beam port and array port surfaces of a Rotman lens.

focus F_2. We assume that the array port receiving elements[3] are positioned on S_a at a regular angular interval $\Delta\theta$ with respect to the primary focus F_2. The beam port at the primary focus F_2 corresponds to a bore-sight beam and the beam ports at the secondary foci F_1 and F_3 correspond to extreme-right and extreme-left beams, respectively (nevertheless, one may opt to place beam ports beyond F_1 and F_3 with larger phase errors). Suppose $2\theta_m$ is the subtended angle of the lens surface S_a at the primary focus. Then for producing three focus beams with no phase error, the radial distances r_1, r_2, and r_3 in Figure 12.18 should satisfy the following three differential equations:

$$\frac{\partial r_1}{\partial\theta} = c \qquad \frac{\partial r_2}{\partial\theta} = 0 \qquad \frac{\partial r_3}{\partial\theta} = -c \qquad (12.43)$$

where c is a constant related to the beam scan angle. In order to deduce the relation between c and other parameters, we assume that the element spacing of the array is Δx as shown in Figure 12.17. Suppose the scan angle corresponding to the left-focus beam with beam port at F_1 is θ_1. Then the phase difference between two elements of the array must be given by

$$\Delta\psi = k_0\,\Delta x\,\sin\theta_1 \qquad (12.44)$$

The above phase difference is realized by the path length difference between two receiving elements on S_a at an angular distance $\Delta\theta$. Therefore, we have

$$k_0\,\Delta x\,\sin\theta_1 = k_0\,\Delta r_1 = k_0\frac{\partial r_1}{\partial\theta}\Delta\theta \qquad (12.45)$$

From (12.43) and (12.45) we write

$$c = \frac{\partial r_1}{\partial\theta} = \frac{\Delta x}{\Delta\theta}\sin\theta_1 \qquad (12.46)$$

---

[3] These receiving elements should not be confused with the antenna elements. The antenna elements are situated on a plane, whereas the receiving elements are situated on the curved surface S_a.

Using the following simple manipulation, c can be expressed as

$$c = \frac{\Delta x}{\Delta \theta} \sin \theta_1 = \frac{N \Delta x}{N \Delta \theta} \sin \theta_1 = \frac{L_{array}}{2\theta_m} \sin \theta_1 \tag{12.47}$$

where N is the number of array elements and L_{array} is the array length.

From (12.47) we notice that, c being a constant, the maximum scan angle θ_1 is not fixed for a Rotman lens but is a function of (almost inversely proportional to) the length of the radiating array L_{array}. Interestingly, (12.47) indicates that the maximum scan angle in terms of the beam-width unit is fixed for a given Rotman lens (see problem 12.6).

We obtain the solutions of the differential equations in (12.43) as

$$r_1 = c\theta + R_1 \qquad r_2 = R \qquad r_3 = -c\theta + R_1 \tag{12.48}$$

where R_1 is the secondary focal length, the distance between A and F_1. Clearly, the three equations cannot be satisfied simultaneously on a single surface because $r_2 = R$ represents a circle, and the other two are linear in θ. Thus, we need to find an equation of a surface that will have minimum deviations from three ideal curved surfaces. To that end we assume that the equation of the surface in polar form is given by

$$r = r(\theta) \tag{12.49}$$

where r is the radial distance on the surface from the principal focus F_2. The first equation in polar form is given by

$$\begin{aligned} r &= d\cos(\alpha + \theta) + \sqrt{r_1^2 - [d\sin(\alpha+\theta)]^2} \\ &= d\cos(\alpha + \theta) + \sqrt{(c\theta + R_1)^2 - [d\sin(\alpha+\theta)]^2} \end{aligned} \tag{12.50}$$

where d is the distance between two foci F_1 and F_2 and α is the angle made by the secondary focus F_1 with the y-axis. Similarly, the second and third equations in polar form are expressed as

$$r = R \tag{12.51}$$

$$r = d\cos(\alpha - \theta) + \sqrt{(-c\theta + R_1)^2 - [d\sin(\alpha - \theta)]^2} \tag{12.52}$$

We thus have three equations. To obtain the "best-fit" curve we simply take the average value[4] of r. Thus, the equation of the surface becomes

$$\begin{aligned} r = \frac{1}{3}\{ &R + d\cos(\alpha - \theta) + \sqrt{(-c\theta + R_1)^2 - [d\sin(\alpha - \theta)]^2} \\ &+ d\cos(\alpha + \theta) + \sqrt{(c\theta + R_1)^2 - [d\sin(\alpha + \theta)]^2}\} \end{aligned} \tag{12.53}$$

---

[4] One may as well consider different weights instead of equal weights for obtaining the best-fit curve.

where the parameters are

$$d = \sqrt{a^2 + b^2} \qquad \alpha = \tan^{-1}\left(\frac{a}{b}\right) \qquad R_1 = \sqrt{d^2 + R^2 - 2Rd\cos\alpha} \qquad (12.54)$$

Notice that the four parameters a, b, R, and c uniquely determine the equation of the surface. These four parameters can be adjusted to obtain the best possible surface for a desired performance of a Rotman lens.

12.5.2 Numerical Results

An array port surface was realized using (12.53) and the theoretical performance of the corresponding lens was evaluated. The following parameters were used: $a = 4\lambda_0$, $b = 2\lambda_0$, $R = 7\lambda_0$, and $\theta_m = 35°$. The beam port surface is considered as a circular arc passing through three foci. Figure 12.19 shows the phase distributions at the array port corresponding to the three focus beams. The amplitude distribution is assumed as the far-field intensity pattern of a beam port radiator. The radiator is considered as a line source of length equal to the port width, which is assumed as $1\lambda_0$ for present computations. Figure 12.20 shows the far-field array patterns of 9 beams associated with 9 beam ports. For the radiating array, 21 omnidirectional radiating elements of $1\lambda_0$ spacing were considered. The highest side lobe was 13.9 dB and that occurred for the edge beams. The maximum scan angle was 15°, which is about five beam widths away from the bore-sight beam.

The four parameters a, b, c, and R can be optimized to design a lens surface. However, it is found that for a given set of a, b, and R, the parameter c must

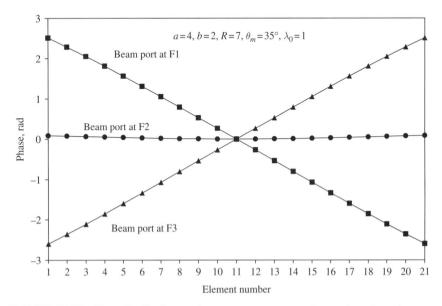

FIGURE 12.19 Phase distributions at the array port corresponding to three focus beams.

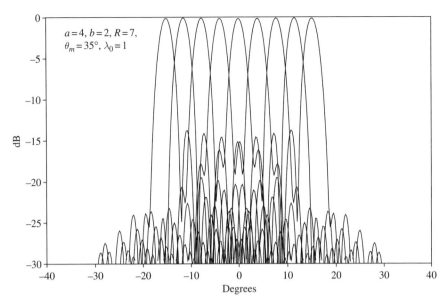

FIGURE 12.20 Far-field array patterns of nine beams produced by a Rotman lens BFN.

be very close to a particular number determined by a, b and R. Otherwise, the three curves in (12.48) would be very different from each other. Consequently, the average curve in (12.53) will have large deviations from the ideal surfaces, causing a large number of phase errors for all three focus beams.

The approximate value of c can be obtained assuming that the lens surface is very close to a circular arc of radius R. Then from Figure 12.18 we obtain

$$\frac{\partial r_1}{\partial \theta} \approx \frac{Rd \sin(\alpha + \theta)}{r_1} \tag{12.55}$$

Notice that $\partial r_1 / \partial \theta$ is a function of θ. The average value of $\partial r_1 / \partial \theta$ gives an estimate of c. One simple way of estimating this parameter is using $\theta = 0$ in (12.55), which gives

$$c \approx \frac{Rd \sin \alpha}{\sqrt{R^2 + d^2 - 2Rd \cos \alpha}} = \frac{Ra}{\sqrt{a^2 + (R - b)^2}} \tag{12.56}$$

From (12.47) and (12.56), the maximum scan angle θ_1 is given by

$$\sin \theta_1 \approx \frac{2\theta_m R a}{L_{\text{array}} \sqrt{a^2 + (R - b)^2}} \tag{12.57}$$

From (12.57) we find that the beam scanning range, θ_1, increases as θ_m increases. However, the phase linearity at the array port deteriorates with an increase in θ_m.

This results in gain loss and side-lobe deterioration. On the other hand, a smaller value of θ_m causes larger spillover loss. The scan angle, θ_1, can also be increased by increasing a. Equation (12.57) can be employed to obtain the effects of various physical parameters on the maximum scan angle. Effects of parameters on lens performance are shown in Table 12.1. The numerical data give an idea for selecting appropriate parameters for a desired performance.

12.5.3 Physical Implementation

In one embodiment, two parallel plates form the cavity of the Rotman lens. The beam ports and the receiving ports are realized by rectangular waveguides supporting the TE_{10} modes, as shown in Figure 12.21. The impedance match is accomplished using

TABLE 12.1 Performance of Rotman Lens with Respect to Different Parameters

a/λ_0	b/λ_0	R/λ_0	θ_m (deg)	Gain Loss (Bore Sight) (dB)	Gain Loss (Maximum Scan Beam) (dB)	Maximum Scan Angle θ_1(deg)	Side-Lobe Ratio for Scan Beam (dB)	Spillover Loss (dB)
4	2	7	30	0.05	0.06	13	14.03	1.79
—	—	—	35	0.13	0.11	15	13.90	1.11
—	—	—	40	0.18	0.22	17	13.45	0.66
4	2	10	30	0.00	0.09	13	13.05	1.79
—	—	—	35	0.01	0.20	15	12.45	1.11
—	—	—	40	0.03	0.35	17	11.76	0.66
5	2	10	35	0.14	0.18	18.5	12.75	1.11

Note: Number of array elements 21, array element spacing $1\lambda_0$, beam port width $1\lambda_0$. Array elements are omnidirectional. The spillover loss is due to the absorbing material.

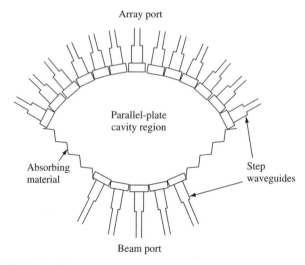

FIGURE 12.21 Rotman lens using rectangular waveguide ports.

steps or tapering the waveguide walls. The reflections from side walls are minimized by installing lossy material. This type of Rotman lens is used for millimeter-wave frequencies to minimize the copper loss. A detailed description with measured data can be found in the literature [6, 7]. A maximum scan angle of about 22° (about seven beam widths) and an insertion loss lower than 2.3 dB are reported. Analysis of multiple reflections due to aperture mismatch is performed invoking a scattering matrix formulation.

Another embodiment employs printed circuit technology. A parallel-plate cavity and printed stripline circuits constitute the lens. Figure 12.22 depicts the center plate and the ports that should be placed between two infinite ground planes. To minimize reflections, some of the dummy ports are match terminated with resistive loads. This configuration is generally used in an active array as a low-level BFN because high loss in the resistive loads reduces the signal strength. The signal loss at the Ka-band frequency typically exceeds 5 dB.

A three-dimensional Rotman lens BFN can be implemented by stacking multiple two-dimensional Rotman len BFNs using the concept depicted in Figure 12.15. It is also possible to design a single lens surface to have beam scanning in two planes [8, 9].

12.5.4 Scattering Matrix

The scattering matrix of a multiport Rotman lens can be determined applying the principle of equivalence. A port aperture at the lens periphery is replaced by equivalent electric and magnetic surface currents that are related to the port current and port voltage. These equivalent currents produce electromagnetic fields inside the lens cavity. A linear relation between the port voltages and port currents leads to the impedance matrix of the lens. The scattering matrix is determined from the impedance matrix. We will present the analysis for the printed lens configuration

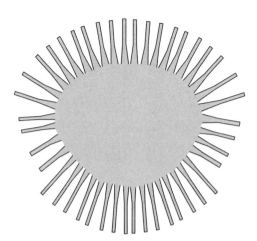

FIGURE 12.22   Center plate (between two ground planes) of a printed Rotman lens.

shown in Figure 12.22. However, the analysis can be easily extended for other configurations.

Suppose $V_1, V_2, \ldots, V_N$ are the port voltages and $I_1, I_2 \ldots, I_N$ are the port currents (current flowing toward the cavity),[5] where N represents the total number of ports (including the loaded ports) of Figure 12.22. We assume that the port apertures are touching each other, so that no space exists between port apertures. If a space exists in reality, one has to define a "virtual port" at that location for the present analysis. The virtual port can be loaded appropriately (for example, with open circuit) after the mutiport scattering matrix is obtained.

Using the principle of equivalence, a port aperture can be replaced by equivalent electric and magnetic surface currents [10] given by

$$\hat{z} I_{sm} = -\hat{z} \frac{I_m}{W_m} \qquad \hat{t}_m M_{sm} = \hat{t}_m \frac{V_m}{h} \tag{12.58}$$

where W_m represents the width of the mth port; h is the port height, which, in this case, is the distance between the center conductor and a ground plane; $\hat{z}$ is a unit vector perpendicular to the cavity plane; and $\hat{t}_m$ is a unit vector along the lens contour at the location of the mth port. The midpoint of the mth port has the coordinate (x_m, y_m). Suppose the z-component of the electric field inside the lens cavity produced by the above surface currents is given by

$$E_{zm}(x, y) = I_m E^I_{zm}(x, y/x_m, y_m) + V_m E^M_{zm}(x, y/x_m, y_m) \tag{12.59}$$

where E^I_{zm} and E^M_{zm}, respectively are the electric fields at (x, y) produced by unit electric and magnetic currents at the mth port. Then, the total electric field inside the cavity would be

$$E_z(x, y) = \sum_{m=1}^N I_m E^I_{zm}(x, y/x_m, y_m) + V_m E^M_{zm}(x, y/x_m, y_m) \tag{12.60}$$

The voltage at the ith port location can be determined from the field inside the cavity. This voltage must be identical with V_i. Thus we can write

$$V_i = -h E_z(x_i, y_i) = -h \sum_{m=1}^N \left[I_m E^I_{zm}(x_i, y_i/x_m, y_m) \right.$$
$$\left. + V_m E^M_{zm}(x_i, y_i/x_m, y_m) \right] \qquad i = 1, 2, \ldots, N \tag{12.61}$$

Equation (12.61) can be rearranged into the following matrix format:

$$[A][V] = [B][I] \tag{12.62}$$

---

[5] Using symmetry, we only consider the cavity region between the center conductor and the lower ground plane without affecting the end results.

where $[A]$ and $[B]$ are square matrices of order $N \times N$ and $[V]$ and $[I]$ are column vectors of order N. The impedance matrix $[Z]$ of the multiport lens is given by

$$[Z] = [A]^{-1}[B] \qquad (12.63)$$

From (12.61) and (12.62), the elements of $[A]$ and $[B]$ are obtained as

$$a(i, m) = \delta_{mi} + hE_{zm}^{M}(x_i, y_i/x_m, y_m)$$
$$b(i, m) = -hE_{zm}^{I}(x_i, y_i/x_m, y_m) \qquad (12.64)$$

The matrix elements will be determined next.

Spectral Domain Formulation The matrix elements in (12.64) are directly related to the electric fields produced by electric and magnetic surface currents of the mth port located at (x_m, y_m). To obtain the fields we employ the spectral domain analysis. To simplify the analysis, we assume that the sources are located at $(0,0)$ instead of (x_m, y_m) and the source plane is on the xz-plane, as shown in Figure 12.23. Later we will do the necessary coordinate adjustment. First we consider the magnetic surface current of length W_m, of width h, and directed along x as shown. We assume that h is sufficiently small compared to λ, the wavelength inside the cavity, so that the equivalent magnetic current is uniform along z. The EM fields produced by a magnetic current source can be obtained from the electric vector potential [10], $\vec{F}$, which satisfies the differential equation

$$\nabla^2 \vec{F} + k^2 \vec{F} = -\hat{x} M_{sm} \delta(y) \qquad (12.65)$$

Notice, the Dirac delta function stands for the surface current on the surface $y = 0$. The source will produce only the x-component of $\vec{F}$ that will satisfy

$$\nabla^2 F_x + k^2 F_x = -M_{sm} \delta(y) \qquad (12.66)$$

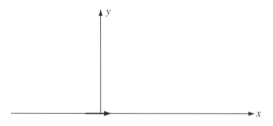

FIGURE 12.23 Equivalent magnetic current source located at (0,0) point.

For small h, the EM fields do not vary with z; thus (12.66) simplifies to

$$\frac{\partial^2 F_x}{\partial x^2} + \frac{\partial^2 F_x}{\partial y^2} + k^2 F_x = -M_{sm}\delta(y) \tag{12.67}$$

We express F_x in terms of its Fourier spectrum as below:

$$F_x(x, y) = \int\limits_{-\infty}^{\infty} \int\limits_{-\infty}^{\infty} \tilde{F}_x(k_x, k_y) \exp(-jk_x x - jk_y y) \, dk_x \, dk_y \tag{12.68}$$

Substituting (12.68) into (12.67) and then taking the inverse FT we obtain

$$\tilde{F}_x(k_x, k_y) = -\frac{1}{4\pi^2(k^2 - k_x^2 - k_y^2)} \iint\limits_{xy} M_{sm}\delta(y) \exp(jk_x x + jk_y y) \, dx \, dy \tag{12.69}$$

Now, $M_{sm} = V_m/h$ for $|x| \le W_m/2$ and $M_{sm} = 0$ for $|x| > W_m/2$; thus, performing the integration with respect to x and y, we obtain

$$\tilde{F}_x(k_x, k_y) = -\frac{V_m W_m}{4\pi^2 h(k^2 - k_x^2 - k_y^2)} \operatorname{sinc}\left(\frac{k_x W_m}{2}\right) \tag{12.70}$$

where $\operatorname{sinc}(p) = \sin(p)/p$. Substituting the expression for $\tilde{F}(k_x, k_y)$ in (12.68) and then performing the integration with respect to k_y, we obtain

$$F_x(x, y) = -j\frac{V_m W_m}{4\pi h} \int\limits_{-\infty}^{\infty} \frac{\operatorname{sinc}(k_x W_m/2)}{\sqrt{k^2 - k_x^2}} \exp(-jk_x x - j\sqrt{k^2 - k_x^2}\, y) \, dk_x \tag{12.71}$$

We use the following identity to obtain the above expression:

$$\int\limits_{-\infty}^{\infty} \frac{\exp(-jk_y y)}{k_y^2 - a^2} \, dk_y = -j\frac{\pi}{a} \exp(-jay) \tag{12.72}$$

The z-component of the electric field is obtained using $\vec{E} = -\nabla \times (\hat{x} F_x)$, yielding

$$V_m E_{zm}^M(x, y) = -\frac{V_m W_m}{4\pi h} \int\limits_{-\infty}^{\infty} \operatorname{sinc}\left(\frac{k_x W_m}{2}\right) \exp(-jk_x x - j\sqrt{k^2 - k_x^2}\, y) \, dk_x \tag{12.73}$$

Recall, E_{zm}^M is the electric field corresponding to a unit voltage source as per (12.60); hence the voltage term V_m appears on the left-hand side of (12.73). Furthermore,

E^M_{zm} being a function of x and y, it is reasonable to use the average value[6] of E^M_{zm} on the ith aperture in (12.61). The average value is given by

$$\bar{E}^M_{zm}(x_i, y_i) = \frac{1}{W_i} \int_{W_i} E^M_{zm}(x, y) \, dl$$

$$= -\frac{W_m}{4\pi h} \int_{-\infty}^{\infty} \text{sinc}\left(\frac{k_x W_m}{2}\right) \text{sinc}(\sigma_{im})$$

$$\times \exp(-jk_x x_i - j\sqrt{k^2 - k_x^2} y_i) \, dk_x \qquad (12.74)$$

where W_i is the width of the ith port-aperture and

$$\sigma_{im} = \frac{1}{2} W_i (k_x \cos\alpha + \sqrt{k^2 - k_x^2}\sin\alpha) \qquad \alpha = \cos^{-1}(\hat{t}_i \cdot \hat{t}_m) \qquad (12.75)$$

Recall that we considered the source at $(0,0)$. If we place the source at (x_m, y_m), then the average electric field at (x_i, y_i) can be obtained by replacing x_i and y_i by $x_i - x_m$ and $y_i - y_m$, respectively. Thus we have

$$\bar{E}^M_{zm}(x_i - x_m, y_i - y_m) = -\frac{W_m}{4\pi h} \int_{-\infty}^{\infty} \text{sinc}\left(\frac{k_x W_m}{2}\right) \text{sinc}(\sigma_{im})$$

$$\times \exp[-jk_x(x_i - x_m) - j\sqrt{k^2 - k_x^2}(y_i - y_m)] \, dk_x \quad (12.76)$$

The matrix element $a(i, m)$ becomes

$$a(i, m) = \delta_{im} - \frac{W_m}{4\pi} \int_{-\infty}^{\infty} \text{sinc}\left(\frac{k_x W_m}{2}\right) \text{sinc}(\sigma_{im})$$

$$\times \exp[-jk_x(x_i - x_m) - j\sqrt{k^2 - k_x^2}(y_i - y_m)] \, dk_x \qquad (12.77)$$

For $i = m$, the integral becomes equal to $1/2$. This is primarily due to the fact that a planar magnetic surface current immersed in an isotropic and homogeneous medium will produce electric fields on both sides of the source plane of equal magnitude and opposite in polarity. Therefore for numerical computation one can use

$$a(i, m) = \begin{cases} -\dfrac{W_m}{4\pi} \int_{-\infty}^{\infty} \text{sinc}\left(\dfrac{k_x W_m}{2}\right) \text{sinc}(\sigma_{im}) \\[2mm] \quad \times \exp[-jk_x(x_i - x_m) - j\sqrt{k^2 - k_x^2}(y_i - y_m)] \, dk_x & i \neq m \quad (12.78) \\[2mm] \dfrac{1}{2} & i = m \end{cases}$$

---

[6] The present formulation is equivalent to Galerkin's MoM formulation with subdomain basis functions. A test with pulse weighting function boils down to taking average values.

Employing a similar procedure the electric field produced by the equivalent electric current can be determined. The magnetic vector potential is determined in this case instead of the electric vector potential. The matrix element $b(i, m)$ is obtained from the average electric field at the ith port aperture. The end result is obtained as

$$b(i, m) = \frac{k^2 h}{4\pi\omega\varepsilon} \int_{-\infty}^{\infty} \frac{1}{\sqrt{k^2 - k_x^2}} \mathrm{sinc}\left(\frac{k_x W_m}{2}\right) \mathrm{sinc}(\sigma_{im})$$

$$\times \exp[-jk_x(x_i - x_m) - j\sqrt{k^2 - k_x^2}(y_i - y_m)]\, dk_x \qquad (12.79)$$

For numerical computation, it is convenient to divide the entire range of integration in three subdomains as

$$\int_{-\infty}^{\infty} (\cdot)\, dk_x = \int_{-\infty}^{-k} (\cdot)\, dk_x + \int_{-k}^{k} (\cdot)\, dk_x + \int_{k}^{\infty} (\cdot)\, dk_x \qquad (12.80)$$

The first and the last integrals on the right-hand side of (12.80) are computed after substituting $k_x = k \tan \gamma$, while for the middle integral the substitution is $k_x = k \sin \gamma$. With these substitutions, the apparent singularity due to the denominator term in $b(i, m)$ is removed, making the numerical computation straightforward. Furthermore, for fast computations asymptotic expressions for the integrals can be used for the ports that are far from each other. The asymptotic expressions for $a(i, m)$ and $b(i, m)$ are deduced as

$$a(i, m) \approx j \frac{W_m}{4} k \frac{y_i - y_m}{\rho_{im}} H_1^{(2)}(k\rho_{im})$$

$$b(i, m) \approx \frac{\omega\mu h}{4} H_0^{(2)}(k\rho_{im}) \qquad (12.81)$$

with $\rho_{im} = \sqrt{(x_i - x_m)^2 + (y_i - y_m)^2}$ and $H_n^{(2)}$ represents the Hankel function of the second kind of order n. It is found numerically that the asymptotic expressions can be used if ρ_{im}/λ is greater than 6 and the port widths are less than $\lambda/4$.

The impedance matrix of the planar lens shown in Figure 12.24 was computed and the scattering matrix was obtained from the impedance matrix. The number of ports were considered as 121 (see Figure 12.24) in order to make the port widths sufficiently small (less than $\lambda/4$). The small port widths are required to satisfy the uniform field assumption that was used to deduce the expressions for the matrix elements. In order to check the consistency of the scattering matrix solution, we compute the following quantity with respect to a given port (say the mth port):

$$P_m = \sum_{i=1}^{N} \frac{|S(i, m)|^2 Y_i}{Y_m} \qquad (12.82)$$

where $S(i, m)$ are scattering matrix elements representing coupling between the ith and mth ports when all ports are match terminated. In (12.82) Y_m represents the characteristic admittance of the transmission line connected with the mth port,

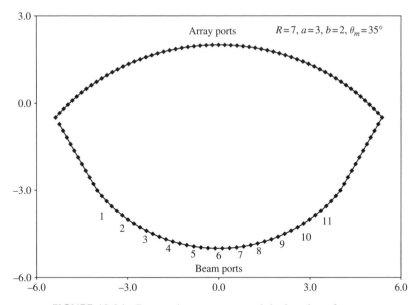

FIGURE 12.24 Rotman lens geometry and the location of ports.

which is assumed as $\sqrt{\mu/\varepsilon}(W_m/h)$. This is the characteristic admittance of a TEM waveguide with magnetic walls. From power conservation, P_m should be unity; however, due to the uniform field assumption and other approximations involved in numerical computations, this quantity differs from unity. It is found that for port widths less than $\lambda/4$, the computed P_m differs from unity by 5% or less. This error increases with the port widths. If the actual port widths are larger (greater than $\lambda/4$, for instance), then more than one virtual port needs to be considered within a physical port for the scattering matrix computation, and the virtual ports matrix elements can be combined to obtain the scattering parameter of the actual ports. For example, if three virtual ports of equal port widths make an actual port, then the formula for the reflection coefficient of a port constituted by $m-1$, m, and $m+1$ virtual ports may be computed as

$$\Gamma = \frac{1}{3} \sum_{i=m-1}^{m+1} \sum_{j=m-1}^{m+1} S(i,j) \tag{12.83}$$

The coupling (or isolation) between two actual ports (specially beam ports) are of interest. The mutual coupling parameter between two consecutive ports (actual) is

$$T = \frac{1}{3} \sum_{i=m-1}^{m+1} \sum_{j=m-1}^{m+1} S(i, j+3) \tag{12.84}$$

For different port widths of the virtual ports, the formula should be modified.

Figure 12.25 shows the return loss of the beam ports (three virtual ports make a beam port in this case) at three different frequencies. The return loss is better

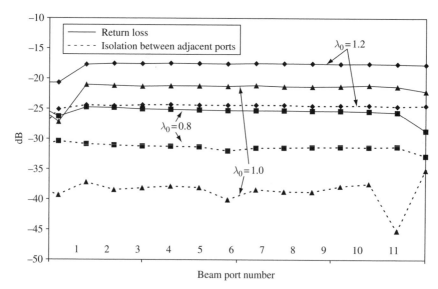

FIGURE 12.25 Return loss and isolation of the beam ports of the Rotman lens in Figure 12.24.

than −17 dB for all beam ports. The lens is capable of operating within 40% bandwidth. The isolation between two adjacent beam ports is also shown. The worst case isolation is 24 dB, which is considered to be good for many applications. The isolation can be improved if more than three virtual ports are combined to make a beam port. With respect to a beam port excitation, the phase distribution at the array port was obtained directly from the scattering matrix elements. The phase distribution is found to be consistent with the intended phase distribution within 5°. The array ports are assumed to be $\lambda/4$ apart for the numerical data.

It is worth pointing out that the spectral domain analysis presented here uses a "voltage match" condition [see (12.61)]. One can use the "current match" condition as well. It is found that both formulations yield similar results if the virtual port widths are sufficiently small. For the present example with 121 ports, the difference in the return loss is within 3 dB at the −20-dB level. If the virtual port widths are wider than $\lambda/4$, then the difference increases. The voltage match spectral domain formulation presented here is equivalent to the contour integral approach [11], except the latter is conducted in the space domain as opposed to the spectral domain.

12.6 DIGITAL BEAM FORMER

A digital beam former (DBF) offers significant flexibility in regard to beam forming and beam reconfiguration. Typically, a digital beam former is used in a receiving array for ultra low side-lobe reception, adaptive pattern nulling, and antenna self-calibration. The principle of operation of a digital beam former differs from that of the passive beam formers, such as the Butler BFN and Rotman lens BFN discussed before. To understand the basic operation of a DBF, we consider an analog equivalent of a DBF. Consider a receiving array of N elements receiving a signal from an angular direction θ, as shown in Figure 12.26. The BFN consists of N phase shifters and a power combiner block. A DBF essentially utilizes a similar concept, but the phase delaying and power combining are conducted in the digital domain instead of the analog domain. To better understand this, let us first consider the operation principle of a digital phase shifter.

12.6.1 Digital Phase Shifter

An analog phase shifter typically utilizes a delay line causing a phase delay or phase shift. A delay line of a given length provides only a fixed amount of phase shift (unless a switching mechanism is introduced). However, by using a $90°$ hybrid and two amplifiers one can generate any amount of phase shift, as shown in the block diagram of Figure 12.27.

The 90° hybrid essentially is a 3-dB power divider and provides 90° phase difference between two output signals represented by cosine and sine functions.

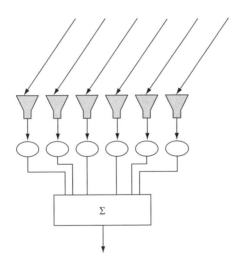

FIGURE 12.26 Receiving array.

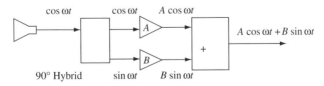

FIGURE 12.27 Schematic of an analog phase shifter utilizing a hybrid, amplifiers, and a combiner.

The output signals pass through two amplifiers with gains A and B, respectively. The last block is a power combiner. By properly choosing A and B, the phase of the output signal can be modified as desired. The output signal can be represented by

$$V_0(t) = A \cos \omega t + B \sin \omega t = \sqrt{A^2 + B^2} \cos(\omega t - \psi) \qquad \psi = \tan^{-1}\left(\frac{B}{A}\right)$$

By selecting A and B, we can introduce any amount of phase shift. Furthermore, A and B also control the magnitude of the output signal that is required for implementing an amplitude taper for a low-side-lobe array pattern.

As mentioned before, the entire operation is conducted in the digital domain in a DBF. The digital equivalence of a 90° hybrid is a "Hilbert transformer (HT)," which converts the signal $\cos \omega t$ to $\sin \omega t$ using the following operation, known as Hilbert transformation:

$$\sin \omega t = \frac{1}{\pi} \int\limits_{-\infty}^{\infty} \frac{\cos \omega \tau}{t - \tau} d\tau$$

The above integral is equivalent to a convolution between $\cos \omega t$ and $1/\pi t$ that can be performed using a digital filter. The amplification and combination in the digital domain are nothing but multiplication and addition that can be performed on a digital computer. The digital equivalence of the analog beam former of Figure 12.27 is shown in Figure 12.28.

A complete DBF consists of digital phase shifter blocks connected with each antenna element and the output of the phase shifters are added together to obtain the final output from a beam, as in Figure 12.26. The number of digital phase shifters connected with an array element should be equal to the number of receiving beams in an array.

12.6.2 System Characteristics

A DBF is generally employed in the case of a narrow-band signal because of the limited processing speed of a digital computer. For example, a signal of 1 MHz bandwidth requires a 2-MHz sampling rate. For a modest array of 400 elements, the number of multiplication operations would be 2×2 MHz $\times 400$ million per second (1600 Mops).

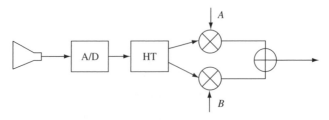

FIGURE 12.28 Digital phase shifter that evolved from Figure 12.27.

For a wide-band signal, this Mops number becomes so high that a modern digital computer cannot handle such an operational speed. Another important factor is the dynamic range (P_{max}/P_{min}) of the system, which is determined by the number of bits (B) of the analog-to-digital (A/D) converter and the number of array elements (N). The dynamic range increases 6 dB per bit for a given number of array elements. The relation is given by Steyskal [12]:

$$\frac{P_{max}}{P_{min}} = 6(B-1) + 10 \log N \qquad \text{dB}$$

We limit our discussion to the fundamental principle of operation of a DBF. A practical DBF may include a few more stages (for instance, A/D conversion may be done at an intermediate frequency level, instead of the carrier level, and thus a down-converter block is needed before the DBF block). Also, a different algorithm, such as the FFT algorithm, may be invoked for simultaneous beams in the digital domain. For details, the reader is encouraged to read Steyskal's paper [12].

12.7 OPTICAL BEAM FORMERS

In an optical beam former (OBF), the RF signal is amplitude modulated with an optical frequency using a nonlinear photodiode. Then the optical signal is propagated through optical fiber transmission lines. The phase sift (time delay) is implemented by adjusting the length of the optical fibers. Figure 12.29 shows a schematic of a received array with an OBF. The major advantage of an OBF is that the fiber links are lightweight and have low optical loss and thus could be useful for an airborne array antenna. Furthermore, an OBF is a true time delay beam former and thus has wide bandwidth. The disadvantage is that the RF-to-optical-frequency conversion

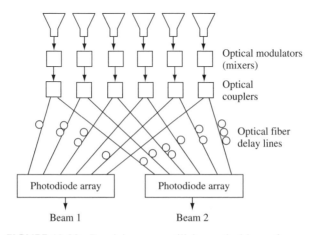

FIGURE 12.29   Receiving array utilizing optical beam former.

incurs a large signal loss; thus the signal-to-noise ratio becomes poor. It is expected that this conversion loss will be minimized with the advent of solid state technology at the optical domain and the OBF may be attractive for future commercial uses. A recent paper on the OBF for a large number of beams is presented by Minasian and Alameh [13].

REFERENCES

[1] M. H. Hayes, *Digital Signal Processing*, Schaum's Outline Series, McGraw-Hill, New York, 1999, Chapter 7.

[2] J. L. Butler and R. Lowe, "Beam Forming Matrix Simplifies Design of Electronically Scanned Antennas," *Electron. Design*, Vol. 9, pp. 170–173, 1961.

[3] R. Levy, "A High Power X-Band Butler Matrix," *Microwave J.*, Vol. 27, p. 153, Apr. 1984.

[4] J. R. Willington, "Analysis, Design and Performance of a Microstrip Butler Matrix," in *Proc. European Microwave Conference*, Brussels, Belgium, Sept. 1973.

[5] S. Mosca, F. Bilotti, A. Toscano, and L. Vegni, "A Novel Design Method for Blass Matrix Beam-Forming Network," *IEEE Trans.*, Vol. AP-50, pp. 225–232, Feb. 2002.

[6] A. F. Peterson and E. O. Rausch, "Scattering Matrix Integral Equation Analysis for the Design of a Waveguide Rotman Lens," *IEEE Trans.*, Vol. AP-47, pp. 870–878, May 1999.

[7] E. O. Rausch, A. F. Peterson and W. Wiebach, "Millimeter wave Rotman Lens," in *Proc. of 1997 IEEE National Radar Conference*, Syracuse, May 1997, pp. 78–81.

[8] G. C. Sole and M. S. Smith, "Design and Performance of a Three Dimensional Rotman Lens for a Planar Antenna Array," *Proceedings of ISAP*, 1985, Vol. 112–1, pp. 369–372.

[9] J. B. L. Rao, "Multifocal Three-Dimensional Bootlace Lenses," *IEEE Trans.*, Vol. AP-30, pp. 1050–1056, Nov. 1982.

[10] R. F. Harrington, *Time-Harmonic Electromagnetic Fields*, McGraw-Hill, New York, 1961.

[11] K. C. Gupta and M. D. Abouzahra, "Planar Circuit Analysis," in T. Itoh (Ed.), *Numerical Techniques for Microwave and Millimeter-Wave Passive Structures*, Wiley, New York, 1989.

[12] H. Steyskal, "Digital Beamforming Antennas," *Microwave J.*, Vol. 30, pp. 107–124, Jan. 1987.

[13] R. A. Minasian and K. E. Alameh, "Optical-Fibre Grating-Based Beamforming Network for Microwave Phased Arrays," *IEEE Trans.*, Vol. MTT-45, pp. 1513–1518, Aug. 1997.

BIBLIOGRAPHY

Butler, J. L., "Digital, Matrix, and Intermediate Frequency Scanning," in R. C. Hansen (Ed.), *Microwave Scanning Antennas*, Peninsula Publication, Los Altos, CA, 1985, Chapter 3.

Chan, K. K., and K. Rao, "Design of a Rotman Lens Feed Network to Generate a Hexagonal Lattice of Multiple Beams," *IEEE Trans.*, Vol. AP-50, pp. 1099–1108, Aug. 2002.

Gagnon, D. R.,"Procedure for Correct Refocusing of the Rotman Lens According to Snell's Law," *IEEE Trans.*, Vol. AP-37, pp. 390–392, Mar. 1989.

Hall, P. S., and S. J. Vetterlein, "Review of Radio Frequency Beamforming Techniques for Scanned and Multiple Beam Antennas," *IEE Proc.*, Vol. 137, Pt. H, No. 5, pp. 293–303, Oct. 1990.

Hansen, R. C. in *Phased Array Antennas*, Wiley, New York, 1998, "Multiple-Beam Antennas" Chapter 10.

Mailloux, R. J., in *Phased Array Antenna Handbook*, Artech House, 1994, "Special Array Feeds for Limited Field-of-View and Wide band Arrays," Chapter 8.

Rotman, W., and R. F. Turner, "Wide-Angle Microwave Lens for Line Source Applications," *IEEE Trans.*, Vol. AP–11, pp. 623–632, Nov. 1963.

Stark, L., "Microwave Theory of Phased Array Antennas—A Review," *Proc. IEEE*, Vol. 62, No. 12, pp. 1661–1701, Dec. 1974.

Yuan, N., J. S. Kot, and A. J. Parfitt, "Analysis of Rotman Lens Using a Hybrid Least Squares FEM/Transfinite Element Method," *IEE Proc.*, Vol. MAP-148, No. 3, pp. 193–198, June 2001.

PROBLEMS

12.1 Develop the eight-point FFT algorithm and construct the signal flow graph.

12.2 Prove that for 2^m-point FFT the number of multiplication operations is $2^m(m-1)+1$.

12.3 Obtain the nine-point FFT and construct the signal flow graph.

12.4 Show that in the modified Butler BFN in Figure 12.13, if two cosine beams are two beam widths apart, then the beams become orthogonal. Construct a hybrid BFN with four array ports and two beam ports with perfect beam isolation.

12.5 Determine the amplitude and phase distributions corresponding to beam A and beam B for the Blass matrix BFN shown in Figure 12.16. Assume identical directional couplers with power coupling values of 0.8 with the straight arm and 0.2 with the side arm, that is, with reference to Figure P12.5,

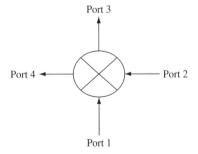

FIGURE P12.5 Schematic of a directional coupler.

$|S_{31}|^2 = |S_{42}|^2 = 0.8$, $|S_{41}|^2 = |S_{32}|^2 = 0.2$, $|S_{21}| = |S_{12}| = 0$. Assume TEM transmission lines, $d = 0.5\lambda$, $\alpha_1 = 30°$, $\alpha_2 = 60°$. Phases of S_{31} and S_{42} are zero, phase of $S_{41} =$ phase of $S_{32} = 90°$. What are the circuit efficiencies for beam A and beam B?

12.6 Using (12.57), deduce an expression for the number of beam widths that can be scanned by a Rotman lens BFN. Use the formula for the 3-dB beam width $\Delta\theta(3 \text{ dB}) \approx \lambda_0/L_{\text{array}}$ radian. Also use $\sin\theta_1 \approx \theta_1$.

12.7 Using the contour integration method, show that for $y > 0$ and $Im(a) < 0$,

$$\int_{-\infty}^{\infty} \frac{\exp(-jk_y y)}{k_y^2 - a^2} dk_y = -j\frac{\pi}{a}\exp(-jay)$$

Hence deduce (12.71).

12.8 Assuming W_m is very small, obtain the asymptotic expression for F_x in (12.71). Hence deduce the asymptotic formulas for $a(i, m)$ and $b(i, m)$ in (12.81).

12.9 Deduce the formulas for the reflection coefficient if three virtual ports having characteristic admittances Y_{01}, Y_{02}, and Y_{03} are combined together to have an actual port.

CHAPTER THIRTEEN

Active Phased Array Antenna

13.1 INTRODUCTION

In modern communication satellites in-orbit reconfigurability has become a neces-
sity due to rapid changes in the marketplace. A phased array is the most appropriate
antenna structure that allows in-orbit reconfigurability. However, a conventional
phased array has only one RF power source; as a result it is susceptible to a single
point failure caused by the breakdown of the power source. In an active phased
array antenna, on the other hand, the RF power source is distributed at the element
level. Thus an active phased array overcomes the limitation of the single point
failure because the antenna remains operational if a small percentage of the elements fail. The radiated power, however, decreases gradually with the number of
failed elements (graceful degradation), which may be preferable than a total loss of
power at once. Major disadvantages of an active array are high fabrication cost, low
solid state power amplifier (SSPA) efficiency causing power dissipation, and the
unwanted intermodulation (IM) product generation that degrades the signal quality
to some extent.

   The purpose of this chapter is to introduce the basic structures and subsystems
of active array antennas and highlight important system design aspects which are
of concern to an antenna system designer. We begin with generic block diagrams
of active arrays and qualitatively explain the functions of the blocks. We then
consider the design of the radiating aperture for a spot beam array. This includes the
procedure for estimating the minimum number of elements and the corresponding
element size with respect to a desired gain within a coverage region. Then we focus
on the active elements behind the radiating aperture, including SSPAs and phase
shifters. We define the important system parameters of the solid state components

Phased Array Antennas.   By Arun K. Bhattacharyya
© 2006 John Wiley & Sons, Inc.

and explain them in the context of antenna performance. The SSPA nonlinearity, which causes unwanted IM product, is studied. A quantitative analysis of an IM beam, including location and power level, is presented. To aid in the array system analysis the noise temperature and noise figure of various components are deduced. A typical example of array system analysis is presented. A discussion of the active array calibration methods is presented at the end.

13.2 ACTIVE ARRAY BLOCK DIAGRAMS

An active array antenna structure consists of passive and active RF devices, DC and digital control circuitries and radiating elements. In order to accommodate these components, two different architectures, namely *brick* and *tile* architectures, are commonly used. In the brick architecture, the DC and digital control circuits and RF devices are mounted on a single panel perpendicular to the aperture plane. An array consists of many such panels. In the tile architecture, the radiating elements, RF devices, and DC and digital control circuits are mounted on different panels located at different layers parallel to the aperture plane. These panels are mechanically integrated using a thermal conducting structure (generally aluminum support lattice) which also acts as a thermal contact between the active modules and the spacecraft. Because a significant amount of heat dissipation occurs in active RF elements (usually SSPAs), efficient thermal design of the array architecture is crucial. For frequencies over 20 GHz, brick architecture is preferred over tile architecture because in a tile architecture a module surface area is limited by the unit cell area of the radiating elements [1]. In order to maximize the EIRP (effective isotropic radiating power), it is necessary to optimize each component, such as the radiating element efficiency, power-added efficiency of the SSPA, and power consumption of the control electronics.

Figure 13.1 shows a generic block diagram of an active array consisting of variable-phase-shifters and power combiners that constitute the beam forming network. A DC circuit controls the phase shift. A variable attenuator (not shown in the figure) at the amplifier output usually controls the amplitude. This configuration essentially provides maximum flexibility in regard to beam location, beam shape, and side-lobe requirement, primarily because the phase and amplitude of the signals at the antenna input can be controlled independently. Each block is represented by rectangles (dashed lines) generally mounted on a single panel.[1] Each panel may have multiple layers of dielectric materials, and the total number of layers depends on the design architecture and function of the panel.

Another configuration utilizing a low-level passive beam former is also possible, as depicted in Figure 13.2. This configuration, as such, is limited to multiple fixed

---

[1] This is true for tile architectures only. For an efficient packaging, one may choose to combine two blocks in a single panel, for instance the SSPA and the BPF blocks, assuming that a printed bandpass filter (BPF) is used.

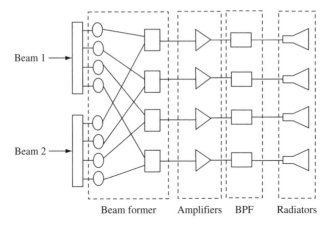

FIGURE 13.1   Generic active transmit array with independent amplitude and phase control.

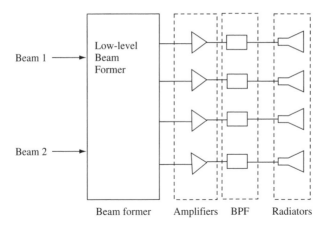

FIGURE 13.2   Active transmit array using low-level beam former.

beams or predetermined shaped beams because independent in-orbit phase control is not possible. A Butler matrix or Rotman lens (see Chapter 12) is typically used as the low-level beam former.

The configurations we presented correspond to an active transmit array antenna. In the case of a receive array, the sequence of the components is different. The amplifier is placed next to the antenna, followed by the phase shifter module, BPF, and power combiner modules. For multiple beams, a power divider module is placed between the antenna and the phase shifter.

In order to minimize the weight and cost, every effort should be made to minimize the number of elements in an active array. In the following sections we will present a procedure to determine the minimum number of array elements and the corresponding element size for a given array gain.

13.3 APERTURE DESIGN OF ARRAY

One of the most important tasks of an active array designer is to select appropriate array size and element size. The array size is generally decided by the beam size and the upper limit of the element size is decided by the grating lobe locations. In most cases, the covering beam edge is defined by the -3-dB-relative-gain contour with respect to the beam peak. It is found that this definition of the coverage region may not yield an optimum aperture design for a spot beam, where a minimum gain needs to be satisfied inside the region. On the other hand, to minimize the cost of the array the number of radiating elements should be as low as possible, because that would subsequently reduce the number of active modules (phase shifter, SSPA, attenuator, and BPF make a module). Active modules are the most expensive parts in an active array, especially for multiple beam applications, where the beam forming modules become very complex from an implementation point of view. In this section we will discuss the methodology for determining the minimum number of radiating elements for a spot beam [2].

13.3.1 Number of Elements and Element Size

In a spot beam array, the edge of the coverage (EOC) directivity at the maximum scan angle is generally specified (for instance, maximum scan angle $8°$ from bore sight, beam diameter $1°$, desired EOC directivity for the $8°$ scanned beam 37 dBi). In order to obtain the number of array elements, we first deduce the directivity of a square array of $n \times n (= N)$ elements with element size $a \times a$ and then study the EOC gain behavior of the array. We assume uniform aperture distributions for the elements (100% aperture efficiency) and uniform array excitation. If the array is phased to scan at a direction (θ_s, ϕ_s), then the array factor becomes

$$\text{AF}(\theta, \phi) = \frac{\sin[k_0 a \sqrt{N}(u_0 - u)/2]}{\sin[k_0 a (u_0 - u)/2]} \frac{\sin[k_0 a \sqrt{N}(v_0 - v)/2]}{\sin[k_0 a (v_0 - v)/2]} \tag{13.1}$$

where

$$u_0 = \sin \theta_s \cos \phi_s \qquad v_0 = \sin \theta_s \sin \phi_s$$
$$u = \sin \theta \cos \phi \qquad v = \sin \theta \sin \phi$$

In the above (θ, ϕ) is the observation direction and k_0 is the wave number in free space. The normalized element pattern of a square element of size $a \times a$ with uniform aperture distribution can be expressed as

$$\text{EP}(\theta, \phi) = \frac{\sin(k_0 a u/2)}{k_0 a u/2} \frac{\sin(k_0 a v/2)}{k_0 a v/2} \tag{13.2}$$

The radiation pattern of the array is the product of the array factor and the element pattern. We now normalize the array pattern with respect to 4π radiated power by introducing a constant term. The final array pattern becomes

$$\mathrm{AP}(\theta, \phi) = \frac{a}{\lambda_0}\sqrt{\frac{4\pi}{N}}\,\mathrm{AF}(\theta, \phi)\,\mathrm{EP}(\theta, \phi) \tag{13.3}$$

In (13.3) λ_0 is the wavelength in free space. The directivity in dBi is given by

$$\mathrm{Directivity\ (dBi)} = 10\,\log|AP|^2 \tag{13.4}$$

To verify, the directivity for the bore-sight beam is $10\log(4\pi Na^2/\lambda_0^2)$ dBi, which is expected from an array of N elements with uniform excitation and uniform aperture distribution of the elements.

In order to obtain the optimum element size and minimum number of elements for a given EOC directivity, we generate a number of parametric curves employing (13.3) and (13.4). Figure 13.3 shows the EOC directivity versus a with N as a parameter. For the plot, we consider the scan location at $\theta_s = 8°, \phi_s = 0°$. The beam diameter is assumed as $1°$; therefore the EOC directivity is computed at $\theta = 8.5°, \phi = 0°$. For a given N, the EOC directivity initially increases, reaches a maximum, and then decreases. The optimum element size corresponds to the peak directivity. For example, with $N = 144$, the peak EOC directivity is 37.7 dBi and the optimum element size $a = 2.9\lambda_0$. The peak directivity increases with increasing N, but the corresponding optimum element size decreases. The broken-line curve is the locus of the peak EOC points. This curve can be utilized to determine the optimum element size for a desired EOC directivity. For a desired EOC directivity of 37 dBi,

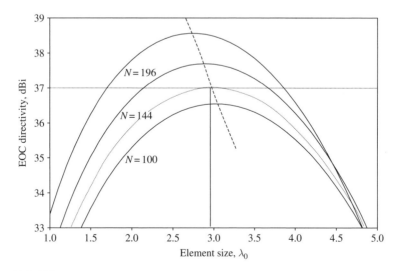

FIGURE 13.3 EOC directivity versus element size a with N as a parameter.

the optimum element size is the abscissa of the point, where the 37-dBi horizontal line meets with the locus curve, which in this case is about $2.95\lambda_0$. The number of elements N would be between 100 and 144, which can be found by interpolation. The element number N should be such that the corresponding parametric curve peaks at 37 dBi. It is found that the parametric curves in Figure 13.3 that are generated for square apertures are valid for circular and hexagonal apertures also with negligible error.

It should be pointed out that we assumed uniform amplitude taper for the array. For nonuniform amplitude taper, commonly applied for low side lobes, one should generate the parametric curves using an appropriate array factor. However, one can still use Figure 13.3 for a nonuniform taper. In this situation a larger value of the EOC directivity should be used to compensate for the loss of the aperture efficiency caused by the nonuniform taper. Table 13.1 yields the taper loss versus edge taper of a circular array aperture. It is found that this procedure is fairly accurate (within ±0.2 dB) for small scan angles (less than $10°$).

Another important point to consider is the actual aperture efficiency of the radiating elements. Figure 13.3 uses 100% aperture efficiency by considering uniform aperture distribution for an element. To compensate for a lower element-aperture-efficiency, a higher EOC directivity should be used particularly for small scan angles from bore-sight.

13.3.2 Radiating Element Design Consideration

In order to minimize the array size and number of array elements, the aperture efficiency of the radiating element should be as high as possible. We have seen that the optimum element size is decided by the EOC directivity. For an array of horns, the element design is straightforward, because a horn radiator can be designed to have the exact aperture dimensions dictated by the EOC directivity. Typical smooth-wall pyramidal horns have aperture efficiency between 70 and 80%. A multimode step horn radiator can be designed to have over 90% aperture efficiency [4].

TABLE 13.1 Side-Lobe Level and Array Aperture Efficiency of a Circular Array for Different Edge Taper

Edge Taper, C (dB)	Side-lobe Level(dB)	Array Aperture Efficiency(dB)
8	24.7	−0.37
10	27.0	−0.57
12	29.5	−0.79
14	31.7	−1.01
16	33.5	−1.23

Note: The array aperture distribution considered is a parabolic taper on a pedestal given by $V(r) = C + (1 - C)[1 - (r/R)^2]^2$, where R = radius of the array, r = distance of an element from center, and C-edge taper.
Source: Form [3]

TABLE 13.2 Aperture Efficiency of Subarray of Patch Elements

Subarray Elements	Directivity of Subarray (dBi)	Aperture Efficiency (%)
1×1	5.81	5.06
2×2	11.41	21.78
3×3	17.8	96.25
4×4	18.67	99.54

Note: Subarray size $2.5\lambda_0 \times 2.5\lambda_0$. patch element size $0.35\lambda_0 \times 0.35\lambda_0$.

The scenario is very different in the case of patch elements. A radiating patch cannot be of any size, because unlike a horn element the patch dimensions are dictated by the resonant frequency. Therefore, in the case of patch elements, multiple resonant patches should fill the optimum element space; otherwise the element efficiency becomes low. An optimum patch array thus is an array of subarrays. The number of elements in a subarray is a compromise between feed complexity within a subarray and the aperture efficiency of the subarray. More patch elements are better for aperture efficiency, but that increases the complexity of the feed distribution network. It has been found that if the patch element spacing (center to center) is less than $1\lambda_0$, the aperture efficiency, excluding the RF feed loss, is very close to 100%. Table 13.2 shows the aperture efficiency versus number of patch elements in a subarray of size $2.5\lambda_0 \times 2.5\lambda_0$. This aperture efficiency is the embedded efficiency in the array environment that includes mutual coupling effects, as discussed in Chapter 4. Exact analysis of an array of subarrays including mutual coupling has been presented in Chapter 5.

In the next few sections we focus our discussion on the components used in a typical active array antenna.

13.4 SOLID STATE POWER AMPLIFIER

An active array aperture design, particularly a transmit array aperture design, strongly depends on the characteristics of available SSPAs. For instance, with respect to a given EIRP requirement, the required array directivity depends on the output power of the SSPAs. In this section we present a brief discussion of various technologies available for an SSPA with respect to compactness, RF power output, and noise level. This will be followed by definitions of system parameters in light of an active array design.

The power amplifier circuit should be compact in order to accommodate it within a unit cell. This is particularly true for high-frequency arrays, because the physical size of a unit cell is proportional to the wavelength of operation. Microwave integrated circuit (MIC) technology and more recently monolithic microwave integrated circuit (MMIC) technology offer the required compactness. In a hybrid MIC discrete components such as resistors, capacitors, and transistors are bonded on a thin dielectric substrate (alumina, Teflon, etc.). The transmission lines are printed onto the same substrate. In an MMIC, the active and passive components are grown on a semiconductor substrate (most common substrate is GaAs; also used are Si and InP). For mass-scale production, the MMIC is cost effective, because the manual labor is eliminated in the manufacturing process. One disadvantage of the MMIC compared to the MIC is that, due to the small size of the MMIC, the amount of the output power is limited by heat dissipation.

Commercial microwave MMIC SSPAs are designed using MESFET (metal–semiconductor field-effect transistor), HEMT (high-electron-mobility transistor), or HBT (heterojunction bipolar transistor) technologies. Gallium arsenide (GaAs) based devices have high electron mobility and high electron saturation drift velocity, resulting in low noise figure and high output power. Recently, Silicon–germanium (SiGe) HBTs are gaining popularity for improved noise performances. The HMET technology offers an improvement over MESFET technology, because the former exhibits lower noise figure and higher gain than the latter. Detailed characterization of the technology is beyond the scope of the book. A comprehensive treatment can be found elsewhere [5].

13.4.1   System Characteristics

The following characteristic features must be considered when selecting an appropriate SSPA for an active array design:

A. Input–output characteristics (power and phase)
B. Power-added efficiency (PAE)
C. Noise figure

Input–Output Characteristics   The input–output characteristics essentially provide the RF signal gain and the maximum RF power that can be obtained from an SSPA. A typical input–output characteristic of an SSPA is shown in Figure 13.4. The output power varies linearly with input power until saturation. The insertion phase also varies with input power. For a single-beam antenna, the operating point is set near saturation in order to maximize the output power, which in turn maximizes the EIRP and minimizes power dissipation. For more than one beam (multibeam array), the operating point is kept in the linear region to reduce the intensities of IM beams that otherwise could degrade the signal-to-noise ratio. The IM beam is analyzed in more detail in Section 13.6.

Power-Added Efficiency   Power-added efficiency is another important parameter of an SSPA. It signifies the percentage of the total DC power converted to RF output. The PAE is defined as the ratio between net output RF power and total DC power drawn from the DC source. Mathematically, PAE can be expressed as

$$\text{PAE} = \frac{P_{\text{out}} - P_{\text{inp}}}{P_{\text{DC}}} \times 100\% \qquad (13.5)$$

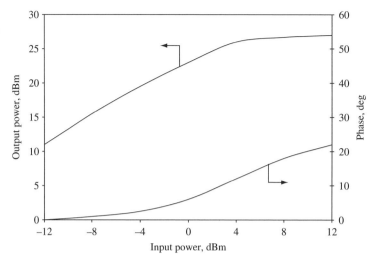

FIGURE 13.4 Input–output characteristics of an SSPA at Ka band.

where P_{inp} and P_{out} respectively are the input and output RF power of the SSPA and P_{DC} is the power drawn from the DC source. From PAE, one can determine the SSPA power dissipation and DC power consumption. Power dissipation data is very important to an active array designer in order to design the heat sink of the antenna system. Power dissipation P_{dis} can be expressed in terms of PAE and the input–output RF power as

$$P_{\text{dis}} = P_{\text{DC}} - P_{\text{out}} + P_{\text{inp}} = P_{\text{out}} \left[1 - \frac{1}{G} \right] \left[\frac{100}{\text{PAE}} - 1 \right] \tag{13.6}$$

where G represents the power gain of the amplifier. Typical PAE of a Ka-band SSPA is about 30%. Thus for an SSPA of 0.5 W output power and with $G = 10$, the power dissipation can be computed as 1.05 W. For an array of 500 elements, the total power dissipation (only for SSPAs) would be over 500 W. A thermal dissipation structure is needed to remove the excess heat.

The PAE of an SSPA decides the total DC power consumption for a given RF output power. These PAE data are utilized to estimate the required DC power source in a satellite with respect to a desired EIRP. Figure 13.5 shows the typical PAE characteristic of an SSPA. The PAE becomes highest near saturation. An SSPA is generally operated near the saturation point to avail of its highest PAE.

Noise Figure The noise figure (NF) of an SSPA is directly related to the quality of the signal radiated by the antenna. The NF of a device is a measure of the noise power introduced by the device to the output signal. It is defined as

$$\text{NF} = \left. \frac{S_i/N_i}{S_o/N_o} \right|_{N_i = kTB} \tag{13.7}$$

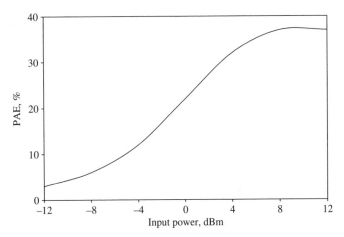

FIGURE 13.5 Power-added efficiency versus input power of an SSPA.

where S_i = signal power at the input of the device (SSPA in this case)
$\quad S_o$ = signal power at the output of the device
$\quad N_i = kTB$ = assumed input noise power to define NF
$\quad N_o$ = noise power at the output of the device
$\quad k$ = Boltzmann constant
$\quad T$ = standard room temperature (290 K)
$\quad B$ = bandwidth of the SSPA

Suppose G is the gain of the SSPA. Then $S_o = GS_i$. Substituting in (13.7) we obtain

$$\text{NF} = \frac{N_o}{GN_i} = \frac{N_o}{GkTB} \tag{13.8}$$

The output noise power is the summation of two uncorrelated noise sources: noise introduced by the SSPA and the "amplified" input noise. Thus the device-introduced noise N_d is given by (with reference to the SSPA input port)

$$N_d = \frac{N_o - GN_i}{G} \tag{13.9}$$

From (13.8) and (13.9) we write

$$N_d = (NF - 1)kTB \tag{13.10}$$

Equation (13.10) yields the noise power introduced by a device. Notice, the device-introduced noise is independent of the input noise. Figure 13.6 shows a block diagram of an SSPA and its equivalent model to facilitate output noise power computation. For many applications, the simplified model in Figure 13.6*b* can be used for signal-to-noise ratio computation at the output for any signal-to-noise ratio at the input.

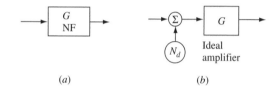

FIGURE 13.6 (*a*) Amplifier. (*b*) Simplified model for *S/N* ratio computation.

FIGURE 13.7 *M* modules in cascade.

In an active array several devices are connected in cascade; therefore, the combined noise figure is of interest. If *M* devices are cascaded in series as shown in Figure 13.7, the overall noise figure is obtained as

$$\mathrm{NF} = \mathrm{NF}_1 + \frac{\mathrm{NF}_2 - 1}{G_1} + \frac{\mathrm{NF}_3 - 1}{G_1 G_2} + \cdots + \frac{\mathrm{NF}_M - 1}{G_1 G_2 \cdots G_{M-1}} \tag{13.11}$$

The noise figure of an SSPA is strongly dependent on the frequency of operation. For example, the typical noise figure of an HEMT MMIC amplifier (that has 10% bandwidth and 15 dB gain) could be less than 0.5 dB near 2 GHz frequency. The noise figure is about 2 dB near 30 GHz frequency. The MESFET devices have higher noise figure than HEMT devices. Around 30 GHz, a typical MESFET amplifier has about 4 dB noise figure. Table 13.3 shows an overview of frequency bands, output powers, PAEs, and noise figures of SSPAs employing various technologies. The noise figures of other passive devices will be considered in Section 13.7.2.

It should be mentioned that the noise figure essentially designates the intrinsic noise of an SSPA. For a multichannel signal, the IM noise must also be included

TABLE 13.3 An Overview of RF Performance of SSPAs Using Various Technologies

Device Technology	Frequency (GHz)	Output power (dBm)	PAE (%)	Noise Figure (dB)
GaAs MESFET	1–18	35–27	65–15	0.4–2.5
GaAs HBT	1–20	33–30	50–10	1.0–4.0
InP DHBT	1–22	35–26	60–40	2.0–5.0
GaAs HEMT	4–100	40–28	65–15	0.5–4.5
InP HEMT	60–195	23–10	40–10	1.7–5.0
GaN HEMT	2–36	46–39	50–10	1.0–4.0

Note: The data is compiled from various sources in the open literature.

in the noise analysis. In this situation, the noise power ratio (NPR) provides the necessary information about the total output noise power. The NPR is considered in Section 13.6.3.

13.5 PHASE SHIFTER

Phase shifters allow an array to scan a spot beam or reconfigure a shaped beam in real time. An ideal phase shifter should have low insertion loss, predictable phase accuracy with respect to the bias control, and minimum amplitude variation with respect to the frequency of operation. Typically three types of phase shifters are used in a phased array: (a) ferrite phase shifters (b) analog phase shifters, and (c) digital phase shifters. A ferrite phase shifter operates based on the principle that the permeability of a ferrite material varies with the intensity of a DC magnetic field bias; hence the phase shift of an RF signal through the material changes with bias field intensity. The analog phase shifter exploits the property of a varactor (variable capacitor) diode where the capacitance varies with the bias voltage. A digital phase shifter relies on two-state RF switches (on/off states) such as positive–intrinsic–negative (PIN) diodes or FET switches implemented in a MMIC design. The insertion phase is altered by appropriately changing the state of the switches using DC bias voltages.

Selection of the type of phase shifter depends on the phased array configuration and the RF power level. For a high-power passive array, ferrite phase shifters are preferable because of their low insertion loss and high power handling capacity. For a moderate-power array, analog phase shifters are used that have somewhat lower insertion loss. For an active array antenna, the accuracy of the phase and the size of the phase shifter are more critical than the insertion loss (because the signal power is amplified after the phase shift); thus digital phase shifters are usually preferred. We will limit our discussion to digital phase shifters only. For further details see [6].

Figure 13.8 shows a conceptual sketch of a four-bit digital phase shifter[2]. A four-bit phase shifter employs four two-state switches that can change states with bias voltages. When a switch is off, the RF signal must travel through the adjacent transmission line section experiencing a phase delay. When the switch is on, the signal finds a straight-through path, and thus no phase delay occurs. Thus two phase states occur corresponding to the two states of the switch. In some phase shifter design, the transmission lines are replaced by lumped circuit elements such as resistors and capacitors. By using proper combination of the on/off states, one can realize any discrete number of phase states between 0 and 360°. The four-bit phase shifter in Figure 13.8 is capable of yielding $2^4 (= 16)$ discrete phase states at $360/16 = 22.5°$ interval. The maximum phase quantization error for this

---

[2] The reader should not be confused with the processor-based "digital phase shifter" presented in Figure 12.28, although we use the same name.

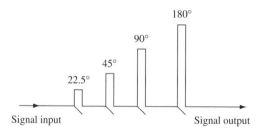

FIGURE 13.8 Conceptual sketch of a 4-bit digital phase shifter.

phase shifter is 11.25°. To minimize phase quantization error, the number of bits, and hence number of switches, should be increased. This will add to the circuit complexity and RF loss. The number of bits is selected based on the desired tolerance interval of the side-lobe level for the far-field pattern of an array. The more the number of bits, the less is the tolerance interval.

The phase shifter circuit is implemented either in MIC or MMIC technologies, typically using MESFET switches. The transmission lines are either microstriplines or coplanar waveguides on a dielectric substrate. Typical insertion loss of an MMIC phase shifter lies between 4 and 15 dB for a signal frequency between 3 and 30 GHz. The insertion loss is larger for higher frequencies.

13.6 INTERMODULATION PRODUCT

An IM product arises when two or more signals pass through a nonlinear device. In the case of a multibeam array, multiple signals with different carrier frequencies pass through SSPAs that have nonlinear characteristics. As a result, the output consists of many other frequency components in addition to the input carrier frequencies. To illustrate we present a simple analysis of two CW (carrier wave) signals. Figure 13.9 shows a simplified model of a unit cell of an active array with two CW signals. The signals have amplitudes P and Q with angular frequencies ω_1 and ω_2 and phases β_1 and β_2, respectively. The phase is responsible for the beam location. We assume

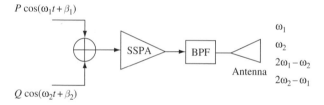

FIGURE 13.9 Unit cell of an array of two beams.

that the nonlinearity of the SSPA at the frequency band of interest can be modeled as a third-degree polynomial[3] as

$$V_{out} = AV_{in} + BV_{in}^2 + CV_{in}^3 \tag{13.12}$$

Substituting $V_{in} = P\cos(\omega_1 t + \beta_1) + Q\cos(\omega_2 t + \beta_2)$ in (13.12) we obtain

$$V_{out} = A[P\cos(\omega_1 t + \beta_1) + Q\cos(\omega_2 t + \beta_2)] + B[P\cos(\omega_1 t + \beta_1)$$
$$+ Q\cos(\omega_2 t + \beta_2)]^2 + C[P\cos(\omega_1 t + \beta_1) + Q\cos(\omega_2 t + \beta_2)]^3 \tag{13.13}$$

With algebraic manipulations we can express V_{out} as a summation of several CW signals with different frequencies. The BPF will reject the signals that are far out of the band, for example signals with frequencies $2\omega_1$, $2\omega_2$, and other higher frequencies. The signal at the output of the BPF consists of only four frequencies as below:

$$C_1 = P[A + 0.75CP^2 + 1.5CQ^2]\cos(\omega_1 t + \beta_1) \tag{13.14}$$

$$C_2 = Q[A + 0.75CQ^2 + 1.5CP^2]\cos(\omega_2 t + \beta_2) \tag{13.15}$$

$$I_1 = \frac{3}{4}CPQ^2\cos(2\omega_2 - \omega_1 + 2\beta_2 - \beta_1) \tag{13.16}$$

$$I_2 = \frac{3}{4}CP^2Q\cos(2\omega_1 - \omega_2 + 2\beta_1 - \beta_2) \tag{13.17}$$

In the above, C_1 and C_2 are the amplified version of the two input signals and I_1, I_2 represent the IM products. Notice, the IM frequencies differ from the input signal frequencies.

In Figure 13.10 we have plotted the input signal power (P in decibels) versus an output carrier power (C_1 in decibels) and an IM (I_1 in decibels) power for $A = 10$ and $C = -0.1$. We assumed $P = Q$ for this plot. The carrier power initially increases with increasing input power and reaches a saturation point. Expectedly, the IM power in decibels varies linearly with the input power in decibels. The slope of the line is 3 for $P = Q$, as apparent from (13.16). The difference in the ordinates indicates the carrier power level with respect to the IM power level. The dashed line tangential to the carrier output power curve is the extension of the low signal gain curve that has a constant slope of unity. These two lines intersect at a point, known as the *IM intersect point*. The significance of the IM intersect point is that the low signal voltage gain (A) of the SSPA and the output IM power can be determined if the location of this point is specified.

It is important to notice that the SSPA nonlinearity causes coupling between the signals, that is, the amplification factor of one signal is dependent upon the

---

[3] Typically, the input–output characteristics are estimated from single-tone measurement data at the operating band.

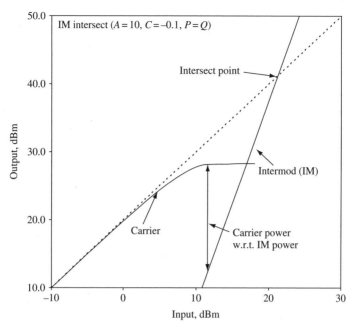

FIGURE 13.10 Input signal power versus output carrier power and IM power for $A = 10$ and $C = -0.1$.

input amplitudes of the other. The SSPA nonlinearity also determines the IM power level. In the next section we will discuss a method that can be used to find SSPA parameters that quantify the nonlinearity.

13.6.1 Estimation of SSPA Parameters from IM Data

From (13.14)–(13.17) we notice that the parameters A and C play major roles in carrier amplification and IM power generation; thus it is important to estimate these two parameters. The ratio between A and C can be determined from the carrier and IM amplitudes given in (13.14) and (13.16). The carrier-to-IM-amplitude ratio is obtained as

$$\frac{\overline{C}_1}{\overline{I}_1} = \frac{A + 0.75CP^2 + 1.5CQ^2}{0.75CQ^2} \tag{13.18}$$

where the bars denote the amplitudes for the signals (ignoring cosine factors). Performing a simple algebraic manipulation we obtain the ratio A/C as

$$\frac{A}{C} = 0.75Q^2 \left[\frac{\overline{C}_1}{\overline{I}_1} - 2 \right] - 0.75P^2 \tag{13.19}$$

Thus the A/C ratio of an SSPA can be determined from the measured input power of both carriers, the output power of one carrier, and the output power of one IM product. From the measured value of the IM output power, the value of C can be estimated directly from (13.16). Thus A can be determined from (13.19).

The A/C ratio measurement technique discussed here is generally referred to as the "two-tone measurement technique" in the literature, because the input signals used are CW signals with zero bandwidths (tone). Typically, these parameters are estimated from the measured data using a single tone at a band of interest.

13.6.2 IM Beam Locations

For an active array producing multiple beams, the IM products coherently produce IM beams called IM lobes. The location of an IM lobe can be determined from the desired carrier beam locations. For the present analysis we use two narrow-band signals with midfrequencies ω_1 and ω_2, respectively, that can be approximated as two single-tone signals. We assume that ω_1 and ω_2 differ by several bandwidths of individual bands such that the IM bands do not interfere with the signal bands.

We assume that intended spot beams are at (θ_1, ϕ_1) and (θ_2, ϕ_2) for the two desired signals of frequencies[4] ω_1 and ω_2, respectively. We first consider the IM product of frequency $2\omega_2 - \omega_1$. The phase at the SSPA input is a function of the element's coordinate, beam location, and signal frequency. For frequency ω_1, the phase at the SSPA input associated with the (m, n) element is

$$\beta_1(m, n) = -k_1 ma \sin\theta_1 \cos\phi_1 - k_1 nb \sin\theta_1 \sin\phi_1 \tag{13.20}$$

The phase of the signal with frequency ω_2 would be

$$\beta_2(m, n) = -k_2 ma \sin\theta_1 \cos\phi_1 - k_2 nb \sin\theta_1 \sin\phi_1 \tag{13.21}$$

where k_1 and k_2 are free-space wave numbers corresponding to the frequencies ω_1 and ω_2 and $a \times b$ is the unit cell size (rectangular grid). From (13.16), the phase of the IM product of frequency $2\omega_2 - \omega_1$ is given by

$$\begin{aligned}\beta_i(m, n) &= 2\beta_2(m, n) - \beta_1(m, n)\\ &= -2k_2 ma \sin\theta_2 \cos\phi_2 - 2k_2 nb \sin\theta_2 \sin\phi_2\\ &\quad + k_1 ma \sin\theta_1 \cos\phi_1 + k_1 nb \sin\theta_1 \sin\phi_1\end{aligned} \tag{13.22}$$

Because the above phase is a linear function of m and n, all the IM signals emanating from the array elements will produce a coherent beam. Suppose the location of the IM beam is (θ_i, ϕ_i). Then we can write

$$\beta_i(m, n) = -k_i ma \sin\theta_i \cos\phi_i - k_i nb \sin\theta_i \sin\phi_i \tag{13.23}$$

---

[4] We loosely use the word "frequency" to imply *midfrequency*.

In the above equation $k_i = 2k_2 - k_1$ is the free-space wave number for the IM product. Comparing (13.22) and (13.23) we obtain

$$k_i \sin \theta_i \cos \phi_i = 2k_2 \sin \theta_2 \cos \phi_2 - k_1 \sin \theta_1 \cos \phi_1 \qquad (13.24)$$

$$k_i \sin \theta_i \sin \phi_i = 2k_2 \sin \theta_2 \sin \phi_2 - k_1 \sin \theta_1 \sin \phi_1 \qquad (13.25)$$

The above two equations can be solved for (θ_i, ϕ_i), the location of the IM beam. Figure 13.11 shows the locations of the two IM beams for a linear array. The carrier frequencies are 9.5 and 10.5 GHz, and the intended beam locations are at 5° and 10° scan angles, respectively. As can be noted, the IM beams do not lie between 5° and 10° scan angles. In fact, it can be proven that in a liner array with two carrier beams θ_i does not lie between θ_1 and θ_2; that is, the IM beam does not fall between the carrier beams. In the case of a planar array with two carrier beams, the locations of IM lobes and carrier beams are collinear on the uv-coordinate plane. Moreover, the IM beams lie outside the line segment between the carrier beam locations, as shown in Figure 13.12.

13.6.3 Multiple-Channel Array: Noise Power Ratio

Thus far we focused on IM products due to two narrow-band signals in an active array antenna. Generally more than two channels are used in a practical array and the channel bands are kept very close to each other in order to utilize the available bandwidth efficiently. Furthermore, the "single-tone" approximation for a channel cannot be utilized for IM analysis because the individual channel bandwidths are

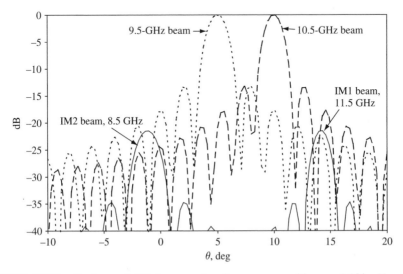

FIGURE 13.11 Carrier and IM beams of a linear array with $a = 1.18$ in, $N = 30$, $A = 10$, $C = -0.1$, $f_1 = 9.5$GHz, $f_2 = 10.5$GHz.

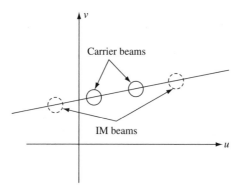

FIGURE 13.12   Locations of IM and carrier beams on $(u-v)$ plane of a 2-D array.

wider and the individual signal spectra are very closely spaced (less than one signal bandwidth). For such cases it is very difficult to determine the IM interference analytically, even though the SSPA nonlinearity can be characterized accurately.

In order to provide a quantitative characterization of an SSPA in terms of IM distortion, the noise power ratio (NPR) is introduced. The NPR data of a nonlinear device essentially provides an estimate for the signal-to-noise ratio at the output terminal due to IM products caused by multiple signal interference. The NPR of an SSPA at a test channel band is defined as the ratio between (a) the mean output noise power measured at the test channel band if the input port of the SSPA is connected with a band pass Gaussian noise source and (b) the mean output noise power measured at the test channel band if the input Gaussian noise source excludes the portion of the noise power that pertains to the test channel band. The block diagram for NPR measurement is shown in Figure 13.13, consisting of a white noise generator, a BPF, a band-reject notch filter, and a power meter. The bandwidth of the BPF should be equal to the total bandwidth of the channels. In the first part of the measurement, the BPF is directly connected to the SSPA input and the noise power at the test channel band is recorded. In the second part, the input noise at the test channel band must be zero, which is enforced by passing the bandpass noise through the notch filter.

In Figure 13.14 we present conceptual sketches of power meter displays for an NPR measurement. The input consists of seven channels (possibly with a guard

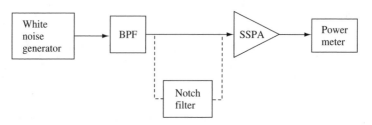

FIGURE 13.13   NPR test set up.

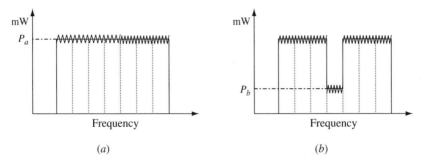

FIGURE 13.14 Sketches of SSPA output with two different input noise required for NPR measurement (a) Noise power output of a band limited noise, (b) Noise power output when the input noise passes through the notch filter.

band between two channels, not shown here). The NPR with respect to the fourth channel from the left is under consideration, which should be given by

$$\text{NPR} = \frac{P_a}{P_b} \tag{13.26}$$

To measure the NPR for other channels, the notch filter needs to set at the respective channel bands.

It should be pointed out that the NPR is dependent on the signal level, because the IM power is strongly dependent upon the operating region of the nonlinear SSPA. Therefore, the NPR data are generally specified with respect to the output (or input) power. In reality, the noise power set for NPR measurement should correspond to the actual signal power level.

The NPR data are commonly used to determine the useful power within a channel of interest. The NPR essentially represents the ratio between the signal plus noise power and the noise power. The signal-to-noise ratio at the output port is given by $(\text{NPR}-1)$. With reference to Figure 13.14, the noise power in the fourth channel is P_b. The total power (signal + noise) is P_a. Therefore, the useful signal power is

$$P_{\text{useful}} = P_a - P_b = P_a \left(1 - \frac{1}{NPR} \right) \tag{13.27}$$

The useful power that determines the EIRP of the array thus can be expressed in terms of the NPR and the channel output power. It is worth mentioning that the NPR includes not only the IM noise but also the intrinsic device noise that is otherwise determined from the noise figure.

It is also worth pointing out that with a multiple-beam array the spreading of energy in IM lobes will tend to improve the NPR of the desired beam at the far field. This NPR improvement factor should be considered when defining the SSPA operating point [7].

13.6.4 AM–PM Conversion

The output phase of an amplifier is a function of the input power lever, as depicted in Figure 13.4. For a single-channel or multichannel amplitude modulated (AM) signal the input power varies with time, making the output phase vary with time and resulting in a phase-modulated (PM) output signal [8]. This phenomenon in a nonlinear amplifier is known as AM–PM conversion. In a typical SSPA, operating in the liner region, the AM–PM conversion effect is not very substantial because the output phase does not vary significantly. For traveling wave tube amplifiers, however, this effect could be substantial.

13.7 NOISE TEMPERATURE AND NOISE FIGURE OF ANTENNA SUBSYSTEMS

Evaluation of the signal quality transmitted or received by an array necessitates array system analysis. A system analysis essentially includes estimation of EIRP and signal-to-noise ratio for the signals. Typically, the EIRP information is of significance for a transmit array because the signal power at the antenna aperture is much stronger than the noise power generated by the antenna subsystems. On the other hand, the signal-to-noise ratio is of concern for a receive array because the power of the receiving signal at the antenna aperture may be comparable with the noise power. This is particularly so in the case of an on-board satellite antenna, because the received signal consists of thermal noise radiated by Earth's surface and by the sun. In order to estimate these quantities one needs to know the device parameters that constitute the antenna system. The device parameters essentially signify the functions and quality of the device. For instance, gain, noise figure, output power, and NPR are the parameters pertaining to an SSPA, while gain and antenna noise temperature are the parameters pertaining to an antenna element. Useful parameters for devices associated with an active array, not defined before, will be considered in the following sections.

13.7.1 Antenna Noise Temperature

A receiving antenna receives black-body radiation emanating from the surrounding sources. This undesired power degrades the SNR of the received signal. The amount of black-body radiation received by an antenna depends on the radiation pattern of the antenna, locations of the black bodies, and their temperatures. The antenna noise temperature of an antenna is a measure of the noise power it receives from the surrounding black bodies.

Consider an antenna to be exposed to an object of temperature T in Kelvin. According to Planck's law of black-body radiation, the object will radiate electromagnetic energy and the radiated power density per frequency per unit projected surface area per solid angle of the object is given by [9]

$$R(f) = \frac{2hf^3}{c^2} \frac{1}{\exp(hf/kT) - 1} \qquad \text{W/m}^2/\text{Hz/rad}^2 \qquad (13.28)$$

where k is the Boltzmann constant $(1.381 \times 10^{-23} \text{J/K})$, f is the frequency of the radiating energy, c is the velocity of light, and h is Planck's constant $(6.625 \times 10^{-34} \text{J} - \text{S})$. At the microwave frequency, $kT \gg hf$; thus (13.28) can be approximated as

$$R(f) = \frac{2kTf^2}{c^2} = \frac{2kT}{\lambda^2} \qquad (13.29)$$

where λ is the wavelength. The total noise power per solid angle radiated by the object within a bandwidth B is

$$P_n \approx R(f_0)A_n B = \frac{2kTA_nB}{\lambda_0^2} \qquad (13.30)$$

where A_n is the projected surface area of the noise source (Figure 13.15) along the direction of the antenna, f_0 is the center frequency of the band, and λ_0 is the corresponding wavelength. The amount of power P_n will radiate isotropically, because the phase of the noise source has a random distribution on the object surface. The power received by the antenna with respect to a given polarization (noise power would contain both polarizations with equal power) would be given by

$$P_a = \frac{P_n/2}{R^2}A_e \qquad (13.31)$$

where R is the distance between the antenna and the object and A_e is the effective aperture area of the antenna in the direction of the noise source. The $\frac{1}{2}$ factor is due to the polarization mismatch. Substituting P_n from (13.30) into (13.31) we have

$$P_a = \frac{kTA_nB}{R^2\lambda_0^2}A_e$$

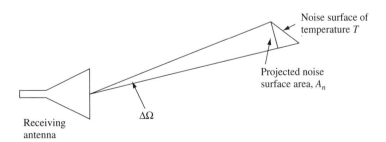

FIGURE 13.15   Noise surface and a receiving antenna.

Now, A_n/R^2 is the solid angle $\Delta\Omega$ of the noise area at the antenna location. Thus we write

$$P_a = \frac{kTB}{\lambda_0^2} A_e \, \Delta\Omega \tag{13.32}$$

The gain of the antenna in the direction of the noise source is

$$G_a = \frac{4\pi A_e}{\lambda_0^2} \tag{13.33}$$

Using (13.33) in (13.32) we finally obtain the received noise power as

$$P_a = \frac{kTB}{4\pi} G_a \, \Delta\Omega \tag{13.34}$$

Noise sources are incoherent. If multiple noise sources exist around the antenna, then the total power is a summation of the individual noise power. For an extended noise source with available temperature profile, the total noise power received by an antenna is

$$P_a^{\text{Total}} = \frac{kB}{4\pi} \iint_\Omega T(\Omega) G_a(\Omega) \, d\Omega \tag{13.35}$$

The antenna noise temperature T_a is defined by

$$P_a^{\text{Total}} = kT_a B \tag{13.36}$$

which yields

$$T_a = \frac{1}{4\pi} \iint_\Omega T(\Omega) G_a(\Omega) \, d\Omega \tag{13.37}$$

For an isotropic antenna $G_a(\Omega) = 1$; thus the antenna temperature becomes equal to the average temperature surrounding the antenna. Notice, the noise power received by an antenna can be determined using (13.36) if the antenna noise temperature T_a is specified.

13.7.2 Noise Temperature and Noise Figure of Resistive Circuits

The noise temperature of a passive circuit of resistive elements is of importance because such circuits are utilized in several components of an array feed network such as power dividers, power combiners, and attenuators. To determine the noise temperature of a circuit, one must be familiar with Thevenin's noise equivalent circuit of a lumped resistor and the definition of the noise temperature. These items will be considered next. This will be followed by derivations of noise temperatures and noise figures of useful circuits in a typical active array feed chain.

FIGURE 13.16 Noise equivalent circuit of a resistor.

Thevenin's Equivalent Circuit and Noise Temperature of Resistor [10] It is shown by Nyquist [11] that the open-circuit rms noise voltage, V_n^{rms}, across a lumped resistor R at temperature T in Kelvin with bandwidth B is given by

$$V_n^{rms} = \sqrt{4kTBR} \qquad (13.38)$$

The mean value for noise voltage, $< V_n >$, is zero. Using Thevenin's theorem, an equivalent circuit of the noise voltage can be constructed as shown in Figure 13.16. The circuit consists of an ideal voltage source of rms voltage V_n^{rms} and a series resistor R. The series resistor acts as the internal resistance of the noise source. This noise equivalent circuit is very effective for obtaining the noise power of a complex resistive network.

The noise temperature is defined as the maximum time-average noise power delivered to an external load divided by kB. The maximum power transfer theorem states that the maximum power transfer takes place if the external load is equal to the internal resistance R of the source. Under such a condition the time-average power to the external load would be

$$< P_{max} > = \frac{(V_n^{rms})^2}{4R} = kTB \qquad (13.39)$$

As per the definition, the noise temperature T_n of the resistor R would be

$$T_n = \frac{< P_{max} >}{kB} = T \qquad (13.40)$$

Expectedly the noise temperature of a single resistor is equal to its physical temperature.

Next, using Thevenin's equivalent circuit of a resistor we will prove an important theorem that is necessary to determine the noise temperature of a complex network.

Theorem of Equivalent Noise Temperature

Statement In a complex resistive circuit, if the component resistors have the same temperature T, then the effective noise temperature of the equivalent resistor is also T.

Proof We first prove the theorem for two basic building blocks, namely a series circuit and a parallel circuit of two resistors. We show that if two resistors have a noise temperature T, then their equivalent resistance for series (or parallel) combination will also have the same noise temperature T. We then argue that the theorem

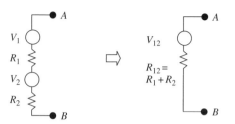

FIGURE 13.17 Thevenin's equivalent circuit of two resistors in series.

must be valid for any resistive network, because any such network consists of a number of series and parallel connections of resistors. Repeated application of the theorem for series and parallel combinations would result in the same noise temperature as that of the individual components.

First we consider a series circuit of two resistors R_1 and R_2. Figure 13.17 shows Thevenin's equivalent circuit of the series resistors and the equivalent noise source. The Thevenin source voltage, V_{12}, should simply be the algebraic summation of individual noise voltages. Thus we have

$$V_{12} = V_1 + V_2 \tag{13.41}$$

The mean-square value of this voltage is

$$< V_{12}^2 > = < V_1^2 > + < V_2^2 > + 2 < V_1 V_2 > \tag{13.42}$$

Now, V_1 and V_2 are uncorrelated. Thus $< V_1 V_2 > = < V_1 > < V_2 >$, which must vanish because $< V_1 > = < V_2 > = 0$. Furthermore, $< V_1^2 > = 4kTBR_1$, $< V_2^2 > = 4kTBR_2$, where T represents the noise temperature of the resistors. Thus we obtain

$$< V_{12}^2 > = 4kTB(R_1 + R_2) = 4kTBR_{12} \tag{13.43}$$

According to (13.39) and (13.40), the equivalent noise temperature becomes

$$T_n = \frac{< V_{12}^2 >}{4kBR_{12}} = T \tag{13.44}$$

We thus prove that the noise temperature of the equivalent resistor R_{12} in a series circuit is identical to that of the individual resistors.

A similar procedure can be followed for the parallel circuit in Figure 13.18. Thevenin's equivalent resistance in this case is

$$R_{12} = R_1 || R_2 = \frac{R_1 R_2}{R_1 + R_2} \tag{13.45}$$

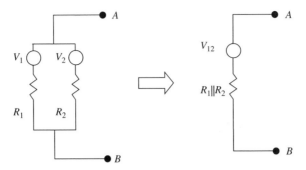

FIGURE 13.18   Noise equivalent circuit of parallel resistors.

Thevenin's open-circuit voltage V_{12} is given by

$$V_{12} = V_1 - \frac{R_1(V_1 - V_2)}{R_1 + R_2} = \frac{V_1 R_2 + R_1 V_2}{R_1 + R_2} \tag{13.46}$$

The mean-square value of V_{12} is

$$<V_{12}^2> = \frac{<V_1^2> R_2 + R_1 <V_2^2>}{R_1 + R_2} \tag{13.47}$$

Recall $<V_1^2> = kTBR_1$, $<V_2^2> = 4kTBR_2$, yielding

$$<V_{12}^2> = \frac{4kTBR_1 R_2}{R_1 + R_2} = 4kTBR_{12} \tag{13.48}$$

The equivalent noise temperature becomes

$$T_n = \frac{<V_{12}^2>}{4kBR_{12}} = T \tag{13.49}$$

We have proven that the noise temperature of the resultant resistor remains unchanged if the individual resistors have identical noise temperatures and the resistors are either in series or in parallel combination. As a corollary, we can state that the theorem is generally true for any circuit of multiple resistive components of identical noise temperatures, because using the rules of series and parallel combinations we can simplify the circuit systematically to a single resistive element. The resultant element will have the same noise temperature as that of the individual components.

Thus far we only considered the resistors with identical noise temperature. Next we demonstrate the procedure for determining the equivalent noise temperature of two series resistors having different noise temperatures.

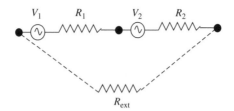

FIGURE 13.19   Noise equivalent circuit of two series resistors.

Equivalent Noise Temperature of Resistive Circuit at Different Temperatures

The equivalent noise temperature of a resistive circuit can be determined from the equivalent circuit of individual resistors. To demonstrate the methodology we consider a simple example of two resistors in series (Figure 13.19) that have different temperatures.

Suppose T_1 and T_2 are the physical temperatures of two resistors R_1 and R_2, respectively, that are connected in series. The equivalent circuit of this combination consists of two ideal voltage sources and two series-connected resistors, as shown in Figure 13.19. The maximum power transfer to the external load R_{ext} takes place if $R_{ext} = R_1 + R_2$. Under such a situation, the power delivered to the external load is

$$P_{max} = \frac{(V_1 + V_2)^2}{4(R_1 + R_2)} \tag{13.50}$$

The average power delivered to the load is given by

$$<P_{max}> = \frac{1}{4(R_1 + R_2)}[< V_1^2 > + 2 < V_1 V_2 > + < V_2^2 >] \tag{13.51}$$

Now $< V_1 V_2 > = 0$, because the two noise voltages are uncorrelated. Furthermore, according to (13.38), $< V_1^2 > = 4kT_1 BR_1$ and $< V_2^2 > = 4kT_2 BR_2$. Thus we obtain

$$<P_{max}> = \frac{kB}{R_1 + R_2}[T_1 R_1 + T_2 R_2] \tag{13.52}$$

The effective noise temperature T_{12} is

$$T_{12} = \frac{<P_{max}>}{kB} = \frac{T_1 R_1 + T_2 R_2}{R_1 + R_2} \tag{13.53}$$

Expectedly, substituting $T_1 = T_2$ in (13.53), we obtain the same noise temperature of the series circuit as the noise temperature of an individual resistor.

Noise Temperature of Resistive Attenuator

Determination of the noise temperature of a resistive attenuator is more involved than that of the previous cases, primarily due to the fact that an attenuator is a two-port device as opposed to

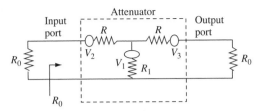

FIGURE 13.20   Equivalent circuit of an attenuator for noise temperature calculation.

a one-port device. For a two-port device, the definition of noise temperature is associated with the noise power delivered to a matched load at the output port while the input port is also terminated by a matched load. A symmetrical two-port passive device can be modeled as a symmetrical T-network; thus we will use a T-network model for the attenuator, as shown in Figure 13.20. The attenuator is perfectly matched with the input and output ports, with characteristic impedance R_0 as shown. To obtain the noise power delivered to the matched load at the output port, we employ Thevenin's model. Notice, we purposely do not include the noise sources associated with the match terminations at the input and output ports because we are interested only in the noise power contributed by the attenuator circuit. For a perfectly matched attenuator, we must have

$$R + \frac{(R+R_0)R_1}{R+R_0+R_1} = R_0 \tag{13.54}$$

The voltage attenuation, A, of the attenuator is given by

$$A = \frac{R_0}{R+R_0} \frac{(R+R_0)R_1}{(R+R_0+R_1)R_0} \tag{13.55}$$

From (13.54) and (13.55) we obtain

$$A = \frac{R_0 - R}{R_0 + R} \tag{13.56}$$

Equation (13.56) can also be expressed as

$$R = \frac{1-A}{1+A}R_0 \tag{13.57}$$

Substituting the expression for R in (13.54) we deduce the expression for R_1 as

$$R_1 = \frac{2A}{1-A^2}R_0 \tag{13.58}$$

Equations (13.57) and (13.58) essentially are the design equations for an attenuator of voltage attenuation A and input–output resistance R_0.

In order to obtain the noise power dissipated at the output load R_0, we first compute the noise current that flows through R_0. The simplest way, perhaps, is to invoke the superposition theorem. Using that, the total current can be expressed as

$$I_0 = \frac{V_1}{R + 2R_1 + R_0} + \frac{V_2}{2R_0}\frac{R_1}{R + R_1 + R_0} + \frac{V_3}{2R_0} \tag{13.59}$$

The noise power dissipation is given by

$$P_0 = I_0^2 R_0 \tag{13.60}$$

The time-average power is obtained by integrating P_0 with respect to time, t. Because the noise voltages are uncorrelated, the average power becomes

$$<P_0> = \frac{<V_1^2> R_0}{(R + 2R_1 + R_0)^2} + \frac{<V_2>^2}{4R_0}\frac{R_1^2}{(R + R_1 + R_0)^2} + \frac{<V_3^2>}{4R_0} \tag{13.61}$$

Now, according to the Nyquist theorem, $<V_i^2> = 4kTBR_i, i = 1, 2, 3$, where T is the noise temperature of the resistors in the attenuator. We use (13.61) and then replace R_1 and R in terms of R_0 and A using (13.57) and (13.58). With a lengthy but straightforward algebraic manipulation we arrive at

$$<P_0> = kTB(1 - A^2) \tag{13.62}$$

In the literature the attenuation parameter L is defined as the reciprocal of $A^2(L > 1)$. Thus we write

$$<P_0> = \frac{kTB(L - 1)}{L} \tag{13.63}$$

The device noise, hence the noise temperature, is expressed with reference to the input port as shown in Figure 13.6b. The noise temperature of the attenuator (as referred to the input) becomes

$$T_n = \frac{<P_0> L}{kB} = (L - 1)T \tag{13.64}$$

The noise figure, NF, of the attenuator can be determined using (13.63) and (13.10). The attenuator introduced output noise power <P_0> which must be equated with $(NF-1)GkT_0 B$ with $T_0 = 290$ K and $G = 1/L$. Thus we obtain

$$NF = 1 + (L - 1)\frac{T}{T_0} \tag{13.65}$$

This completes the derivation of the noise temperature and noise figure of an attenuator.

The equivalent noise temperature of an attenuator can also be understood intuitively by invoking the equivalent noise temperature theorem discussed before. To

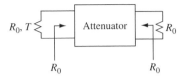

FIGURE 13.21 Attenuator with input-output terminations with matched loads.

that end we consider an attenuator and terminate both ends with R_0 (Figure 13.21). We assume that the terminated resistor R_0 at the input end has a noise temperature T. Then by virtue of the theorem, the noise temperature of the equivalent resistance looking from the attenuator output (Figure 13.21) should be T because all the resistive elements to the left have an identical noise temperature. Furthermore, the attenuator has an output resistance R_0, and thus the combined noise power contributed by the attenuator and the input port resistor R_0 is kTB. Of this amount of power, the input resistor contributes kTB/L, because kTB is the incident noise power from R_0 to attenuator's input and L is the power attenuation of the attenuator. The contribution of the attenuator is therefore $kTB(1-1/L)$, which is the same as deduced in (13.63).

Noise Temperature of Power Divider/Combiner The noise temperature of Wilkinson's power divider (Figure 13.22) can be determined from (13.64). The power divider can be considered as an attenuator if only two ports, for instance, port 1 and port 2, are used while the third port is match terminated. Under such a condition $L = 2$, because for a Wilkinson divider $|S_{21}|^2 = \frac{1}{2}$. Thus using (13.64) we obtain that the noise temperature of a Wilkinson divider is the same as its physical temperature, T. Because $S_{21} = S_{12}$, the noise temperature remains the same if the input–output ports switch. The noise figure of the Wilkinson divider using (13.65) becomes

$$NF = 1 + \frac{T}{T_0}. \tag{13.66}$$

For a $1 : N$ power divider, $L = N$. Thus the noise temperature becomes $(N - 1)\, T$ subject to the condition that *one input port and one output port are*

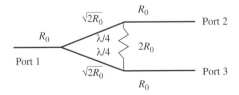

FIGURE 13.22 Wilkinson's 1:2 power-divider. Thick lines represent transmission line segments with characteristic impedances written by their sides.

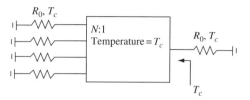

FIGURE 13.23   Circuit for determining the noise power introduced by a power combiner.

used and the remaining ports are match terminated. The noise figure in this case becomes

$$\text{NF} = 1 + (N-1)\frac{T}{T_0} \tag{13.67}$$

It must be emphasized that the noise figure in (13.67) is valid if, out of $N+1$ ports, only two ports are utilized while the other ports are match terminated. In other words, the divider must be used as a two-port device. If, however, all of the $N+1$ ports are used (one input port and N output ports), then the device-generated noise becomes zero because the total output noise does not exceed the input noise.

The output noise power of an $N:1$ combiner (N input ports, one output port) depends on its circuit configuration and the nature of the noise sources (uncorrelated or correlated). Figure 13.23 shows a combiner. A typical combiner operates perfectly (no loss of power) if the signals at the input ports are in phase. On the other hand, the noise powers from input ports do not add up if the noise sources are uncorrelated. To illustrate this we match terminate all the input ports and assume that the terminated loads are at temperature T_c. Then looking from the output port, the resultant noise temperature of the device would be T_c per the theorem of equivalent noise temperature in Section 13.7.2, making the noise power delivered to the output load kT_cB. Thus one can say that each matched load at the input ports transfers only kT_cB/N amount of noise power to the output port, making the total output noise power $N(kT_cB/N) = kT_cB$. Therefore, effectively an *attenuation of uncorrelated noise* occurs in a combiner. On the contrary, no such attenuation occurs if the noise sources at the input ports are correlated and in phase. This point must be carefully considered in the system analysis of an array antenna.

13.8   ACTIVE ARRAY SYSTEM ANALYSIS

System analysis is one of the most important tasks when examining the quality of a signal. This is particularly so for a receive array antenna because the signal power is very low and often comparable to the noise power received by the antenna and the intrinsic noise power of the active components. In the case of a transmit array with multiple beams, the IM products generally dominate the noise power. The NPR data of the SSPAs can be utilized to estimate such a noise power. The signal-to-noise ratio (SNR) at the antenna aperture can be estimated in a straightforward manner.

However, the SNR at the far-field point could be significantly different from that at an aperture. The noise voltages (excluding the IM noise portion) produced by various array elements being uncorrelated, the noise phase is randomly distributed over the array elements. This results in an omnidirectional far-field noise pattern. However, the IM noise forms coherent beams (IM lobes) as discussed before. For a dual-channel signal, the IM beam locations differ from the channel beam locations. This is not generally true for a multichannel signal. If an IM beam interferes with a channel beam, then the SNR should be estimated from the NPR data. The system analysis of a transmit array should be performed on a case-by-case basis knowing the NPR for each beam and the locations of the channel beams and IM beams.

In the case of a receive array, the total noise power is contributed by the antenna noise and the intrinsic noise of the phase shifter, low-noise amplifier (LNAs), power attenuator, and power combiner circuits. Figure 13.24 shows a simplified block diagram of a receive array. The phase shifter, LNA, and attenuator in a feed chain are combined in a single block. The gain and the noise figure of the block can be determined using the chain rule. The antenna noise voltages received by the antenna elements combine coherently at the combiner output because the array elements receive noise power from a common source (Earth, for example, in the case of a satellite uplink array[5]). Contrary to this, the device noise sources are uncorrelated, and therefore the noise voltages combine incoherently. The IM noise is generally absent because the LNAs are operated in the linear region.

Suppose G and F respectively are the gain and the noise figure of the block in Figure 13.24. The noise power received by an array element is given by

$$P_a = kT_aB \qquad (13.68)$$

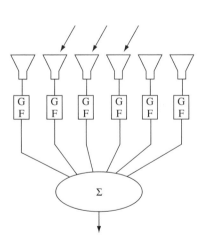

FIGURE 13.24 Simplified block diagram for SNR computation of a receive array.

---

[5] This estimation is somewhat conservative, because the entire thermal noise power does not appear from a single point, but rather is distributed within a solid angle, so a portion of the power combines incoherently.

where T_a is the antenna temperature of an array element and B is the signal bandwidth. The block that exists before the power combiner amplifies this noise power. The $N:1$ power combiner coherently combines this antenna noise because the phase of the noise signal would be uniform at the input ports of the combiner.[6] Thus the output noise would be

$$P_{Na} = NP_a G = NkT_a BG \qquad (13.69)$$

where N is the total number of elements and G is the power gain of an amplifier block. We assumed uniform taper in this case. The device noise (of the amplifier block) at one input terminal of the power combiner is (see the discussion of noise figure in Section 13.4.1)

$$P_d = kT_0 B(F-1)G \qquad (13.70)$$

with $T_0 = 290$ K. The transmitted noise at the output port of the combiner is

$$P_{do} = \frac{kT_0 B(F-1)G}{N} \qquad (13.71)$$

The total uncorrelated noise output power due to N input port noise is

$$P_{do}^T = NP_{do} = kT_0 B(F-1)G \qquad (13.72)$$

The noise power introduced by the $N:1$ power combiner is zero, because we consider here an ideal power combiner. Thus the total noise is equal to $P_{Na} + P_{do}^T$. The signal power at the output of the combiner is

$$S_o = NGS_a \qquad (13.73)$$

where S_a is the signal power received by an array element. Notice, unlike the device noise power, the signal being coherent, no signal power is lost in the power combiner. The SNR at the output of the combiner becomes

$$\left(\frac{S}{N}\right)_{out} = \frac{S_0}{P_{Na}+P_{do}^T} = \frac{S_a}{kT_a B + kT_0 B(F-1)/N} \qquad (13.74)$$

The RF power received by an array element is

$$S_a = a^2 \eta P_0 \qquad (13.75)$$

---

[6] This is true only under the assumption that the antenna noise source and the signal source (the transmitter) are colocated (as in a satellite array receiving a signal from Earth). If the signal power and the antenna noise power approach from two different angles (for instance, if the solar noise dominates), then the antenna noise power will be incoherent; thus the phase of the antenna noise voltage received by the antenna elements should be taken into consideration.

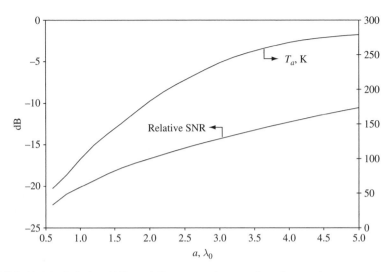

FIGURE 13.25 Relative SNR and T_a versus element size of a receive array of square elements with N=100.

where a^2 is the aperture area of an element (square aperture), η is the aperture efficiency in the direction of the incident field, and P_0 is the power density (watts per unit area) at the element's aperture. The SNR thus can be expressed as

$$\left(\frac{S}{N}\right)_{out} = \frac{P_0\eta}{kB}\left[\frac{a^2}{T_a+T_0(F-1)/N}\right] \tag{13.76}$$

where term inside the square brackets is of interest because that term is responsible for the SNR improvement with respect to the antenna parameters. We see that the SNR at the output improves as N increases. Apparently, for a fixed N the SNR increases as the element size, a, increases. However, an increase in a results in an increase in the antenna noise temperature T_a, particularly in the case of a satellite antenna. Thus the SNR does not increase linearly with the element size. Figure 13.25 shows a plot of relative SNR versus the element size of an array. Also plotted is the antenna temperature of an array element looking at Earth from a geostationary orbit. For the antenna temperature computation Earth temperature is assumed 290 K and the atmosphere temperature is assumed to be 4 K. The SNR improves by about 8 dB when the element size increases from $1\lambda_0$ to $4\lambda_0$.

13.9 ACTIVE ARRAY CALIBRATION

In an on-board satellite active array antenna the amplitudes and the phases of SSPAs and phase shifters drift over time. This happens primarily because (a) the DC supply voltage reduces due to aging of the solar cells, (b) the junction capacitance of the solid state devices changes due to temperature variations, and (c) the gain of an SSPA changes due to temperature variations. The performance of an active array thus slowly degrades over time even before the end of the satellite's life. Array calibration essentially is an attempt to detect and correct the amplitude and phase drifts of the array modules in order to maintain the performance of the array as much as possible. In some situations calibration may be required even before the installation of the array with the satellite bus for necessary adjustments. For an on-board satellite array antenna, calibration is generally conducted from a ground station by measuring the amplitude and phase of the array signals at different known conditions. Direct measurement of element amplitude and phase is not possible because that would require turning off all but the test element. Turning off an array element is not permissible because an amplifier dissipates a large amount of power when the input RF signal is switched off, increasing the temperature of a module to an unacceptable limit. The only parameter that can be varied for calibration purposes is the phase of a module. Three procedures are generally followed to measure the amplitude and phase of the RF power radiated by modules of an active array antenna. We will present them next.

13.9.1 Two-Phase-State Method

This is a very simple and straightforward method to determine the amplitude and phase of an element. In this method the entire array is excited with respect to a reference condition, for instance with respect to a desired beam. The amplitude and phase of the carrier signal emanating from the array are measured at a station on Earth with respect to a reference signal. Then a 180° phase shift is applied on the test element, while keeping phases of the remaining elements unchanged. The amplitude and phase of the signal are now measured from the same station. From the difference between the two complex field intensities, one can easily figure out the field intensity radiated by the test element.

To illustrate, suppose $e_1, e_2, \ldots, e_N$ are the complex field intensities at the measuring station radiated by the elements with respect to a reference state. Then, the measured field intensity of the array would be

$$E_1 = e_1 + e_2 + \cdots + e_m + \cdots + e_N \tag{13.77}$$

Suppose the mth element is the test element. We now shift the phase of the mth element by 180°. With this change, the array field intensity become

$$E_2 = e_1 + e_2 + \cdots - e_m + \cdots + e_N \tag{13.78}$$

From (13.77) and (13.78) we find e_m as

$$e_m = \frac{1}{2}(E_1 - E_2) \tag{13.79}$$

Notice e_m is the field intensity of the mth element at the location of measurement. The actual amplitude and phase at the aperture can be found after dividing the data by the element pattern and the space factor.

This method is very simple conceptually. The phase is measured by mixing the RF signal with a highly stable local oscillator, as in a synchronous receiver. The data are collected at a fast rate to avoid phase drift of the local oscillator.

13.9.2 Multiple-Phase Toggle Method

In order to avoid the array phase measurement, the phase of the array test element is toggled in four states. The power of the array is measured with respect to the four phase states of the test element. From the measured power data, one can estimate the amplitude and phase of the test element. Suppose P_0, P_1, P_2, and P_3 are the magnitudes of the power received from a station if the test element's phase increases by 0°, 90°, 180°, and 270°, respectively. Suppose E is the magnitude of the field intensity of the array excluding the test element and e is the magnitude of the field intensity of the test element. Suppose ϕ is the relative phase of the test element's field with respect to the resultant phase of the remaining elements. Then we can write

$$(E + e\cos\phi)^2 + (e\sin\phi)^2 = P_0 \tag{13.80}$$

Simplifying,

$$E^2 + e^2 + 2eE\cos\phi = P_0 \tag{13.81}$$

For 90°, 180°, and 270° phase toggles, the relations become

$$E^2 + e^2 - 2eE\sin\phi = P_1 \tag{13.82}$$

$$E^2 + e^2 - 2eE\cos\phi = P_2 \tag{13.83}$$

$$E^2 + e^2 + 2eE\sin\phi = P_3 \tag{13.84}$$

Solving (13.81)–(13.84) we find

$$\phi = \tan^{-1}\left(\frac{P_3 - P_1}{P_0 - P_2}\right) \tag{13.85}$$

$$E = \frac{1}{2}\left[\left(\frac{P_0 + P_2}{2} + \frac{P_0 - P_2}{2\cos\phi}\right)^{1/2} + \left(\frac{P_0 + P_2}{2} - \frac{P_0 - P_2}{2\cos\phi}\right)^{1/2}\right] \tag{13.86}$$

$$e = \frac{1}{2}\left[\left(\frac{P_0 + P_2}{2} + \frac{P_0 - P_2}{2\cos\phi}\right)^{1/2} - \left(\frac{P_0 + P_2}{2} - \frac{P_0 - P_2}{2\cos\phi}\right)^{1/2}\right] \tag{13.87}$$

From (13.85) and (13.87) we obtain the amplitude and phase of the test element. Notice that with four phase toggles we could estimate the amplitude and phase of a test element from the power data only. A block diagram of the test setup is shown in Figure 13.26.

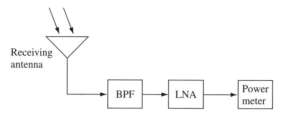

FIGURE 13.26 Calibration by power measurements only.

It is important to point out that due to the presence of noise, there is an error involved in the power measurement, and hence the amplitude and phase data may be inaccurate. The error caused by the random noise can be reduced by taking the average value of multiple measurements. Statistical analysis shows that for 5° phase accuracy four repetitive measurements are necessary for each phase state if the noise power is 10 dB below the element signal power [12]. For a lower signal power, more than four measurements would be required for a similar phase accuracy.

13.9.3 Simultaneous Measurement: Hadamard Matrix Method

In the previous two approaches, we noticed that the amplitude and phase of the array elements are measured one element at a time. For a large array this process is very time consuming. In order to expedite the measurement process, one can perform the measurements using fast digital processors that are installed at the array unit and at the measuring station. In one such scheme, 180° phase shifts for a particular group of array elements are implemented and then the complex field intensity of the array is measured. By selecting N different groups, this process is repeated N times, where N is the number of array elements. From the measured complex intensities, one can estimate the amplitude and phase of each array element.

To accelerate the processing speed, the phases of the array elements are set according to the row elements of a very special orthogonal matrix called a Hadamard matrix [13]. The elements of the Hadamard matrix are either +1 or –1. The phase change of an array element should be zero if the matrix element is +1 and 180° if the matrix element is –1. The $2n$th-order Hadamard matrix is recursively related to the nth-order matrix as shown:

$$[H_{2n}] = \begin{bmatrix} [H_n] & [H_n] \\ [-H_n] & [H_n] \end{bmatrix} \tag{13.88}$$

where $[H_m]$ represents a square matrix of order $m \times m$. We set $[H_1] = [1]$. The higher order matrices can be constructed using (13.88). For instance, a 4×4 Hadamard matrix can be found as

$$[H_4] = \begin{bmatrix} 1 & 1 & 1 & 1 \\ -1 & 1 & -1 & 1 \\ -1 & -1 & 1 & 1 \\ 1 & -1 & -1 & 1 \end{bmatrix} \tag{13.89}$$

Notice, the rows are mutually orthogonal; thus the inverse of the matrix becomes its transpose multiplied by the reciprocal of its determinant. To demonstrate the principle of operation, we consider a four-element array, and thus the corresponding Hadamard matrix is $[H_4]$, as in (13.89). Suppose the far fields radiated by the elements for nominal excitations are e_1, e_2, e_3, and e_4 and the corresponding far-field array intensity is E_0. Changing the phase of two selective elements by 180°, in accordance with the rows of the Hadamard matrix, we obtain the four different array field intensities as the product of two matrices:

$$
\begin{bmatrix}
1 & 1 & 1 & 1 \\
-1 & 1 & -1 & 1 \\
-1 & -1 & 1 & 1 \\
1 & -1 & -1 & 1
\end{bmatrix}
\begin{bmatrix}
e_1 \\ e_2 \\ e_3 \\ e_4
\end{bmatrix}
=
\begin{bmatrix}
E_0 \\ E_1 \\ E_2 \\ E_3
\end{bmatrix}
\tag{13.90}
$$

The nominal element fields can be determined easily from (13.90) by taking the transpose of the Hadamard matrix, which becomes

$$
\begin{bmatrix}
e_1 \\ e_2 \\ e_3 \\ e_4
\end{bmatrix}
=
\frac{1}{4^2}
\begin{bmatrix}
1 & -1 & -1 & 1 \\
1 & 1 & -1 & -1 \\
1 & -1 & 1 & -1 \\
1 & 1 & 1 & 1
\end{bmatrix}
\begin{bmatrix}
E_0 \\ E_1 \\ E_2 \\ E_3
\end{bmatrix}
\tag{13.91}
$$

Thus from the measured complex array intensities E_0, E_1, E_2, and E_3, one can determine the element field intensities. The excitation coefficients can be determined from e_1, e_2, e_3, and e_4.

It must be pointed out that the element number of the array should be an integer power of 2 (2^n, where n is an integer). If the number of array elements differs from 2^n, for instance a six-element array, then one should add a few dummy elements with zero nominal excitations to make up the total number of elements as 2^n and carry on the process.

In order to implement this approach, one needs to change the element phase by means of an on-board digital processor. The on-board digital processor generates N frames (N = number of elements in the array) and each frame contains N binary pulses. The pulse signals control the phase shifters to change the phase of a module in two states, namely 0° and 180°. The received array signal is time synchronized with the on-board digital processor. The amplitude and phase of the RF signal of each frame correspond to the complex array field intensity.

13.10 CONCLUDING REMARKS

The objectives of this chapter were to introduce the active array concept, define technical terms, and present the most common concerns from the antenna design perspective. Toward these goals, generic block diagrams of active arrays were shown and system characteristics of various components in a feed chain module were presented. A considerable amount of emphasis was given to SSPA nonlinearity and IM products, which play an important role in active array performance, particularly for a multiple-beam active array. Noise figures of passive components were derived from basic principles. The system analysis of a simplified receiving array was presented. Finally, the most common array calibration procedures were explained.

It an active array design a considerable amount of effort goes to the structural design of an active array, including thermal radiator design, because an active array generates a significant amount of heat due to the low PAE of an SSPA. Furthermore, phase shifters, digital control circuits, SSPA bias circuits, and their packaging are important issues to be dealt with in a real design process. Suggested references are listed for interested readers.

REFERENCES

[1] D. E. Reimer, "Packaging Design of Wide-Angle Phased-Array Antenna for Frequencies Above 20 GHz," *IEEE Trans. Antennas Propagat.*, Vol. 43, No. 9, pp. 915–920, Sept. 1995.

[2] A. K. Bhattacharyya, "Optimum Design Consideration for Multiple Spot Beam Array Antennas," in *AIAA Conference Proceeding*, Vol.1, pp. 424–431, Monterey, CA, May 2004.

[3] W. L. Stutzman and G. A. Thiele, *Antenna Theory and Design*, Wiley, New York, 1981.

[4] A. K. Bhattacharyya and G. Goyette, "A Novel Horn Radiator with High Aperture Efficiency and Low Cross-Polarization and Applications in Arrays and Multi-Beam Reflector Antennas," *IEEE Trans. Antennas Propagat.*, Vol. AP-52, No. 11, pp. 2850–2859, Nov. 2004.

[5] S. Y. Liao, *Microwave Devices and Circuits*, 3rd ed., Prentice-Hall, Englewood Cliffs, NJ, 1990.

[6] N. Fourikis, in *Phased Array-Based Systems and Applications*, Wiley, New York, Sons, 1997, "Transmit/Receive Modules," Chapter 4.

[7] A. Cherrette, E. Lier, and B. Cleaveland, "Intermodulation Suppression for Transmit Active Phased Array Multibeam Antennas with Shaped Beams," U.S. Patent No. 6,831,600, Dec. 14, 2004.

[8] O. Shimbo, "Effects of Intermodulation, AM-PM Conversion and Additive Noise in Multicarrier TWT Systems," *Proc. IEEE*, Vol. 59, No. 2, pp. 230–238, Feb. 1971.

[9] J. D. Kraus, in *Radio Astronomy*, 2nd ed., Cygnus Quasar Books, 1986, "Radio-Astronomy Fundamentals," Chapter 3.

[10] C. Kittel, in *Elementary Statistical Physics*, Wiley, New York, 1958, p. 141.

[11] H. Nyquist, "Thermal Agitation of Electric Charge in Conductors," *Phys. Rev.*, Vol. 32, pp. 110–113, 1928.

[12] R. Sorace, "Phased Array Calibration," *IEEE Trans.*, Vol. AP-49, pp. 517–525, Apr. 2001.

[13] A. Hedayat and W. D. Wallis, "Hadamard Matrix and Their Applications," *Ann. Stat.*, Vol. 6, pp. 1184–1238, 1978.

BIBLIOGRAPHY

ACTIVE ARRAY SYSTEMS

Ingvarson, P., S. Hagelin, and L. Josefsson, "Active Phased Array Activities in Sweden During COST 245," in *Proceedings, ESA Workshop on Active Antennas*, ESTEC, Noordwijk, The Netherlands, ESA-WPP-114 June 27–28, 1996, pp. 303–310.

Jacomb-Hood, A., and E. Lier, "Multibeam Active Phased Arrays for Communication Satellites," IEEE *Microwave* Magazine, Vol.1, No. 4, pp. 40–47, Dec. 2000.

Mailloux, R. J., "Antenna Array Architecture," *IEEE Proc.*, Vol. 80, No. 1, pp. 163–172, Jan. 1992.

Mailloux, R. J.,"Recent Advances and Trends in the USA on Active Array Antennas," in *Proceedings, ESA Workshop on Active Antennas*, ESTEC, Noordwijk, The Netherlands, June 27–28, 1996, pp. 269–276.

McQuiddy, Jr., D. N., R. L. Gassner, P. Hull, J. S. Mason, and J. M. Bedinger, "Transmit/Receive Module Technology for X-Band Active Array Radar," *IEEE Proc.*, Vol. 79, No. 3, pp. 308–341, Mar. 1991.

Sanzgiri, S., D. Bostrom, W. Pottenger, and R. Q. Lee, "A Hybrid Tile Approach for Ka Band Subarray Modules," *IEEE Trans. Antennas Propagat.*, Vol. 43, No. 9, pp. 953–958, Sept. 1995.

Tang, R., and R. W. Burns, "Array Technology," *IEEE Proc.*, Vol. 80, No. 1, pp. 173–182, Jan. 1992.

Whicker, L. R., "Active Phased Array Technology Using Coplanar Packaging Technology," *IEEE Trans. Antennas Propagat.*, Vol. 43, No. 9, pp. 949–952, Sept. 1995.

SSPAS AND PHASE SHIFTERS

Ali, F., and A. Gupta (Eds.), *HMETs and HBTs: Devices, Fabrications and Circuits*, Artech House, Norwood, MA, 1991.

Bahl, I. J., and P. Bhartia., *Microwave Solid State Circuit Design*, Wiley, New York, 1988.

Chang, K., *Microwave Solid State Circuits and Applications*, Wiley, New York, 1994.

Chang, W-J., and K-H. Lee, "A Ku-Band 5-Bit Phase Shifter Using Compensation Resistors for Reducing the Insertion Loss Variation," *ETRI J.*, Vol. 25, No. 1, pp. 19–24, Feb. 2003.

Ellinger, F., R. Vogt, and W. Bachtold, "Compact Reflective-Type Phase-Shifter MMIC for C-Band Using Lumped-Element Coupler," *IEEE Trans. Microwave Theory Tech.*, Vol. 49, No. 5, pp. 913–917, May 2001.

Gonzalez, G., *Microwave Transistor Amplifiers*, Prentice-Hall, Englewood Cliffs, NJ, 1984.

Hung, H. A., T. Smith, and H. Huang, "FETs: Power Applications," in K. Chang (Ed.), *Handbook of Microwave and Optical Components*, Vol. 2, Wiley, New York, 1990, Chapter 10.

IEEE Trans. Microwave Theory and Techniques, Special issue on microwave solid state devices, Vol. 30, No. 10, Oct. 1982.

Kim, H-T., D-H. Kim, Y. Kwon, and K-S. Seo, "Millimeter-Wave Wideband Reflection-Type CPW MMIC Phase Shifter," *Electron. Lett.*, Vol. 38, No. 8, pp. 374–376, Apr. 11, 2002.

Megej, A., and V. F. Fusco, "Low-Loss Analog Phase Shifter Using Varactor Diodes," *Microwave Opt. Tech. Lett.*, Vol. 19, No. 6, pp. 384–386, Dec. 20, 1998.

Mondal, J., J. Geddes, D. Carlson, M. Vickberg, S. Bounnak, and C. Anderson, "Ka-Band High Efficiency Power Amplifier MMIC with 0.30 μm MESFET for High Volume Applications," *IEEE Trans. Microwave Theory Tech.*, Vol. 40, No. 3, pp. 563–566, Mar. 1992.

Schwierz, F., and J. J. Liou, *Modern Microwave Transistors—Theory, Design and Applications*, Wiley, New York, 2002.

Shigaki, M., S. Koike, K. Nogatomo, K. Kobayashi, H. Takahashi, T. Nakatani, N. Tanibe, and Y. Suzuki, "38-GHz-Band High-Power MMIC Amplifier Module for Satellite On-Board Use," *IEEE Trans. Microwave Theory Tech.*, Vol. 40, No. 6, pp. 1215–1222, June 1992.

Sze, S. M., *Physics of Semiconductor Devices*, 2nd ed., Wiley, New York, 1981.

INTERMODULATION PRODUCT AND ANTENNA NOISE

Carvalho, N. B. D., and J. C. Pedro, "Compact Formulas to Relate ACPR and NPR to Two-Tone IMR and IP3," *Microwave J.*, Vol. 42, pp. 70–84, Dec. 1999.

Hassun, R., "Noise Power Ratio Measurement Techniques," Agilent Technologies, Santa Rosa, CA, Sept. 26, 2002.

Hemmi, C., "Pattern Characteristics of Harmonic and Intermodulation Products in Broad-Band Active Transmit Arrays," *IEEE Trans. Antennas Propagat.*, Vol. 50, No. 6, pp. 858–865, June 2002.

Lee, J. J., "G/T and Noise Figure of Active Array Antennas," *IEEE Trans. Antennas Propagat.*, Vol. 41, No. 2, pp. 241–244, Feb. 1993.

Safier, P. N., et al., "Simulation of Noise-Power Ratio with the Large-Signal Code CHRISTINE," *IEEE Trans. Electron Device*, Vol. 48, No. 1, pp. 32–37, Jan. 2001.

ARRAY CALIBRATION

Aumann, H. M., A. J. Fenn, and F. G. Willwerth, "Phased Array Antenna Calibration and Pattern Prediction Using Mutual Coupling Measurements," *IEEE Trans. Antennas Propagat.*, Vol. 37, No. 7, pp. 844–850, July 1989.

Hampson, G. A., and A. B. Smolders, "A Fast and Accurate Scheme for Calibration of Active Phased-Array Antennas," *IEEE APS Symp. Dig.*, Vol. 2, pp. 1040–1043, Aug. 1999.

Ng, B. C., and C. M. S. See, "Sensor-Array Calibration Using a Maximum-Likelihood Approach," *IEEE Trans. Antenna Propagat.*, Vol. 44, No. 6, pp. 827–835, June 1996.

Solomon, I. S. D., et al., "Receiver Array Calibration Using Disparate Sources," *IEEE Trans. Antennas Propagat.*, Vol. 47, No. 3, pp. 496–505, Mar. 1999.

PROBLEMS

13.1 For a spot beam array with a maximum scan angle of 8° and a beam size of 1° (diameter), the estimated number of elements is 196 for a desired EOC directivity of 38.5 dBi (see Figure 13.3). If the total DC power available for the active array is 1 kW, estimate the EIRP of the beam. Assume that the PAE of an SSPA is 25% and the total RF loss is 1.5 dB. Assume three simultaneous beams of equal power.

13.2 Deduce (13.11).

13.3 In a dual-beam array prove that an IM lobe does not fall between two carrier beams. Also, show that in a two-dimensional array, the two carrier beams and the IM lobes are collinear.

13.4 Deduce the effective noise temperature of two resistors R_1 and R_2 with noise temperatures T_1 and T_2, respectively, connected in parallel.

13.5 Deduce the noise figure of a 90° hybrid if it is used as a 90° phase shifter with two ports match terminated. Assume that the hybrid RF loss is 0.5 dB.

13.6 Determine the signal-to-noise ratio at the amplifier output for the two-element receive array shown in Figure P13.6. Assume that the antenna noise temperature of a radiating element is 150 K, signal midfrequency is 12 GHz, signal bandwidth is 1 GHz, amplifier gain is 20 dB, antenna element directivity is 12 dBi along the direction of the receiving signal, and signal power density is −30 dBm/in.^{2} The noise figure of the amplifier is 4. Element spacing, d, is such that $k_0 d \sin \theta = \pi/2$, θ being the angle of the receiving signal measured from the bore sight. Assume that the antenna noise source is located along the bore sight of the array.

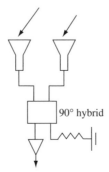

FIGURE P13.6   A simple two-element receive array.

13.7 Prove that the value of the determinant of the Nth-order Hadamard matrix, $[H_N]$, is $N^{N/2}$.

CHAPTER FOURTEEN

Statistical Analysis of Phased Array Antenna

14.1 INTRODUCTION

In a phased array antenna position and shape of a beam determine the amplitude and phase distributions. In an active array power amplifiers and attenuators implement the desired amplitude distribution, while the phase shifters implement the phase distribution. Because of the manufacturing tolerance, bias instability, and other factors, the magnitude of the output voltage of an amplifier has uncertainty, typically represented by a probability density function. A phase shifter also has a probability density function representing its phase uncertainty. The statistical nature of the element voltage and element phase makes the array pattern nondeterministic. This chapter is concerned with the statistics of the far-field pattern of an active array antenna.

In the first part we formulate the probability density function of the far-field intensity in terms of the element amplitude and phase uncertainties. It must be emphasized that we deal with the statistics of the far-field intensity, which is equivalent to the EIRP of an array. We first obtain the statistics of the real and imaginary parts of the array factor and then apply the central limit theorem to deduce the respective probability density functions in closed form. The concept of joint probability is then invoked to deduce the probability density function of the magnitude of the far-field intensity. Simplified expressions of the probability density function at important far-field locations such as at the beam peak, nulls, and peak side lobes are obtained. This is followed by approximate formulas for the 95% confidence boundaries. In the last part, we incorporate the effects of element failure to the array statistics. The numerical results with and without element failure

Phased Array Antennas. By Arun K. Bhattacharyya

are presented. Effects of amplitude and phase uncertainty and element failure on the side-lobe level and the null depth are shown. A phase adjustment model for incorporating element position uncertainty is presented at the end.

14.2 ARRAY PATTERN

Figure 14.1 shows a simplified block diagram of a phased array antenna. The amplifiers control the amplitude distribution and the phase shifters provide the necessary phase distribution to produce a beam pattern. The array pattern[1] of a linear phased array[2] with N elements can be expressed as

$$F(u) = \sum_{n=1}^{N} A_n \exp[j(\phi_n + nud)] \tag{14.1}$$

where A_n = real amplitude of the voltage source for the nth element
$\qquad \phi_n$ = phase of the voltage source
$\qquad d$ = element spacing
$\qquad u = k_0 \sin\theta$, k_0 being the wave number and θ the observation angle measured from bore sight

We assume isotropic radiating sources. The amplitude, A_n, and the phase, ϕ_n, of an element are realized by two independent components; therefore they can be treated as independent variables. It is also assumed that the mean amplitudes and phases are the corresponding desired values. Furthermore, it is also assumed that the phase variance does not differ, while the amplitude variance differs from element

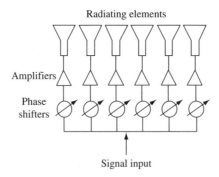

FIGURE 14.1   Simple block diagram of a phased array antenna.

---

[1] The statistical analysis is performed on the far-field intensity or the EIRP, which is very relevant for practical applications.
[2] For a planar array, nud in (14.1) and elsewhere is replaced by $x_n u + y_n v$, where (x_n, y_n) is the location of the nth element, $u = k_0 \sin\theta \cos\phi$ and $v = k_0 \sin\theta \sin\phi$, and (θ, ϕ) is the observation angle.

to element. The following symbols are used to represent the mean and variance of the random variables:

$$< A_n > = \bar{A}_n = \text{expected value of } A_n$$
$$< \phi_n > = \bar{\phi}_n = \text{expected value of } \phi_n$$
$$< (A_n - \bar{A}_n)^2 > = \sigma_{An}^2 = \text{variance of } A_n (\text{varies with } n)$$
$$< (\phi_n - \bar{\phi}_n)^2 > = \sigma_{\phi n}^2 = \text{variance of } \phi_n (\text{assumed identical for all values of } n)$$

We now proceed to determine the statistics of the array pattern. Notice that each term in (14.1) is a complex quantity which is a random variable because A_n and ϕ_n are random variables. It is difficult to deal with a complex random variable; therefore we decompose them into real and imaginary parts and treat them separately. We then determine the statistics of the absolute value of the radiated field. Thus we express the array pattern as below (the element pattern part is omitted here because that is unimportant with respect to the present context):

$$F(u) = \sum_{n=1}^{N} A_n \cos(\phi_n + nud) + j \sum_{n=1}^{N} A_n \sin(\phi_n + nud)$$
$$= R + jI \tag{14.2}$$

We denote the real part of $F(u)$ as R and the imaginary part of $F(u)$ as I for brevity. Each term under a summation sign of (14.2) is a random variable. As mentioned, our ultimate objective is to obtain the probability density function of the magnitude of the far intensity $|F(u)|$ at a given u. For that we do not need the distribution function of each term under the summation signs, but we must know the mean and variance of each term. This will be considered in the following section.

14.3 STATISTICS OF R AND I

In this section we will obtain the statistical parameters of the real and imaginary parts, R and I, of the array factor $F(u)$ given in (14.2). First we determine the parameters term by term under the summation signs and then apply the procedures for multiple independent variables. The mean or expected value of the nth term of R in (14.2) is given by

$$< A_n \cos(\phi_n + nud) > = < A_n > < \cos(\phi_n + nud) > \tag{14.3}$$

To obtain the mean of the cosine term, we assume that ϕ_n is normally distributed with mean $\bar{\phi}_n$ and variance σ_ϕ^2. The expected value of the cosine term can be expressed as

$$< \cos(\phi_n + nud) > = \frac{1}{\sqrt{2\pi}\sigma_\phi} \int_{-\infty}^{\infty} \cos(\phi_n + nud) \exp\left[-\frac{(\phi_n - \bar{\phi}_n)^2}{2\sigma_\phi^2}\right] d\phi_n \tag{14.4}$$

The above integral can be evaluated analytically using the following identity:

$$\frac{1}{\sqrt{2\pi}\sigma_\phi} \int_{-\infty}^{\infty} \exp(jk\phi_n)\exp\left[-\frac{(\phi_n-\bar{\phi}_n)^2}{2\sigma_\phi^2}\right]d\phi_n = \exp(jk\bar{\phi}_n)\exp\left(-\frac{k^2\sigma_\phi^2}{2}\right)$$

$$(14.5)$$

The final expression for $<\cos(\phi_n+nud)>$ becomes

$$<\cos(\phi_n+nud)> = \cos(\bar{\phi}_n+nud)\exp(-\frac{1}{2}\sigma_\phi^2) \qquad (14.6)$$

From (14.3), (14.4), and (14.5) we obtain the expected value of $A_n\cos(\phi_n+nud)$ as

$$<A_n\cos(\phi_n+nud)> = \bar{A}_n\cos(\bar{\phi}_n+nud)\exp\left(-\frac{1}{2}\sigma_\phi^2\right) \qquad (14.7)$$

In a similar fashion it can be shown that the expected value of $A_n\sin(\phi_n+nud)$ is

$$<A_n\sin(\phi_n+nud)> = \bar{A}_n\sin(\bar{\phi}_n+nud)\exp\left(-\frac{1}{2}\sigma_\phi^2\right) \qquad (14.8)$$

In order to determine the variance of $A_n\cos(\phi_n+nud)$, we first find the variances of A_n and $\cos(\phi_n+nud)$ individually and then use the variance formula for the product of two random variables. The variance of A_n is assumed as σ_{An}^2 and the variance of $\cos(\phi_n+nud)$ is given by

$$\text{Var}[\cos(\phi_n+nud)] = <\cos^2(\phi_n+nud)> - <\cos(\phi_n+nud)>^2 \qquad (14.9)$$

The expected value of the cosine-square term is given by

$$<\cos^2(\phi_n+nud)> = \frac{1}{\sqrt{2\pi}\sigma_\phi}\int_{-\infty}^{\infty}\cos^2(\phi_n+nud)\exp\left[-\frac{(\phi_n-\bar{\phi}_n)^2}{2\sigma_\phi^2}\right]d\sigma_\phi$$

$$(14.10)$$

Employing (14.5) one derives the following closed-form expression:

$$<\cos^2(\phi_n+nud)> = \frac{1}{2}[1+\cos\{2(\bar{\phi}_n+nud)\}\exp(-2\sigma_\phi^2)] \qquad (14.11)$$

Using the results of (14.6) and (14.11) in (14.9) we obtain the variance of the cosine function as

$$\text{Var}[\cos(\phi_n+nud)] = \frac{1}{2}[1+\cos\{2(\bar{\phi}_n+nud)\}\exp(-2\sigma_\phi^2)]$$

$$-\cos^2(\bar{\phi}_n+nud)\exp(-\sigma_\phi^2) \qquad (14.12)$$

We now employ the following formula for the variance of the product of two independent random variables X and Y:

$$\text{Var}(XY) = \text{Var}(X)\text{Var}(Y) + \text{Var}(X)\bar{Y}^2 + \text{Var}(Y)\bar{X}^2 \qquad (14.13)$$

We set $X = A_n$ and $Y = \cos(\phi_n + nud)$ and obtain

$$\text{Var}[A_n \cos(\phi_n + nud)] = \frac{\bar{A}_n^2 + \sigma_{An}^2}{2}[1 + \cos\{2(\bar{\phi}_n + nud)\}\exp(-2\sigma_\phi^2)]$$
$$- \bar{A}_n^2 \cos^2(\bar{\phi}_n + nud)\exp(-\sigma_\phi^2) \qquad (14.14)$$

Similarly, the variance of $A_n \sin(\phi_n + nud)$ is deduced as

$$\text{Var}[A_n \sin(\phi_n + nud)] = \frac{\bar{A}_n^2 + \sigma_{An}^2}{2}[1 - \cos\{2(\bar{\phi}_n + nud)\}\exp(-2\sigma_\phi^2)]$$
$$- \bar{A}_n^2 \sin^2(\bar{\phi}_n + nud)\exp(-\sigma_\phi^2) \qquad (14.15)$$

From (14.2), the mean of R can be expressed as

$$\bar{R} = <R> = <\text{Real}\{F(u)\}> = \sum_{n=1}^{N} <A_n \cos(\phi_n + nud)> \qquad (14.16)$$

Using (14.6) we obtain

$$\bar{R} = \exp\left(-\frac{\sigma_\phi^2}{2}\right)\sum_{n=1}^{N} \bar{A}_n \cos(\bar{\phi}_n + nud) \qquad (14.17)$$

Similarly, the variance of R is expressed as

$$\sigma_R^2 = \text{Var}(R) = \sum_{n=1}^{N} \text{Var}[A_n \cos(\phi_n + nud)] \qquad (14.18)$$

Using (14.14) we obtain

$$\sigma_R^2 = \sum_{n=1}^{N} \left[\frac{\bar{A}_n^2 + \sigma_{An}^2}{2}[1 + \cos\{2(\bar{\phi}_n + nud)\}\exp(-2\sigma_\phi^2)] \right.$$
$$\left. - \bar{A}_n^2 \cos^2(\bar{\phi}_n + nud)\exp(-\sigma_\phi^2) \right] \qquad (14.19)$$

This can be simplified to an alternate expression as

$$\sigma_R^2 = \frac{1}{2}\sum_{n=1}^{N} \sigma_{An}^2 + \frac{1}{2}[1 - \exp(-\sigma_\phi^2)]\sum_{n=1}^{N} \bar{A}_n^2$$
$$+ \frac{1}{2}\sum_{n=1}^{N} \cos\{2(\bar{\phi}_n + nud)\}[(\bar{A}_n^2 + \sigma_{An}^2)\exp(-2\sigma_\phi^2) - \bar{A}_n^2 \exp(-\sigma_\phi^2)] \qquad (14.20)$$

Similarly, the mean and variance of I are

$$\bar{I} = \exp\left(-\frac{\sigma_\phi^2}{2}\right) \sum_{n=1}^{N} \bar{A}_n \sin(\bar{\phi}_n + nud) \tag{14.21}$$

$$\sigma_I^2 = \frac{1}{2} \sum_{n=1}^{N} \sigma_{An}^2 + \frac{1}{2}[1 - \exp(-\sigma_\phi^2)] \sum_{n=1}^{N} \bar{A}_n^2$$

$$- \frac{1}{2} \sum_{n=1}^{N} \cos\{2(\bar{\phi}_n + nud)\}[(\bar{A}_n^2 + \sigma_{An}^2)\exp(-2\sigma_\phi^2) - \bar{A}_n^2 \exp(-\sigma_\phi^2)] \tag{14.22}$$

We compute the standard deviations σ_R and σ_I deduced in (14.20) and (14.22), respectively, for various observation angles of a linear array of 20 elements. The element spacing is one wavelength ($d = \lambda_0$), and the elements are excited uniformly in amplitude and phase to generate a bore-sight beam. The nominal value for the element amplitude is 1 V. The standard deviations for the amplitude error and phase error are 0.2 V and 0.087 rad, respectively. Figure 14.2 shows the variations of σ_R and σ_I with respect to the observation angle at the far field. Also plotted in the figure is the nominal radiation pattern of the array. As can be noted, the two standard deviations (SD) closely follow each other at most observation angles, except at the bore sight and at selective angles (at 30° off bore sight, which is a null location for this example). Observe that the first two terms of (14.20) and (14.22) are identical, while the third term differs by a sign. At the bore sight, $\bar{\phi}_n + nud = 0$ for all n; therefore, the third term is nonzero, causing significant difference between

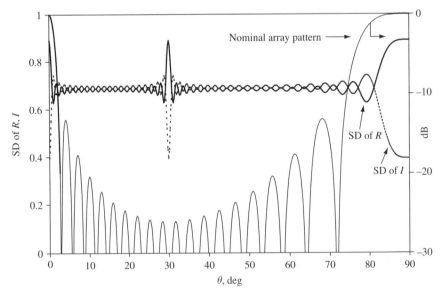

FIGURE 14.2 Nominal array pattern of a uniform array and magnitudes of σ_R and σ_I at various locations of the far field for $N = 20$, $\sigma_A/\bar{A} = 0.2$, $\sigma_\phi = 0.087$ rad, $d = \lambda_0$.

σ_R and σ_I. In general, this difference happens if $\bar{\phi}_n + nud = m\pi$, where m is an integer, which is satisfied at the main lobe, grating lobes, and near selective null locations. For the other observation angles, however, $\bar{\phi}_n + nud$ is neither zero nor an integral multiple of π. Therefore the cosine functions in (14.20) assume various values between -1 and $+1$. As a result, the contribution of the third term of (14.20) becomes small because it is a summation of several positive and negative numbers. This makes σ_R and σ_I close to each other. This is particularly true when the number of elements is large.

As we mentioned before, we assume σ_ϕ independent of n, although the present analysis is not limited to that constraint. This assumption is generally valid because the phase error is primarily due to phase quantization error in a digital phase shifter and/or to fabrication tolerances. On the other hand, σ_{An} cannot be assumed independent of n, because this standard deviation is very much dependent on the magnitude of the excitation voltage. However, in order to simplify the analysis, it is generally assumed that σ_{An} is proportional to $\bar{A}_n$, so that the ratio $\sigma_{An}/\bar{A}_n$ remains constant for all excitation sources.

14.4 PROBABILITY DENSITY FUNCTION OF $|F(u)|$

In order to estimate the standard deviation and confidence intervals of the far-field intensity at a given observation angle, it is necessary to obtain the probability density function (PDF). The PDF of the magnitude of the far-field intensity $|F(u)| = \sqrt{R^2 + I^2}$ can be determined from the PDFs of R and I. Recall from (14.2) that R is a summation of N independent random variables; therefore, according to the central limit theorem R has a Gaussian PDF with mean $\bar{R}$ and variance σ_R^2, assuming N is sufficiently large. Similarly, I would have a Gaussian PDF with mean $\bar{I}$ and variance σ_I^2. The PDF of $|F(u)|$ can be expressed in terms of the PDF of R and I. Interestingly, by means of the central limit theorem we are able to obtain the PDFs of R and I without knowledge of the PDFs of individual terms in (14.2). Before deducing the PDF of $|F(u)|$ we first give a proof for the central limit theorem to convince ourselves the validity of such a powerful concept in statistical theory.

14.4.1 Central Limit Theorem

The central limit theorem (CLT) states that if $x_1, x_2, \ldots, x_N$ are N independent random variables, then their sum $z = x_1 + x_2 + \cdots + x_N$ has a Gaussian PDF provided N is sufficiently large. In order to prove the CLT, we need to understand the joint probability of two independent variables. Suppose x_1 and x_2 are two independent random variables with PDFs $f(x_1)$ and $g(x_2)$, respectively. Then it can be shown by applying a concept of fundamental probability theory that the random variable $z = x_1 + x_2$ has the PDF $h(z)$ given by [1]

$$h(z) = \int_x f(x)g(z-x)dx \qquad (14.23)$$

Notice, $h(z)$ is the convolution of $f(x)$ and $g(x)$. In the Fourier domain, the above relation can be represented by

$$H(k) = F(k)G(k) \tag{14.24}$$

where $H(k)$, $F(k)$, and $G(k)$ are FTs of $h(x)$, $f(x)$, and $g(x)$, respectively. The FT of a function $p(x)$ is defined as

$$P(k) = \int_x p(x)\exp(jkx)\ dx \tag{14.25}$$

Notice, unlike the usual definition of the FT that is used throughout this book, we purposely exclude the $1/2\pi$ factor here. However, we will make necessary adjustments while taking the inverse transformation.

Using the chain rule, the relation in (14.24) can be extended for a sum of N independent random variables. Accordingly, the FT of the PDF of $z = x_1 + x_2 + x_3 + \cdots + x_N$ can be expressed as

$$H(k) = F_1(k)F_2(k)F_3(k)\cdots F_N(k) \tag{14.26}$$

To prove the CLT we first consider a simple case of N independent random variables $x_1, x_2, \ldots, x_N$. We assume that the variables have zero mean with variances as $\sigma_1^2, \sigma_2^2, \ldots, \sigma_N^2$ respectively. For nonzero mean it is a matter of a linear shift of the PDF. The FT of the PDF of x_n is given by

$$F_n(k) = \int_{x_n} f_n(x_n)\exp(jkx_n)dx_n \tag{14.27}$$

where $f_n(x_n)$ is the PDF of x_n. Since the PDF $f_n(x_n)$ is always greater than zero, it can be easily seen that $|F_n(k)|$ is maximum at $k = 0$. In the vicinity of $k = 0$, $F_n(k)$ can be approximated by the following Taylor series with three terms:

$$F_n(k) \approx F_n(0) + kF_n'(0) + \frac{1}{2}k^2 F_n''(0) \tag{14.28}$$

From (14.27) we see that $F_n(0) = 1$ and the first two derivatives of $F_n(k)$ are

$$F_n'(0) = j\int_{x_n} x_n f_n(x_n)\ dx_n = j\bar{x}_n = 0 \text{ (by assumption)} \tag{14.29}$$

$$F_n''(0) = -\int_{x_n} x_n^2 f_n(x_n)\ dx_n = -\sigma_n^2 \tag{14.30}$$

Substituting (14.29) and (14.30) into (14.28) we obtain (in the vicinity of $k = 0$)

$$F_n(k) \approx (1 - \frac{1}{2}k^2\sigma_n^2) \approx \exp(-\frac{1}{2}k^2\sigma_n^2) \tag{14.31}$$

Substituting (14.31) in (14.26) we can write

$$H(k) = \prod_{n=1}^{N} F_n(k) \approx \exp\left(-\frac{k^2}{2} \sum_{n=1}^{N} \sigma_n^2\right) = \exp\left(-\frac{k^2\sigma^2}{2}\right) \tag{14.32}$$

where $\sigma^2 = \sum_{n=1}^{N} \sigma_n^2$. One may think that the above expression for $H(k)$ is valid only in the vicinity of $k = 0$, because the expression of $F_n(k)$ in (14.31) is derived for small $|k|$ only. However, we argue that even though (14.31) is valid for small values of $|k|$, (14.32) is valid for all k, provided N is sufficiently large. This argument is based on the fact that $F_n(k) = 1$ for $k = 0$ and $|F_n(k)| < 1$ for $|k| > 0$ for all n; thus the product of N such functions remains unity at k=0 while decaying rapidly away from the $k = 0$ region. The lower limit of N for the validity of (14.32) for all values of k depends on the individual PDFs. However, as we can see from (14.32), the limit, in general, depends on the value of $\sum_{n=1}^{N} \sigma_n^2$. The higher the value, the better is the approximation. For more details the reader is referred to the classic text by Papoulis [2].

The PDF of z is the inverse FT of $H(k)$. Thus we can write the PDF as

$$h(z) = \frac{1}{2\pi} \int_{-\infty}^{\infty} H(k) \exp(-jkz)\, dk \tag{14.33}$$

Substituting the expression of $H(k)$ from (14.32) we obtain

$$h(z) = \frac{1}{2\pi} \int_{-\infty}^{\infty} \exp(-jkz) \exp\left(-\frac{k^2\sigma^2}{2}\right) dk \tag{14.34}$$

Using the result of (14.5) we obtain

$$h(z) = \frac{1}{\sqrt{2\pi}\sigma} \exp\left(-\frac{z^2}{2\sigma^2}\right) \tag{14.35}$$

Thus we prove that $h(z)$ becomes Gaussian with zero mean and variance σ^2. The zero mean is primarily due to the zero-mean assumption of all the variables $x_1, x_2, \ldots, x_N$. For a nonzero mean of x_n, $h(z)$ will have a simple coordinate shift. The corresponding Gaussian PDF would be given by

$$h(z) = \frac{1}{\sqrt{2\pi}\sigma} \exp\left(-\frac{(z-\bar{z})^2}{2\sigma^2}\right) \tag{14.36}$$

where

$$\bar{z} = \sum_{n=1}^{N} \bar{x}_n \qquad \sigma^2 = \sum_{n=1}^{N} \sigma_n^2$$

In the above $\bar{x}_n$ represents the mean of the random variable x_n. This completes the proof of the CLT.

14.4.2 PDF of *R* and *I*

The CLT can be utilized to obtain the PDF of R because R is the summation of N independent random variables $A_n \cos(\phi_n + nud)$, $n = 1, 2, \ldots, N$, as given in (14.2). For a sufficiently large value of N, the PDF of R is given by

$$f_R(R) = \frac{1}{\sqrt{2\pi}\sigma_R} \exp\left[-\frac{(R - \bar{R})^2}{2\sigma_R^2}\right] \qquad (14.37)$$

In the above, $\bar{R}$ and σ_R^2 are given in (14.17) and (14.20), respectively. Similarly, the PDF of I is given by

$$f_I(I) = \frac{1}{\sqrt{2\pi}\sigma_I} \exp\left[-\frac{(I - \bar{I})^2}{2\sigma_I^2}\right] \qquad (14.38)$$

where $\bar{I}$ and σ_I^2 are given in (14.21) and (14.22), respectively.

14.4.3 PDF of $\sqrt{R^2 + I^2}$

In order to deduce the PDF of the random variable $V = \sqrt{R^2 + I^2}$, we use the following assumptions:

1. Both R and I have Gaussian PDF.
2. Both R and I are independent.

By virtue of the CLT, the first assumption is valid if the number of elements N is large. Furthermore, R and I can be treated as independent random variables because the covariance of R and I becomes very small for large N (see problem 14.5).

To obtain the PDF of V we first obtain the joint PDF of R and I. The joint PDF of two independent variables R and I is given by

$$f_{RI}(R, I) = f_R(R)f_I(I) \qquad (14.39)$$

Suppose we want to determine the probability of R and I lying inside the annular region on the RI-plane shown in Figure 14.3. The probability P that R and I lie inside the ring is

$$P = \iint_{\text{Ring}} f_{RI}(R, I)dR \, dI \qquad (14.40)$$

We now use the following substitutions:

$$R = V \cos\alpha \qquad I = V \sin\alpha \qquad (14.41)$$

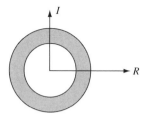

FIGURE 14.3   Annular region on the RI-plane.

Then, (14.40) becomes

$$P = \int_V^{V+\Delta V} \int_0^{2\pi} f_{RI}(V\cos\alpha, V\sin\alpha)V\, d\alpha\, dV \tag{14.42}$$

We assume that the width of the ring in Figure 14.3 is very small. Thus the probability P can be expressed as

$$P = f_V(V)\, \Delta V = \int_0^{2\pi} f_{RI}(V\cos\alpha, V\sin\alpha)V\, d\alpha\, \Delta V \tag{14.43}$$

The PDF with respect to the random variable V thus becomes

$$f_V(V) = \int_0^{2\pi} f_{RI}(V\cos\alpha, V\sin\alpha)V\, d\alpha \tag{14.44}$$

Using (14.39) in (14.44) we finally obtain

$$f_V(V) = V \int_0^{2\pi} f_R(V\cos\alpha)\, f_I(V\sin\alpha)\, d\alpha \tag{14.45}$$

PDF at Beam Peak   Now consider the PDF of the far-field intensity at the beam peak location. At the beam peak, $\bar{I} = 0$. Using the Gaussian PDF for R and I, we write $f_V(V)$ in (14.45) as

$$f_V(V) = \frac{V}{2\pi\sigma_R\sigma_I} \int_0^{2\pi} \exp\left[-\frac{(V\cos\alpha - \bar{R})^2}{2\sigma_R^2} - \frac{V^2\sin^2\alpha}{2\sigma_I^2} \right] d\alpha \tag{14.46}$$

Notice, the magnitude of the integrand becomes maximum if the following two conditions are simultaneously satisfied:

$$V\cos\alpha - \bar{R} = 0 \qquad V\sin\alpha = 0 \tag{14.47}$$

The above conditions are satisfied at $V = \bar{R}$, $\alpha = 0$. Therefore, for small values of σ_R and σ_I compared to $\bar{R}$, the major contribution of the integral comes from the $\alpha = 0$ region. In that region, the cosine and sine functions can be approximated as

$$\cos \alpha \approx 1 - \frac{1}{2}\alpha^2 \qquad \sin \alpha \approx \alpha \qquad (14.48)$$

Using the above approximation in (14.46) we obtain

$$f_V(V) = \frac{V}{2\pi\sigma_R\sigma_I} \int_{-\pi}^{\pi} \exp\left[-\frac{[V(1-\alpha^2/2) - \bar{R}]^2}{2\sigma_R^2} - \frac{V^2\alpha^2}{2\sigma_I^2} \right] d\alpha \qquad (14.49)$$

Notice, we also changed the limits of the integration, but that change will have no effect on the final result because the integrand in (14.46) is a periodic function of α. Upon simplification and ignoring terms of higher order than order 2, we obtain

$$f_V(V) = \frac{V}{2\pi\sigma_R\sigma_I} \exp\left[-\frac{(V-\bar{R})^2}{2\sigma_R^2} \right] \int_{-\pi}^{\pi} \exp\left[-\frac{\alpha^2\{m^2V^2 - V^2 + V\bar{R}\}}{2\sigma_R^2} \right] d\alpha \quad (14.50)$$

with $m = \sigma_R/\sigma_I$. We now make the following assumptions in order to have a closed-form expression for the integral:

$$m^2V^2 - V^2 + V\bar{R} > 0 \qquad (14.51)$$

$$\frac{\sigma_R}{\sqrt{m^2V^2 - V^2 + V\bar{R}}} \ll \pi \qquad (14.52)$$

With these assumptions, the limits of the integral can be extended from $-\infty$ to ∞ without affecting its value significantly. Now let us examine the validity of these assumptions. From (14.50) we notice that $f_V(V)$ dominates in the vicinity of $V = \bar{R}$, and in that vicinity (14.51) is satisfied in a small range of V. The relation in (14.52) is also satisfied in that small range of V subject to the condition that $\sigma_I/\bar{R} \ll \pi$. This is a valid condition because at the beginning we assumed that σ_R and σ_I are small compared to $\bar{R}$. For $m < 1$ and $V \gg \bar{R}$, (14.51) is not satisfied. In that situation, one needs to evaluate (14.46) numerically. In practice that may not be needed, because for $V \gg \bar{R}$, $f_V(V)$ itself is very small, as is apparent from (14.50). Thus for all practical purposes (14.51) and (14.52) can be regarded as valid conditions. In the vicinity of $V = \bar{R}$, $f_V(V)$ will have a closed form expression as

$$f_V(V) = \frac{V \exp[-(V-\bar{R})^2/2\sigma_R^2]\sqrt{2\pi}\sigma_R}{2\pi\sigma_R\sigma_I\sqrt{m^2V^2 - V^2 + V\bar{R}}} \qquad (14.53)$$

Upon simplification, the final expression for $f_V(V)$ becomes

$$f_V(V) = \frac{1}{\sqrt{2\pi(\sigma_R^2 - \sigma_I^2 + \sigma_I^2\bar{R}/V)}} \exp\left[-\frac{1}{2\sigma_R^2}(V-\bar{R})^2 \right] \qquad (14.54)$$

Notice, the PDF of the field intensity at the beam peak is slightly different from a Gaussian function. For $\sigma_I \ll \sigma_R$, the PDF can be approximated as a Gaussian function with mean $\bar{R}$ and standard deviation σ_R. Typically, that happens if σ_ϕ becomes very small. In other situations $f_V(V)$ deviates from a Gaussian PDF.

In Figure 14.4 we plot $f_V(V)$ versus V at the beam peak of an array of 20 uniformly excited elements. The standard deviation of the element amplitude, σ_{An}, is assumed as 0.8 (80%) and the standard deviation of the element phase, σ_ϕ, is assumed as 0.67 rad. We plot three different PDFs. The solid line corresponds to (14.46) and the dashed line (nondistinguishable) corresponds to (14.54). This essentially verifies the accuracy of the approximation in (14.54). The dots correspond to a standard Gaussian PDF with mean $\bar{R}$ and standard deviation σ_R. Evidently, $f_V(V)$ in (14.54) is a better approximation than the Gaussian PDF.

PDF for $\sigma_R = \sigma_I$ We now consider another special case. We notice from Figure 14.2 that, except at bore sight and at a few other locations, σ_R and σ_I are very close to each other. If we assume $\sigma_R = \sigma_I$, the integral of (14.45) reduces to a closed form. Using $\sigma_R = \sigma_I$ in (14.45) and with straightforward algebraic manipulation, we obtain

$$f_V(V) = \frac{V}{\sigma_R^2} \exp\left[-\frac{V^2 + \bar{R}^2 + \bar{I}^2}{2\sigma_R^2}\right] \frac{1}{2\pi} \int_0^{2\pi} \exp\left[\frac{V\sqrt{\bar{R}^2 + \bar{I}^2}}{\sigma_R^2} \cos(\alpha - \psi)\right] d\alpha$$

$$(14.55)$$

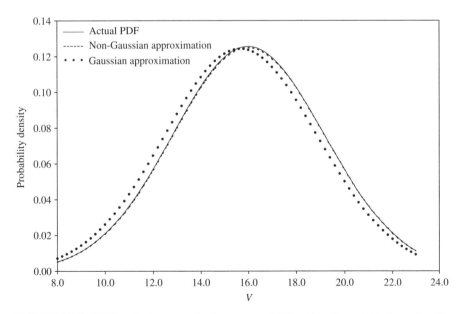

FIGURE 14.4 PDF at the beam peak of an array of 20 uniformly excited elements with $\sigma_A/\bar{A} = 0.8$, $\sigma_\phi = 0.67$ rad, $d = \lambda_0$, $\bar{A} = 1$.

where $\psi = \tan^{-1}(\bar{I}/\bar{R})$. The integral can be expressed in terms of a zeroth-order modified Bessel function of the first kind [3]. The final expression for $f_V(V)$ becomes

$$f_V(V) = \frac{V}{\sigma_R^2} \exp\left[-\frac{V^2 + \bar{R}^2 + \bar{I}^2}{2\sigma_R^2}\right] I_0 \left(\frac{V\sqrt{\bar{R}^2 + \bar{I}^2}}{\sigma_R^2}\right) \tag{14.56}$$

The above PDF is known as a Ricean PDF [4]. The above PDF can be used for the far-field intensity *except at the bore-sight, grating lobe, and other selective regions,* as discussed in the previous section. At null locations, the PDF assumes a closed-form expression that is presented next.

PDF at Null Locations   We consider another special case $\bar{R} = \bar{I} = 0$ that happens at the null locations. Using $\bar{R} = \bar{I} = 0$ in (14.45) and with algebraic manipulations we can show that

$$f_V(V) = \frac{V}{\sigma_R \sigma_I} \exp\left[-\frac{V^2}{4}\left(\frac{1}{\sigma_I^2} + \frac{1}{\sigma_R^2}\right)\right] I_0\left[\frac{V^2}{4}\left(\frac{1}{\sigma_I^2} - \frac{1}{\sigma_R^2}\right)\right] \tag{14.57}$$

Therefore, at each null the PDF of the field intensity follows (14.57).

Figure 14.5 depicts PDFs of far-field intensities at two nulls located at 2.886° and 30°, respectively. Figure 14.6 is the PDF at the first side-lobe location, near 4.103°. For these computations we assume $\bar{A}_n = 1$, $\bar{\phi}_n = 0$, $\sigma_{An} = 0.2$, and $\sigma_\phi = 0.087$

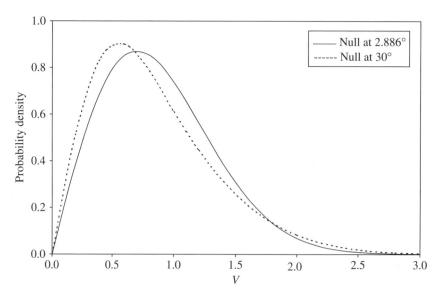

FIGURE 14.5   PDFs of the far-field intensities at two different nulls. Array parameters are the same as in Figure 14.2.

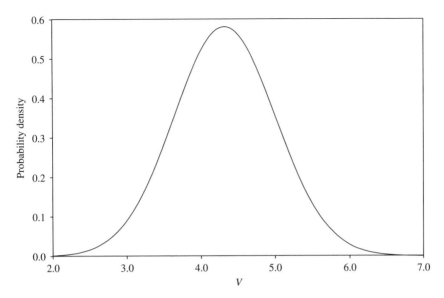

FIGURE 14.6 PDF of the far field at the first side-lobe location near 4.103°. Array parameters are the same as in Figure 14.2.

rad for all n. The number of elements $N = 20$. The PDF for the side-lobe peak approximately follows a Ricean PDF as discussed before. However, the mean intensity at the side-lobe peak being larger compared to its standard deviation, the Ricean PDF almost appears as a Gaussian PDF. For the null at 2.886°, σ_R and σ_I are very close to each other, while these quantities differ for the null at 30°. As a result, the two PDFs significantly differ from each other, as seen in Figure 14.5.

14.5 CONFIDENCE LIMITS

The mean and standard deviation of the field intensity are calculated at each far-field point using the general PDF in (14.45). Figure 14.7 show the error-free pattern of an array. Figures 14.8a–c show mean and 1σ boundaries of the far-field intensity for three different sets of parameters. The mean performance deteriorates with increasing uncertainties in amplitude and phase. In addition, the 1σ range becomes wider as the amplitude and phase uncertainties increase.

An important parameter is the 95% confidence limit for the far-field intensity. At the beam peak location this limit is defined as the "threshold intensity" where the array has 95% probability to exceed the threshold. This definition of confidence limit changes for side lobes and nulls. For a side lobe (or null), this limit is defined as the threshold intensity where the array has a probability of 95% *not to* exceed this threshold. We will consider the 95% confidence limits for three important cases.

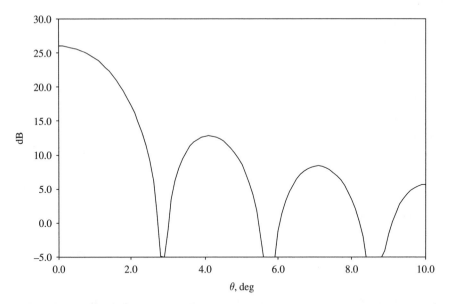

FIGURE 14.7   Nominal far-field intensity of the array in Figure 14.2.

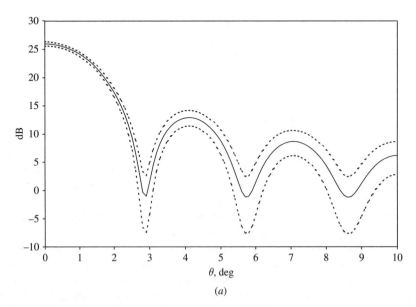

(a)

FIGURE 14.8   Mean and 1σ boundaries of the far-field intensity of a uniform array with $N = 20$, $\bar{A} = 1$, and $d = \lambda_0$: (a) $\sigma_A/\bar{A} = 0.2$, $\sigma_\phi = 0.087$ rad; (b)$\sigma_A/\bar{A} = 0.2$, $\sigma_\phi = 0.17$ rad; (c)$\sigma_A/\bar{A} = 0.4$, $\sigma_\phi = 0.17$ rad.

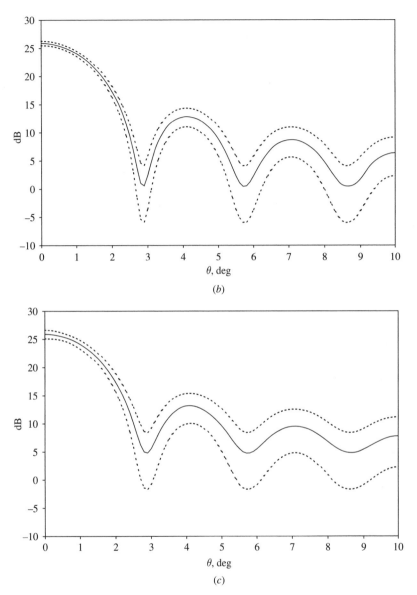

FIGURE 14.8 (Continued).

14.5.1 Beam Peak

The 95% confidence limit for the field intensity at the beam peak location can be determined analytically. For $\bar{R} >> \sigma_I, \sigma_R$ and in the vicinity of $V = \bar{R}$, $f_V(V)$ in (14.54) can be approximated as

$$f_V(V) = \frac{1}{\sqrt{2\pi}\sigma_R} \exp\left[-\frac{1}{2\sigma_R^2} \left(V - \bar{R} - \frac{\sigma_I^{\,2}}{2\bar{R}} \right)^2 \right] \qquad (14.58)$$

The right-hand side of (14.58) is a Gaussian PDF with mean $\bar{R} + \sigma_I^2/2\bar{R}$ and standard deviation σ_R. The 95% confidence limit (lower limit) is situated at a distance $1.645\sigma_R$ below the mean. Accordingly the desired 95% confidence limit becomes

$$V_{95} = \bar{R} + \frac{\sigma_I^2}{\bar{R}} - 1.645\sigma_R \tag{14.59}$$

14.5.2 Side Lobe

As mentioned before the upper limit is of interest for a side lobe. The side lobe follows the PDF given in (14.56). The magnitude of $\sqrt{\bar{R}^2 + \bar{I}^2}$ is much larger than σ_R for the highest side lobe. For example, if $\bar{A}_n = 1$, $\bar{\phi}_n = 0$, $\sigma_{An} = 0.2$, and $\sigma_\phi = 0.087$ rad, the value of $\sqrt{\bar{R}^2 + \bar{I}^2}$ at the highest side-lobe peak at 4.103° is about 4.4, with $\sigma_R = 0.68$. Therefore, under the assumption that $\sqrt{\bar{R}^2 + \bar{I}^2} \gg \sigma_R$, the modified Bessel function in (14.56) can be replaced by its asymptotic form, leaving the right-hand side of (14.56) as

$$f_V(V) \approx \frac{1}{\sigma_R} \sqrt{\frac{V}{2\pi P}} \exp\left[-\frac{(V-P)^2}{2\sigma_R^2}\right], \tag{14.60}$$

with $P = \sqrt{\bar{R}^2 + \bar{I}^2}$. Because $P \gg \sigma_R$, the function $f_V(V)$ is dominant in the region where V is very close to P. In that region, the right-hand side of (14.60) can be approximated as a Gaussian PDF with mean P and standard deviation σ_R. The 95% confidence limit for the peak side lobe thus is given by

$$V_{95} \approx P + 1.645\sigma_R = \sqrt{\bar{R}^2 + \bar{I}^2} + 1.645\sigma_R \tag{14.61}$$

14.5.3 Nulls

The PDF of the field intensity at a null location is deduced in (14.57). Most of the nulls that are close to the main beam have identical values for σ_I and σ_R. Using $\sigma_I = \sigma_R$ in (14.57) one obtains the PDF as

$$f_V(V) = \frac{V}{\sigma_R^2} \exp\left[-\frac{V^2}{2\sigma_R^2}\right] \tag{14.62}$$

The above PDF is a Rayleigh PDF, typically used for noise analysis in communication systems [5]. The 95% confidence limit can be calculated by solving V_{95} from the following integral equation:

$$\int_0^{V_{95}} \frac{V}{\sigma_R^2} \exp\left[-\frac{V^2}{2\sigma_R^2}\right] dV = 0.95 \tag{14.63}$$

The integral can be expressed in closed form. The result is

$$1 - \exp\left[-V_{95}^2/2\sigma_R^2\right] = 0.95 \tag{14.64}$$

TABLE 14.1   Comparison Between Exact Confidence Limits and Estimated Confidence Limits at Three Important Far-field Locations of 20-Element Array

Far-Field Location (deg)	$\sigma_{An} = 0.2, \sigma_\phi = 0.0873$ rad		$\sigma_{An} = 0.2, \sigma_\phi = 0.1745$ rad	
	Computed V_{95} (V)	Estimated V_{95} (V)	Computed V_{95} (V)	Estimated V_{95} (V)
0 (beam peak)	18.50	18.47	18.28	18.27
2.866 (first null)	1.65	1.68	2.00	2.05
4.103 (first side-lobe)	5.47	5.48	5.67	5.69

Note: $\bar{A}_n = 1 (n = 1, 2, \ldots, 20)$, $d = \lambda_0$, beam peak at $0°$.

which yields

$$V_{95} = 2.448\sigma_R \tag{14.65}$$

It should be mentioned that this confidence limit is strictly valid for the nulls near the main-beam region for which the equality condition $\sigma_I = \sigma_R$ is satisfied. If this condition is not satisfied at a null location (for example the null at $\theta = 30°$), then (14.57) should be used to determine the above confidence limit.

The 95% confidence limits for three important locations are computed numerically using (14.45) and the results are compared with the estimated values obtained using (14.59), (14.61), and (14.65). The results are tabulated in Table 14.1. The agreement is within 2% error in most cases. It should be pointed out that (14.65) is exact; therefore the numerical result should match exactly with that obtained using (14.65). The difference is within the accuracy of the numerical integration.

14.6   ELEMENT FAILURE ANALYSIS

There is a finite probability, however small, that an element will completely fail. It is usually assumed that when an element fails, the excitation voltage becomes zero. This failure probability can be incorporated into the PDF of an element's amplitude. Suppose the PDF of an element's amplitude, A, nominally follows a Gaussian distribution with mean $\bar{A}$ and standard deviation σ_A and assume F is the probability of failure of the element. Then the resulting distribution, including element failure, can be expressed as

$$f_F(A) = F\delta(A) + (1 - F)\frac{1}{\sqrt{2\pi}\sigma_A}\exp\left[\frac{-(A - \bar{A})^2}{2\sigma_A^2}\right] \tag{14.66}$$

The resulting distribution is sketched in Figure 14.9.

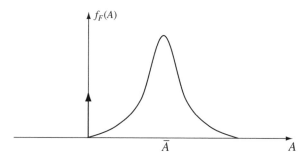

FIGURE 14.9   PDF of an element amplitude including failure probability.

We now determine the mean and variance of the resulting distribution $f_F(A)$ in (14.66). Suppose the mean value with failure is denoted as $\bar{A}^F$. We obtain the resultant mean as

$$\bar{A}^F = \int_{-\infty}^{\infty} A f_F(A)\, dA = (1 - F)\bar{A} \tag{14.67}$$

The resultant standard deviation σ_A^F that includes element failure can be determined similarly. The variance $(\sigma_A^F)^2$ is given by

$$(\sigma_A^F)^2 = \int_{-\infty}^{\infty} (A - \bar{A}^F)^2 f_F(A)\, dA = (1 - F)[\sigma_A^2 + F\bar{A}^2] \tag{14.68}$$

The standard deviation of the amplitude with failure thus becomes

$$\sigma_A^F = \sqrt{(1 - F)(\sigma_A^2 + F\bar{A}^2)} \tag{14.69}$$

The ratio of σ_A^F and $\bar{A}^F$ is obtained as

$$\frac{\sigma_A^F}{\bar{A}^F} = \sqrt{\frac{F + (\sigma_A/\bar{A})^2}{1 - F}} \tag{14.70}$$

From (14.70) we see that the ratio of the standard deviation and the mean increases with the failure rate. Equation (14.70) also indicates that if the ratio of the standard deviation and mean excitation does not vary from element to element (independent of n), then this ratio, including the failure rate, will follow the same characteristic.

In order to obtain the array statistics, including the element failure into consideration, we need to use $\bar{A}_n^F$ and σ_{An}^F instead of $\bar{A}_n$ and σ_{An} in the formulation. The most important parameters are the mean and variance of R and I, defined in (14.2). These parameters are expressed in (14.17), (14.20), (14.21), and (14.22).

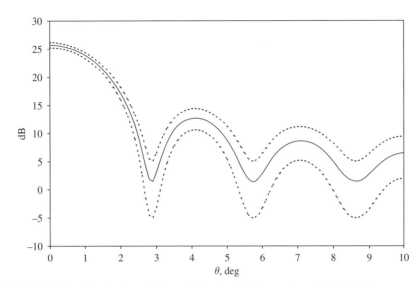

FIGURE 14.10 The 1σ boundaries of the far-field intensity with 2% failure rate and $\sigma_A/\bar{A} = 0.2$, $\sigma_\phi = 0.17$ rad, $N = 20$, $d = \lambda_0$.

For the 20-element linear array, we computed the mean and variance of $V = \sqrt{R^2 + I^2}$ using the PDF in (14.45). Figure 14.10 shows the 1σ boundaries of the far-field intensity of the array with 2% failure rate and with $\bar{A}_n = 1$, $\sigma_{An} = .2$, and $\sigma_\phi = 0.17$ rad. It is found that the 2% element failure changes the boundaries. In particular, the peak side lobes and the null depths deteriorate because of element failure.

Figures 14.11 and 14.12 show the effects of amplitude and phase errors and the element failure on the peak side-lobe level and the depth of the first null of the 20-element array. The peak side lobe occurs at 4.103° and the error-free level is 12.87 dB, which is 13.15 dB below the beam peak. Figure 14.11 shows the upper 1σ boundary (with reference to the error-free level) of the side lobe for five different cases. The significance of the upper 1σ boundary is that the odds are about 84% that the side lobe lies below that boundary. Expectedly, the boundary level increases with σ_{An} (the standard deviation of A_n, represented by the horizontal axis). The boundary level also deteriorates with an increase of σ_ϕ. The element failure, F, has significant effects for smaller values of σ_{An} and σ_ϕ. For larger σ_{An} and σ_ϕ, the element failure has minimal effect on the boundary level.

Figure 14.12 shows the upper 1σ boundary for the null depth that nominally occurs at 2.866°. The null depths are computed with respect to the nominal beam peak level. The null depth boundary deteriorates with increasing σ_{An} and σ_ϕ. The element failure rate has a significant effect on the boundary level as depicted in the plot.

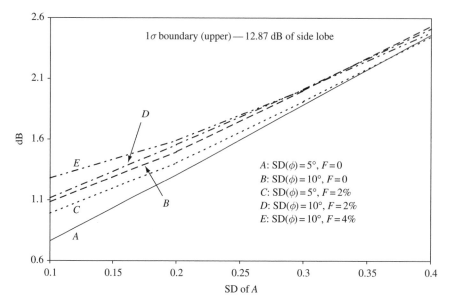

FIGURE 14.11 Effects of amplitude and phase errors and element failure effects on the peak side-lobe level of a uniform array with $\bar{A} = 1$, $N = 20$, $d = \lambda_0$.

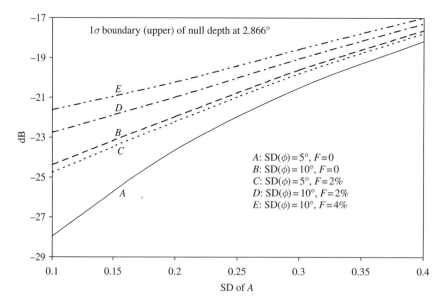

FIGURE 14.12 Effects of amplitude and phase errors and element failure on the depth of the first null of the 20-element array of Figure 14.11.

14.7 CONCLUDING REMARKS

The primary objective of this chapter was to introduce a statistical analysis of the array far-field intensity due to amplitude and phase uncertainties of the radiating elements. The PDF of the far field intensity was deduced using the fundamental principles of statistics. It was shown that element failure can be incorporated in the analysis through a modified amplitude distribution. Approximate analytical expressions for the PDFs at important far-field locations were also obtained.

An important aspect of a phased array, namely the element location uncertainty due to fabrication tolerance, was not considered explicitly. However, the present methodology can easily incorporate the above uncertainty. A simple way to incorporate the element location uncertainty is through an effective element phase while considering ideal element locations. To elaborate, consider the nth term on the right-hand side of (14.1). With location uncertainty, the term is modified as

$$T_n = A_n \exp[j(\phi_n + nud + u\,\Delta x)] \tag{14.71}$$

where Δx represents the deviation of the nth element's position from its intended position. Since Δx is a random variable with zero mean and standard deviation σ_x, we can define the new effective element phase as

$$\phi'_n(u) = \phi_n + u\,\Delta x \tag{14.72}$$

Notice, the effective phase now becomes a function of the far-field location, which is not surprising. The mean and variance of the modified phase variable become

$$\bar{\phi}'_n(u) = \bar{\phi}_n \qquad \sigma'^2_{\phi_n} = \sigma^2_{\phi_n} + u^2 \sigma^2_x \tag{14.73}$$

The above statistics of the effective phase can be implemented in the theory to obtain the PDF of the far-field intensity.

REFERENCES

[1] W. B. Davenport, Jr., *Probability and Random Process*, McGraw-Hill, New York, 1970.

[2] A. Papoulis, *Probability, Random Variables, and Stochastic Processes*, McGraw-Hill, New York, 1984.

[3] M. Abramowitz and I. E. Stegun, *Handbook of Mathematical Functions*, Applied Mathematical Series, Washington, DC, National Bureau of Standards, 1964.

[4] S. O. Rice, "Mathematical Analysis of Random Noise," Monograph B-1589, Bell Telephone System, pp. 104–112.

[5] J. C. Hancock, *The Principles of Communication Theory*, McGraw-Hill, New York, 1961.

BIBLIOGRAPHY

Ashmead, D., "Optimum Design of Linear Arrays in the Presence of Random Errors," *IRE Trans.*, Vol. AP-4, No. 1, pp. 81–92, Dec. 1952.

Hendricks, W. J., "The Totally Random Versus the Bin Approach for Random Arrays," *IEEE Trans.*, Vol. AP-39, No. 12, pp. 1757–1762, Dec. 1991.

Lo, Y. T., and V. D. Agrawal, "Distribution of Sidelobe Level in Random Arrays," *Proc. IEEE*, Vol. 57, No. 4, pp. 1764–1765, 1969.

Mailloux, R. J., "Statistically Thinned Arrays with Quantized Element Weights", *IEEE Trans.*, Vol. AP-39, No. 4, pp. 436–447, Apr.1991.

Rondinelli, L. A., "Effects of Random Errors on the Performances of Antenna Arrays of Many Elements," *IRE National Convension Record*, P.1, pp. 174–189, 1959.

Ruze, J., "Antenna Tolerance Theory — A Review," *Proc. IEEE*, Vol. 54, No. 4, pp. 633–640, April 1966.

PROBLEMS

14.1 Evaluate the following infinite integral and show that the integral has the following closed-form expression:

$$\frac{1}{\sqrt{2\pi}\sigma_\phi} \int_{-\infty}^{\infty} \exp(jk\phi_n)\exp\left[-\frac{(\phi_n - \bar{\phi}_n)^2}{2\sigma_\phi^2}\right] d\phi_n = \exp(jk\bar{\phi}_n)\exp\left(-\frac{k^2\sigma_\phi^2}{2}\right)$$

Using the above result, deduce (14.6).

14.2 Deduce (14.13)

14.3 Deduce (14.15) from (14.14), replacing $\phi_n + nud$ by $\phi_n + nud - \pi/2$ in (14.14).

14.4 Prove (14.26) for N variables.

14.5 The covariance of two random variables x and y is defined as

$$\sigma_{xy} = <(x - \bar{x})(y - \bar{y})> = <xy> -\bar{x}\bar{y}$$

Show that $\sigma_{xy} = 0$ for $m \neq n$ and σ_{xy} is proportional to $\sin(2\bar{\phi}_m)$ for $m = n$, where $x = \cos\phi_m$ and $y = \sin\phi_n$ and ϕ_m and ϕ_n are independent random variables with Gaussian PDFs [hence convince yourself that for a large value of N, the covariance of R and I defined in (14.2) is negligibly small].

14.6 Show that, for $\bar{I} = 0$, $\bar{R} >> \sigma_R$ and, for $\sigma_R = \sigma_I$, (14.54) and (14.56) are consistent in the vicinity of $V = \bar{R}$.

14.7 Deduce (14.57)

14.8 Using $V - \bar{R}$ small in (14.54) expand the denominator in binomial series to deduce (14.58). Assume $(\sigma_I/\bar{R})^2$ very small.

Appendix

A.1 SHANNON'S SAMPLING THEOREM

Statement of the Theorem   If a function $f(x)$ exists in the interval $0 < x < a$ and zero outside the interval, then inside the interval $0 < x < a$ $f(x)$ can be reproduced from the "samples" of its Fourier spectrum.

Proof   Suppose $F(k)$ is the Fourier transform (or spectrum) of $f(x)$. Then by definition

$$F(k) = \int_0^a f(x) \exp(jkx)\, dx \tag{A1.1}$$

Then, $f(x)$ can be recovered from its spectrum as

$$f(x) = \frac{1}{2\pi} \int_{-\infty}^{\infty} F(k) \exp(-jkx)\, dk \tag{A1.2}$$

Notice, the recovery process requires $F(k)$ at every point in the k-domain. Now let us define an auxiliary function $g(x)$ which is periodic with periodicity a such that

$$g(x) = \sum_{n=-\infty}^{\infty} f(x - na) \tag{A1.3}$$

Observe that $g(x) = f(x)$ for $0 < x < a$. Since $g(x)$ is a periodic function, $g(x)$ can be expanded in a Fourier series as

$$g(x) = \sum_{n=-\infty}^{\infty} A_n \exp\left(\frac{-j2nx\pi}{a} \right) \tag{A1.4}$$

Phased Array Antennas.   By Arun K. Bhattacharyya
© 2006 John Wiley & Sons, Inc.

The unknown coefficients are obtained using orthogonality of the complex exponential functions under the summation sign. This yields

$$A_m = \frac{1}{a} \int_0^a g(x) \exp\left(\frac{j2mx\pi}{a}\right) dx \qquad (A1.5)$$

By definition, $g(x) = f(x)$ inside the interval $0 < x < a$. Thus from (A1.1) and (A1.5), A_m can be expressed as

$$A_m = \frac{1}{a} F\left(\frac{2m\pi}{a}\right) \qquad (A1.6)$$

Thus $g(x)$ in (A1.4) can be expressed as

$$g(x) = \frac{1}{a} \sum_{n=-\infty}^{\infty} F\left(\frac{2n\pi}{a}\right) \exp\left(\frac{-j2nx\pi}{a}\right) \qquad (A1.7)$$

Since $f(x) = g(x)$ for $0 < x < a$, $f(x)$ is given by

$$f(x) = \frac{1}{a} \sum_{n=-\infty}^{\infty} F\left(\frac{2n\pi}{a}\right) \exp\left(\frac{-j2nx\pi}{a}\right) \qquad \text{for } 0 < x < a \qquad (A1.8)$$

The above expression for $f(x)$ should not be used outside the interval $0 < x < a$ because outside the interval $f(x)$ and $g(x)$ are not identical. Thus we see that $f(x)$ is recovered from the samples of its spectrum at a regular interval. The interval in the k-domain is given by

$$\Delta k = \frac{2\pi}{a} \qquad (A1.9)$$

The function $g(x)$ could be defined with a larger periodicity than a. In that situation, Δk would be smaller than $2\pi/a$. Thus the general rule for the sampling interval becomes

$$\Delta k \leq \frac{2\pi}{a} \qquad (A1.10)$$

A.2 Proof of $\sum_{n=-\infty}^{\infty} \delta(x-na) = \frac{1}{a} \sum_{n=-\infty}^{\infty} \exp(j2nx\pi/a)$

We know that a periodic function $g(x)$ of periodicity a can be expressed in terms of a Fourier series as

$$g(x) = \sum_{n=-\infty}^{\infty} A_n \exp\left(\frac{j2nx\pi}{a}\right) \qquad (A1.11)$$

The unknown coefficient A_m can be determined multiplying both sides by $\exp(-j2mx\pi/a)$ and then integrating in a full period of $g(x)$. This yields

$$\int_{-a/2}^{a/2} g(x)\exp\left(\frac{-j2mx\pi}{a}\right) dx = aA_m \tag{A1.12}$$

We define $g(x) = \sum_{n=-\infty}^{\infty} \delta(x-na)$. Thus for $-a/2 < x < a/2$, $g(x) = \delta(x)$. Using this in (A1.12), we obtain

$$A_m = \frac{1}{a}\int_{-a/2}^{a/2} \delta(x)\exp\left(\frac{-j2mx\pi}{a}\right) dx = \frac{1}{a} \tag{A1.13}$$

Thus from (A1.11) and (A1.12) we have

$$\sum_{n=-\infty}^{\infty} \delta(x-na) = \frac{1}{a}\sum_{n=-\infty}^{\infty} \exp\left(\frac{j2nx\pi}{a}\right) \tag{A1.14}$$

The identity in (A1.14) has an alternative form that has been used in many places of the text. The alternative form is obtained by replacing x by k and b by $2\pi/a$ in (A1.14). This yields

$$\sum_{n=-\infty}^{\infty} \delta\left(k-\frac{2n\pi}{b}\right) = \frac{b}{2\pi}\sum_{n=-\infty}^{\infty} \exp(jnkb) \tag{A1.15}$$

Index

Phased Array Antennas.   By Arun K. Bhattacharyya
© 2006 John Wiley & Sons, Inc.